Dictionary of Renewable Resources

Edited by Hans Zoebelein

Dictionary of
Renewable Resources

Edited by Hans Zoebelein

Weinheim · New York
Basel · Cambridge · Tokyo

Editor:
Dr. Hans Zoebelein
Waldschmidtstraße 4A
D-83727 Schliersee

This book was carefully produced. Nevertheless, authors, editor and publisher do not warrant the information contained therein to be free of errors. Readers are advised to keep in mind that statements, data, illustrations, procedural details or other items may inadvertently be inaccurate.

Executive Editor: Dr. Hans-Joachim Kraus
Editorial Director: Karin Dembowsky
Production Manager: Dipl.-Wirt.-Ing. (FH) H.-J. Schmitt

Library of Congress Card No.: applied for

British Library Cataloguing-in-Publication Data:
A catalogue record for this book
is available from the British Library

Die Deutsche Bibliothek – CIP-Einheitsaufnahme:
Dictionary of renewable resources / ed. by Hans Zoebelein. –
Weinheim ; New York ; Basel ; Cambridge ; Tokyo : VCH, 1997
  ISBN 3-527-30075-9
NE: Zoebelein, Hans [Hrsg.]

© VCH Verlagsgesellschaft mbH, D-69451 Weinheim (Federal Republic of Germany), 1997

Printed on acid-free and chlorine-free paper

Composition: Filmsatz Unger & Sommer, D-69469 Weinheim
Printing: strauss offsetdruck GmbH, D-69509 Mörlenbach
Bookbinding: Großbuchbinderei J. Schäffer, D-67269 Grünstadt
Printed in the Federal Republic of Germany.

Foreword

This Dictionary of Renewable Resources (RR) is designed to function as a bridge of understanding between the various groups of people interested in this old, yet highly modern subject: Farmers, agronomists, botanists, plant breeders, seed producers, ecologists, chemists in industry and universities, economists, politicians, societies and authorities dealing with various aspects of RR, and last but not least, the interested and educated layman.

This book attempts to give a complete survey of the subject. It is not limited to the use of RR as chemical feedstocks, but also covers the topics of "energy and fuel" as well as the direct (functional) uses without chemical modification, e. g., as construction and insulation materials, textile fibers and resins for coatings and adhesives.

The use of these resources in nutrition is not included extensively. However, food additives that are the result of chemical modifications and/or sophisticated formulations are fully covered. Plant ingredients and their application in medical, pharmaceutical, and cosmetic uses are limited to the most important due to the enormous number of special uses and the unclear borderline between scientifically proven effects and more homeopathic, not always clearly defined efficiencies. For similar reasons only the main fragrances, their sources and products isolated from them, are treated in this dictionary.

The book describes
- sources (plants and animals),
- technologies of isolation,
- resulting chemicals and their properties,
- chemical derivatizations and the operations involved,
- most important areas of application,
- economic significance and
- aspects for future development.

Modern, highly actual subjects are treated as well as old, almost forgotten ones, which today experience a revitalization or may be considered again in the future.

In the Preface "Renewable Resources: History, Definitions, Status and Outlook", a comprehensive review of the subject and all the pro and con arguments, guesses and opinions used in social, political and technical discussions today are compiled. Thus, the dictionary section itself could be limited to scientific, technical and economical facts.

The literature given at the end of a keyword is not a complete survey but leads to important publications, which supply further details and ongoing literature. Frequently cited literature (e. g., encyclopedias) and their abbreviations are compiled in the adjacent list.

Most of the agricultural production figures are taken from PS&D View, Users Manual and Database, Alan Webb and Karl Gudmunds, USDA Economic Research Service.

General abbreviations used throughout the dictionary are explained in a list at the end of this foreword.

The editor wants to thank his team of authors for the friendly and constructive cooperation:

Dr. Volker Böllert, Hamburg, who contributed the cosmetic-related keywords, Prof. Dr. René Csuk, Halle, who took care of the medical and pharmaceutical aspects, Dr. Frank Hirsinger, Düsseldorf, who compiled most of the biological and agricultural keywords, Dr. Friedrich Schierbaum, Potsdam, who worked out all keywords related to starch, saccharides and special polysaccharides, and Barbara Zoebelein, New York, who compiled all the fragrance information.

The editor also wants to thank the Henkel KGaA in Düsseldorf, a company for which he worked for more than 30 years and which is one of the few chemical companies that processes and uses RR on a broad industrial scale, for its general support and the permission to make use of the expertise of the following Henkel experts:

Dr. F. Bongardt (lubricants), G. Egeland (paper additives), Dr. D. Feustel (polymerization additives), Dr. C.P. Herold (oilfield chemicals), Dr. K. Hill (alkylpolyglucosides), Dr. R. Höfer (polymerization additives, defoamers), D. Köppl (mining chemicals), Dr. R. Mathis (fibers and fiber finishes), U. Nagorny (coatings and additives), Dr. K. Schlüter (textile auxiliaries), Dr. K.H. Schmid (surfactants), Dr. H.Tesmann (surfactants and cosmetic ingredients), Dr. B. Wegemund (plastics additives), Dr. R. Zauns-Huber (leather auxiliaries). Their contributions are highly acknowledged, with thanks also to Dr. G. Klement, who coordinated and organized the cooperation with these experts.

The editor also wants to acknowledge the input (food additives, fire fighting chemicals, textile printing auxiliaries and fatty acid/protein condensates) of Dr. H.G. Smolka, W. Adams and A. Sanders of Chemische Fabrik Grünau GmbH, Illertissen, a subsidiary of Henkel KGaA.

A special word of thanks is given to Dr. L. H. Princen, Peoria, USA, who acted as a lectorer. He did not only spent a lot of his time to correct the English but also, due to his professional background, gave many hints and additional information to the subject itself.

The authors would like to thank the publisher, who had the original idea to establish a dictionary of RR and provided all necessary support to the project during the work.

Schliersee, Oktober 1996   Hans R. Zoebelein
                                                                    Editor

List of Authors

Dr. Volker Böllert
Husumer Straße 37
D-20249 Hamburg

Prof. Dr. René Csuk
Institut für Organische Chemie
Martin-Luther-Universität
Weinbergweg 16
D-06120 Halle/Saale

Dr. Frank Hirsinger
Henkel KGaA
D-40191 Düsseldorf

Dr. Friedrich Schierbaum
Kantstraße 4
D-14471 Potsdam

Barbara Zoebelein
c/o Givaudan Roure
1775 Windsor Road
Teaneck, NJ 07666
USA

Dr. Hans Zoebelein
Waldschmidtstraße 4A
D-83727 Schliersee

Abbreviations/Symbols used in this Dictionary

($\rightarrow$ = Definition in the Dictionary)

$\rightarrow$	"see" = hint to another keyword
*	frequently cited literature $\rightarrow$special list
10^6	million
10^9	billion (USA); milliard
[...]	CAS (Chemical Abstracts Service) number
a (/a)	year (/a = per year)
AGU	$\rightarrow$anhydroglucose unit
a.m.	above mentioned
a.v.	acid value
B.C.	before Christ
b.p.	boiling point (°C)
C18:0	characterization of $\rightarrow$fatty acids (C number of C-Atoms, : number of double bonds)
d.b.	dry basis
d.	decomposition
d (/d)	day (/d = per day)
DE	$\rightarrow$dextrose equivalent
d.m.	dry matter
DP	degree of polymerization
DS	$\rightarrow$degree of substitution
d.s.	dry substance
dt	decitonne (0.1 mt = 100 kg)
EC	European Community
EC-number	$\rightarrow$Enzyme Classification
ed(s).	editor(s)
e.g.	for example
et al.	and others
F.	French keyword
ffa	free fatty acid
FSU	former Soviet Union
G.	German keyword
h (/h)	hour (/h = per hour)
ha	hectare (10 000 m^2 = 2.471 acre)
HLB	$\rightarrow$HLB (Hydrophilic/Lipophilic Balance) value
i.v.	iodine value
Lit.	literature
MDG	mono/diglycerides ($\rightarrow$glycerides)
m.p.	melting point (°C)
MS	molecular substitution ($\rightarrow$cellulose ethers)
min (/min)	minute (/min = per minute)
mt	metric ton
m.w.	molecular weight
NY	New York (NY)
OH-v.	hydroxyl value
O/W	oil in water $\rightarrow$emulsion
r.h.	relative humidity (%)

RR	→Renewable Resources (→preface)
r. s.	relative sweetness (→sucrose = 100)
r. t.	room temperature (°C)
sp.	species
spp.	species (plural)
ssp.	subspecies
s. v.	saponification value
Syn.	synonym of a keyword
T_g	glass transition temperature of polymers
v/v	volume per volume
W/O	water in oil →emulsion
w/v	weight per volume
w/w	weight per weight

In its description a keyword is abbreviated with the first letter.

Frequently Cited and Partly Abbreviated Literature

(marked under "Lit.:" with an asterisk *)

Arctander*
Steffen Arctander, "Perfume and Flavor Materials of Natural Origin" (published by the author), Elisabeth NJ. (1960) or Steffen Arctander "Perfume and Flavor Chemicals" (published by the author), Montclair NJ. (1969)

Blanchard*
P. H. Blanchard, "Technology of Corn Wet Milling and Associated Processes" Elsevier Amsterdam/London/NY/Tokyo (1992)

Cobley*
L. Cobley "An Introduction to the Botany of Tropical Crops" Longman, London (1976)

Eggersdorfer*
M. Eggersdorfer, S. Warwel, G. Wulff (eds.) "Nachwachsende Rohstoffe, Perspektiven für die Chemie" VCH Verlagsgesellschaft Weinheim/NY/Basel/Cambridge/Tokyo (1993)

Encycl.Polym.Sci. Engng.*
"Encyclopedia of Polymer Science and Engineering" H. F-Mark, N. M. Bikales, C. G. Overberger, G. Menges, J. I. Kroschitz (eds.) 2. edition, A Wiley-Interscience Publication, John Wiley & Sons, NY/Chichester/Brisbane/Toronto/Singapore (1985–1989)

Falbe*
J. Falbe "Surfactants in Consumer Products" Springer Verlag Berlin/Heidelberg/NY/London/Paris/Tokyo (1987)

Fehr/Hadley*
W. R. Fehr, H. H. Hadley (eds.) "Hybridization of Crop Plants" ASA & CSSA Publishers, Madison (1980)

Franke*
W. Franke, "Nutzpflanzenkunde", Thieme, Stuttgart/NY (1981)

Gildemeister*
E. Gildemeister, F. Hoffmann "Die ätherischen Öle" Akademie Verlag, Berlin (1966)

Gunstone*
F. D. Gunstone, J. L. Harwood, F. B. Pradley "The Lipid Handbook" Chapman & Hall London /NY (1994)

Kirk Othmer*
"Kirk-Othmer Encyclopedia of Chemical Technology" J. I. Kroschwith, M. Howe-Grant (ed.) 3. edition: 1978–1984, 4. edition: since 1991; A Wiley-Interscience Publication, John Wiley & Sons, NY/Chichester/Brisbane/Toronto/Singapore

Levring*
T. Levring, H. A. Hoppe, O. J. Schmidt "Marine Algae: A Survey of Research and Utilization" Cram, DeGruyter & Co, Hamburg (1969)

Lichtenthaler*
F. W. Lichtenthaler (ed.) "Carbohydrates as Organic Raw Materials" VCH Verlagsgesellschaft, Weinheim/NY/Basel/Cambridge (1991)

Martin/Leonhard*
J. H. Martin, W. H. Leonhard "Principles of Field Crop Production" Macmillan NY (1967)

Purseglove*
J. W. Purseglove "Tropical Crops, Dicotyledons" Longman, London (1974)

Rehm/Espig*
S. Rehm, G. Espig "Die Kulturpflanzen der Tropen und Subtropen" Ulmer Verlag Stuttgart (1995) or the English edition: "The Cultivated Plants of Tropics and Subtropics", Ulmer Verlag (1994)

Rehm/Reed*
H.-J. Rehm and G. Reed (eds.) "Biotechnology, a Comprehensive Treatise in 8 Volumes", Verlag Chemie, Weinheim, since 1981

Röbbelen*
G. Röbbelen, R. K. Downey, A. Ashri (ed.) "Oil Crops of the World" McGraw-Hill, NY (1989)

Ruttloff*
H. Ruttloff (ed.) „Lebensmittelbiotechnologie, Entwicklungen und Aspekte", Akademie-Verlag, Berlin (1991)
and
H. Rutloff (ed.) „Industrielle Enzyme", Behr's Verlag, Hamburg (1994)

Rymon-Lipinski/Schiweck*
G. W. von Rymon-Lipinski, H. Schiweck (eds.) „Handbuch Süßungsmittel, Eigenschaften und Anwendung", Behr's Verlag Hamburg (1991)

Salunkhe/Desai*
D. K. Salunkhe, B. B. Desai "Postharvest Biotechnology of Sugar Crops" CRC Press Inc., Boca Raton, Florida (1988)

Schenk/Hebeda*
F. W. Schenk, R. W. Hebeda (eds.) "Starch Hydrolysis Products, Worldwide Technology, Production and Applications" VCH Publishers, Inc. NY/Weinheim/Cambridge (1992)

Tegge*
G. Tegge „Stärke und Stärkederivate" Behr's Verlag, Hamburg (1988)

The H&R Book*
"Guide to Fragrance Ingredients" English edition Johnson publication Ltd., 130 Wigmore Street, London

The Merck Index*
"An Encyclopedia of Chemicals, Drugs and Biochemicals" 11th ed. Merck & Co., Inc. Rahway (1989)

Ullmann*
"Ullmann's Encyclopedia of Industrial Chemistry" B. Elvers, St. Hawkins, W. Russey (eds.) 5. edition: since 1985 VCH Verlagsgesellschaft mbH, Weinheim/Basel/Cambridge/NY/Tokyo

Van Beynum/Roels*
G. M. A. van Beynum, J. A. Roels (ed) "Starch Conversion Technology" Marcel Dekker Inc. NY/ Basel (1985)

Renewable Resources

History, Definition, Status and Outlook

I. History

"Renewable Resources" (RR) is a modern term for an old subject. It arose in the time between 1973 and 1979 when the price of crude oil, the major base for fuels, energy and chemical feedstock, was lifted by the Organization of Petroleum Exporting Countries (OPEC) in two steps from about 2 US$/barrel to 30 US$/barrel and more. This motivated politicians, economists and industrial leaders to consider the possibility to intensify the use of RR.

RR were the only sources of carbon that the organic chemical industry and its predecessors (craftsmen, small businesses) had available before the middle of the 19th century. The use of products from nature, not only for food but also as "functional" products, is one of the criteria that sets early human beings apart from animals. The use of wood as fuel and construction material, or the making of soap from fats and oils with the help of the (pot)ash from the campfire are early examples of functional uses. During the long history of human development, man learned to tan leather, to make glue from bones, to dye textiles, to make paint for decoration of his cave walls, to ferment carbohydrates to alcohol, to make vinegar, to make coatings to protect wood structures, to make paper, etc. All these developments were based on RR exclusively.

Early scientists claimed that organic compounds, based on carbon, can only be produced with the help of the so-called *"vis vitalis"*. With his successful experiment to make urea from ammonium cyanate, Wöhler (1834) made this theory obsolete. From then on, organic chemistry *in vitro* was investigated intensively and developed rapidly.

With the growing steel industry in the middle of the 19th century, increasing amounts of coke were in demand. The manufacturing of this coke yielded increasing amounts of coal tar, which proved to be a valuable by-product, useful for the synthesis of colors and dyes.

For energy production, wood was used but was gradually substituted by coal. The starting automobile industry used petroleum-based gasoline in increasing amounts and later created the markets for the by-products fuel oil, naphta and bitumen.

The generation of electricity by water power or coal-fueled power stations was the next step. In the early years of the 20th century, discussions were held either to bring electricity to the households by wire or to provide energy in the form of calcium carbide as "stored electricity", whereby acetylene is generated in the home by the addition of water and used for illumination and fuel. When further development favored the wire alternative, there was already a rather large CaC_2 capacity in operation or under construction. An alternative use of this potential was found with the development of acetylene chemistry, which made coal an important resource for organic chemistry by substituting or complementing RR-based chemistry. Acetaldehyde, acetone, acetic acid and dimethylbutadiene were among the first new products manufactured on an industrial scale.

In the 1920's and 1930's, the term "chemurgy" was coined in the USA, initiated by agricultural surpluses and the desire of some economists to be less dependant from imports during war times. Chemurgy was understood as a move, still modern today, to intensify the cooperation between agriculture and chemical industry.

During and especially after the second world war, naphta derived from crude oil became the number one raw material of organic chemistry in tremendous volumes. It was inexpensive, available in large quantities and easy to handle and process, and petrochemistry became the predominant factor in the chemical industry. Traditional resources lost importance, research was focusing only on the new raw material and the chemistry connected with it. Old and new markets and applications were sucessfully conquered by petrochemistry. Development seemed to be unlimited. Only the mentioned price increases for crude oil in

1973 and 1979 ("oil crisis") made the vulnerability of this base visible and stimulated alternative thinking. The lack of consensus and discipline in the OPEC cartel has led to a decrease of the crude oil price in the past few years. Today, the price level is higher than before the crisis years but only half of what it was during those years.

II. Definition

To define what RR are, it is necessary to have a brief look at the sources available for mankind for nutrition, for energy, for functional and technical applications and for chemical conversion.
   The following scheme may be useful:

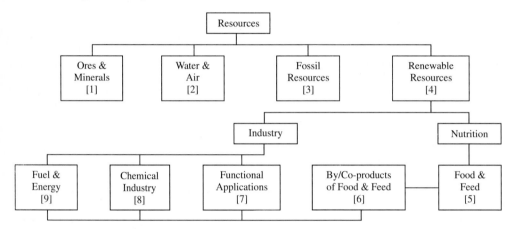

   [1] Ore and minerals are the sources for inorganic chemicals, metals and construction materials. Especially, some of the metals are not available in abundance. Their use in the future must be kept under tight control because otherwise these metals will be used and spread around in concentrations too low for any economic recovering. Restrictions in use, strong waste management and the application of modern recycling technologies are the only possibility to avoid severe shortages in industrial applications and important markets.
   [2] Water and air are important resources for the biological reaction called →photosynthesis:

$$6\,CO_2 + 6\,H_2O + \text{sunlight} \longrightarrow C_6H_{12}O_6 + 6\,O_2$$

This reaction, also called assimilation, is the fundament of all life and generates →biomass and all our RR. The reverse reaction is called oxydation, respiration or combustion:

$$6\,C_6H_{12}O_6 + 6\,O_2 \longrightarrow 6\,CO_2 + 6\,H_2O + \text{energy}$$

Water and air are also raw materials, used directly or as secondary products, in many chemical processes, e. g., hydrogen, oxygen, nitrogen, carbon dioxide. To maintain quality and availability is an important issue for our modern society.
   [3] Fossil resources (i. e., coal, peat and lignite, crude oil and natural gas) are, according to the prevalent theories, derived from plants and/or animals, which were taken out of the biological cycle millions of years ago. These resources are limited in volume, especially because our consumption today is much faster than their generation. This means, that the present rate of exploitation of these resources can never be in agreement with a concept of sustainability and responsibility for future generations. Regardless of all ups and downs of their prices, their limited availability precludes an upward trend in their utilization in the long term. Resources are still huge and new occurrences are found, which allows to postpone decisions for a while, but a growing demand will make it necessary to exploit sources that are decreasingly accessible and need ever more sophisticated and expensive techniques.

In the near future, fossil resources will remain the most important sources for energy, fuel and organic feedstock for the chemical industry. The latter is the most valuable use and will remain so, even if oil is too expensive for energy and fuel. It is said, that the world's last drop of oil will run through an ethylene cracker!

[4] An interesting alternative to [3] is the subject of this dictionary: renewable resources (RR). This subject is discussed in more detail below.

[5] Food for humans, feed for animals (which in turn serve as food for humans) are the main reason for first collecting and later breeding and crop production. Today, this is still the dominant motive for modern agriculture.

Some food products are also permanently used industrially, others only in times of surpluses or low price levels [7–9]. For example →coconut oil, →soybean oil, →corn and →potato starch are used industrially as well as for food. Thus, ≈ 14 % of all →fats and oils produced worldwide are used in →oleochemistry.

[6] It is highly desirable, also from an ecological point of view, if wastes and materials that do not meet the specifications of nutrition are utilized industrially. These co-and by-products of food and feed production, processing and purification are often well accepted for industrial use. Examples in the area of meat production are →wool, hides, bones and fats (e. g., →tallow), while plants may supply →cotton linters, →straw from grain, →bagasse, →fatty acids from soapstocks (foots), →tocopherol and phytosterols (→steroids) from deodorizer distillate from edible oil refining.

It depends on the manufacturer's intention whether wheat starch (→wheat starch production) or →wheat gluten is the "main product". Other examples are →soybean oil and soybean meal or →cotton and →linseed oil. In cases like this, it may be better to use the term "co-product".

[7]–[9] Industrial crops, i. e., plants that are only grown for industrial applications, are used for fuel/energy, functional applications and chemical feedstocks. As mentioned above, food and feed products [5], i. e., products grown and designed for nutrition, and the by-products of food processing [6], are also serving all these areas.

The only energy crop [9] used in substantial quantities is →wood, especially in less industrialized countries.

For functional products [7], wood is an important material for construction. Two-thirds of the wood harvested is used for this purpose. The remainder is used for paper pulp and as chemical raw material.

Other plant products used in functional applications, with almost no chemical but physical processing, are →cotton, →silk, →wool, →flax. Sometimes, their by-products (e. g., cotton linters) are used as chemical feedstock.

Plants that are only grown to gain products for chemical manufacturing include →castor oil and →tung oil. High-erucic →rapeseed is another example. Several plants are grown for medical use or application as fragrances. Most such plant ingredients need further chemical or biotechnological modifications.

The number of specific industrial crops is increasing, and many of them are available in two versions: one designed for food use and another tailored to industrial needs.

From the above, the following definition for RR can be formulated:

RR are products, originated from plants and animals, used for industrial purposes (energy, functional applications and chemical modification) and include also food products that are not used for nutrition, as well as wastes and co-products of food processing.

Another term used frequently in context with RR is →"biomass", which has different, somewhat unclear definitions:

– the total organic matter in our ecological system;
– the total plant mass produced by photosynthesis;
– the organic waste available for production of cellular substance via fermentation.

In general: Biomass is nonspecific and large in volume, RR are specific raw materials for industrial and functional uses.

There is a tremendous volume of "biomass" (total plant mass produced by photosynthesis) that is available theoretically: $170-200 \times 10^9$ mt grow annually on our planet, and only less than 3% is used by men, mostly for nutrition. Out of this, again less than 3% is used industrially.

III. Status and Outlook for Fuel

Status:

Wood and wastes from food production have been sources for energy for heating and cooking since prehistoric times. In many nonindustrialized countries and economies without sufficient fossil energy "carriers", they are still in use. Today, modern societies with their high per-capita consumption depend almost exclusively on fossil or nuclear energy. However, there are many development projects for alternatives, including the use of biomass as a source of energy. Because RR carry some O-functionality (they are partly oxidized), their energy content must always be lower than that of hydrocarbons.

Somewhat different is the status of fuel for transportation. The first steamship and railroad engines used wood as fuel. Coal was the successor. During the second world war, German cars (especially trucks) had built-in gas generators that ran on wood chips. This inconvenient substitute for petroleum-based fuel was abandoned at the end of the war. With the oil crisis of the 1970's, the use of ethanol, made by fermentation of carbohydrates, as a partial substitute (max. 25%) added to normal gasoline and used in normal engines, was considered (Gasohol project in the USA). In addition, Brazil started to manufacture car engines that operate on 100% fermentation alcohol. Fermentation technology was improved in terms of yield and ecology. Many pilot plants were established worldwide. Wastes can be used for biogas production.

Another idea is the use of plant oils (palm oil in Malaysia, rapeseed oil in Europe) as a substitute for diesel fuel. When degummed oils (trigycerides) are used, the diesel engine has to be modified and adjusted to this fuel. Normal diesel engines require the methyl esters derived from these oils. This makes an additional chemical modification (→transesterification) necessary. Large-scale trials have shown, that in principle both products, the oil and the ester, are running well technically but need further developmental effort.

There are also some projects that focus on the thermal cracking of biomass in the presence of hydrocarbons. This may stretch petroleum feed. Also, some trials are running to use biomass as a source for synthesis gas, which can be hydrogenated to fuel components.

For further details: →fuel alternatives.

Outlook:

Projects to use biomass as an energy source are numerous. Only those that make use of by-products/ wastes of certain bio-based processes within the "own fence" appear to be realistic in terms of economics. Examples are wood wastes, such as saw dust used in the power stations of wood processing plants (saw and paper mills); lignin and hemicellulose wastes in paper pulping plants; slop (vinage), which is used to make biogas in fermentation plants; straw or other wastes of farming.

If biomass is produced with the primary intention to use it as an energy source, its economic value is much more difficult to calculate. →China grass (*Miscanthus sinensis*) and fast-growing trees (poplar and eucalyptus) are being considered, but none of these projects has been commercialized. One exception is eucalyptus, grown in Brazil in huge plantations, the wood of which is used to make charcoal on a large scale for steel manufacturing. The main problem is the energy density, i.e., the tremendous volume of these materials in relation to their energy "concentration". Logistic problems are evident.

"Fuel for cars" is much more in demand than the mere energy use of biomass. As already stated, the use of ethanol as complete or partial substitute of gasoline has proven its technical maturity on a large scale. Somewhat less experience has been collected with diesel substitutes. However, there is no doubt that all these alternative fuels work well technically, although some improvements are necessary. The major problem is economy. If not subsidized by governments, none of these alternatives could reach the present economic level at the current price of petroleum-based fuels. In most economies, petroleum-based fuels are a substantial tax income, while biomass-based alternatives would be a tax consumer, at least unless the crude oil price reaches a level that is far above that of today. There are many advantages and disadvantages connected with the use of biomass as fuel base, which are difficult to evaluate as to their future significance.

Some of these factors are as follows:

Pro:
- Saves fossil resources for more valuable use than burning;
- Generates (with some limitations due to use of fossil energy for harvesting, fertilizer production, process energy) as much CO_2 during burning as was taken out of the atmosphere during growing, thus it does not contribute to the climate problem;
- In many countries, it provides independence from imports (crisis situations, hard currency problems);
- It creates opportunities for otherwise obsolete agricultural areas and labor;
- It extends the possibilities of crop rotation.

Con:
- Not yet economic; need for subsidies;
- It requires high investments (fermentation plants, transesterification units), which make the move to other interesting alternatives difficult (cementing of a technology that might remain or become uneconomic in the long term);
- It provides only weak economic contributions of by-products (glycerol, rapeseed meal);
- It may require irrevocably adjusting or exchanging of engines and motors;
- There are ecological problems in production (effluent of fermentation with high oxygen demand during biodegradation);
- Production is more labor-intensive than crude oil-based materials;
- There are high losses of biomass in the case of gasohol (half of the biomass is transferred to CO_2) and energy demands to concentrate the broth;
- It may require huge agricultural monocultures.

Summary:
All this leads to the following outlook: Ethanol as a major fuel is presently not under consideration and will not be in the near future. For the Gasohol project in the USA today, ethanol is seen as an anti-knocking additive, which might be of interest as a substitute for methyl-tertiary butyl ether (MBTE), which is already expected to eventually cause ecological problems. In Brazil, there is some tendency to reduce alcohol production for economical and ecological reasons, but social considerations and the investment in 100% alcohol-fueled cars (once subsidized by government) do not allow a fast termination.

Triglycerides may be preferred for diesel substitution because methyl esters need an additional cost-increasing step that requires substantial investments and professional handling (methanol). Triglyceride oils demand the introduction of special engines, which can be used only for this fuel (see problems in Brazil mentioned above).

IV. Status and Outlook for Functional Use

Status:
Wood for construction and natural fibers (cotton, silk, wool, flax) are still in great favor. It took the petrochemical industries decades to convince the consumer that their products are at least equivalent to RR and to get rid of the aftertaste of a "wartime substitute". It is, however, obvious in many keyword entries of this dictionary that RR lost ground. Their volume is rather stable, but the strong increases of many markets are served by petro-based products. Arguments to save the woods worldwide and to reduce extreme exploitation have contributed to the reduction of uses of wood in furniture, lumber window frames, etc.

Outlook:
There is a tendency for a change again. If one ensures that trees are planted at the same rate as they are being cut (sustainable forestry), there is also no objection to use wood for construction. Natural fibers are

in fashion again, and new methods of growing and harvesting flax and hemp may lead to a renaissance of these good old fibers. Modern technologies in cellulose pulping and spinning give also optimism for stronger use of cellulosic fibers.

V. Status of RR in Chemical Applications

Status:

Today, RR are in an active stand-by. Nevertheless, 10% of the organic carbon processed, modified and used in the chemical industry in Western Europe is of RR origin. This is equivalent to all organic chemicals based on coal. Their use has been increasing gradually since the two price shocks in the 1970's and has lead to a moderate revitalization of the RR scenario.

Restrictions are that presently available crops are mainly designed and grown for nutritional purposes and not for industrial use. Basic and applied research in universities has neglected RR for the last 40 years. Therefore, the use of RR-based organics has been limited to certain areas over many years.

On the other hand, attempts of petrochemical synthesis to conquer areas where RR dominate did not lead to much success: Synthesis of fatty acids from petrochemical resources has never materialized, with the exception of paraffin oxidation in Germany during the war and in Russia thereafter. Fatty acids were always the prominent domain of RR. Glycerol synthesis via propylene is almost entirely abandoned today, and natural glycerol is in use. Fatty alcohol synthesis based on olefin chemistry, which was and is a large-scale operation, is stagnant today. One petrochemical producer is running a synthetic alcohol plant in combination with a hydrogenation unit for RR-based feedstock in order to have maximum flexibility. Water-soluble synthetic polymers, such as polyvinylalcohol and polyacrylamide, once seen as substitutes for cellulose- and starch-based polymers, are limited to certain applications, and the RR-based polymers have regained or opened new markets with new biopolymers.

There are areas in which petrochemical-based products are dominant and others where RR-based materials have their domain; there is a large overlap in between, which is a battlefield of price and performance, with one gaining today and loosing tomorrow.

While hydrocarbons have no functionality and can be used in principle for the synthesis of almost all organic compounds, RR contain some O-functionality. Therefore, RR-derived products are preferred if the synthetic input of "mother nature" can be utilized.

Outlook:

The use of RR as chemical feedstocks is presently stagnant with a tendency to slight increases. This indicates that markets, easy to access with the present spectrum of RR, are covered.

RR will make stronger inroads into entirely new positions or those now held by petrochemicals, if one or more of the following events will happen:

- an increase in price of crude oil and naphta feedstock
  There is an inherent long-term tendency for price-increases if a raw material is finite. More expensive techniques are necessary to maintain the present production level. One has to consider that prices of the various crude oil fractions are the result of a mixed calculation, where the chemical use of the naphta segment is small in volume but has the highest incremental value for the end products. A few cents more for gasoline gives tremendous price flexibilty for the naphta fraction. The price increase for crude oil may not only come from normal market forces but could also be implemented by legislation and artificial measures due to environmental or other protective considerations (e. g., CO_2 tax).

- a decrease in price or an increase in quality of RR
  RR must and can be tailored to chemical use in yield and composition. This will strongly enhance applications and create additional markets for agricultural products. For example, lower prices for fats and oils have favored the use of alkyd resins in industrial coatings during the last years compared to petrochemical resins.

Natural sources always supply composite materials with a mixture of a "main product", many "co-and by-products" and some minor components, potentially of high value. This creates a problem because the ratio of all these components never correlates with the ratio desired by the markets. For a successful business, all components must be used at their maximum value.

Modern plant breeding, supported by recombinant DNA techniques, is a tool to accomplish modifications in order to meet the ratio of the components desired.

The chemical industry is not buying rapeseed oil as such, like the food industry does, but as a source of →erucic acid. Increasing the content of this acid in the seed oil (today ≈ 50%) by plant breeding will give it added value for industrial applications.

There are some starch sources with high amylose content and others with a high percentage of amylopectin. Again, plant breeding can help to modify the contents of both components or to adjust a certain ratio for a specific use.

Many projects that focus on this concept are under way and will bring new impacts and ideas to RR in order to make them competitive in new areas by normal market mechanisms.

A price reduction by public subsidies should not be seen as a desirable alternative. Such support is acceptable for R+D and pilot projects only.

- More activities in nonindustrialized countries
  Developing countries may use their opportunities, their cheap labor, their land and their favorable climate and pick up ideas and projects, which are not or cannot be accomplished in the countries where they were developed. It is a reasonable concept of economic development to base it on local opportunities and go down-stream later.

- More basic and applied research for RR
  New, cheaper and more ecologically oriented processes for growing, isolation (e.g., →wood pulping), modification and derivatization (e.g., better accessibility of the →cellulose molecule) as well as new areas of application will promote the use of RR. Modern methods (e.g., biotechnology) applied to an old, scientifically long neglected area will create new opportunities. In this effort, the development of new RR-based specialties is more promising than the attempt to substitute petrochemical commodities.

- Increasing importance of ecological advantages of RR-based products
  RR-based products are natural products that often fit better in biological degradation cycles than an entirely synthetic product. This argument can be used by sales promotion of the manufacturer, convincing the consumer to buy products made from RR, even if they are a little more expensive. Legislation can support this tendency. A high degree of derivatization reduces this advantage of RR-based products and makes them ecologically comparable to petrochemical alternatives.

- For RR-based products, "sustainable concepts" can be developed.
  Sustainability means that products grown, produced, processed and used may leave no economical or ecological burden to following generations.

VI. A General Comparison of RR versus Fossil Raw Materials

The following synopsis compares fossil (petroleum) and RR-based products in practical use:

Petroleum:	RR:
homogeneous	heterogeneous/large variety
concentrated/compact	diluted/large volume
limited availability	unlimited availability/use of fallow land
no sustainable concept	fits in modern sustainable concepts
only a few suppliers	almost every country can be producer
established technology of the rich	high development potential for poor countries
high internal energy	low energy density
climate problems (CO_2)	less climate problems
low waste problems in production	some waste problems (by- products) in production
fully used at high value	only parts of high value, many low-value by-products
established, sophisticated technology involved	only farming and isolation are sophisticated technologies (food)
limited chances of further development	chances of improvment by further development (e. g., plant breeding)
high flexibility to produce almost all organics	makes use of the synthetic work already done by nature; but limited flexibility

One can assume from this synopsis that RR have advantages

- in their unlimited, regenerative availability;
- in reduced CO_2 output (climate aspect);
- in bringing jobs and economical impacts to developing countries;
- in their high potential for improvement (e. g., plant breeding);
- in new crops, new chemistry and technology and applications.

A change in crude oil price, environmental legislation, the necessity to use fallow land in countries with surplus agro-production and specific development of industrial crops may support the tendency to make more use of RR.

A

Abacá
→Manila Hemp

Abietic Acid
→Rosin

Abrasin Oil
→Tung Oil

Absolute
G.: Absolute; F.: essence absolue

A. is a highly concentrated, entirely alcohol-soluble, normally liquid perfume and flavor raw material, reflecting the entire hydrocarbon-soluble odorous principle of a clearly defined (species/origin) plant material. It is obtained by alcohol →extraction of →concretes, →resinoids or →pommades. These products are mixed and stirred in alcohol at r.t. or under gentle heat. The well-mixed alcoholic extract is then chilled under stirring for a considerable time (up to several weeks). Waxes, sesqui→terpenes and most odorless matters precipitate and are removed by filtration or centrifugation. Subsequent recovery of the alcohol by evaporation yields a.

Absorbents, Water-
→Superslurper

7-ACA
→7-Amino-cephalosporanic Acid

Acacia
Acacia senegal (L.) Willd.
Leguminosae
G.: Akazie, Gummi arabicum Baum;
F.: acacia à gomme

A. are wild and semi-cultivated trees in the Sudan in areas with a rainfall as low as 300 mm/a. However, a. is widespread in the dry savannas of tropical Africa and extends to the Red Sea and Eastern India. The spiny tree is up to 12 m tall with fragrant creamy-white flowers borne on spikes 5–10 cm long. The gum called →gum arabic is gained from cuts in the bark of the tree, and the exudate is allowed to dry on the bark.

The Sudan has exported the gum to Europe and the Middle East for over 2000 years.
Lit.: Cobley* (1976)

Acacia senegal
→Acacia

Acarbose
G.: Acarbose; F.: acarbose

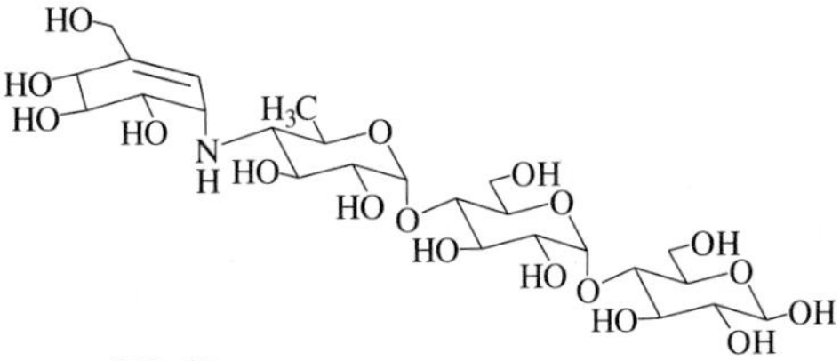

m.w.: 645.63

A. is a pseudotetrasaccharide containing an unsaturated cyclitol moiety. A. is an α-glucosidase inhibitor that reduces sugar uptake in the gastrointestinal tract. It is isolated from strains of *Actinoplanes*, is used as an antidiabetic, and has potential use in the prophylaxis of dental caries.

Lit.: J. Creutzfeld (ed.) "Acarbose for the Treatment of Diabetes mellitus", Springer, Berlin, (1988) [56180–94-0]

Acetem
→Acetic Acid Esters of Mono/Diglycerides

Acetic Acid Esters of Mono/Diglycerides
Syn.: AMG; Acetem; Acetylated Mono/Diglycerides
G.: Aceto-mono/diglyceride;
F.: esters acétiques de mono/diglycérides

$CH_2-O-CO-R^1$
$CH-OR^2$ R^1: C_{15-17} alkyl
$CH_2-O-CO-CH_3$ R^2: H or $-CO-R^1$

s.v.: 260–340; m.p.: 32–46 °C

The physical (including s.v. and m.p.) and practical properties of a. depend on the degree of acetylation (normally 50–90%), the ratio of mono- to diglycerides and the kind of fatty acid used. The formula above is schematic.

A. are more stable against heat and →hydrolysis than the →lactic acid esters and →citric acid esters of mono/diglycerides.

A. are produced by reaction of →mono/diglycerides with acetic acid anhydride or by →transesterification of triglycerides (→fats and oils) with triacetin (→glyceryl triacetate) in the presence of additional →glycerol.

Uses are in cakes and other baked goods. The excellent film formation, flexibility and elongation are used to protect food articles by dipping them into molten a. They are also used in shortenings, cakes and toppings and as an approved food lubricant.

US consumption (1981) was 450 mt. European consumption (1993) was 2000 mt.

Lit.:   G.Schuster, "Emulgatoren in Lebensmittel" Springer Verlag Berlin/Heidelberg/NY/Tokyo (1985)
N.J.Krog "Food Emulsions" K.Larsson, S.E. Friberg (eds.), Marcel Dekker NY/Basel (1990)

Acetin
→Glyceryl Monoacetate

Acetone
→Acetone/Butanol Fermentation

Acetone/Butanol Fermentation
G.: Aceton/Butanol-Fermentation;
F.: fermentation acétone/butanol

Acetone and butanol, two important organic chemicals, are nowadays almost exclusively made by chemical synthesis. During the second world war, however, a fermentation process was used in many countries for the production of a mixture of acetone, butanol and ethanol. Some members of the *Clostridium* genus are able to transfer cornmeal or molasses into the desired mixture of these chemicals, the ratio of which is determined by the kind of raw material and the bacteria species used. The maximum overall concentration obtainable is 2%. The process was abandoned after the war in most countries but returned into discussion when oil prices increased in the 1970's. It is not very likely that this process will be used again unless the technology is improved and raw material prices are favorable.

Lit.: Microbiol.Rev. **50**, 484–524 (1986)

Acetylated Mono/Diglycerides
→Acetic Acid Esters of Mono/Diglycerides

Acetylated Tartaric Acid Esters of Mono/ Diglycerides
→Diacetyl Tartaric Acid Esters of Mono/Diglycerides

Acetylcellulose
→Cellulose Acetate

Acid Modified Starches
→Thin-boiling Starch

Activated Carbon
G.: Aktivkohle; F.: charbon actif

A. is a material with many fine pores, the surfaces of which have a small amount of chemically bound oxygen and hydrogen. Commercial products have an inner surface area of 500–1500 m^2/g.

The properties depend largely on raw material used, processing, pore size and distribution.

Raw materials are wood (→charcoal), saw dust, nut shells (esp.→coconut shells) and fossil sources, e. g., peat, lignite, bituminous coal.

It is manufactured either by carbonization of carbonaceous materials with simultaneous chemical activation or by gas treatment of already carbonized material (e. g., charcoal). Chemical treatment is carried out at 400–1000 °C with phosphoric acid or zinc chloride and gas activation at 800–1000 °C in the presence of water or carbon dioxide. Several types of kilns and furnaces are in use.

The absorbent property of a. is used in gas and water purification, in solvent recovery, for decoloration and purification in food, chemical and pharmaceutical industries.

Well known is also its use in gas masks and treatment of diarrhea. A. also serves as a catalyst and carrier for catalysts and is an important →mining chemical in gold recovery.

The world capacity is estimated to be 380 000 mt/a.

Lit.: Ullmann* (5.) **A5,** 124
[7440-44-0]

Acylamino Alkane Sulfonates
Syn: Taurates; Taurides
G.: Tauride; F.: taurides

$$R-CO-N-CH_2-CH_2-SO_3Na$$
$$|$$
$$CH_3 \qquad\qquad R: C_{11-17} \text{ alkyl}$$

A. are made by reaction of fatty acids or their chlorides with N-methyltaurine at 20–30 °C. Aqueous solutions of 40% are obtained, which can be spray-dried to recover a powder.

The →surfactants obtained are skin-friendly, insensitive to water hardness and have good foaming
and soil-suspending properties. They are used in
the textile industry, in →cleaners and in shampoos
(→hair preparations) and toothpastes (→oral hygiene products).

Lit.: Ullmann* (5.) **A25,** 747
        Kirk Othmer* (3.) **22,** 354

Acyloxy Alkane Sulfonates
Syn: Isethionates
G.: Isethionate; F.: acides iséthioniques

R–CO–O–CH$_2$–CH$_2$–SO$_3$Na   R: C$_{11-17}$ alkyl

For manufacturing, sodium isethionate and fatty
acids or their chlorides are reacted at $100-120\,°C$.
A friable solid is obtained.

The products are used as →surfactants with good
foaming and emulsifying properties. Because they
are nonirritating to the skin, they are used in cosmetics and →cleaners. Technical uses are in the
textile industry.

Lit.: Ullmann* (5.) **A25,** 747
        Kirk Othmer* (3.) **22,** 354

N-Acylsarcosinates
→Sarcosinates

Adeps Lanae
→Lanolin

Adhesives
Syn.: Glues
G.: Klebstoffe; F.: adhésifs, colles

An adhesive is a material, that binds two similar or
different substrates firmly together. Cohesive forces
give strength to the adhesive layer itself, while adhesive forces are responsible for the bond strength
to the surface of the substrate. Aside of this basic re-
quirement, there are many other criteria important
in formulating and selecting of a., which depend on
the type of the final application.

Early a., such as animal glue (→glue, animal) and
→sealing wax, were based on RR exclusively.
Modern adhesives make use of a broad variety of
synthetic polymers and additives, especially in
structural a. used in construction, automotive assembly and aircraft industry.

Uses of RR-based a. (the following is only focusing on these) are limited, with some exceptions, to
less demanding and traditional but still large-volume applications:

Starch-based a.:
Starch of different origin (corn, potato, wheat,
manioc) is still a major a. raw material (→starch
industrial applications). Starches modified by different physical or chemical treatments (→pre-
gelatinized starch, →starch derivatives, →dex-
trins), are used in the packaging industry for the
manufacture of corrugated board, paper and board
lamination, bag and envelope manufacturing, top-
flap gumming of envelopes, carton and case sealing, board tube winding, cigarette manufacturing
and packaging, bottle labeling and as wallpaper
paste.

Cellulose-based a.:
Mainly →carboxymethyl cellulose and →methyl
cellulose are used in wall paper paste. In some
countries starch-based a. are preferred for this application.

Protein-based a.:
→Casein is used in glues (frequently combined
with starch) for icewater-proof bottle labeling.

The glues (→glue,animal) made from bones and
hides -formerly the only wood glue- are substituted
mainly by synthetic a. These collagen- (→proteins)
based a. are still used in the manufacture of abrasives, gummed paper tapes, book binding, match
heads and boxes.

Natural rubber-based a.:
Pressure-sensitive a. are still based on natural rubber (→rubber, natural) either in solution or latex
form. Its high tack properties are appreciated, but
the sensitivity to thermal and oxydative degradation is detrimental to some applications.

Polyamides based on oleochemicals:
→Dimer fatty acid is the basis for some →poly-
amides used as hot-melt a. in the shoe industry,
cable-joint a. and for other special applications
such as textile interlinings or nonwovens. Poly-
aminoamides are used as hardeners in epoxy a.

Rosin and resin-based A.:
Wood →rosin and natural resins (→resins, natural)
and terpene resins (→terpenes) are used as ingredients in hot-melt a.

Castor oil in polyurethanes:
In polyurethane a., →castor oil and polyols derived
from it are used as reactive components, e.g., for
film laminating.

Other RR-based raw materials:
→Tetrahydrofuran derived from furan is one of the
few solvents for PVC and is used in PVC (pipe) a.,
which temporary dissolves and swells the surface
which forms strong bonds after fixing and evaporation.

→Sealants and →putties have also a. functions.
→Films of →cellulose acetate are used in adhesive tapes as base material.

The global adhesive market (1993) is estimated to be in the range of 4.8×10^6 mt. This figure covers all kinds of a. from low-price starch to expensive specialties. There are no figures available about the volume of RR-based a.

The trends show a preference to environmentally oriented products (biodegradable, no solvent). This may at least stabilize the market share of RR-based products.

Lit.: M.Skeist (ed.): "Handbook of Adhesives" (3.) Van Nostrand-Reinhold, NY (1990)
Kirk Othmer* (4.) **1**, 445
Ullmann* (5.) **A1**, 221

Agar (-Agar)

Syn.: Gelose
G.: Agar(-Agar); F.: agar-agar
m.w.: 110 000–160 000

A. is made up of the jellifying cell-wall heteropolysaccharides of red algae (→marine algae), also called red seaweed (some *Gelidium*-and *Gracilaria*-species, Rhodophyceae) and is a mixture of →agarose ($\approx 70\%$) and agaropectin ($\approx 30\%$).

A. is soluble in hot water but insoluble in cold water. Its structure is believed to be a complex mixture of polysaccharide chains having alternating α-1,4- and β-1,3-linkages. A. is transparent, odorless and tasteless, and is marketed in form of strips, granules and powder. Aqueous solutions containing even less than 1% of a. form solid gels (m.p. 80–100 °C, resolidifying at 45 °C).

A is gained after a labor intensive process, by hot water extraction of the dried algae material, followed by several purification steps (freezing). The main producer is Japan, followed by Taiwan, Korea, Morocco, Chile, Portugal and USA.

A. is used mainly as a nutrient medium for microbiological cultures and as a gelling agent, stabilizer and emulsifier in food (→food additives), especially in the Asian kitchen. It is also used in cosmetic gels, in the production of medicinal encapsulations and ointments, as a dental impression mold base and as a carrier for immobilized enzymes. It shows some cathartic and laxative properties and is not metabolized in the gastrointestinal tract.

Due to its high price, a. is more and more substituted by cheaper gums in many applications.

World production is ≈ 5000 mt/a.

Lit.: Kirk Othmer* (3.) **12**, 45
Ullmann* (5.) **A11**, 570
[9002–18-0]

Agarose

Syn.: Agarose Gel
G.: Agarose; F.: agarose

D-Galactose   3,6-Anhydro-L-galactose

A. is a neutral, jellifying polysaccharide (heterogalactane) consisting of alternating β-1,3-linked D-galactopyranose and α-1,4-linked 3,6-anhydro-L-galactopyranose units; it is obtained from →agar. A. is used as a matrix in chromatographic procedures, for the immobilization of enzymes or of whole cells, as well as in electrophoresis.

Lit.: Methods Enzymol. **152**, 61–87 (1987)
Ullmann* (5.) **B3**, 11–18 and **A9**, 360
[9012–36-6]

Agathis australis
Agathis borneensis
→Resins, natural

Agave sisilana
→Sisal

Agricultural Chemicals

G.: Agrochemikalien, Agrarchemie;
F.: produits chimiques pour l'agronomie

The term a. describes any kind of chemical input into agricultural production, including herbicides, insecticides, nematocides, fungicides, and fertilizers.

Its use is in opposition to organic farming practices, although compromises exist in the application of a., e. g., by integrated pest management or the application of natural enemies of phytopathogenic insects. The whole idea of a. is to increase the return of harvested materials by application of fertilizers and the protection of plants, animals, etc. from massive infestations or damage by →pests and →diseases with pesticides.

Other a. are used as seed coats to obtain a homogeneous seed size as well as uniform emergence in the field and to add fungicides and pesticides to the seed coat. Such coats normally consist of →cellulose ethers (→sugar beet).

Several other applications of a. deal with the improvement of soils by means of soil conditioners, either to improve air or water permeability of the soil, to prevent the loss of water to the subsoil, or to increase aggregation of the soil particles, thus preventing splashing of soil onto the produce. Often such soil conditioners consist of →surfactant-like substances.

See also: →Agronomy, →pesticides, →fertilizer

Agronomy

G: Pflanzenbau; F: agronomie, production végétale

A. is the biological science that deals with the principles of crop production, such as the biology of crop plants and their agricultural methods of production, such as soil fertility, tillage, fertilizer application, sowing, canopy formation, weed eradication, plant protection, harvesting, utilization of by-products, manure utilization and crop rotation.

Lit.: Martin/Leonard* (1967)

Agro-refinery
→Whole Crop Harvesting

Aleurites cordata
Aleurites fordii
Aleurites montana
→Tung

Aleurites moluccana
Aleurites trisperma
→Candlenut Oil

Algae
→Marine Algae

Algin
→Sodium Alginate

Alginates
→Sodium Alginate

Alginic Acid
→Sodium Alginate

Alizarin
→Dyes, natural

Alkaloids
G.: Alkaloide; F.: alcaloïdes

A. occur mainly in plants and are natural products (more than 10 000 are known), which contain at least one, predominantly heterocyclic, nitrogen substituent in the molecule. Many a. show marked pharmacological effects.

The following are of importance:
→Atropine, →Berberine, →Codeine, →Morphine, Nicotine (→Tobacco, →Insecticides), →Noscapine, →Opium, →Quinine, →Saponins, →Thebaine.

Alkoxylation
Syn.: Ethoxylation/Propoxylation
G.: Alkoxylierung; F.: alcoxylation, éthoxylation

In the context of RR, a. is the reaction of →fatty alcohols, →fatty amines, →fatty acids, →fatty acid ethanolamides and fatty acid esters of poly-hydroxy compounds such as mono/diglycerides or sorbitol, with ethylene oxide (EO) or propylene oxide (PO).

Another group of compounds made by a. are →hy-droxyethyl cellulose and →hydroxypropyl cellulose. Starch is modified also by ethoxylation (→starch ethers), and →alginic acid is propoxy-lated for some food applications.

The exothermic reaction is carried mostly out batchwise at elevated temperature ($\approx 150\,°C$) and $5-10$ bar ($0.5-1$ MPa) pressure with alkaline (sodium or potassium hydroxide or methylate) or acidic catalysts:

$$R-OH + n\,CH_2-CH_2 \longrightarrow R-O-(CH_2-CH_2-O)_n-H$$
$$\underset{O}{\diagup}$$

R: C_8-C_{18-22} alkyl

n: depends on the reaction conditions and can be adjusted to the desired properties

Normally, alkaline catalysts are used because there are less by-products and corrosion of the equipment. Alkaline catalysis however, leads to a broad spectrum of homologous polyglycol ethers (n = $0-\approx 12$), which may cause problems later in detergent manufacturing (pluming, which is the formation of aerosols in the exhaust of the spray tower). Sometimes they have advantages in emulsification behavior. Acidic catalysis leads to a narrower distribution of EO units but can yield undesired by-products, e.g., dioxane. Many attemps have been made to improve alkaline as well as acidic catalysis. There is some practical use of Mg/Al hydroxy-carbonate (hydrotalcid) or alkaline-earth salts of modified fatty acids, which give narrow-range al-kyoxylates with less undesirable by-products, less odor, better solubility, lower volatility (pluming) and better thickening properties.

Production safety requires respect for the toxic and explosive character of EO.

The resulting products are the most important class of nonionic →surfactants. EO is the major reagent, PO is used to modify the hydrophilic/hydrophobic balance (→HLB value) of the surfactant.

For product details: →fatty alcohol ethoxylates and propoxylates →fatty amine ethoxylates and →fatty acid ethoxylates.

The a. of →cellulose and starch is carried out at 30–80 °C, and resulting products are used as more or less water-soluble thickeners (→starch ethers, →hydroxyethyl cellulose and →hydroxypropyl cellulose). →Guar gum is also derivatized by a.

Lit.: F.E.Bailey, J.V. Koleske "Alkylene Oxides and Their Polymers" Marcel Dekker Inc. NY (1991)

Alkyd Resins

G.: Alkydharze; F.: résines alkydes

A. are polyesters based on the following basic constituents:

-O-R-O-   = a polyol component, frequently
   | →glycerol but also pentaerythritol and
   O   trimethylolpropane. Most of the OH
   | groups are esterified with the acid
              components.

-OC-A-CO- = a dibasic acid component mainly
              phthalic acid but sometimes, among
              others, →azelaic or →sebacic acid or
              →dimer acid.

F-   = a →fatty acid

From this, the following schematic structure of an a. results:

-OC-A-CO-O-R-O-OC-A-CO-O-R-O-OC-A
         | |
         O   O
         | |
         F   F
   -O-R-O-OC-A-CO-O-R-O-OC
   | |
   O   O
   | |
   F   H

The properties of a. can by tailored easily to specific needs because there are many parameters available for adjusting:

- m.w., branching and amount of free hydroxyl or carboxyl groups by changing the ratio and kind of polyol and dibasic acid.
- flexibility, hydrophobicity and hardness by changing the dibasic acids, e.g., long-chain acids such as →azelaic and →sebacic acid instead of phthalic acid.

- amount of F– (called "oil length") which ranges between 30% (short oil) and 70% (long oil) of fatty acid in the total resin. Oil length has influence on the compatibility with other resins, and on viscosity, drying time, film hardness, gloss, brushability, durability and solvent requirements.

- the type of F– provides many possibilities for variation in physical and chemical characteristics. Normally, fatty acid mixtures directly derived from the trigyceride by →hydrolysis are used. Only in rare cases is fractional distillation needed. Unsaturated acids make a. reactive for crosslinking (drying) by air oxydation. A higher degree of unsaturation, conjugation and *trans* fatty acids bring stronger drying characteristics into an a. compared with less unsaturated and unconjugated *cis* fatty acids. In the alcoholysis process, the following oils are used instead of fatty acids: →linseed, →safflower, →oiticica, →tung, →sunflower, →cottonseed, →castor (dehydrated), →soybean, →coconut and many other oils. An important starting material, not derived from a trigyceride, is →tall oil fatty acid.

- modified a. are made by grafting vinyl monomers (styrene, methacrylates, etc.) by a radical mechanism onto unsaturated sites of the resin. Free hydroxyl groups of the a. are used for reactions with silicones and isocyanates. →Polyamides based on →dimer acid are also used to modify a. Cationic polymerization of →alkyloxazolines is another method of modification.

Production: There are two principle possibilities for manufacturing. The first starts with the preformed fatty acids, which react with the other two components in a one-step/one-pot reaction. The other process, called alcoholysis, starts with the oil (→fats and oils), which is reacted at 220–260 °C (PbO used as catalyst) with additional glycerol to a statistical monoglyceride. This is reacted in a second stage with the dibasic acid(s) or its anhydride. This second reaction type is more popular and economic because the oil provides all the fatty acid and part of the polyol.

Both reaction types can be carried out — either by fusion, i.e., the direct heating of all reaction components (an inert gas is blown through the melt to eliminate water from the reaction) — or by a solvent process, which is superior and avoids many disadvantages of the fusion technique. Small amounts (3–10%) of xylene are used, which carries the water out of the reactor azeotropically.

Uses: A. are the workhorses of surface →coatings (paints, enemals, lacquers and varnishes). They are the most versatile and economic coating binder in architectural, industrial and specialty applications.

Nondrying a. (low unsaturation, short oil, free OH groups) are designed for elevated-temperature curing by mixing with aminoplasts (melamine- or urea-formaldehyde resins) or as a polyol component in isocyanate curing formulations. They are used in many industrial applications (automotive industry, coil coating, factory wood finishing, metal furniture). The use in automotive coatings is however limited because a. cannot be used for the modern clear coats on metallic finishes due to unsufficent light and weather resistance. Nondrying a. are also used as plasticizer in lacquers based on →cellulose nitrate (furniture finishes). Combinations of a. with other binders (e. g., acrylics) are widely used for improvement of performance properties.

Drying a. (→drying oils) are used at room or slightly elevated temperature. The reaction with oxygen is accelerated by addition of →dryers (Co, Mn, Pb salts of naphtenic, →rosin, →caproic, →caprylic, →linoleic acid). Soybean and tall oil-based a. dry well. However, much better drying performance is obtained with linseed or fish oil-based resins. Maximum in drying are those a., which contain oils with highly conjugated double bonds, such as tung, oiticica or dehydrated castor oil. Short oil types dry fast by evaporation of the solvent, but crosslinking is limited. Long oil a. dry slower, but their final durability is much better due to better crosslinking. Drying a. are mainly used in trade sales products or architectural coatings.

Modified a. are used widely in many applications where higher weatherability, durability, faster drying and gloss are desired.

To reduce the amount of solvent (air pollution, fire hazard, transport regulations, health considerations) there is a tendency to high-solids or water-reducible a. High solids a. consist of low molecular backbones and selected diluents achieve low viscosities and high solids contents. Water reducibles are made by either converting the resin into an emulsion (→emulsifiers) or by incorporating water-soluble and crosslinking groups, e. g., carboxyl groups neutralized with ammonia or reactive amines. In such cases, trimellitic acid is used instead of phthalic acid.

There are some small applications outside the coating area, with →inks being the major outlet.

The US market for a. is estimated to be in the range of 300000 mt/a (1989), which is ≈30% of the total US coating-resin volume. The US coating market is split into architectural 42%, industrial 40% and specialties 18%.

The consumption of a. in Western Europe was (1989) in the range of 500000 mt/a (≈40% of the total coating resin market) with 55% going into architectural and construction applications. The Japanese market was 154000 mt in 1987 with a growing tendency. Main area of application (80%) is industrial.

During the 1960's and 1970's, a constant decline in alkyd use could be observed. Because the price level of natural fats and oils declined in the 1980's, a. gained importance again. The future can be seen as positive because a. are versatile, convenient in use, cheap, easily adaptable to the modern trends of "high solids" and "water reducability". A significant part of the raw materials for their production is renewable and therefore available in the long term.

Lit.: Ullmann* (5.) **A1**, 409
      Kirk Othmer* (4.) **2**, 53
      Encycl.Polym.Sci.Engng.* (2.) **11**, 315

Alkyl Amines
→Fatty Amines

Alkyl Ether Sulfates
→Fatty Alcohol Ether Sulfates

Alkyl Ether Phosphates
→Fatty Alkohol Ether Phosphates

Alkyloxazolines

Syn.: Oxazolines

G.: Alkyloxazoline; F.: alkyloxazolines, oxazoles

A. have the structure

$$\begin{array}{c} CH_2\!-\!CH_2 \\ | \qquad\quad | \\ N \qquad\; O \\ \diagdown\;\diagup \\ C \\ | \\ R \end{array} \qquad R: C_{7-21} \text{ alkyl}$$

and are formed by →dehydration of fatty acid monoethanolamides (→fatty acid ethanolamides) at 170–230 °C with $Ti(OR)_4$ as catalyst.

They are able to undergo cationic polymerization and can be used to modify →alkyd resins

Lit.: U.Eiken, U.Nagorny Farbe und Lack **99**, 911 (1993)

Alkyl Polyglucosides

Syn.: APG

G.: Alkylpolyglucoside; F.: polyglucosides d'alkyle

A. are modern nonionic →surfactants entirely based on RR. They are acetals of →fatty alcohol and →dextrose units:

R: C_{8-14} alkyl

The alcohol chainlengths normally vary between C_8 and C_{14}. They have an →HLB value of >10. Longer chains (e. g., C_{16}) yield more hydrophobic products. The glucose part may have an average DP of 1.3.

Normally, aqueous solutions ($\approx 50\%$) are sold.

There are two synthetic routes for acetalization, a direct heterogeneous reaction of glucose with the alcohol ($120\,°C/2 \times 10^3$ Pa) or a two-step process in which a short-chain alcohol (e. g., butanol) is reacted first with either glucose ($115\,°C$/no pressure) or starch ($140\,°C/4 \times 10^5$ Pa under depolymerization). The resulting glucoside is further reacted ($120\,°C/2 \times 10^3$ Pa) in homogeneous phase with the fatty alcohol. The reduced pressure enhances the removal of water. In both routes, acid catalysis is used. The excess alcohol is separated by →distillation.

A. are readily biodegradable. They are foaming well (even with high concentrations of fatty dirt) and not skin-irritating, and they are therefore used in manual dishwashing detergents as such or -with synergistic effects- in combination with surfactants presently used in dishwashing (LAS and →fatty alcohol sulfates). Due to their capability to adjust viscosity of detergent formulations based on →fatty alcohol (ether) sulfates, they could be used as substitutes of →fatty acid alkanolamides (eventually forming nitrosamines).

Another widening field of application is the light-duty detergent market. Advantages are viscosity control, freeze stability and less use of hydrotropes.

In the →cosmetic industry, a. are used in the formulation of shampoos, hair conditioners (→hair preparations), bath products and facial and other skin cleaners (→skin preparations). Hydrophobe types (C_{16}) find use as emulsifiers in cosmetics and antifoamers.

Technical applications focus on strong alkaline cleansers in the food and beverage industry, where normal nonionic or anionic surfactants cannot be used (insolubilty and strong foaming). In combination with terminally blocked →fatty alcohol ethoxylates, a. are able to form low-foaming or even antifoaming detergents based on 50% sodium hydroxide solutions.

A. are now available in more than pilot-plant volume. Production plants with a capacity up to 25 000 mt/a each exist in the USA and Germany. The excellent environmental and consumer-friendly properties of these first totally RR-based nonionic surfactants will allow a bright outlook for the future.

Lit.: B.Salka: Cosmetics & Toiletries **108**, 89–94 (1993)
K.Hill: "Carbohydrates as Organic Raw Materials" II, Gérard Descotes (ed.), p. 163 VCH, Weinheim (1993)

Alkyl Polyglycol Ethers

→Fatty Alcohol Ethoxylates
→Fatty Alcohol Propoxylates

Alkyl Sulfates

→Fatty Alcohol Sulfates

Alligator Pear

→Avocado

Aluminum Soap

→Metallic Soaps

AMG

→Acetic Acid Esters of Mono/Diglycerides

Amide Soaps

→Fatty Acid Cyanamides

Amino Acids

G.: Aminosäuren; F.: acides aminés

A. are the building blocks of →proteins and are also contributing to the structure of nucleic acids. About 20 a. are of importance (see table):

$$R-\underset{\underset{H}{|}}{\overset{\overset{NH_3^+}{|}}{C}}-COO^-$$

All a. have high m.p. (>180 °C). With the exception of glycine a. show stereo→isomerism. In nature the L-form occurs almost exclusively.

A. are produced:
- by acidic or enzymatic →hydrolysis of proteins, followed by isolation through crystallization or

Name	Code	R
Aliphatic a.		
Glycine	Gly	H–
Alanine	Ala	CH_3-
Valine	Val	$(CH_3)_2-CH-$
Leucine	Leu	$(CH_3)_2-CH-CH_2-$
Isoleucine	Ile	$\begin{array}{c}CH_3-CH_2\\ \rangle CH-\\ CH_3\end{array}$
Acidic a.		
Aspartic acid	Asp	$HOOC-CH_2-$
Glutaminic acid	Glu	$HOOC-CH_2-CH_2-$
Asparagine	Asn	$H_2N-CO-CH_2-$
Glutamine	Gln	$H_2N-CO-CH_2-CH_2-$
A. with O–, S– or NH-function		
Serine	Ser	$HO-CH_2-$
Threonine	Thr	$CH_3-CHOH-$
Cysteine	Cys	$HS-CH_2-$
Methionine	Met	$CH_3-S-CH_2-CH_2-$
Lysine	Lys	$H_3N^+-CH_2-CH_2-CH_2-CH_2-$
Arginine	Arg	$\begin{array}{c}H_2N\\ \rangle C-NH-CH_2-CH_2-CH_2-\\ H_2N^+\end{array}$
Histidine	His	$\begin{array}{c}NH\\ \| \rangle-CH_2-\\ HN^{\text{+}}\end{array}$
Aromatic and heterocyclic a.		
Phenylalanine	Phe	⬡–CH_2-
Tyrosine	Tyr	HO–⬡–CH_2-
Tryptophan	Trp	(indole ring)$-CH_2-$
Proline	Pro	Iminoacid, includes the amino-group in a five-membered ring

precipitation techniques (possible sources are →soybean or other →hydrolized vegetable proteins);
– by chemical synthesis; however, only racemates (DL) are obtained;
– by microbial or enzymatic processes; especially immobilized-enzyme techniques show increasing importance.

Almost 675 000 mt/a of a. are produced worldwide. Economically, the most important are:
- →L-Glutamic acid, produced by fermentation and used as flavor enhancers in food (→food additives), represents half of this volume.
- DL-Methionine as a supplement to animal feed in a volume of $\approx 250\,000$ mt/a is produced by

chemical synthesis because animals are able to make use also of the D-isomer.
- L-Lysine is produced by fermentation in a volume of $\approx 70\,000$ mt/a also as a animal feed supplement.
- L-Phenylalanine and L-aspartic acid are of increasing importance because l-aspartyl-L-phenylalanine methyl ester is a powerful sweetener (r.s. 200).

Methionine and lysine are the most important out of a group of nine so-called essential a., which can't be produced by mammals and have to be taken up by food or as supplement.
Small volumes of a. are used in medicinal, cosmetic and technical applications.

Lit.: Kirk Othmer* (4.) **2**, 504–571
　　Rehm/Reed* **3**, 479

7-Amino-cephalosporanic Acid

Syn.: 7-ACA
G.: 7-Amino-cephalosporansäure;
F.: acide 7-amino-céphalosporanique

m.w.: 240.22

A. is an important intermediate and starting material for the production of semisynthetic cephalosporins. It is produced either by enzymic cleavage (e. g., with cephalosporin-acylase) or by treatment of biotechnologically produced cephalosporin C with nitrosyl chloride.
Production is $\approx 10\,000$ mt/a.

Lit.: Vandomme "Biotechnology of Industrial Antibiotics" p. 205 Marcel Dekker Inc. NY (1984)
　　Ullmann* (5.) **A2**, 470 + **A18**, 130
　　[957–68–6]

6-Amino-penicillanic Acid

Syn.: Penicine; 6-APA
G.: 6-Amino-penicillansäure;
F.: acide 6-amino-pénicillanique

m.w.: 216.28; m.p.: 209–210 °C (d.)

A. is a central building block of penicillins and is produced either by fermentation (*Penicillium chry-*

sogenum or *Pleurotis astreatus*) in the absence of suitable side-chain precursors or by chemical or enzymatic cleavage of penicilline G (phenoxymethylpenicilline).

Penicillin G

For the enzymatic cleavage, immobilized penicillin acylase is used. Alternatively, the fresh enzyme or resting cells can be used.
Production ≈ 10 000 mt/a.

Lit.: The Merck Index* (11.) 470
    Ullmann* (5.) **A2,** 469, 473, 504, 511
    Kirk Othmer* (4.) **3,** 1
    [551–16-6]

Amphidiploid
→Cytology

Amphoteric Surfactants
Syn.: Betaines
G.: Amphotenside; F.: tensioactifs amphotères

A. are specialty →surfactants:

$$R-\underset{\underset{H}{|}}{\overset{\overset{H}{|}}{N^+}}-CH_2-COO^-$$

These compounds, called ampholytes or betaines, form a cationic compound by addition of H^+ (acidic media) and an anionic species by addition of OH^- (alkaline media). This behavior is called amphoteric.
Better known und more used are however the betains where the two H atoms at the N-atom are substituted by two CH_3-groups. They are produced by reaction of tertiary amines with chloroacetic acid and sodium hydroxide.
Betaines are insensitive to water hardness, almost nontoxic, nonskin-irritating and have antimicrobial properties. They are therefore used in cosmetic formulations. They are compatible with other surfactants and have good washing and foaming performance.
There are two other types of a. of minor importance: the sulfobetains (the anionic group is a sulfo acid instead of a carboxylic acid) and imidazolinium betains, which are made by reaction of imidazolines (→imidazoline derivatives) with chloroacetic acid.
US production of a. in 1980 was 12 000 mt.
Lit.: Ullmann* (5.) **A25,** 747

Amylases
Syn.: Glucosidases
G.: Amylasen; F.: amylases

Group of carbohydrate hydrolyzing →enzymes (hydrolases), which are specialized on splitting the α-glucosidic bonds of the →starch polysaccharides as well as of →malto-oligosaccharides with preference for positions α-1,4 and α-1,6 (α-1,3) (→hydrolysis, glucosidic linkages). A. are classified, according to their mode of action, as endo- or exoamylases as well as by their substrate and linkage position specifity : α-a., β-a., α-glucosidases, glucoamylases (γ-amylases), maltases, oligases.
The systematic names of all characterized a. are standardized with respect to their action pattern and specificity by the enzyme nomenclature of the International Union of Biochemistry, (EC-numbers), e. g., α-amylase = 1,4-α-D-glucane-glucanohydrolase, EC 3.2.1.1.

Lit.: Ruttloff* 144–146 + 577–594 (1994)
    [9000-92-4]

Amylocorn
→Amylomaize Starches
→High-amylose Starches

Amylolysis
G.: Amylolyse; F.: amylolyse
A. is the depolymerization (→hydrolysis, glucosidic linkages) of →starch polysaccharides by →amylases.

A. is a catalytic process, performed by single amylases or by combinations or sequences thereof. The conditions are highly specific with respect to enzyme-substrate relations, pH, concentration, activators and inhibitors. Amylolytic processes are run industrially in continuous or semi-continuous devices. A powerful step ahead in continuous a. was reached by the development of immobilized →enzymes or enzyme combinations with multiple utilization instead of the former batch processes with a simple enzyme.
Enzymes of technical significance are currently of microbial origin, produced by breeding of microorganisms such as fungi (moulds) or bacteria. Upgrading the yield of enzymes is performed by genetic engineering. Amylases from higher plants,

such as barley (malt), soybean and sweet potatoes, are still dominating the β-amylases for producing →maltose syrups and →maltose as well as in the brewery.

Starch and other carbohydrate-modifying enzymes hold the second place in world enzyme production behind detergent enzymes.

Lit.: Ruttloff* 577–645, (1994)

Amylomaize Starches

Syn.: High-Amylose Starches
G.: Amylomaisstärke;
F.: amidon amylosique du maïs, amyloamidon du maïs

→Corn starch from *Zea mays* L. (amylose extender, a homozygote mutant) with →amylose contents of 50–85% has gained technical significance as a source of enriched amylose, which needs no costly isolation from regular starch. Beyond an amylose content of 80%, a. behave like pure amylose in practical use.

Isolation from corn kernels by →wet-milling is performed in the same way as described for →corn starch processing,especially in the USA.

Composition: →Starch granules are round or deformed, 3–24 μm in diameter (12 μm average; normal corn starch 15μm), crystalline x-ray polymorph type A (cereal starches), crude lipids 0.4%, minerals 0.2%, m.w. ($\times 10^6$) a. 16.7, a.-amylopectin 68.9, a.-amylose 2.45 [1].

The rheological properties during gelatinization, cooking and cooling deviate from regular corn starch: onset of gelatinization at >67 °C; low viscosity gain by swelling until 95–99 °C; low swelling power at 95 °C, 6 g water/g ; total gelatination and dissolution need pressure cooking at >140 °C; on cooling rapid clouding and set back to rigid white gels occurs (→retrogradation starts at 60–70 °C).

In the food industry, these properties are used in production of quickly setting sweets, jellies, edible foils and coatings.

A. solutions are well suited to be cast to biodegradable films and fibers from the unmodified state, in which the functional properties are enhanced with increasing amylose content. After derivatization with fatty acid (DS >1) the esters are soluble in organic solvents and show properties similar to cellulose esters.

Therefore, a. are highly interesting raw materials for nonfood applications as biodegradable plastics. Moreover, a. solutions can form inclusion compounds with hydrophobic molecules (such as →cyclodextrins) for the retention, protection and retarded liberation of drugs in pharmaceutical products and inclusion of flavors.

At present, research is directed on breeding of more effective mutants (higher yields on crop, starch and amylose content).

Commercial products are delivered with 50% and 70% amylose.

A. is not suited for cultivation in the European climate. As a rich-amylose variant, the breeding and cultivation of pea varieties (→pea-starches) is attempted.

Lit.: van Beynum/Roels* 20–45 (1985)
    Tegge* p. 31, 158–160 (1988)
    Stärke im Nichtnahrungsbereich*, 95–106, (1990)
[1] Th. Aberle et al.: starch/stärke **44**, 329–335. (1994)

Amylopectin

G: Amylopektin; F: amylopectine

A. is one of the polysaccharide constituents of starch with a branched molecular structure and an industrial product, which can be obtained either by separation from a starch solution or from one of the waxy varieties of starch-bearing plants.

A. is the major compound of the starch polysaccharides in →starch granules, about 65–85%, of the "normal" starches from maize, wheat, potato, manioc, rice. →Waxy starch varieties are carriers of nearly pure amylopectin, such as waxy maize, waxy sorghum, waxy barley and waxy rice (98–100%). Other varieties of maize, "amylomaize or amylocorn", are enriched in amylose and the a. content is reduced to 30–50%. Wrinkled pea contains 22–50% a. The m.w. of a., $(C_6H_{10}O_5)_n$, largely depends on the starch origin:
($\times 10^6$) (methodical differences possible):

Maize	112.2	Waxy maize	76.9
Potato	60.9	Wheat	77.6
Amylomaize	68.0	Wrinkled pea	77.9
Smooth pea	53.8		

A. is constituted from n AGU (→ Anhydro Glucose Unit) according to the structure

Amylopectin

The AGU are linked together by α-1,4-glucosidic (94–96%) and by α-1,6-glucosidic (4–6%) bonds. The latter are responsible for the branched structure with only one reducing end group for about 5000 nonreducing end groups positioned at the exterior branches of the macromolecule. The C-chain with the only reducing end group is branched by α-1,6-glucosidic bonds into numerous B-chains, which are further ramified by α-1,6-glucosidic bonds into outer A-chains, which carry the nonreducing end groups. The generally accepted "cluster-model" of the branched starch →polysaccharide looks like

of a. are highly stable in freezing and thawing. Viscous, gel-like structures may be formed only at high concentrations, at low temperatures, after long-time storing. These gels belong to the thermally reversible type of starch gels, in which →retrogradation may be overcome simply by heating.

A. behaves like starch or amylose in reactions such as →hydrolysis, →derivatization or physical →modification. Application of debranching →enzymes (pullulanase and iso-amylase) opens the possibility of total splitting of the α-1,6-linkages, thus enhancing hydrolysis as well as producing

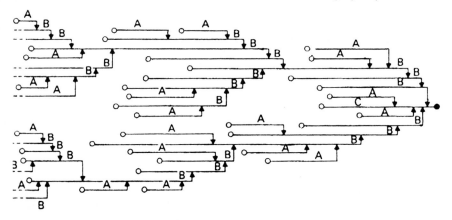

The average length of the α-1,4-glucosidic chains are: outer A-chains 20–25 AGU, the shorter segments between branching points 5–9 AGU, the longest B-chains 40–50 AGU.

Differences in the degree of branching, in A/B-chain relations and in chainlength determine the specific behavior of different starch types.

Considering the linear chains and chain segments of a. as short-chain (low molecular) →amylose, similar interactions like those of amylose are possible. The crystalline regions of the →starch granules are formed by elementary cells of double helices, which are joined together by antiparallel unwinding. The main share of this starch crystallinity is due to a.

On swelling of starch granules in aqueous liquids, a. remains behind as the higher water-binding, insoluble residue, whereas amylose is eluted. Pure a. results, after complete solubilization of the starch and precipitation of the amylose by complexing, as a soluble residue in the mother liquor and may be isolated after solvent-precipitation. These preparations are soluble in aqueous solvents either on heating or at pH >8. Solutions

short-chain amyloses from a. on a commercial scale.

The application in food as well as in the nonfood industry is based on its functional properties: solubility, high water-binding capacity, viscosity increase, clearness and stability of solutions (nonsetting), stabilizing effects, cohesiveness, film forming, coating and binding, digestibility, biodegradability. A. is as easily available as normal starches; waxy cereals are a source of a. with high mass potential. A. preparations from normal starches are used merely in structure research as well as for studying interactions and hydrolyzing and other degrading reactions.

Lit.: T.Galliard (ed.) "Starch: Properties and Potential" 55–78 John Wiley & Sons, Chichester/NY/Brisbane/Toronto/Singapore, (1987)
A.Imberty et al. starch/stärke **43**, 375–384 (1991)
Th. Aberle et al.starch/stärke **46**, 329–335 (1994)

Amylopectin Starches
→Waxy Starches

Amylose

G: Amylose; F: amylose

A. is one of the polysaccharide constituents of starch with a predominantly linear molecular structure. It is an industrial product that can be obtained either by aqueous elution of starch granules or by fractionation of a starch solution.

A. represents the lesser amount of the starch constituents in the →starch granules, about 15–35 % in the "normal" starches from maize, wheat, potato, maniok and rice. Special varieties of plants exhibit enhanced contents of a.: amylomaize (50–80 %) and →wrinkled pea (50–78 %) are the most promising carriers of "high-amylose starch" for industrial utilization.

The m.w. of a. largely depends on starch origin: ($\times 10^6$) (methodical differences possible)

Maize	2.09	Amylomaize	2.45
Potato	19.6	Wrinkled pea	2.62
Wheat	8.51	Smooth pea	5.45

A. is constituted from n AGU, joined together by α-1,4-glucosidic linkages according to the structural scheme:

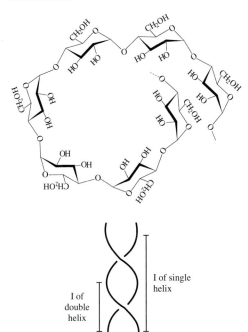

In its primary structure, the AGU are existing in the conformation c (chair conformation). The valence angles between the AGU are favoring a helical conformation, formed by 6–8 AGU, as the energetically most suitable state. The normal state in solution is that of a disturbed helix. An ideal stable helix conformation is formed and stabilized when hydrophobic molecules (iodine, aliphatic alcohols and acids) are allowed to penetrate into the molecular channel. The formation of such inclusion complexes is a typical property of a. and may be compared best with the inclusion behavior of →cyclodextrin. Insoluble complexes with organic solvents are used to precipitate amylose from starch solutions during fractionation.

A. is the more soluble part in the native starch granule and belongs to the amorphous fraction. When starch granules are heated in aqueous liquids a. will be leached out from the onset temperature of starch granule swelling until gelatinization, whereas amylopectin remains in a highly hydrated state. Solutions of a. are unstable. In the dissolved state at temperatures of <50 °C, the molecules associate with each other, leading either to precipitates or to turbid gels. This behavior is also shown in a.-containing starch systems and is called "→retrogradation".

Such retrograded a. are insoluble in water below 150 °C. They are structured in elementary cells of double helices which exhibit the typical crystalline x-ray diffraction pattern of the B-polymorph. This type is found in native starches of roots and tubers and always in gelatinized and retrograded systems as the crystalline component. When retrogradation is forced above 50 °C, the A-polymorph is formed.

Association of linear α-1,4-linked polyglucan chains into double-helix structures is the functional origin of the typical behavior of a. as well as a.-containing starches:

turbidity and instability of solutions, set-back (structure forming), changes in viscosity, insolubility, decrease of digestibility as well as of enzymatic susceptibility. The tendency for aggregation depends on the chainlength and has its maximum at 80–120 AGU.

The most prominent functional property of a. is its ability to form films, foils and fibers, resulting from its linear structure. Such products may be achieved from unmodified a. in aqueous systems as well as from chemical derivatives (acetylated a.) in organic solvents in which they are soluble (like acetyl-cellulose). Applications have been successful for paper coatings and layers as well as for textile fabrics. Edible or soluble coatings, capsules, films and packing materials have been developed for the food and pharmaceutical industries. Encapsulation of flavors and active pharmaceutical agents by spray drying or complexing with a., using inclusion mechanisms comparable with those of →cyclodextrins, have been studied extensively, and a high state of knowledge has been achieved.

This starting position makes a. one of the most important raw materials in the production of biodegradable →starch plastics as well as a component in starch-filled synthetics or in →starch-containing plastics. In such systems, firmness, elasticity and flexibility is enhanced with rising a. content whereas the amylopectin component imparts brittleness and lower firmness. Therefore, a high a. content is demanded for most applications of starchy materials in biodegradable plastics as well as in many other applications as RR.

In addition to the described applications, the following are noteworthy: lacquer bases, degradable films, foils and other packaging materials.

Plant breeding is expected to deliver more high-amylose species (amylomaize, wrinkled pea) with stable starch properties and high yields. This approach is more economical than isolating a. from normal starches.

Lit.: Tegge* 29–41 + 157–160 (1988)
A.Imberty et al. starch/stärke **43**, 375–384 (1991)
Th. Aberle et al.starch/stärke **46**, 329–335 (1994)
Eggersdorfer* 311–322 (1993)

Amylum

G.: Amylum; F.: amylum, fécule

The term for →starches, derived from the Greek (a mylos, the cereal substance that cannot be further crushed by milling). The term a. is used frequently for pharmaceutically approved starches, always in connection with the botanical term of the starch source:

Solani amylum — potato starch
Maidis amylum — maize starch
Tritici amylum — wheat starch

Anetto

→Dyes, natural

Anhydro Glucose Unit

Syn.: AGU
G.: Anhydro Glucose Einheit (AGE);
F.: unité d'anhydroglucose (UAG)

The monomeric unit constituting oligo- and polysaccharides, joined together by glucosidic bonds in numerous α- and β-glucans.

$[C_6H_{10}O_5]_n$

Its formation may be considered as a poly-addition process of n molecules of glucose under elimination of n−1 molecules of water. The inverse process, splitting of n−1 glucosidic bonds with binding of n moles of water and setting free n AGU, is characterized as →hydrolysis.

Anionic Starches

G.: Anionische Stärke;
F.: amidon (fécule) anionique

Modified starches that contain groups capable of conferring on the macromolecules a negative charge in aqueous solutions under appropriate pH conditions are called a.

The term a. comprises: →starch ethers and inorganic and organic →starch esters with substituents carrying free acidic groups such as R−COO′, −(PO_4)″, −(PO_4H)′ and −(SO_4)′ as well as →oxidized starches in which primary or secondary OH-groups of the AGU-backbone are converted to carboxylic groups. Besides their anionic character, substances of this group exhibit diminished or total lack of →retrogradation such as shown by changes in turbidity, precipitation, set back, syneresis and freeze-thaw instability. Moreover, with increasing DS, the temperatures of swelling and gelatinization of the →starch granules decreases until they swell in cold water.

Typical a. are: →carboxymethyl starch (→starch ethers), starch adipate, starch citrate, starch maleate, starch phosphate, starch succinate, starch sulfate, starch xanthate (→starch esters), →oxidized starches.

Lit.: Ullmann* (5.) **A25**, 17

Annual

G: Einjährig; F: annuel

A. plants have a life cycle of one year only, meaning that sowing and harvesting of the fruits will have to take place within 12 months. Examples are most agricultural crops like →wheat, →rapeseed or →corn. A. crops are segregated into winter a. which are planted in the fall and harvested in the summer of the following year such as winter →wheat or w. →rapeseed or →meadowfoam, and spring a., which are planted in the spring and harvested in the fall of the same year (e.g., →corn, →soybean or →cotton). Some crops appear to have an a. life cycle, like →sugarbeet (which is a →biennual crop because it bears fruits only in the second year; roots, however, which comprise the harvested organ, mature already in the first year) or →sugarcane (which bears fruits in the first year but is considered a →perennial crop).

Anthocyanidins
→Dyes, natural

Antibiotics
G.: Antibiotika; F.: antibiotiques

A. are low-molecular weight, nonenzymic, secondary metabolites or substances which result from (bio)synthetic or semisynthetic processes. They inhibit the growth of other microorganism even at low concentrations.
→7-Amino-cephalosporanic Acid, →6-Amino-penicillanic Acid, →Kanamycin, →Tetracyclin

Antistatic Agents
Syn.: Antistats
G.: Antistatika; F.: antistatiques

Some materials (e.g., synthetic fibers, plastic and rubber goods) show the ability to generate strong electrostatic charges when being rubbed (tribo-charging) or by induction (transfer from other charged materials). Nasty or even dangerous effects may result: Textile fabrics may "stick" to each other, soil may be attracted to plastic articles making them look unattractive, uncomfortable electric shocks may be experienced by people walking on a synthetic fiber carpet and may even be responsible for explosions at sudden discharging. It is therefore highly desirable to control this static electricity by a. This can be done in some cases by adding conductive substances (metal, carbon black, graphite) to the polymer mass. In most applications it is accomplished by adding polar organic substances (→surfactants) to the batch prior to extrusion or molding. A. can also be applied as more or less durable finishes to the surface by padding or spraying. In general, a. act by creating a film of higher humidity on the surface, and water has excellent antistatic properties due to its high dielectric constant.
A. for fibers (→textile auxiliaries) are sulfonates and phosphates, →fatty amines, →fatty acid amides and their ethoxylates, →quaternary ammonium compounds as well as →fatty alcohol ethoxylates, fatty acid esters and their derivatives. For more durable finishes, polyamine resins cross-linked on the fiber are used.
In plastic articles (→plastic additives), →fatty acid amides, the ethyleneglycol esters of →oleic acid, →fatty amine ethoxylates, →fatty acid diethanol-amides and mono/diglycerides are used.
Lit.: Kirk Othmer* (4.) **3**, 540

Antistats
→Antistatic Agents

6-APA
→Amino-penicillanic Acid

APG
→Alkylpolyglucosides

Araban
→Hemicelluloses

L-Arabinose
→Pentoses

Arabinogalactan
→Hemicelluloses

Arabinoxylan
→Hemicelluloses

Arachic Acid
→Arachidic Acid

Arachidic Acid
Syn.: Arachic Acid; Eicosanoic Acid; C20 : 0
G.: Arachinsäure; F.: acide arachidique

$CH_3-(CH_2)_{18}-COOH$

m.w.: 312.5; m.p.: 75.4 °C; b.p.: 246 °C (1.33 kPa/10 mm)

A. is insoluble in water and soluble in organic solvents.
A. is contained in →peanut oil and in waxes, together with other long-chain acids. It can be produced by →hydrogenation of →eicosenoic acid, which is contained in →camelina and →meadow-foam in larger quantities.
A. is used in organic synthesis and in lubricants. Zn soaps (→metal soaps) are greases in metal drawing (→metalworking fluids).

Lit.: Ullmann* (5.) **A10**, 247
       [506–30-9]

Arachidonic Acid

Syn.: 5,8,11,14-Eicosatetraenoic Acid; C20:4
(Δ 5,8,11,14)

G.: Arachidonsäure; F.: acide arachidonique

$CH_3-(CH_2)_4-(CH=CH-CH_2)_4-(CH_2)_2-COOH$

m.w.: 304.46; m.p.: $-49.5\,°C$

A. is contained in small quantities in animal fats
and organs, particularly in fish oils. All double
bonds are *cis*. Together with other long-chain
highly unsaturated fatty acids ($\rightarrow$linoleic and
$\rightarrow$linolenic acid) it represents an essential part of
human nutrition and diet ($\rightarrow$linolenic acid).

Enzymatic oxidation yields numerous compounds
called eicosanoids like $\rightarrow$prostaglandines, thromboxanes, leucotrienes etc., which are highly important biochemicals.

Lit.: Ullmann* (5.) **A10**, 248
      [7771-44-0]
      [27400-91-5]

Arachis hypogaea

$\rightarrow$Peanut

Arachis Oil

$\rightarrow$Peanut Oil

L-Ascorbic Acid

Syn.: Vitamin C

G.: L-Ascorbinsäure; F.: acide ascorbique

```
  ┌─OH
  ├─OH
  ┌O
  │   ╲
  │    O
HO    OH
```

m.w.: 176.12; m.p.: $192\,°C$; $[\alpha]_D^{20}$: $+23°$

A. crystallizes in colorless and odorless platelets,
which are easily soluble in water or ethanol. It possesses a strong acidic taste and a high reducing
power (di-enolic constitution). Aqueous solutions
are rapidly oxidized by air. From 4 possible
stereoisomers ($\rightarrow$isomerization), only the L-threo
configuration has vitamin C potency.

A. is naturally widespread in plants but is excep-
tionally enriched in the following fruits and organs
(>100mg/100g fresh weight): West Indian cherry,
rose hips, paprika fruit, pine needles, guava, black
currant, parsley, broccoli, green pepper, kiwi fruit.

Traditionally a. is produced (Reichstein synthesis)
from $\rightarrow$starch, via D-glucose, $\rightarrow$D-sorbitol, $\rightarrow$D-
sorbose by a combination of chemical and biotech-

nological (e.g., biological oxidation of $\rightarrow$sorbitol
with *Acetobacter suboxydans*) processes.

Approaches to short-cut and improve this traditional
process of 14 steps by different catalytic pathways as
well as by a genetically engineered microorganism
are promising. In addition, fermentation of $\rightarrow$D-glu-
cose with a recombinant strain of *Erwinia herbicola/
Corynebacterium* sp. resulted in the formation of 2-
oxo-L-gulonic acid, a direct precursor of a.

Main applications are based on its action as vitamin C in food products, in nutrition and in pharmaceuticals. It is used for treatment of vitamin C de-
ficiency (scurvy). Its reducing action on $\rightarrow$wheat
gluten protein disulfide bonds leads to improvement of baked goods. Moreover, numerous applications to prevent autoxidation of lipids in food
processing operations and in storage depend on its
strong reducing properties. Color stability is
achieved by prevention of nonenzymatic browning
(Maillard reaction of sugars and amino acids), as
well as of the enzymatic phenoloxidase reaction in
fruit and vegetables.

The sodium salt [134-03-2] is a white or slightly
yellow crystalline powder, practically odorless and
used as an antioxidant.

Enhanced environmental sensitivity will favor utilization of a. instead of sulfur dioxide and its salts
in special fields of application.

30 000 – 40 000 mt/a are produced worldwide.

Lit.: Lichtenthaler* 267 – 288 (1991)
      Kirk Othmer* (3.) **24**, 8
      Ullmann* (5.) **A11**, 563
      [50-81-7]

Assimilation

$\rightarrow$Photosynthesis

Atropa belladonna

$\rightarrow$Atropine

Atropine

Syn.: D,L-Hyoscyamine

G.: Atropin; F.: atropine

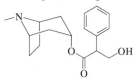

m.w.: 289.4; m.p.: $114 - 116\,°C$

The parasympatholytic alkaloid is isolated from
Atropa belladonna L., *Datura stramonium* L. and
other Solanaceae. During extraction, partial race-

mization of the l-hyoscyamine takes place, which is completed by treatment with dilute alkali on heating in chloroform solution. The dried and finely divided plant materials are extracted with ethanol; evaporation of the crude extract and reextraction with water and 0.5% sulfuric acid, ether and chloroform gives an alkaloidal fraction, which is then converted to salts and purified by recrystallization. A. forms long orthorhombic prisms. Intoxication with a. causes blurred vision, suppressed salvination, vasodilatation, hyperpyrexia, excitement, agitation and delirium. In human therapy, it is used as an anticholinergic and mydriatic agent.

Lit.: Ullmann* (5.) **A1**, 360–361
       Kirk Othmer* (4.) **1**, 1046–1050
       [51-55-8]

Avena sativa
→Oats

Avocado
Syn.: Alligator Pear
Persea americana Mill.
Lauraceae
G.: Avocado; F.: avocat

A. is a tropical tree of 12–15 m height with pear-shaped fruits with soft light-green, cream-like meat in the mesocarp and a hard kernel.

The name "avocado" is derived from the word "aguacate", which the Spanish conquerors heard from the Aztec Indians of Latin America. It spread from the tropics to the subtropical winter-rain areas.

Main producing areas are in Mexico, California, Florida, Brazil, Argentina, North Australia, India, South Africa, Israel, and several Mediterranean countries. A new a. plantation will bear fruits after 4–6 years. Yields are 5–12 mt fresh fruits per ha per year.

The mesocarp of the fruit has the following composition: 59–86% water, 5–32% oil, 0.8–4.4% protein, 1–10% carbohydrates, 1.5% crude fiber. The typical fatty acid composition of the oil is:

C14:0	C16:0	C18:0	**C18:1**	C18:2
1–3%	8–26%	0–1%	**65–82%**	6–12%

A. is a typical vegetable of high value. Fruits that have become too mature for transportation are manufactured to vegetable oil. A special application of a. oil are high-value cosmetics.

Lit.: Purseglove* (1974)

Azadirachta indica
→Neem

Azelaic Acid
Syn.: Nonanedioic Acid; 1,7-Heptane Dicarboxylic Acid
G: Azelainsäure; F: acide azélaïque

$HOOC–(CH_2)_7–COOH$

m.w.: 188.2; m.p.: 106 °C; b.p.: 286 °C (13.3 kPa/100 mm)

A. is soluble in hot water, alcohol and other organic solvents.

It results from oxydative cleavage of →oleic acid and is produced in commercial quantities by →ozonolysis.

Technical products contain 80% azeleic acid and 15% other dibasic acids. Higher purities (90%) are also available.

The acid is used to modify →alkyd resins and to produce →polyamides (nylon 6/9). The esters are used in →lubricants, →hydraulic fluids and plasticizers (→plastics additives)

Lit: Ullmann* (5.) **A8**, 531
       Kirk Othmer* (3.) **7**, 623
       [123-99-9]

B

Babassú

Orbignya martiana Barb.-Rodr., *O. oleifera* Burret
O. speciosa
Palmae
G.: Babassu; F.: attalée, babassu

The b. palm is one of the few wild oil-bearing palms that have become sources of commercially important quantities of lauric oils. Nevertheless, the development of oil production from the b. palm has been disappointing in relation to the quantities of fruit produced by the $5-25 \times 10^9$ trees growing in northern Brazil (states of Maranhao and Piaui). The palm bears large bunches that contain 200–600 fruits or nuts. The nut is egg-shaped, 7×10 cm long and has a tough outer fibrous portion. Most of the weight of the b. nut consists of a hard thick shell that encloses 1–6 oblong kernels (4×1 cm long).

The main obstacle to exploit the b. palm is the difficulty to crack the hard shell of the nut as well as the comparatively low weight percentage (12–17%) of kernels per nut. The nuts are mostly cracked by hand with low efficiency of about 3–6 kg of kernels from 25 kg of nuts per day and person. This is done by holding the nuts with one hand across the sharp edge of an axe and beating

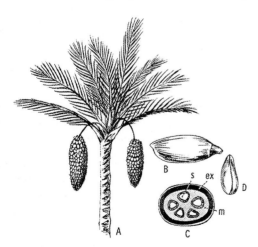

Babassu palm; A: Habit of palm tree with fruit bunches.
B: Fruit. C: Cross-section of fruit: (ex = exocarp plus mesocarp, m = endocarp, s = seed with seed coat) D: Seed (from Franke*, 1981, with permission).

on the nut with a wooden club so that the nut may be split along its longitudinal axis. The single kernels can then be pulled out of the wooden shell of the nut. The by-product woody shell is hard and yields a valuable →charcoal, suitable for special purposes such as steelmaking or the manufacture of →activated carbon. Another by-product of the b. nut is the starch from the outer fibrous layer of the nut, which has not been utilized intensively. It is known that rodents feed on this outer starch layer.

The kernel contains about 65% →babassú oil.

Several mechanical cracking methods have been developed and tried in Brazil to reach better productivity. However, none have been able to compete with the oil quality of hand-cracked nuts, and all are limited to the natural obstacle of having to cope with huge amounts of low value by-products. Local people tend to crack nuts only if they are short of cash or if they have no other jobs. Nuts are being exchanged in a primitive way in local food-stores for groceries. This exchange system is still dominating in the rural areas of northern Brazil, where no other sources of income exist for poor people. On the other hand, the exchange trade and the local oilmills in North-East Brazil are being owned by a few rich families, which have a special interest in keeping this system of dependency alive.

In 1980 about 250000 mt of b. kernels were produced in Brazil, but in 1984, only 150000 mt. It is believed to have dropped to less than 100000 mt in 1992. The b. kernels are normally crushed in oil-mills without solvent extraction. The oil is used mostly as raw material for local →soap production and as cooking oil. The meal is applied as feed.

Lit.: P.H.May et al. "Subsistence Benefits from the Babassú Palm". Economic Botany **39**, 113–129 (1985)

Babassú Oil

G.: Babassuöl; F.: huile de orbignya
s.v.: 242–253; i.v.: 13–16; m.p.: 22–26 °C

B. is the oil of the kernels of the babassú nut (→babassú).

The fatty acid composition is similar to that of →coconut oil:

C8:0	C10:0	**C12:0**	**C14:0**
4–5%	6–8%	**44–45%**	**15–18%**

C16:0	C18:0	C18:1
8–10%	2–5%	12–16%

It is used almost exclusively for soap making and nutrition, e. g., for margarine in Brazil internally. Technical use, e. g., as babassú oil fatty acids and derivatives, as well as exports are of minor importance. The annual production was $\approx 160\,000$ mt (1980) and is declining.

Lit.: Ullmann* (5.) **A10**, 221

Babassú Oil Fatty Acid
→Babassú Oil

Bacterial Alginates
→Microbial Gums

Bacterial Fructans
→Fructans

Bacterial Polysaccharides
→Microbial Gums

Bagasse
G.: Bagasse; F.: bagasse

The solid, fibrous residues remaining after the expression and extraction of ground sugar cane or sweet sorghum (→sugar cane processing, →sugary sorghum processing) consists of $\approx 40-60\%$ →cellulose, $\approx 20-30\%$ →hemicelluloses and $\approx 20\%$ →lignin. It is either burned as a solid fuel, e. g., in the sugar industry directly, or is used as a raw material for →paper and cardboard production. B. is sometimes used as a soil conditioner. After reduction of the lignin content, followed by mixing with →sucrose molasses and a suitable nitrogen-containing material, b. serves as animal feed.

The pentosane content is the basis for commercial production of →furfural and →levulinic acid.

Inhalation of bagasse dust may cause pneumonitis and asthma.

Annual volume derived from sugar production is $\approx 60-70 \times 10^6$ mt.

Lit.: Ullmann* (5.) **A25**, 345
     Kirk-Othmer* (3.) **3**, 434–438

Barium Soap
→Metallic Soaps

Barley
Hordeum vulgare
Gramineae, Poaceae
G.: Gerste; F.: orge

A cereal with dense, bearded spikes of flowers, each made up of three single seeded spikelets (→cereals).

Probable centers of origin of b. are Asia and Ethiopia. During the stone age, b. was cultivated by the Swiss lake dwellers. Cultivated b. probably originated from *H. agriocrithon*, a wild 6-row b. from Tibet or from *H. spontaneum*, a 2-row type. B. is grown throughout the temperate regions. It is the most dependable cereal crop with good adaptation to cool climate, summer frost, drought and alkali soils. It will stand more heat under semiarid than under humid conditions. The warmer climates enable fall- or winter-sown b. Typical botanical characteristics of b. are the awns that are formed at the end of the seeds.

Production in 1995 ($\times 10^6$ mt)	Barley seed
FSU	49.23
Germany	11.50
Canada	13.00
France	7.80
Turkey	7.50
Australia	5.50
China	4.00
World	160.17

B. grain is used in food and derived products, especially for making barley malt which is converted to beer. B. grains are also used in soups and as a feed for animals, especially for nonruminant animals such as hogs.

Lit.: Fehr/Hadley* (1980)

Basil Oil
G.: Basilikumöl; F.: essence de basilic

B. is steam-distilled from the flowering tops of *Ocimum basilicum* (Labiatae) cultivated in France, USA, Italy, eastern Europe, Egypt, Madagascar, Reunion, Seychelles, and Comores Islands.

It is a pale yellow, mobile liquid with an intensely fresh, green, slightly minty and distinctly anise-like topnote that blends into a sweet-spicy, herbal note of medium tenacity. Characteristic constituents are linalool and estragole.

B. is used in flavors for meat sauces, culinary seasonings, canned soups and dishes and liquors. In perfumery, it is used as a freshness element in Eau de Colognes and men's fragrances, as an interesting topnote in floral or green fragrances or as a contrast in heavier perfume types, such as →chypre or →orientals.

Lit.: Arctander*
 The H&R Book*

Beef Tallow
→Tallow (Beef-)

Beeswax
Syn.: Cera Alba; Cera Flava
G.: Bienenwachs; F.: cire d'abeille
s.v.: 87–102; i.v.: 8–10; m.p.: 62–65 °C

The wax of the honeycomb of the bee (*Apis mellifica*) has been known and used for hundreds of years.
It is soluble in polar and nonpolar solvents, especially at elevated temperatures.
The wax is harvested by removing the honey and melting in hot water. Further purification is done by filtration or centrifugation. It is yellow and can be bleached by sunlight or peroxides to a white wax.
The composition varies depending on geographical origin. It consists (70–80%) of the esters of →palmitic acid, →melissic acid and →cerotic acid with alcohols of chainlengths of $C_{24}-C_{32}$. It contains also some free fatty acid (10–20%) and hydrocarbons (10%).
Largest markets are high quality, decorative →candles, cosmetics (→skin preparations) and pharmaceuticals (ointments, →suppositories). Smaller volumes are incorporated in →polishes. Industrial applications are limited because of price.
In 1981, 10000 mt were produced worldwide.

Lit.: Kirk Othmer* (3.) **24**,466
 [8012-89-3]
 [8006-40-4]

Beet Sugar
→Sugar Beet
→Sucrose

Behenic Acid
Syn.: n-Docosanoic Acid; C22:0
G.: Behensäure; F.: acide béhénique

$CH_3-(CH_2)_{20}-COOH$

m.w.: 340.57; m.p.: 80.7 °C; b.p.: 306 °C (8 kPa/60 mm)

B. is contained in →peanut oil and can be produced commercially by →hydrogenation of →erucic acid

or by →splitting of hydrogenated, high-erucic rapeseed oil with subsequent fractionation. Purities of up to 94% are commercially available.
B. is used in cosmetics and was used in large quantities as foam control (→defoamers) in machine →detergents. Esters and amides are used in lubricants for PVC and antiblocking agents for polyolefin films (→plastics additives).

Lit: Ullmann* (5.) **A10**, 245
 [112-85-6]

Behenyl Alcohol
Syn.: 1-Docosanol
G.: Behenylalkohol; F.: alcool béhénique
m.w.: 326.6; m.p.: 71 °C; b.p.: 241 °C (1.33 kPa/10 mm)

Properties are similar to →stearylalcohol. Commercial quantities are produced by →hydrogenation of the corresponding methyl esters. B. is transformed to behenyl acrylate, which, as a polymer, finds use in crude oil processing (→oilfield chemicals).

Lit.: [661-19-8]

Benni Oil
→Sesame Oil

Benzoin Resinoid
G.: Benzoe Resinoid; F.: benjoin

Two types of b. are commercially available, b. Siam and b. Sumatra. The Siam type is produced by hydrocarbon solvent extraction of the tree resin of *Styrax tonkinensis,* Styraceae, which grows wild in Thailand, Laos, Cambodia and Vietnam. A related tree, *Styrax benzoin,* gives the starting material for b. Sumatra. The resins are collected by making incisions in the tree bark and later collecting the accumulated sun-dried material. This raw b. is then processed to →resinoid, and in an additional purifying step to →absolute, or it is immersed and macerated in alcohol to produce a b. tincture.
B. Siam is the more sought-after and expensive product, a medium- to dark-brown viscous product with a rich, balsamic, slightly spicy-cinnamic and vanilla-like note of excellent tenacity.
Main constituents are coniferyl benzoate, benzoic acid, cinnamic acid, benzoresinyl benzoate, vanillin and styrene. B. Siam lends long-lasting richness and sweetness to floral, oriental and chypre-type (→odor description) fragrances and is widely used in colognes and fragrances for cosmetics and body-wash products.

B. Sumatra is similar in aspect and odor type but less fine in quality, more harsh-cinnamic and less vanilla-chocolate. Its uses in fragrance types are similar, and due to its lower price, it is be used in all kinds of applications.

Lit.: Arctander*
     The H&R Book*

Berberine

Syn.: Umbellatine
G.: Berberin; F.: berbérine

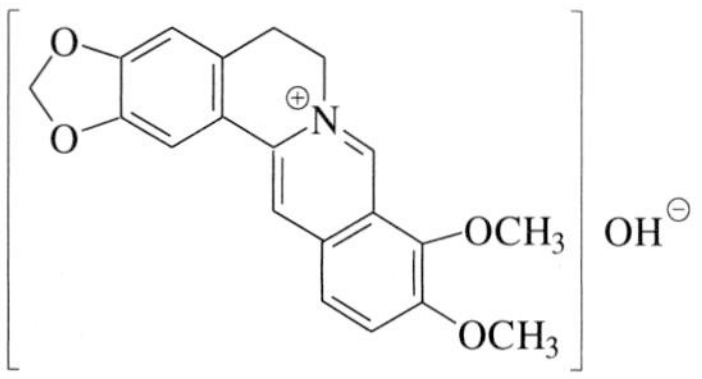

m.w.: 353.36; m.p.: 147–148 °C; (d. at 160 °C)

B. is an isoquinoline →alkaloid mainly found in Berberidaceae. Higher doses of b. extracts cause asphyxiation, whereas small doses stimulate breathing. B. shows antibacterial, fungicidal and antipyretic properties and acts as an antimalarial agent; several derivatives of b. show antitumor activity.

B. is produced by cell culture technology.

Lit.: Ullmann* (5.) **A1**, 374
     The Merck Index* (11.) 1169
     [2086-83-1]

Bergamot Oil

D.: Bergamotteöl; F.: essence de bergamote

B. is produced by expression from the peels of the nearly ripe fruits of the small →bergamot tree. Most b. used today is subjected to vacuum distillation to yield the terpeneless oil, free of phototoxic substances such as bergaptene, which might cause skin reactions when exposed to sunlight. Expressed b. is a green or olive-green mobile liquid, whereas terpeneless b. is practically colorless. The odor profiles are fairly similar, with a stronger, fresher topnote present in the expressed b. The typical odor is zesty and fresh, slightly bitter, gradually fading into a sweet herbaceous note of medium tenacity. Main constituents are linalol, linalyl acetate (about 60–70%), citral, methyl anthranilate, traces of aliphatic aldehydes and coumarine. B. is the main constituent of Eau de Cologne-type fragrances and important in oriental and floral types (→odor description) as a lively fresh topnote. Fla-

vor uses are extensive, ranging from tea flavorings (Earl Grey) to candles, beverages and tobaccos.

Lit.: Arctander*

Bergamot Tree

Citrus bergamia
Rutaceae
G.: Bergamotte Baum; F.: bergamotier

The small tree is exclusively cultivated for the production of →bergamot oil gained from the peels of its fruits, which are nonedible. The b. fruit is almost perfectly round and its color ranges from yellow to green. Main cultivation area is Southern Italy (Calabria). Other plantations are in Spain, West Africa and California.

Lit.: Gildemeister/Hoffmann*
     The H&R Book*

Betaines

→Amphoteric Surfactants

Betanin

→Dyes, natural

Beta vulgaris

→Dyes, natural

Biennial

G: Zweijährig; F: bisannuel

B. plants form flowers only in the summer or fall of the second year after sowing/planting. In the first year, b. plants will only form leaves, rosettes or roots (such as →sugarbeets, →evening primrose, carrots). After a temperature shock during the winter season (vernalization), such plants will form shoots or stems in the second year followed by flower and fruit formation.
→annual; → perennial

Lit.: P.Raven et al. "Biology of Plants" (5.) Worth Publish. NY (1992)

Biodegradable Polymers

→Degradable Polymers
→Starch-Containing Plastics
→Starch Plastics

Biodiesel

→Fuel Alternatives

Bioethanol

→Fuel Alternatives

Biomass

G.: Biomasse; F.: biomasse

There are several definitions for b.:
1. the complete living, organic matter in our eco-logical system, (volume/nonspecific).
2. the plant material constantly produced by →photosynthesis with an annual growth of $1.7-2 \times 10^{11}$ mt (marine plants excluded).
3. the cell-mass of plants, animals and microor-ganisms used as raw materials in microbiologi-cal processes.

B. constituents are →cellulose, →hemicellulose and →lignin.
They are the primary products of reduction of CO_2. Within plants, they are further transformed to specific ingredients (secondary products).
B. has always been used and is considered for more intensive use as a source for chemical raw materials (RR). It can also be used for energy gen-eration (simple burning), or as starting material for large-scale thermal decomposition[1] (cracking) with fuel oil as a resulting product (→whole-crop harvesting).
B. is also directly used as starting material for fer-mentation processes.

Lit.: Ullmann* (5.) **A4**, 99

[1] F.Goudriaan, D.G.R.Perferoen "Liquid Fuel from Bio-mass Via a Hydrothermal Process Chem.Engineering Service **45**, 2729 (1990)

Biopolymers

G.: Biopolymere; F.: biopolymères

The term b. is used for two categories of polymers:
1. All polymers occurring in nature, such as →proteins, →polysaccharides, nucleic acids and, in even more extented definitions, also derivatives thereof such as →cellulose acetate.
2. Synthetic, mostly biotechnologically produced, polymers(→microbial gums), that have structures and/or properties similar to natural polymers (→dextrans, →levans, →polyhydroxybutyric acid, →polylactic acid, →pullulan, →xanthane).

Lit.: Encycl.Polym.Sci.Engng.* (2.) **2**, 286
Kirk Othmer* (4.) **4**, 184

Biosurfactants

G.: Biotenside; F.: biosurfactants

There are several surface-active compounds that ocurr naturally (→saponins). However, the term b. include mainly surfactants generated by microor-ganisms, where the hydrophobic part derives from →fatty acids or hydrocarbons and the hydrophilic part consists mainly of a polyol, polysaccharide or peptide.
There are numerous patents dealing with this sub-ject. However, not many b. have reached practical application.

Lit.: Falbe* 118–124 (1987)

Bixa orellana
→Dyes, natural

Bixin
→Dyes, natural

Black Currant

Ribes nigrum L.
Saxifragaceae
G: Schwarze Johannisbeere; F: cassis

B. originated in northern Europe. Bushes are grown for their aromatic berries in France, UK, Germany, Scandinavia, Russia and Southern Europe. As a by-product, the seeds of b. contain ca. 30% of fatty oil. The oil is composed of 50% linoleic, 14% α- and 20% γ-linolenic acid (→lin-olenic acid).
Other sources for γ-linolenic acid are→evening primrose and →borage.

Lit: H.Traitler, "Fractionation of Black Currant Seed Oil". J.Amer.Oil Chem.Soc. **65**, 755, (1988).

Bladderpod
→Lesquerella

Blubber Oil
→Whale Oil

Bodynote
→Odor Description

Boehmeria nivea
→ Ramie

Bombyx mori
→Silk

Borage

Borago officinalis L.
Boraginaceae
G.: Borretsch, Gurkenkraut; F.: bourrache

Annual species with a blue (white is rare) 5-petaled crown (appearance of flower is similar to that of nightshades such as →potato). Typical for b. are rough hairs on almost all parts of the upper plant. Seeds have a typical oblong shape like a small insect, ca. 5 mm in length, with a small white spot at one end. Its origin is the Mediterranean area. First cultivated in Turkish Asia and Syria, but later cultivated in Spain by Arabian people during the Moorish occupation. Today, it has spread all over Europe as a weed, and it is mostly used as a spice.

Its economic importance is limited to the application of the leaves as a kitchen spice. Seeds are mostly produced in small quantities for selling them in the home gardening business. Seed yields range from 2–10 dt/ha. Seeds of b. contain 30–40% oil with about 25–40% γ-linolenic acid, which is also occurring in higher amounts in the seeds of →black currant and of →evening primrose.

There are no reports about special breeding goals or agronomic work with respect to b.

Lit.: R. Kleiman et al., "Search for New Industrial Oils XI. Oils of *Boraginaceae*" J.Amer.Oil Chem. Soc. **41**, 459–460 (1964)

Borago officinalis
→Borage

Brake Fluids
→Hydraulic Fluids

Branched Oligosaccharide Syrups
G : Sirup verzweigter Oligosaccharide
F.: sirop d'oligo-saccharides ramifés

B. are purified, concentrated solutions of nutritive saccharides obtained by enzymatic hydrolysis of starch polysaccharides with simultaneous transglucosidase action. They consist of oligosaccharides $[C_6H_{10}O_5]_n$ containing α-1,6-glucosidic and α-1,4-glucosidic linkages.

Typical saccharide composition of a commercial syrup is :

Glucose 40.5%, →maltose 6.7%, →isomaltose 16.9%, other disaccharides 4.9%, →maltotriose 0.8%, →panose 12.5%, →isomaltotriose 3.4%, other trisaccharides 2.3%, tetraoligomers 8.9%, pentaoligomers 3.3%.

B. are manufactured on an industrial scale by liquefying a starch slurry (≈30%) with bacterial α-amylase to DE 5–10%, followed by combined action of β-amylase from soybean and transglucosidases from *Aspergillus niger*.

After purification, the syrup is made up to 75% d.s. Higher enrichment of branched oligosaccharides (≈ 85% d.s.) by separation from the glucose is achieved by cation-exchange-resin chromatography.

The functional properties are:

relatively low viscosity of the solutions, high moisture-retaining properties, low moisture activity, which retards microbial growth, mild sweetness, no fermentability by yeast but entire digestibility by intestinal enzymes, physiological energy content 17.2 kJ/g.

Important physiological effects include:

increase of the *Bifidobacterium* count in human intestines, improvement of colon conditions and consistency of the feces, inhibition of synthesis of water-insoluble glucans from glucose, thus contributing to noncariogenity of such sweeteners.

Applications in food and pharmaceutical industries include:

confectionary, soft drinks, sake making, bakery seasonings, medicated syrups, energy supply, liquid diets.

Lit : Schenk/Hebeda* (1992)

Brasilian Wax Palm
→Carnauba

Brassica campestris
→Turnip Rape

Brassica napus
Brassica oleifera
→Rapeseed

Brassica spp.
→Mustard

Brassidic Acid
Syn.: *trans*-13-Docosenoic Acid; C22:1 (Δ 13)
G.: Brassidinsäure; F.: acide brassidique

$CH_3-(CH_2)_7-CH=CH-(CH_2)_{11}-COOH$

m.w.: 338.56; m.p.: 61.5 °C; b.p.: 256 °C (1.33 kPa/10mm)

B.is the *trans* isomer of →erucic acid

Lit.: Kirk Othmer* (4.) **5**, 150
[506-33-2]

Brassylic Acid

Syn.: Tridecanedioic Acid; 1,13-Undecandicar-boxylic Acid

G.: Brassylsäure ; F.: acide brassylique

HOOC–(CH$_2$)$_{11}$–COOH

m.w.: 244.33; m.p.: 114 °C

B. is produced by →ozonolysis of →erucic acid. Fermentation of tridecane has been investigated recently in Japan. Commercial qualities range around 80%. Pure acid (99%) is also available.

B. is used in the manufacture of →polyamides (nylon 6/13 and 13/13) and esters, which are employed as low-temperature plasticizers (→plastic additives) and as →lubricant components. In organic synthesis it can be used as starting material for synthetic musk.

Lit: Ullmann* (5.) **A8**, 532 Kirk Ottmer* (3.) **7**, 624
[505-52-2]

British Gums

→Dextrins

Broad Bean Starch

Syn.: Horse Bean Starch

G.: Ackerbohnenstärke; F.: amidon de fève

B. granules are deposited in the cotyledons of various bean cultivars (*Vicia faba*, L.) as reserve polysaccharides together with oligosaccharides and proteins.

Industrial or pilot-plant products are obtained by integrated →wet-milling and refining of ripe fababeans or bean meal.

The raw material may contain 35–50% starch, 28–35% crude protein, 6–8% oligosaccharides, 2% lipids and 2–8% crude fiber.

B. is isolated after dry dehulling and milling, by steeping of bean meal, removing the dissolved proteins and separating insoluble protein, raw starch refining, dewatering and drying.

Starch composition and properties:

Granules are of the ellipsoid-shaped type with a granule size of 12–48 µm (variation is due to different cultivars); x-rays show C-polymorph; crude protein 0.35–0.50%, minerals 0.06–0.25%, amylose content 31.5–34.5%. Granule swelling occurs between 56–81 °C with retarded increase of viscosity at 70 °C; low viscosity level after full granule swelling; remarkably high consistency stability against cooking and mechanical forces. This behavior is similar to that of other legume starches as well as heat/moisture-treated granular potato

starches. On cooling, b. pastes rapidly set to rigid white gels.

The world production of broad beans in 1988 has been estimated at 4.72×10^6 mt. Starch production is <10 000 mt/a.

Production of b. will be economic if the protein can be isolated and used in food protein enrichment. B. may be successfully applied as rapidly gelling agents for reducing thickener input in food preparations and in the paper industry.

Lit.: F.Schierbaum et al.: Nahrung/Food **29**, 267–278, (1985)
N.U.Haase et al.: starch/stärke **43**, 205–208 (1991)

Buffalo Gourd

Cucurbita foetidissima Kunth ex H.B.K.

Cucurbitaceae

G: Büffelkürbis; F: courge

B. has been studied for several years as a potential source of vegetable oil and starch in the USA. It is a →biennial crop and accommodating to arid climates, being indigenous to the southern USA, northern Mexico, Lebanon and India. The habit of the plant is vine-like with a typical long fleshy, pumpkin-like fruit that bears flat seeds. Experimental plantings in Texas yielded 300–900 kg of seed per acre. The oil content of the seed is 30–40%.

Fatty acid pattern:

C16:0	C18:0	C18:1	**C18:2**
7–10%	3–4%	21–30%	**57–68%**

Lit.: Gunstone* (1994)
J.A.Vasconcellos et al. "Buffalo Gourd" in "New Sources for Fats and Oils" L.H.Princen et al. (ed.), Amer.Oil Chem.Soc. p. 55 (1981)

Butanedioic Acid

→Succinic Acid

Butanol

→Acetone/Butanol Fermentation

Butenedioic Acid

→Fumaric Acid

Butterfat

G.: Butterfett; F.: graisse de beurre

s.v.: 218–235, i.v.: 25–38, m.p.: 37 °C

B. is contained at 80% in butter, which is derived from milk by churning and has the following fatty acid composition:

C12:0	C12:0	C14:0	C16:0	C18:0
7–9%	2–5%	8–14%	24–32%	9–13%

C18:1	C18:2	C18:3
19–33%	1–4%	2–6%

Butter is used exclusively for food purposes. Industrial uses have been and are being discussed during times of surplus production, however, the broad and unspecific spectrum of fatty acids prevents such intentions. See also →ghee fat.

Production in 1993 ($\times 10^6$ mt)

India	1.11
FSU	1.10
USA (1992)	0.61
Germany	0.48
France	0.44
New Zealand	0.27
Netherlands	0.19
Poland	0.17
Ireland	0.14
Australia	0.13
World	3.90

Lit.: Kirk Othmer* (3.) **15**, 522

Butyl Stearate
→Fatty Acid Esters

Butyrospermum parkii
→Shea Butter

C

Cacao Butter
→Cocoa Butter

Cadmium Soap
→Metallic Soaps

Cajeputol
→Eucalyptol

Calcium Soap
→Metallic Soaps

Calcium Stearoyl Lactate
→Lactylic Esters of Fatty Acids

Calendula officinalis
→Calendulic Acid

Calendulic Acid
Syn.: Calendic Acid; *trans,trans,cis*-,8,10,12-Octa-
decatrienoic Acid; C18:3 (Δ 8,10,12)
G.: Calendulasäure; F.: acide calendulique

CH$_3$–(CH$_2$)$_4$–(CH=CH)$_3$–(CH$_2$)$_6$–COOH

m.w.: 278.2; m.p.: 40–40.5 °C

C. derives from *Calendula officinalis* L., Compositae, grown mostly as an ornamental or as a medicinal plant for its →essential oils. The seeds contain
40–46% of an oil with 50–55% highly conjugated c. and 28–30% of the nonconjugated →lin-
olenic acid as main components.

Lit.: Gunstone* (1994)
    M.J. Chisholm, C.Y.Hopkins, Can.J.Chem. **38**, 2500
    (1960)
    [28872-28-8]

Callitris quadrivalvis
→Resins, natural

Camelina
Syn.: Gold of Pleasure; False Flax
Camelina sativa (L.) Crtz.
Cruciferae
G.: Leindotter; F.: caméline

C. has a typical cruciferous morphology with four
small, yellow petals and a mostly hairless structure
of all organs. The fruit is a small pod with yellow

to red-brown seeds, which appear similar to wheat
seeds but are only 1–2.5 mm long.
C. is a typical example of a "forgotten" oil crop,
which lost its importance with modern agricultural
production methods. Oldest findings date back to
2000 B.C. During the 15th century, c. was planted
in larger areas; later it was reported as a weed in
→flax. During the last centuries c. was known as
an undemanding and comparatively fast growing
oil crop, that was adapted to light soils. Today, c.
has only moderate economical importance in the
former SU and in Poland. Small experimental plots
were grown in the 1950's in Sweden and in Germany and later in Canada and the USA for testing
its adaptation as a RR for industrial application.
Seed yields reach 15–30 dt/ha.
The seeds yield 33% protein and 40% of an oil of
the following composition:

C16:0	C18:1	C18:2	**C18:3**	C20:1
3–8%	16–18%	18–22%	**35–45%**	15–20%

Lit.: W.Radatz, W.Hondelmann, Landbauforschung Völ-
    kenrode, 31/4, 227–240 (1981)

Camellia oleifera
Camellia sinensis
→Teaseed

Camphor
Syn.: 1,7,7-Trimethylbicyclo [2.2.1]-2-heptanone
G.: Kampfer; F.: camphre

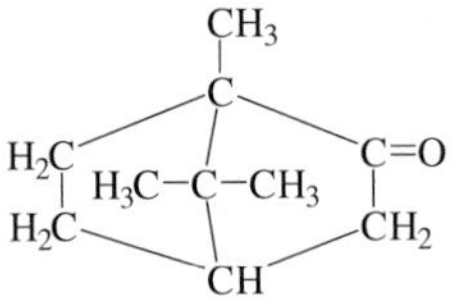

m.w.: 152.24; m.p.: 179–180 °C; b.p.: 208 °C

C. is available as crystalline, translucent mass or as
grainy crystals with an ethereal diffusive odor of
minty and medicinal quality, and a warm bitter,
then cool taste.
C. is naturally isolated from distilled camphor
wood oil (*Cinnamomum camphora*) or synthetically made from pinene and finds use in artificial
→essential oils of the →lavandin family as well as
in other fresh, green or herbal perfume types to

give lift and freshness. In flavors it reinforces minty notes.

Lit.: Arctander*

Cananga odorata
→Ylang Ylang Tree

Cananga Oil
→Ylang Ylang Oil

Canarium luzonicum
→Resins, natural

Candelilla Wax
G.: Candelillawachs; F.: cire de candelilla

C. is a hard wax (→waxes), which melts at 70 °C. Solubility is similar to →carnauba wax. It is harder than bees wax but less hard than carnauba wax.

It is harvested from the shrubs of *Euphorbia cerifera* and *E. antisiphilitica* (Zuccarini), as well as from *Pedilanthus pavonis* (Boissier) and *P. aphyllus* (Boissier), Euphorbiaceae, which grow in northern Mexico, Arizona and Texas. The entire plants are uprooted and boiled in 0.2% sulfuric acid. The wax floats to the surface and is collected, filtered and sometimes bleached and degummed. C. consists almost 50% of paraffins, the other half is a mixture of waxesters of C_{30}–C_{32} acids with alcohols of the same chainlength.

The largest market in the past was as a component in chewing gums. Today, cosmetic applications such as lipsticks are dominating (→skin preparations). Other outlets are identical to →carnauba wax.

In 1974, 3000 mt were produced. Production is declining.

Lit.: Kirk Othmer* (3.) **24**, 468
    [8006-44-8]

Candles
G.: Kerzen; F.: bougies

A candle consists of a wick (cotton) in the center and the waxy (→waxes) fuel around it.

The use of c. goes back to prehistoric times. Until the middle of the 18th century, they were made of →beeswax, →tallow, →spermaceti and →stearin. The main raw materials today are paraffin and other hydrocarbon waxes, but products based on RR, especially beeswax (in high value c. and for making wax sheets as decorative cover) and stearin as opacifier, are still in use. For colored c., dyes are added. Numerous formulations, to tailor the performance of a c. to the specific need, are in use.

Manufacturing techniques reach from repeated dipping of the wick in the wax melt to the use of drawing machines and extruders.

In 1979, 100 000 mt were produced in USA and 165 000 mt in Western Europe.

Lit.: Ullmann* (5.) **A5**, 29

Candlenut Oil
→Candlenut

Candlenut Tree
Syn.: Lumbang
Aleurites moluccdna (L.) Willd., *A. trisperma* Blanco
Euphorbiaceae
G.: Lichtnußbaum, Kerzennußbaum;
F.: noyer de bancoul, noyer des moluques

C. seeds come from a tropical tree widely distributed in Australia, Sri Lanka, Hawaii, Indonesia and Malaysia. It is grown in plantations, e. g., in Sumatra. The oil gets its name from the custom of stringing the nuts on bamboo and using them for lighting. The oil is not appreciably commercialized, but there has been a significant interest in the oil over a number of years because of its high degree of unsaturation. The kernels contain 57–69% oil, which is similar to linseed oil but of slightly lower i.v.

The fatty acid composition of c. oil is:

16:0	18:0	18:1	**18:2**	18:3
5.5%	6.7%	10.5%	**48.8%**	28.5%

Lit.: Gunstone* (1994)

Cane Sugar
→Sugar Cane
→Sucrose

Cannabis sativa
→Hemp

Canola
→Rapeseed

Caper Spurge
→Euphorbia

Capric Acid
Syn.: n-Decanoic Acid; C10:0
G.: Caprinsäure;
F.: acide décanoïque, acide caprique

CH_3–$(CH_2)_8$–COOH

m.w.: 172.26; m.p.: 31.3 °C; b.p.: 270 °C

C. is soluble in most organic solvents, insoluble in water. It is derived from →coconut oil and other medium-chain sources (→fats and oils) by →hydrolysis and →fractionation. Technical qualities range up to 90%.

C. is used in synthesis of →fragrance raw materials (aldehyde), plasticizers (→plastic additives) and →resins. →Triglycerides find use as mild, well-spreading components in →skin preparations. The name derives from goat (lat. capra) because of its unpleasant odor.

Lit.: Ullmann* (5.) **A10**, 245
[334-48-5]

Capric Alcohol
Syn.: 1-Decanol; n-Decyl Alcohol
G.: Caprinalkohol, Decylalkohol;
F.: alcool caprique

$CH_3-(CH_2)_8-CH_2OH$

m.w.: 158.3; m.p.: 7 °C; b.p.: 230 °C

C. is a colorless, oily liquid with flowery odor. It is gained from lauric acid oils (→fats and oils) by →hydrolysis and →fractional distillation.

C. is used as intermediate for making →surfactants, plasticizers (→plastic additives) and →fragrance raw materials (aldehydes).

Lit.: Ullmann* (5.) **A1**, 292
[112-30-1]

Caproic Acid
Syn.: n-Hexanoic Acid; Capronic Acid; C6:0
G.: Capronsäure, Hexansäure;
F.: acide caproïque, acide hexanoïque

$CH_3-(CH_2)_4-COOH$

m.w.: 116.16; m.p.: -3.9 °C; b.p.: 205 °C

C. is soluble in alcohol and ether, slightly soluble in water.

Derives from →fermentation of butyric acid. It can also be isolated from →coconut oil or other lauric acid oils (→fats and oils) as well as from rancid fats by →hydrolysis and →fractional distillation. It is commercially available up to 99.8% purity.

The acid is used in synthesis of →rubber chemicals, pharmaceuticals and dryers for coatings (→drying oils).

For origin of name →capric acid.

Lit.: Ullmann* (5.) **A10**, 245
[142-62-1]

Capronic Acid
→Caproic Acid

Caprylic Acid
Syn.: n-Octanoic Acid; C8:0
G.: Caprylsäure; F.: acide caprylique

$CH_3-(CH_2)_6-COOH$

m.w.: 144.2; m.p.: 17 °C; b.p.: 237.9 °C

C. is soluble in alcohol and slightly soluble in water. It is derived from →coconut oil and other medium-chain sources (→fats and oils) by →hydrolysis and →fractionation. Technical grades range up to 99%.

It is used in organic sythesis (drugs, fragrances, fungicides) and in ore separation (→mining chemicals). C. is also used to manufacture driers for coatings (→coating additives).

→Hydrogenation yields →caprylic alcohol (octanol), which is used for making plasticizers (→plastic additives). →Triglycerides find use as mild, well-spreading components in →skin preparations.

For origin of name →capric acid

Lit.: Ullmann* (5.) **A10**, 245
[124-07-2]

Caprylic Alcohol
Syn.: 1-Octanol; n-Octyl Alcohol
G.: Caprylalkohol; F.: alcool caprylique

$CH_3-(CH_2)_6-CH_2OH$

m.w.: 130.2; m.p.: -16 °C; b.p.: 195 °C

C. is a colorless liquid with a pleasant odor. It is derived from →lauric acid oils and available in high purity.

C. is used in →fragrances, as a solvent and as an intermediate to make plasticizers (→plastic additives)

Lit.: Ullmann* (5.) **A1**, 290
[111-70-6]

Caramel Color
G.: Zuckercouleur; F.: couleur de caramel

C. is a food dyestuff and taste-giving agent, exclusively obtained by controlled heating of edible sugars, either in the pure state or in the presence of one or more chemical compounds; the products are amorphous, dark-brown, water-soluble substances or mainly syrups.

Production of c. is normally based on →sucrose, →invert sugar, →dextrose and dried glucose syrups. Heating is performed in steam-heated, stirred stainless-steel equipment at 120–180 °C for 5–10 h depending on the nature of the applied catalysts. The process is highly empirical and determined by the final application. If the sugars are heated by themselves, a material is formed that is used for flavoring. In the presence of catalysts (ammonium, sodium, potassium hydroxide or their salts of acetic, citric, phosphoric or carbonic acid) color production is increased. Liquid c. types are prepared with 65–75 % d.s., arbitrary glucose content 44–61 % DE and pH 2–7. C. is a mixture of molecular and colloidal substances.

Use is concentrated on coloring in food products (→food additives), such as alcoholic and nonalcoholic beverages, vinegar, sweets, convenience foods, dry soups, puddings and bakery products.

Lit.: J.Schorrmüller (ed.): "Handbuch der Lebensmittelchemie", V, 1. "Kohlenhydratreiche Lebensmittel" 683–685, Springer-Verlag Berlin (1967)
J.W.Knight: "The Starch Industry", 118–122, Pergamon Press, Oxford (1969)

Carboxymethylated Fatty Alcohol Ethoxylates

Syn.: Polyalkoxy Carboxylates
G.: Carboxymethylierte Ethoxylate;
F.: carboxylates d'alcool gras éthoxylé

$R-O-(CH_2-CH_2O)_n-CH_2-COO^- \ Na^+$
R: C_{12-18} alkyl

The acids of c. are clear, mobile liquids while the salts are viscous pastes.

C. are made by reaction of fatty alcohol ethoxylates with chloroacetic acid in the presence of caustic soda.

The anionic →surfactants show good resistance to water hardness and good water solubility. They are nonskin-irritating and good dispersants and emulsifiers and are therefore used in →detergents and →cosmetics. They are also good industrial emulsifiers.

Lit.: Ullmann* (5.) **A25,** 747

Carboxymethylcellulose

Syn.: CMC
G.: Carboxymethylcellulose;
F.: carboxyméthylcellulose

C. is the most common cellulose ether because it is versatile and can be manufactured easily. For details concerning structure →cellulose ethers. DS values range from 0.3–1.2.

C. is anionic, and its aqueous solutions show pseudoplasticity, thixotropy and a correlation between pH and viscosity. At high (12) and very low (1) pH, precipitation occurs, while the maximum viscosity is reached at a pH of 6–7. Heavy-metal salts, trivalent cations and cationic detergents and some polymers cause precipitation. Increasing substitution leads to better solubility and biodegradation. Films are sensitive to humidity because they absorb a large amount of water. Slight crosslinking reduces this sensitivity.

The most important mixed ether is carboxymethyl hydroxyethyl cellulose (CMHEC). Even low substitution (carboxy-methyl-DS 0.3 and hydroxyethyl-MS 0.4) leads to clear solution in water.

There are many different techniques and equipment to produce c. A slight excess of alkali is necessary when the sodium chloroacetate reacts at 25–70 °C (exothermic). If the free chloroacetic acid is used as reagent, another mole of alkali is necessary. The reaction may be carried out without or with small (wetting) or large (slurry) amounts of solvents such as acetone or isopropanol. More uniform substitution, more economic use of reagents and better solubility result by applying solvents. Reagents may be mixed into the cellulose first by spraying, followed by the addition of alkali, or vice versa. The crude, neutralized and dried product (40 % NaCl) needs no purification for many applications (oil fields, →detergents, paper). If decreased viscosities are desired hydrogen peroxide is added to the reaction mixture prior to neutralization. In uses where pure types are necessary (e.g.,food), NaCl is extracted with alcohol/water mixtures. Mixed ethers (CMHEC) are made in one or two steps by the same technology as used for CMC or HEC. Crosslinking may be accomplished by adding alum, formaldehyde, bifunctional reagents or simply by heating the finished c. at 80 °C for 12 hours.

C. is used, like other cellulose ethers, in the building industry as additive in plasters, mortars and cements, in wallpaper pastes (→adhesives) and as thickener and suspension aid in paints (→coating additives) and in water-based →adhesives. It is used as plasticizer for electrical porcelain and for vitrous enamels in →ceramics. Core binders (→foundry industry) contain c. as well as →pencil lead. A main application for c. is in →detergent formulations to avoid soil redeposition. Finishes, warp sizes and printing in the →textile industry are other big outlets for c. Large volumes are consumed by the →paper industry for increasing dry strength, as sizes and for coatings. One of the largest consumer of c. is the petroleum industry for the formulation of drilling muds (→oil field chemicals). CMHEC is also widely used in modern drilling techniques because it combines the properties of both substituents. In → agriculture, c. is applied as soil aggregant. Purified c. is approved for →food applications and used in ice cream formulations as crystallization inhibitor. The pharmaceutical industry makes use of the property of c. to be insoluble in acidic media (stomach) and soluble in alkaline environments (intestine). Other uses are for dental impressions and in toothpaste. Crosslinked c. is used as a most powerful water absorbent in hygenic applications such as diapers and tampons. The sodium and calcium salts are used as suspending and viscosity-increasing agents in pharmaceutical applications. They act also as tablet binders as well as tablet disintegrating agents. CMMC is an effective film former in the manufacture of →tobacco sheets

Total US consumption is estimated at 35 000 mt (1984), and Western Europe capacity at 160 000 mt/a.

Lit.: Ullmann* (5.) **A5**, 477
Encycl.Polym.Sci.Engng.* (2.) **3**, 239 + **7**, 593
[9004-32-4] and [9000–11-7]

Carboxymethyl Guar
→Guar Gum

Carboxymethyl Starch
Syn.: Starch Glycolic Acid; CMS
G.: Carboxymethylstärke;
F.: carboxyméthyl amidon,

C. is a →starch ether in which some or all of the available hydroxyl groups of the starch have been etherified by carboxymethyl groups (→anionic starch).

The starch-O-carboxymethylether is easily formed by alkaline reaction of granular starch with monochloric acetic acid or with glycolic acid lactone:

$$AGU-OH + Cl-CH_2-C=O + NaOH \longrightarrow$$
$$\qquad\qquad\qquad\qquad |$$
$$\qquad\qquad\qquad ONa$$

$$AGU-O-CH_2-C=O + NaCl + H_2O$$
$$\qquad\qquad\qquad |$$
$$\qquad\qquad\quad ONa$$

The reaction will proceed in aqueous suspension, but the ether formed becomes swellable in cold water if DS becomes >0.1. Therefore, the process is mostly run in aqueous organic solvents such as methanol, ethanol, isopropanol, acetone. After the reaction, the unswollen and undissolved c. is

filtered off, washed and dried. Usually DS is 0.2–0.6.

Carboxymethylation enhances the hydrophilic character of →starch polysaccharides (→anionic starch).

C. rapidly absorbs large amounts of water, which swells and disintegrates the starch granules. The pastes formed are clear and highly viscous, without tendency of →retrogradation. On heating and stirring, as well as in the presence of ions, more or less thinning or breakdown of the paste occurs. This draw-back may be overcome by →crosslinking.

C. is preferably used in nonfood and pharmaceutical applications as water-binding agent (tooth pastes, emulsions).

Lit.: Tegge*, 187-188 (1988).

Carmin
→Dyes, natural

Carnauba
Syn.: Brazilian Wax Palm
Copernicia prunifera (Mill.) H.E. Moore; *C. cerifera* Mart.
Palmae
G.: Karnaubapalme; F.: palmier carnauba

C. is a palm tree native to northern Brazil, about 15 m tall. It is not used as a cultivated crop but in its natural growing habitats. The c. wax is formed on both sides of the up to 2 m long fronds as protection against water loss due to transpiration. For the c. palm, this is a good adaptation to dry climates. Harvesting is done by carefully cutting every other month 6–8 fronds per palm. This may only be done three times during the dry season. The fronds are laid on mats. During drying of the fronds, they will shrink and the wax scales will be uncovered. For getting hold of the wax, the fronds will be shaken, and the wax is also scraped from the dry leaf tissue. By this process, 5–8 g of wax are gained per frond, which is 120–180 g per tree and year. Experimental machine harvesting is said to deliver higher wax yields.

Lit.: E.Taube "Carnauba Wax-product of a Brazilian Palm". Econ. Bot. 6. 379–401 (1952)

Carnauba Wax
G.: Karnaubawachs; F.: cire de carnauba
s.v.: 78–88

C. is the most important vegetable wax. It is hard and melts at 82–86 °C. It is soluble in hydrocarbons and other nonpolar solvents but not in alcohol. It resists water and is compatible with most other waxes.

It derives from the →carnauba palm tree. There are four categories of quality (yellow, light fatty gray, fatty gray and chalky), depending on the intensity of the various purification steps. C. increases the melting point, hardness and gloss of other waxes and is therefore frequently used in combination.

Chemically, c. is mainly a wax ester derived from acids with average chainlength of C_{26} and alcohols of C_{32}. There are also some hydroxy- and aromatic acids.

Due to its high gloss, hardness and resistance to harsh environment, it is used in →polishes for floors, shoes, furniture and cars. Smaller applications are in candles, lipsticks and in leather manufacturing. It is also used for coatings of citrus fruits (→food additives), to polish candies and for sugar-coated tablets and sustained-release pharmaceutical formulations in combination with →stearyl alcohol. Its ability to retain oils and solvents makes it highly useful in the manufacture of ink ribbons and carbon paper.

In 1981, 13 000 mt were produced. Almost 80% is exported.

Marketing is controlled by the Brazilian government.

Lit.: Kirk Othmer* (3.) **24**, 470
[8015-86-9]

Carob Tree

Syn.: Locust Bean Tree; Saint John's Bread Tree
Ceratonia siliqua L.
Leguminosae/Caesalpinioideae
G.: Johannisbrotbaum F.: caroubier

C. is a → dioecious, evergreen tree that grows preferentially on poor soils in the Mediterranean region (Spain, Portugal, Sicily, Turkey, Morocco). It blooms one year and bears fruits the following year. The fruit is a pod 20 cm in length and 2 cm in width, which carries comparatively small seeds. The pulp of the fruits may be used as feed and, after toasting as food or a substitute for coffee and cocoa, or as a flavoring agent. It contains 30–40% sugar, 35% starch, 7% protein and 0.5% oil.

The seeds, which have almost always precisely the weight of 0.18 g each, are the origin of the jewelry and gemstone unit "Karat" (derived from its genus name "Ceratonia").

The hulls are removed from the kernel by an acid process or by roasting followed by a washing step. The germs are sifted off from the endosperm, which are ground, split and cleaned to produce c. seed meal (→locust bean gum).

Lit.: W.N.L.Davies "The Carob Tree and its Importance in the Agricultural Economy of Cyprus", Econ. Bot. **24**, 460 (1970)

Carob Seed Gum
→Locust Bean Gum

Carotene
→Dyes, natural

Carrageenan

Syn.: Irish Moss; Irish Gum
G.: Carrageen; F.: carragheen, chondrus
m.w. $10^5–10^6$

A group of →seaweed extracts, c. is constituted from sulfated galactan polysaccharides of different repeating units, such as D-galactose-sulfate and 3,6-anhydro-D-galactose-sulfate.

Sources for commercial isolation of c. are the red seaweeds (→marine algae, Rhodophyceae), species *Chondrus crispus* and several species of the genus *Gigartina, Eucheuma, Hypnea* and *Iridaea*. C. is the most important RR of the seaweed extracts. The algae, after harvesting and cutting, are washed and then extracted by alkaline solutions at 130 °C, yielding a crude extract, which is further purified by sieving, carbon treatment and filtration; it is then vacuum-concentrated to 3% d.s. The high viscosity does not allow further concentration. Precipitation by alcohol and vacuum- or drum-drying leads to crude c. Commercial products are yellow-white powders, soluble in water and insoluble in organic solvents. C. is not digested by α-amylases and is therefore noncaloric. In solution c. behaves as an anionic polymer. Three major c.-fractions have been identified in commercial products. The relationship between them varies widely depending on botanical and geographical origin:

R = OH   → κ-Carrageenan
R = OSO$_3^-$ → ι-Carrageenan

.3)-β-D-Galactose-4-sulfate(1.4) α-D-3.6-anhydro-galactose-(1.

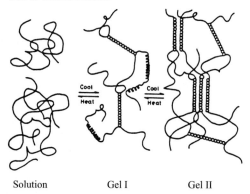

λ-Carrageenan

.3)-β-D-Galactose-2-sulfate(1.4)α-D-galactose-2.6-disulfate-(1.

Lambda-c.:
is an α-1,3-glucosidically linked linear-chain molecule esterified with sulfuric acid groups at all primary hydroxyl groups and of 70% of the hydroxyl groups in the C-2 position of both galactosyl residues. The sulfate content is 32–39% (d.s.). Lambda-c. is the only nongelling fraction.

Kappa- and Iota-c.:
are both α-1,3-linked linear polymers, constituted from the dimer carrabiose as repeating unit. Kappa-c. has a sulfate content of 25–30%, the hydroxyl groups in position C-4 of the galactosyl residue are esterified. In iota-c., additionally, the hydroxyl group in C-2 of the anhydro-galactosyl residue is esterified; sulfate content is 28–35% (d.s.). Both types form thermally reversible, clear aqueous gels in the presence of cations.

Such gels result from the formation and aggregation of double helices:

Solution   Gel I   Gel II

The gelling mechanism of carrageenan
(according to Rees)

Gelling is influenced differently by cations: kappa-c. needs K^+-cations, iota-c. needs Ca^{++}-cations. These cations, as well as added sucrose, increase the gel strength. Interactions by direct association between kappa-c. and -in the pure state- nongelling galactomannans (→locust bean gum) leads to the formation of elastic gels with properties for upgraded applications. The kappa- as well as the

iota-c. are reactive with proteins, leading to gels or precipitates. C. are used as food additives (15 000 mt/a) as well as in nonfood applications (2000–3000 mt/a) for emulsifying, thickening, stabilizing, suspending, gelling and coating. Some food applications are baked goods, dairy products, dessert gels, low-calory jellies, reconstructed meat and poultry products. Nonfood applications are found in the pharmaceutical industry (emulsions, tablet binding, cough drops and syrups). Medically it is used for therapy of ulcers as well as for diseases of the blood vessels; because it is indigestible by the carbohydrases of the human intestine c. acts as a soluble dietetic fiber. In cosmetics, it is used for water regulation in skin ointments, creams, pastes, hair tonics and foam stabilizers. Other technical industries utilize c. for stabilizing paints and pigments, for polishing agents, textile printing inks, leather and paper processing, as emulsifying agent for wax emulsions, graphite suspensions and latex.

Lit.: Levring* 319–324 (1969)
   J.M.V. Blanchard, J.R.Mitchell, "Polysaccharides in Food" 188–189, Butterworths, London/Boston/Sydney/Wellington/Durban/Toronto (1979),
   Ullmann* **A25**, 40–45,

Carthamus tinctorius
→ Safflower

Casein
G.: Casein; F.: caséine
m.w.: 34 000–375 000

C. is the →protein contained in cow's milk (≈3%). There are two types: acid and rennet. The term c. applies normally to the acid form. Rennet is only used in the food industry.

The main →amino acids of c. are →glutamic acid (22.4%), proline (11.3%) and leucine (9.2%). C. has a rather high content of phosphate groups.

C. is a granular, pale yellow powder, which has a milky odor. It is insoluble in water but forms micellar structures and dispersions in acid or alkaline, salty and organic media.

The production of c. starts with precipitation of skim milk by acid. Washing, pressing, drying and sifting give 3 kg of c. from 100 kg milk.

70–80% of the world production of c., also in the form of its Ca-salt, is used in food applications, where it gives firmness to bread, smoothness to creams and acts as a binder in milk substitutes and pet food.

Industrial uses are in →adhesives, →paper additives, paper and leather coatings (→leather auxiliaries) and in decorative plaster wall paints (coatings). Plastics made of c. (→galalith) have lost importance. The same happened to the c. fibers, which were marketed in 1930–1950. C. has been proposed as a ion exchanger for cleaning chromium-containing effluents from electroplating and tanning plants.

Production (1994) ($\times 10^3$ mt)

New Zealand	75
Poland	8
Ireland	33
France	30
Netherlands	10
Denmark	14
Germany	28
World	205

Lit.: Encyl.Polym.Sci.Engng.* (2.) **2**, 685
P.F.Fox "Development in Dairy Chemistry" Applied Science Publications, London (1982)

Cassava

Syn.: Manioc, Tapioca
Manihot esculenta Crantz
Euphorbiaceae
G.: Maniok; F.: manioc

C. is a tropical root crop, domesticated in South and Central America. Wide distribution in many tropical countries in Africa and Asia. C. is a shrubby, woody, short-lived →perennial, 3 m in height, which is easily propagated from stem cuttings. Leaves are large, deeply lobed and dark green.
Stem cuttings root easily and produce an extensive fibrous root system. Tubers are developed by secondary thickening of the roots near the base of the stem about 18 months after planting. Normally, they will be harvested individually as needed, be-

cause of poor storage behavior. They are about 50 cm long and 10 cm in diameter, sometimes branched and about 1m long. They are to a certain extent poisenous due to the cyanogenic glucoside linamarin and may only be eaten after removal of the outer cell layers and special boiling and sqeezing methods. They may be boiled, fried or ground to flour.
Peeled tubers contain about 65% water, 35% carbohydrate and <2% protein. Besides its importance as food in the tropics, there are several industrial applications that made c. →starch a world commodity.
Tapioca is prepared from fine c. starch, obtained from the tubers after washing and pressing. The starch grains are heated on plates until they swell. They are then pelletized as so-called tapioca. During the heating, some of the starch is converted into sugar. C. starch is also being applied for adhesives as well as for the manufacture of sugars, alcohol and acetone.

Lit.: Cobley* (1976)

Castor

Syn.: Ricinus
Ricinus communis L.
Euphorbiaceae
G.: Rizinus; F.: ricin

The c. plant is an easily adaptable crop of tropical, subtropical and even some temperate regions.
The seeds are up to 25 mm long, 50 mm wide and 5 mm thick with a typical brown and white pattern; they resemble the shape of a beetle.
The crop has been known for more than 2000 years, and its origin is believed to be in Ethiopia. It belongs to the spurge family (→caper spurge),

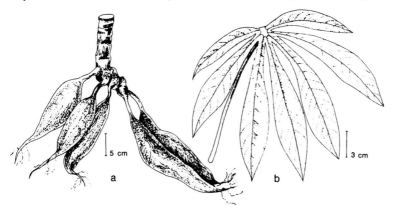

Cassava; roots (a) and leaf (b) (from Rehm and Espig*, 1995, with permission).

with only one species in the genus *Ricinus*. Today, c. is spread all over the tropical and subtropical areas, with main production in Brazil, India, China (these three represent 80% of the world production) and Thailand. There are attempts to revitalize production in the USA after it was abandoned in the 1970's. Most of the world c. crop is derived from beans collected from "semi-wild" plants and not from commercially cultivated plantations. Harvesting of the crop can be successfully handled manually or mechanically at any farming system. Due to the harvesting mode of "collection", it may be considered a typical cash crop of the tropical areas. Typical seed yields are around 5–12 dt/ha. Leaves of the →perennial crop may be used for feeding silkworms.

The seed contains 40–50% oil (→castor oil) with 14–22% crude protein, 15–18% crude fiber, and 2–3% ash. Important for further use of the protein are two types of →alkaloids: 0.03–0.15% ricinin and about 3% ricin, the toxicity of which is higher than that of potassium cyanide. This means that, after oil →expelling and →extraction, the residue is not to be used for any feeding purposes, unless it is treated further to destroy toxic and allergenic substances. It may, however, be used as fertilizer.

There are also recent genetic efforts (Costa Rica, Europe, USA) to increase oil content and quality and to reduce the toxic effects of the plant and seeds.

Lit: D.Atsmon "Castor" in Röbbelen* 438–448 (1989)
     Rehm/Espig* (1991)

Castor Oil

Syn.: Ricinus Oil
G.: Rizinusöl; F.: huile de ricin
s.v.: 177–187; i.v.: 82–90 OH–; v.: 160–168;
m.p.: –10 °C

C. is a yellowish, viscous liquid, soluble in alcohol and hot hexane.

The oil is gained from the seeds of →castor tree by →expelling and →extraction and is further treated (→fats and oils) by degumming, refining, deodorizating and decoloring. Various grades of c. are commercially available.

Fatty acid composition is characterized by a content of 87–91% →ricinoleic acid, an unsaturated C_{18} hydroxy acid. The remainder is 2% C16:0; 2% C18:0; 4% C18:1; 4% C18:2.

C. is one of the few triglycerides with a fatty acid of high functionality (OH, double bond and carboxylic group), which makes it an interesting starting material for many oleochemical modifications.

Another similar oil derives from →lesquerella seed, which contains →lesquerolic acid.

C. is used in medicine (laxative) and as lubricant in technical applications (textile, leather, metal). The oil itself as well as its alkoxylated form is extensively used as a polyol component in polyurethanes (coatings and foams).

→Hydrolysis yields castor oil fatty acids, which consist of 87–91% →ricinoleic acid an interesting intermediate for oleochemistry. It can be used as such or is purified by →distillation and →fractional distillation. It is the starting material for making →ricinenic acid, esters, →sebacic acid, →undecylenic acid, nylon 11 (→polyamides) and →hydroxy stearic acid.

Dehydration at 240–260 °C under reduced pressure in the presence of 0.1–0.2% sulfuric acid as catalyst yields a →drying oil called dehydrated c. [64147–40-6], which contains 40% conjugated and 60% nonconjugated linoleic acid (→ricinenic acid). Uses are nonyellowing →alkyd resins and →coatings.

Partial →hydrogenation gives a waxy material, which is used in antiperspirants (deodorants), leather coatings, roll leaf foil manufacturing and as a component in polyurethanes.

→Sulfonation yields turkey red oil (→sulfated fats and oils), which is used in →textile auxiliaries.

Alkali fusion of c. dissolved in octane at 180–200 °C yields methyl hexyl ketone and 10-hydroxy decanoic acid. At 250–275 °C, octanol-2 and →sebacic acid are obtained.

→Alkoxylation with ethylene oxide of the OH groups of c. leads to a group of nonionic →surfactants, used as →foamers and defoamers in detergents, lubricating oils and cutting fluids. Propoxylated c. serves in lubricants and as a polyurethane polyol. In the USA, 6500 mt of alkoxylated c. were produced in 1988.

Oxidized (blown) c.is used as a plasticizer for cellulose-, polyvinylbutyral- and polyamide-resins and also for natural and synthetic rubber. It is also an excellent tackifier in adhesives, a pigment grinding medium and a base for →inks, →lubricating oils and →hydraulic fluids.

Worldwide production of c. is in the range of 447000 mt/a (1994), 35000 mt of which are transformed to nylon 11.

Lit.: Kirk Othmer* (4.) **5**, 301
     [8001-79-4]

Castor Oil Fatty Acid

→Castor Oil

Cationic Starches

G.: Kationische Stärke;

F.: amidons cationiques

→Modified starches, containing positively charged groups in aqueous solutions under suitable pH conditions, are called c.

They are one of the most important classes of →starch ethers because of their large use in the paper industry.

There are two different groups of c.:

(A) tertiary amino-alkyl ethers and (B) quaternary alkyl-ammonium ethers.

Type (A):

$$A G U-O-CH_2-CH_2-\overset{H}{N^+}-(C_2H_5)_2 \ Cl^-$$

Type (B)

$$A G U-O-CH_2-CH(OH)-CH_2-N^+-(CH_3)_3 \ Cl^-$$

Cationizing is achieved by reacting →starch granules of any kind at <70 °C with either (A) chlorotrialkylamines or (B) quaternary epoxy amines in aqueous suspension (pH>10), followed by washing and drying. Reactions in the dry state have been patented and lead to products with a DS of 0.005–0.05.

Gelatinized products are prepared either by pumping the washed reaction slurry on steam-heated roller drums, grinding and sieving or by reacting highly concentrated solutions of gelatinized starches without further purification.

Compounds of type (A) are effective in applications at pH <8.0; they lose their cationic character at higher pH, while compounds of type (B) are cationic at any pH.

Cold-water solubility of both types depends on →DS and becomes perceptible at DS 0.07. Viscosity and its stability as well as clarity of the dispersions are enhanced likewise.

They are used at the wet-end of the paper machine and at the size-press (→paper additives), for textile sizing and as flocculants.

Advantages are:

– improvement of mechanical strenght,
– better retention of fillers,
– faster drainage,
– less water pollution,
– saving of additives and better environmental compatibility.

On average, every sheet of paper contains 0.5 g of c. Increasing applications are in textile sizing (→textile auxiliaries), in flocculation of waste waters and in breaking emulsions.

Annual production of c. for the markets in E.C. and USA amounts to 220000 mt (for making 50×10^6 mt of paper) and 225000 mt, respectively.

Lit.: van Beynum/Roels*, 94–96 (1985)
    Blanchard*, 320–321 (1992)

Cattle

G.: Rind; F.: bœuf

C. is an important RR. The main reason for breeding c. is the production of beef for nutrition. There are also many functional and chemical uses of the by-products: →tallow as an important source for fatty acids and derivatives; hides for making →leather; bones for producing adhesives (→glue, animal). Bile is used as a source for →steroids, and the pancreas is used for →insulin production.

Production in 1994 ($\times 10^6$)	Cattle, total slaughter inventory
USA	35.2
FSU-12	35.0
Brazil	22.4
India	21.3
China	19.0
Argentina	12.3
Australia	8.3
France	6.3
World	220.1

CD

→Cyclodextrins

Cedarwood Oil

G.: Zedernholzöl; F.: essence de cèdre

Various species of different origin are used to produce c. through steam-distillation of saw dust and wood wastes. The most well-known quality, c. Virginia is distilled from wood of *Juniperus virginia*, whereas c. Texas is made from *Juniperus mexicana*. Chinese c. is made from wood of *Cypressus funebris*. All of the above are in the family Cypressaceae, whereas Atlas c.(also known as Morocco c.) is distilled from the wood of a tree belonging to the Pinaceae family, *Cedrus atlantica*.

The odors of the c. from the cypress-family are fairly similar and typically described as "sharpened pencil", a soft, extremely woody note with certain sweetness. Atlas c. is more camphoraceous and has an almost medicinal note in addition to the woody-cedar character. All oils are slightly viscous, pale yellow to yellow in color and may separate a crystalline part when stored at cool temperature due to the presence of cedrol as a main ingredient.

C. is extensively used in low-cost fragrances wherever a long-lasting smooth woody note is required and as an enhancer of the notes of more expensive woody notes, such as →sandalwood.

Lit.: Arctander*
      The H&R Book*

Cedrus atlantica
→Cedarwood Oil

Cellobiose
→Cellulose

Cellophane
G.: Cellophan, Zellglas; F.: cellophane

C. is regenerated cellulose and is made from wood pulp by the same process principles as used for viscose fiber (for process details: →viscose).

The viscose solution (7% cellulose) flows through adjustable slots into an acidic precipitation bath that contains salts, such as sodium and ammonium sulfate. Dyes may be added. Cellulose is regenerated, and the continuous film is running through several treatment baths, one of which is →glycerol. The foil is dried on hot cylinders (60–80 °C) and finally wound up. It contains 8% water and 12% glycerol, both acting as plasticizers.

C. is a strong, flexible, glossy, transparent and, if no dyes are added, colorless film of 12–45 μm, which does not melt but looses strength at 150 °C. C. is biodegradable and can easily be printed and dyed. It is highly permeable to water vapor and sensitive to water but resistant to fats and oils. To achieve better water resistance, the film is sometimes coated with →cellulose nitrate or is laminated with other film material (polyethylene, polyvinylidene chloride, aluminum).

The film is approved for packaging in the food industry where it is widely used. Other applications are in packaging of textile goods, cigarettes, soap and books. Adhesive tapes are another outlet.

In 1970, the market was estimated to be 700000 mt worldwide. but now, 70–80% of this volume has been replaced by oriented polypropylene.

Lit.: Ullmann* (5.) **A11**, 96
      [9005-81-6]

Celluloid
Syn.: Pyroxylin
G.: Celluloid; F.: celluloïd

C. is a solid "solution" of 70–75% low-nitrated cellulose (→cellulose nitrate) in 25–30% →camphor. Sometimes, other plasticizers and a flame retarder (e. g., ammonium phosphate) are added.

The transparent hornlike material is thermoplastic at 80 °C and can easily be transferred into sheets, rods, tubes, films and molded goods.

For manufacturing, the cellulose nitrate (DS 1.9) is kneaded with a concentrated solution of camphor in alcohol at 40–50 °C. Colorants and other additives may be added. The resulting mass is filtered, rolled and cut, whereby most of the alcohol evaporates. The rest of the alcohol (15%) remains in the mass during further processing and is eliminated by drying the final product.

Many traditional applications of c. are nowadays replaced by synthetic resins. However, there are many uses remaining, such as spectacle frames, combs, dolls, buttons, ping-pong balls and office supplies.

Production is declining due to fire hazard, and applications such as photographic films have disappeared completely; 40000 mt were produced worldwide in 1929 with only 5000 mt/a remaining nowadays.

Lit.: Ullmann* (5.) **A5**, 434
      [8050-88-2]

Cellulose
G.: Cellulose; F.: cellulose

Cellulose is the most widespread organic compound and is contained in almost all plants. The annual yield of c. produced by photosynthesis is estimated to be in the range of 1.3×10^9 mt.

The amount of total biomass annually produced of which c. is the predominant part is roughly equal to all known resources of petroleum and natural gas. Thus, c. is available in abundance and reproduced constantly, in contrast to the fossil sources.

Only a few sources consist of pure cellulose (called α-c.) e. g., the hairs of cotton seeds (better than 90%). The large-volume source of c. is wood (40–50%), mainly from pine trees. C. has to be separated from its by-products (→lignin, →hemicellulose) by a pulping process. Mechanical pulping yields a material that is used in papermaking. For chemical utilization content of more than 90% α-cellulose is necessary and can be obtained only by chemical pulping (→paper) or by the use of linters,

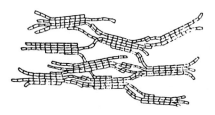

which is the by-product of cotton manufacture: After separation of the long seed hairs by ginning (→cotton), the seed corn is covered with short, coarse fibers called linters. These are cut off and used for production of regenerated c. or derivatives.

C. is a white, taste- and odorless polymeric material of mostly fibrous structure. It is the isotactic β-1,4-polyacetal of cellobiose, which is 4-O-β-D-glucopyranosyl-D-glucose (see formula above).

The quality is defined by the content of α-c. and by the average degree of polymerization (DP), which is the number of glucose units per molecule. DP depends on the type and origin of the source plant as well as on the kind of pulping technology used. C. forms secondary structures. There are four different lattice structures (cellulose I-IV). More important for further modification (derivatization, regeneration) are the crystallite structures. The OH groups of c. are able to form hydrogen bonds, which are, together with the stiff chain structure of c., responsible for the tendency to form crystallites. Therefore, in native as well as in regenerated c., there are crystalline and amorphous segments: 24–40 glucose units form cyrystallites that are separated by amorphous areas.

C. forms also supermolecular structures (texture) called fibrils, which are visible in an electrone microscope.

All these structural variations strongly influence the physical properties, solubility and reactivity of solid c.

C. is nonmelting, and thermal decomposition starts at 180 °C. It is hygroscopic and absorbs 8–14% water at 20 °C and 60% r.h. It dissolves in strong acids under severe degradation of the chains. Caustic causes strong swelling under formation of alkali cellulose. Molecules with lower m.w. will dissolve.

All substitution reactions on c. deal with the three free OH-groups on each glucose unit. Substitution can be partial or complete. The →degree of substitution (DS) is a value between >0 and 3. Because of its supermolecular structure, c. is reacted mostly in a heterogenous mode. All reactions start at the amorphous regions of the elementary fibrils. Longer reaction times, harsher conditions and an excess of reagent forces the reaction slowly to take place also in the crystalline zones, which disappear at a DP of ≈2.5. This course of reaction explains why a partially substituted c. derivative consists of a mixture of fully and unsubstituted glucose units as well as partially substituted molecules with a nonuniform distribution of substituents (blocks) along the chain. As a result, one observes sometimes inhomogenous solutions of the derivative and some fluctuation of properties, e. g., with →c.-ethers. It is therefore necessary to "overderivatize" the amorphous segments to compensate for non-substituted crystalline areas. A high degree of swelling (in water), strong alkalization (with some unwanted degradation of the chain) or the addition of solvents (such as benzene) enhance reactivity. This behavior is also the reason for many attempts to find proper solvents for c. that allow a (more) homogenous reaction.

The following types of reactions are important:

1. Acid degradation: The β-glucoside linkages between the glucose units are split by acids. In heterogeneous hydrolysis, the reaction proceeds rather fast (depending on pH and temperature). It stops almost completely when the DP level of the crystallites (25–100) is reached. Only homogeneous reaction yields cellobiose and glucose.

2. Alkaline degradation: Alkaline hydrolysis requires higher temperatures (>150 °C) and starts at the reducing end of the chain, leading to oxidized fragments by a zipperlike reaction.

3. Microbiological degradation: Cellulase →enzymes also cause hydrolysis of the β-glucoside linkages. This possibility has found increasing interest to use cheap biomass as a feed medium for biotechnological processes.

4. Oxidation reactions: Some oxidative reagents attack c. nonspecificly (chlorine, hypochlorite and chromic acids). All kinds of "oxycellulose" can be

produced. Nitrogen oxides oxidize more specificly the OH-group on C-6 to a carboxylic group, while periodate splits the glycol configuration on C-2 and C-3 to yield a dialdehyde. Most oxidation reactions also result in substantial degradation of the chain.

5. Esterification: →cellulose esters, →c. nitrate, →c. acetate

6. Etherification: →cellulose ethers, →methyl c., →ethyl c., →carboxymethyl c., →hydroxyethyl c. and →hydroxypropyl c.

7. Regeneration: →cellulose regenerate, →viscose and →cellophane.

8. Graft copolymerization: Side chains can be grafted on c. in the presence of a peroxide or a redox system. Thus, styrene, acrylic acid or esters can be added. However, industrial applications of these polymers are limited up to now.

The largest volume of cellulose, after partial removal of noncellulosic material, goes into paper, paperboard and nonwovens (150×10^6 mt/a). The long hairs of cotton are used in textile applications (15×10^6 mt/a). Only 7×10^6 mt/a is used in the form of high α-celluose wood pulp or linters for the production of regenerated fibers (→viscose) or films (→cellophane) and for the manufacture of c. derivatives, especially esters and ethers.

The use of c. in existing and new areas of application can be enhanced if:

– new crops for easy-pulping c. with a high α-content are developed;
– new pulping processes (improved economy and ecology) would exist;
– higher-value uses for by-products (lignin) can be found;
– more homogeneous reaction conditions are possible (solvents);
– regeneration is modernized (less air and water pollution);
– enzymatic hydrolysis is improved.

Lit.: Ullmann* (5.) **A5**, 375
    Encycl.Polym.Sci.Engng.* (2.) **3**, 90
    Kirk Othmer* (4.) **5**, 476
    R.A.Young, R.M. Rowell "Cellulose: Structure, Modification and Hydrolysis" Wiley-Interscience Publ. NY (1986)
    [9004-34-6]

Celluloseacetate

Syn.: Acetylcellulose

G.: Celluloseacetat; F.: acétate de cellulose

Linters (→cellulose) is the preferred raw material for the production of c. However, highly purified wood cellulose (α-cellulose > 90%) can also be used. Acids react slowly; therefore, acid anhydrides are the main reactants. Sulfuric acid is still (since 1879) the preferred catalyst. In any case, it is necessary to form the triester (DS ≈ 3) first.

There are in principle three process types:

– the acetic acid process:
   Glacial acetic acid acts as the solvent in an excess of 10–40%. Sulfuric acid (2–15% based on cellulose) catalyzes the exothermic reaction. Temperature is kept below 50 °C. A highly viscous solution of the triester is obtained.

– the methylene chloride process:
   Cellulose is suspended in CH_2Cl_2. Lesser amounts of sulfuric acid (1%) and anhydrid (5–20%) are necessary. The temperature is controlled by the b.p. of CH_2Cl_2 (41 °C), which is a good solvent for the triester. Corrosion problems, however, require expensive alloys for the equipment.

– fiber acetate process:
   The fiber structure of cellulose is maintained. Large amounts of CCl_4 are necessary to suspend the cellulose. Perchloric acid is used as a catalyst.

Mixed esters such as cellulose acetopropionate and cellulose acetobutyrate are produced by one of the two first-mentioned processes when the anhydrides of propionic or butyric acid are added.

To obtain lower DS, →hydrolysis is carried out by adding water or dilute acid after the reaction. Cellulose esters hydrolyse and degrade rapidly. The reaction is controlled by temperature, acid and water concentration. Thus, the so-called secondary acetate is formed. To get a stable c., it is necessary to wash all traces of acid out of the final product.

The recovered anhydride and acid are recycled.

The viscosity of the final product is determined by the DP (controlled degradation) of the molecule, while the DS is responsible for the solubility in different solvents. The kind of acid used influences melting characteristics and hydrophobicity.

Uses of different c. types and mixed esters:

triacetate	photo films,
(DS: 2.8–2.9)	films and foils
sec. acetate	molded goods,
(DS: 2.3–2.5)	fibers, films, coatings
acetate/propionate	molded goods
(DS: 0.3–2.3)	
acetate/butyrate	molded goods, films
(DS: 2.1/0.5)	
(DS: 0.6/2.3)	coatings

Films are made by casting from methylene chloride solution.

Triacetate shows reduced sensitivity to moisture compared to the secondary acetate.

Fiber production starts from an acetone solution of secondary acetate (20–30%), which is filtered and pressed through spinnerets with up to 100 holes. For cigarette filter tow manufacturing 1000 holes are usual. The dry spinning technique is used almost exclusively. The fiber is formed in a spinning column, 4–6 m long, where the solvent is evaporated in a countercurrent of air at 80–100 °C. The fiber is finally stretched in order to increase strength. Triacetate fibers are manufactured by a wet spinning process.

One major outlet for cellulose acetates is textile fibers (280000 mt/a). Much secondary acetate and less triacetate is used.

For cigarette filter tows, secondary acetate fibers are used (370000 mt/a) (→tobacco industry). These two markets take up 80% of total c. production.

Triacetate (60000 mt/a) for photographic films is substituted by polyester. Secondary acetate and mixed esters are used in molded goods, →coatings and films.

Uses of mixed esters in molded goods and in →coatings are declining, because they are not sufficiently biodegradable. Modern investigations focus on improvements by searching for ether or ester groups or combinations thereof, that give thermoplasticity at lower DS and leave enough OH groups to ensure biodegradation.

Lit.: Ullmann* (5.) **A5**, 438
    Encycl.Polym.Sci.Engng.* (2.) **3**, 158
    [9004-35-7]

Cellulose Acetobutyrate
→Cellulose Acetate

Cellulose Acetopropionate
→Cellulose Acetate

Cellulose Esters
G.: Celluloseester; F.: esters cellulosiques

Reaction of →cellulose with inorganic or organic acids yields cellulose esters (→esterification):

Cell–OH + HX ↔ Cell–X + H_2O
X = –ONO$_2$; –OSO$_2$OH ; –O–CO–CH$_3$;
    –O–CO–(CH$_2$)$_2$–CH$_3$
(Cell-: for detailed structure →cellulose)

Cellulose can be partly or almost completely esterified. The →degree of substitution (DS) ranges from >0–<3.0. Also mixed esters (→cellulose acetate) are possible and in use.

Inorganic esters:
Only →cellulose nitrate is of commercial importance.

Cellulose sulfate is well investigated but it tends to become unstable due to hydrolysis and chain degradation. Many attempts have been made to produce and commercialize a sodium salt of semiester as thickener for paints and inks and for paper applications. Use as ion exchanger has also been considered; however, nothing seems to be in practical use.

Some other esters such as phosphate, phosphite, halogenide, borate, and titanate, are described in the literature.

The tri-nitride is easily formed by reaction of cellulose with nitrosyl nitrate or chloride and has been considered as a substitute for xanthate in the manufacturing of →cellulose regenerate.

The use of expensive solvents, however, hampered the realization of such ideas.

Cellulose xanthate is an important intermediate for large scale production of →cellulose regenerate.

Organic esters:
Cellulose can theoretically form a large variety of organic esters. However, due to low reactivity with most acids and the properties of the final products, only a few esters of short-chain aliphatic acids are of commercial interest.

Formic acid esters are easily formed, but they are unstable and therefore not in use.

A small volume of a mixed ester of acetic and phtalic acid is used in tablet coating.

Only →cellulose acetate and the mixed c. of acetic/propionic acid and acetic/butyric acid are in large-scale production and have many uses.

Lit.: Ullmann* (5.) **A5**, 419
    Encycl.Polym.Sci.Engng.* (2.) **3**, 139 + 158

Cellulose Ethers
G.: Celluloseether; F.: éthers de cellulose

C. are derivatives of →cellulose in which the hydroxyl groups are completely or partially replaced by ether substituents (→etherification). Cellulose is reacted with either alkyl chlorides, ethylene oxide or propylene oxide according to the formulae:

I: Cell–OH + RX ⟶ Cell–OR +HX
     R = –CH$_3$; –CH$_2$–CH$_3$; –CH$_2$–COOH(Na); X: Cl

II: Cell–OH + CH$_2$–CH–R ⟶
              \O/   Cell–O–CH$_2$–CHOH–R

     R: H, CH$_3$
     (Cell-: for detailed structure →cellulose)

The →degree of substitution (DS) may range from >0 − <3. There is a slight preference in etherification of the C2 and the C6 hydroxy groups. Short-chain molecules are alkylated easier than those with a higher m.w. This causes problems due to higher consumption of expensive reagent and higher solubilty of these short chains, which leads to a higher oxygen demand in waste water during work-up procedures. A low hemicellulose content is therefore essential.

Mixed ethers are also possible and are used where a combination of properties is desired. They are produced by →etherification with a mixture of reagents or in a two step process.

Substitution by reagents that generate a new OH group (e. g. hydroxyethyl-), which itself can be etherified, is defined by the molecular substitution (MS).This value indicates the average number of moles of this reactant per anhydroglucose unit.

C. are easy in handling, economic and efficient, low in toxicity and can be tailored to certain applications.

Most c. are →water-soluble polymers that produce aqueous systems with properties that promote adhesion, thickening, film-formation and defined rheology.

C. of high commercial importance are:
→methyl cellulose (MC)
→ethyl cellulose (EC)
→carboxymethyl cellulose (CMC)
→hydroxyethyl cellulose (HEC) and
→hydoxypropyl cellulose (HPC).

They are described seperately. (Mixed ethers can be found under the keyword of the c. with the dominating substituent).

Less or not yet important c. are:

2-(N,N-Diethylamino-)ethyl cellulose (DEAEC): made by reaction of alkaline cellulose with hydrochloride of the amine and used as a weakly alkaline chromatography material or ion exchanger. Strongly basic exchangers are available by tranforming the amino group into the quaternary ammonium compound. These products are also used in shampoo formulations (→hair preparations).

Cyanoethyl cellulose (CNEC): results from a reaction of acrylonitrile with alkaline cellulose. Small quantities are used in paper for improving sheet strength.

Carboxyethyl cellulose (CEC): obtained by reaction of acrylamide with alkaline cellulose and subsequent →hydrolysis. Properties are similar to CMC.

Other cellulose ethers are sodium 2-sulfoethyl cellulose (SEC) and phosphonomethyl cellulose

(PMC). The mixed ether of the latter with hydroxyethylated cellulose (HEPMC) shows unusual gelling properties with multivalent cations at different pH values.

Lit.: Ullmann* (5.) **A5**, 461
        Encycl.Polym.Sci.Engng.*(2.) **3**, 226 + **7**, 593+600

Cellulose Fibers

Syn.: Cellolosic Fibers, Cellulosics
G.: Cellulosefasern; F.: fibres cellulosiques

Cellulose fibers can be classified into different categories according to the degree of treatment:

1.) Natural cellulosic fibers:

These are fibers that occur in nature and are used as such, mostly after a process of purification. Examples are →cotton, which is the most important, and →flax, →jute, →ramie, →hemp, →sisal, →kenaf, coir (→coconut) and →manila hemp. →China gras may get some importance in the future.

2.) Modified but not derivatized fibers:[1]

There are modern processes in development and have already passed the pilot stage to produce fibers by dissolving cellulose in a mixture of N-methyl morpholine oxide (NMMO) and water. The solution is filtered, spun and precipitated in an aqueous bath. NMMO can be recycled with a yield of 99%. The resulting fiber has the highest dry and wet strength of all c. Fibrillation is still a problem. There could be large outlets for such a fiber, which would attack not only the cotton but also the polyester market.

3.) Regenerated fibers:

Cellulose is dissolved by formation of an intermediate derivative and is finally precipitated to form c.

For details →viscose and Cuoxam (→cellulose regenerate).

4.) Derivatized fibers:

Cellulose is derivatized and remains chemically modified after spinning and in use (→cellulose acetate).

Lit.: Ullmann* (5.) **A5**, 391

[1] Chemiefaser/Textilindustrie **43**./95.p. 25 (1993)

Cellulose, Microcrystalline

→Microcrystalline Cellulose

Cellulose Nitrate

Syn.: Nitrocellulose; Gun Cotton; Collodion
G.: Cellulosenitrat, Collodium;
F.: nitrate de cellulose

C. is the only inorganic cellulose ester of commercial importance. It was produced already in the middle of the 19th century. The western world capacity is estimated to be in the range of 150 000 mt/a.
→Cotton linters is the preferred starting material. Wood cellulose can also be used in purified form. Lower quality is tolerable in the production of highly nitrated types. The degree of polymerization (DP) of the starting material is, among other parameters, responsible for the DP of the final product.
The nitrating acid is, just like a hundred years ago, a mixture of nitric and sulfuric acid and water (normally 21.4:66.4:12.2). The ratio may be changed according to the DS and the properties desired. A large excess of acid (20:1–50:1) is used to keep the reaction mixture stirrable. The reaction runs at 10–36 °C for 30 min. in stainless-steel reactors. Continuous processes are known and practiced. After the reaction, the acid is centrifuged off. Further steps deal with the fast and complete elimination of the acid from the product because acids reduce quality and yield by degradation and hydrolysis.
A modern semicontinuous process increases safety, reduces personnel costs, provides ecological advantages and increases the uniformity of the final product.
There are in principle four classes of final product:
– low nitrated (10.5–11.2% N; DS ≈1.9), for thermoplastics (molded goods; →celluloid) and →coatings;
– middle nitrated (11.2–11.7% N; DS ≈2.1), for →adhesives and coatings;
– higher nitrated (11.8–12.2% N; DS ≈2.3), for →adhesives and →coatings;
– almost completely nitrated (11.8–13.7% N, DS ≈2.6), for →explosives.
The various →coating types differ in DS, DP, solubility, kind of application (dipping, spraying, brushing, casting) and go into different fields of application, e.g., coatings for wood, metal, paper, plastic films, →leather, and in printing →inks.
Uses in photographic films and as fiber (Chardonnet silk) are abandoned because of its high inflammability.
For safety reasons, none of the c. types are delivered in pure, undiluted form but as water-wet or phlegmatized products.

Lit.: Ullmann* (5.) **A5**, 421
Encycl.Polym.Sci.Engng.* (2.) **3**, 139
[9004-70-0]

Cellulose Powder

G.: Pulvercellulose; F.: cellulose pulvérisée
m.w.: ≈243 000

Powdered cellulose is manufactured by the mechanical processing of α-cellulose obtained as a pulp from fibrous plant materials. It is a white, odorless, tasteless powder of variable fineness, and it ranges from a free-flowing, dense powder to a coarse, fluffy, nonflowing material. It is insoluble in water, dilute acids and most organic solvents. It is slightly soluble in sodium hydroxide solution. It is used as a tablet or capsule diluent, as a sorbent and a suspending agent.

Lit.: R.W.Chilamkurti, C.T.Rhodfes, J.W.Schwartz, Drug Development and Industrial Pharmacy **8** (1), 63–86 (1982).

Cellulose Regenerate

G.: Celluloseregenerat; F.: cellulose régénérée

Due to secondary and supermolecular structures (hydrogen bonds, lattices and crystallites, fibrils), →cellulose is insoluble in water and other solvents. There is, however, a possibility by forming certain derivatives or complexes to circumvent this property. For example, c. forms a xanthogenate (with CS_2 and NaOH), which is soluble in aqueous systems and can be regenerated to c. by precipitation in acid media after the formation of fibers or films. For details of this important process: →viscose.
Another possibility to convert c. into a fiber, called Cuoxam, is the cuprammonium process. Linters or high α-cellulose pulps dissolve in a mixture of copper hydroxide or a basic copper salt and highly concentrated ammonia. The solution is filtered and contains 4–11% cellulose, 4–6% Cu^{++} and 6–10% NH_3. This solution is extruded through relatively large spinning holes into a vertical waterstream and is simultaneously stretched. Copper is extracted, the fiber is neutralized, finish is applied, and the fiber is finally dried. Filament (and staple fibers), hollow fibers, nonwovens and membranes can be made by this process.
The fiber is similar to →viscose but is superior in gloss and silklike in appearance, but it is inferior in strength.
Production of cuoxam fiber began to decline in the early 1960's due to the success of synthetic alter-

natives, outdated equipment and unresolved ecological problems (copper). Research was then focussed successfully on membranes and hollow fibers, which are now the standard materials worldwide for blood dialysis in artificial kidneys. There
are only a few production plants serving this important specialty market.

There have been many trials to develop other regeneration processes that are economically and
ecologically acceptable and lead to high-quality fiber.

Lit.: Ullmann* (5.) **A5**, 400

Cellulose Sulfate
→Cellulose Esters

Cellulose Xanthate
→Viscose

Cellulosic Fibers
→Cellulose Fibers

Centranthus macrosiphon
→α-Eleostearic Acid

Cephalosporines
→Amino Cephalosporanic Acid

Cera alba/flava
→Beeswax

Ceramics
G.: Keramik; F.: céramique

The production of c. needs auxiliaries. Depending
on the various technologies used in this industry,
there are applications of RR-based products as
plasticizers and lubricants (→stearic acid and their
→metallic soaps, →oleic acid and →glycerol), deflocculants (mainly →oleic acid, but also →stearic
acid, →tartaric acid), →surfactants and foam depressants (stearates,→metallic soaps,→fatty alcohols).

The largest volumes are represented by organic
binders where →sodium alginates, →gum arabic,
→tragacanth, →sucrose molasses, →lignosulfo-
nate, →methyl cellulose (also mixed ethers),
→carboxymethyl cellulose and →starch are used.

A large market for methyl cellulose is the use as
binder for manufacturing the ceramic bodies for
afterburner catalysts for cars.

Lit.: D.W.Richerson,"Modern Ceramic Engineering"
   Marcel Dekker Inc. NY (1992)
   T.Morse "Handbook of Organic Additives for Use
   in Ceramic Body Formulation" Montana Energy
   and MHD Res. and Dev. Inst. Butte Mont.(1979)

Ceratonia siliqua
→Carob

Cereals
G: Getreidearten; F: céréales

C. are any of a variety of mostly →monocotyle-
donous crops of the gramineae family used for
food and as RR such as →wheat, →oats, →barley,
→rice, rye, →corn and sorghum (→sugary
sorghum processing). Their main product is
→starch (→wet milling, →dry milling). Corn and
oats are or may be used also for fat production.
Baking properties of the meal depend on the proportion of certain proteins (→gluten cereals).
Pseudocereals are starch-producing plants of the

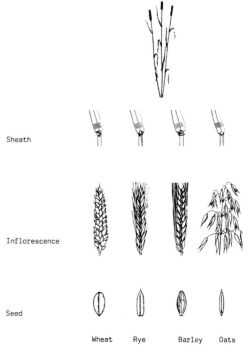

Sheath

Inflorescence

Seed

Wheat   Rye   Barley   Oats

The most important cereal crops and their morphological
differences.

→dicotyledonous crops, such as buckwheat and amaranth.

Lit.: S.R. Chapman,L.P.Carter "Crop Production,Principles and Practices" Freeman and Co.,San Francisco (1976)
H.H.Leonard,J.H.Martin "Cereal Crops" Macmillan NY (1968) Ullmann* (5.) **A6**, 93

Cerotic Acid

Syn.: Hexacosanoic Acid; C26:0
G.: Cerotinsäure; F.: acide cérotique

$CH_3-(CH_2)_{24}-COOH$

m.w.: 396.7; m.p.: 87.7 °C; b.p.: 271 °C (0.53 kPa/ 4 mm)

C. forms white odorless crystals or is a powder. It is derived from →bees wax, wool wax (→lanolin) or →carnauba wax.

Lit.: Kirk Othmer* (4.) **5**, 147
[504-46-7]

Cetaceum

→Spermaceti

Cetyl Alcohol

Syn.: Hexadecanol; n-Hexadecyl Alcohol; Palmitoyl Alcohol
G.: Cetylalkohol; F.: alcool cétylique

$CH_3-(CH_2)_{14}-CH_2OH$

m.w.: 242.5; m.p.: 49 °C; b.p.: 194 °C (2.67 kPa/ 20 mm)

C. is a white crystalline solid formerly produced by →saponification or →hydrogenation with metallic sodium of →spermaceti, which is mainly the cetyl ester of →palmitic acid.

C. is produced today in various grades of purity (up to 99 %) via →hydrogenation of fatty acids (esters).

Uses are in cosmetics, textile auxiliaries and as an agent to prevent water evaporation in dry areas.

Lit.: [36653-82-4]

Cetylic Acid

→Palmitic Acid

C₄ Plants

G.: C_4 Pflanzen; F.: C_4 plantes

Several mostly →monocotyledonous plants have a different plant physiology and mesophyll-cell morphology in which the first product of →photosynthetic action is a four-carbon chain derivative ("C_4").

Normally in plants, the first assimilated product is a C_3 derivative. A typical characteristic of c. is better photosynthetic productivity under intensive light conditions; however, they require also higher temperatures, more light and use water more efficiently. C. are believed to have developed in the subtropical zones where water is scarce and light is available at high levels.

Typical representatives are →corn and millet species, miscanthus (→China grass) and other tropical grasses. With →genetic engineering techniques, it has been attempted to transfer this characteristic to C_3 → cereals. Examples of c. within the →dicotyledonous classes of plants are *Atriplex* and *Falveria*, which actually include both C_3 and C_4 species.

Lit.: N.W.Simmonds,"Evolution of Crop Plants" Longman London/NY (1976)

Charcoal

G.: Holzkohle; F.: charbon de bais

C. is 78–90 % carbon of a fine crystalline graphite type.

It is produced from dry →wood by thermal decomposition (400–500 °C) in kilns with controlled access of air and in retorts in the absence of air. Saw dust, fruit stones, nut shells and corn cobs may be used also. By-products of pyrolysis are acetic acid, methanol, wood creosote and many others.

Bone c. is made from animal bones and is used as black pigment in coatings.

C. is used for barbecuing and for industrial applications, such as metallurgy (iron manufacturing e.g. in Brazil) and production of →activated carbon.

Production in 1984 worldwide was estimated to be 18×10^6 mt. While in USA 95 % is used in barbecuing and 5 % in industrial applications, just the opposite is true in Brazil; in Europe and Japan, the uses are divided half and half.

Lit.: W.Emrich "Handbook of Charcoal Making" D.Reidel Publ.Co., Dordrecht,Boston,Lancaster (1985)
Ullmann* (5.) **A6**, 157

Chardonnet Silk

→Cellulose Nitrate

Chemurgy

G.: Chemiurgie; F.: chimiurgie

C. is defined as a branch of applied chemistry devoted to the industrial utilization of organic raw materials, especially from farm products. In modern terms it is the use of RR (biomass), usually of plant or microbial origin, for use as raw material in the chemical industry, for functional and energy uses. It was almost a social movement during the 1920's and 1930's when large agricultural surpluses caused economic problems in USA. The foundation of four Regional Research Laboratories by the United States Department of Agriculture (USDA) was a result of this movement.

Lit.: Kirk Othmer* (4.) **5**, 902

Chicory
→Fructans

China Grass
Syn.: Elephant Grass; Silver Grass
Miscanthus sinensis L.
Gramineae, Poaceae
G.: Chinaschilf, Elefantengras, Stielblütengras;
F.: miscanthe

The origin of c. is in Northern China and Japan. The name "elephant grass" was given to it because it grows fast, up to a height of 3 m in the first year (depending on the soil). Its comparatively high productivity is due to the →C_4 photosynthetic pathway, which enables better utilization of incoming (more intensive) light, (higher) temperature, and water. It is therefore best adapted to hot, sunny and dry locations (→whole crop harvesting).

In Europe (→Denmark and Germany), c. is being tested as a RR, especially for energy production and as an alternative source of pulp for →paper production. There are also possibilities to use c. as a source of fibers (→cellulose fibers), which might be used for the manufacture of fleeces, insulation material and as a reinforcement of plastics and gypsum. Thus it represents a substitute for wood.

Its root system is extensive, thus enabling a good utilization of →fertilizer without eluation of nitrate to ground water. In 1994, about 130 ha of c. were planted in Germany. Their d.m. yield reached about 30 mt/ha/a. Other so-called fast growing wood sources, such as poplar and willow trees, yield no more than 16 mt of d.m. They may only be harvested every 3 years. First results with c. have indicated little adaptation to the middle-European conditions; they especially miss winter hardness.

Lit.: F.Rexen "The Development of New Crops". In: New Crops for Temperate Regions. K.R.M Anthony et al. (ed.) Chapman & Hall, London (1993)
D.Werner, E.Köhler, Spektrum der Wissenschaft (Scientific American) 6/94, 102–105

Chinese Tallow
Syn.: Stillingia
Sapium sebiferum (L) Roxb.
Euphorbiaceae
G.: Chinesischer Talgbaum; F.: arbre de suif

C., a tree of the tropics and tropical highlands, has been cultivated for centuries in India. Today, it is of commercial interest only in China and India. Introduced as an ornamental tree to the Southern United States, it has gained interest as an oil crop. There are some experimental plots reported from Texas. Source of c. oil is the mesocarp (= outer fruit wall with 62% C16:0 and 27% C18:1). The oil is being used for soap and candle manufacturing as well as for food. The seeds (tallow berry) contain stillingia oil, a fast →drying oil for →coatings and varnishes with the following fatty acid pattern:

C18:1	C18:2	**C18:3**
20%	25–30%	**40%**

Lit.: B.S.Jeffrey, F.B.Padley, J. Amer. Oil Chem. Soc. 68, 123–127. (1991)
E.W. Eckey "Vegetable Fats and Oils", 597–600 Reinhold Publ.Corp. NY (1954)

Chinese Wood Oil
→Tung Oil

Chitin and Chitosan
G.: Chitin und Chitosan F.: chitine et chitosane

Chitin is an ubiquitous polysaccharide found as skeletal material in marine and terrestrial invertebrates and in lower forms of the plant kingdom. The chemical structure of this unbranched polysaccharide is formed by the following units, linked together by β-glucoside bonds:

Chitin   Chitosan

Chitosan is made by deacetylation of chitin by treating it with aqueous 40–50% sodium hydroxide at 110–115 °C for several hours.

Most commercial products are partially deacetylated (chitin 10% and chitosan 60%). Complete deacetylation is reached by repeating the →hydrolysis reaction several times. Enzymatic deacetylation yields a less degraded product.

Both products are white amorphous solids, insoluble in water. Chitin is soluble in formic and methanesulfonic acids, while chitosan dissolves in aqueous organic acids such as formic and acetic acids. Chitosan is one of the few naturally occurring cationic polyelectrolytes.

Shells of crab, shrimp, prawn and krill waste material from the seafood processing industry are the best sources of raw material for chitin. The raw shells are treated with warm hydrochloric acid whereby the calcium carbonate contained in the shells dissolves. Crude chitin is obtained after the extraction of the proteins by dilute sodium hydroxide.

The quality of the product varies with the DP, DS of the amino group, manufacturing technique and the origin of the raw material.

Most uses of c. are in the experimental stage. The main outlet is as coagulant and dewatering agent in waste water treatment. Almost 90% of present production goes into this application. Other uses include adsorbents for metals (from seawater) and lectins (toxic plant proteins), textile finishes, artificial skin and vehicles for controlled release of flavors, pharmaceuticals and pesticides. Chitosan finds increasing interest in cosmetic applications, especially in skin and hair care products.

World production was 400 mt in 1985, with Japan and USA being the main producers.

Lit.: R.A.A. Muzzarelli et al.(ed.):"Chitin in Nature and Technolgy", Plenum Press NY, London (1986)
J.P. Zikakis: "Chitin, Chitosan and Related Enzymes", Academic Press London (1984)
Encycl.Polym.Sci.Engng.* (2.) **3,** 436
[1398-61-4] (Chitin); [9012-76-4] (Chitosan)

Chitosan
→Chitin

Chlorophyll
→Dyes, natural

Chlorophyta
→Marine Algae

Cholesterin
→Cholesterol

Cholesterol
Syn.: Cholesterin
G.: Cholesterin, Cholesterol;
F.: cholestérol, cholestérine

m.w.: 386.64; m.p.: 148.5 °C; b.p.: 360 °C

C. is a principle sterol (→steroids) of all higher animals and is found in all body tissues, especially in the brain and the spinal cord, in animal fats and oils. It is the main constituent of gallstones. It is the precursor in the synthesis of steroidal hormones (sexual hormones) via the androstadiene-estrone route with *Mycobacterium phlei* and of vitamin D_3. It is commercially prepared from the spinal cord of cattle by extraction with petroleum ether of the nonsaponifiable portion; it is also produced from wool grease and the residues of animal derived fatty acid distillation. Purification is normally accomplished by repeated bromination. C. is a nearly odorless and faintly yellow solid which may become yellow to tan upon exposure to light and air; it is practicable insoluble in water. It is used as an emulsifying and solubilizing agent and in varying concentrations in oil/water emulsions or absorption ointment bases.

Esters with →oleic acid, →stearic acid and aromatic acids are used in liquid crystals for displays.
Production (1980) 100 000 mt.

Lit.: Ullmann* (5.) **A25,** 309
Kirk Othmer* (3.) **21,** 645
The Merck Index* (11.), 2204
[57-88-5]

Chromosome
→Cytology

Chrysanthemum coccineum
Chrysanthemum cinerariifolium
→Pyrethrum Plant

Chypre
→Odor Description

Cigarettes
→Tobacco Industry

Cigar Flower
→Cuphea

Cigars
→Tobacco Industry

Cinchona officinalis
→Quinine

Cineole
→Eucalyptol

Cinnamomum camphora
→Camphor

Cinnobar Flower
→Cuphea

Cistus ladaniferus
→Labdanum

Cistus Oil
→Labdanum

Citral
Syn.: Geranial; 3,7-Dimethyl-2,6-octadienal (*trans*-α-Citral), Neral (*cis*-β-Citral)
G.: Citral; F.: citral

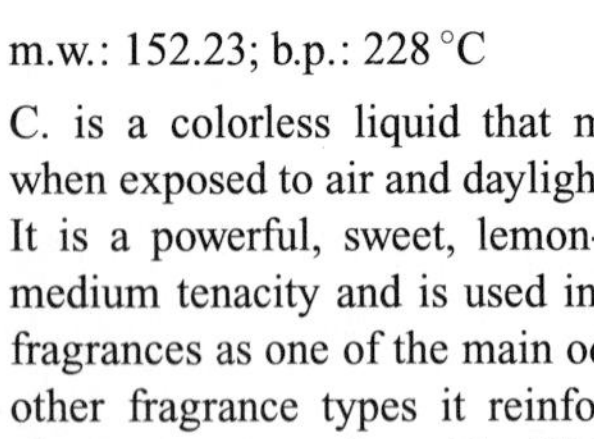

Geranial   Neral

m.w.: 152.23; b.p.: 228 °C

C. is a colorless liquid that may turn yellowish when exposed to air and daylight.

It is a powerful, sweet, lemon-peel type odor of medium tenacity and is used in all types of citrus fragrances as one of the main odor contributors. In other fragrance types it reinforces fresh, natural citrus topnotes and sparkle. Widespread use in citrus flavors and other fruit flavors, such as apple, cherry or grape.

C. is produced from →lemongrass oil or *Litsea cubeba* oil or by synthesis from isoprene via methylheptenone and dihydrolinanool.

Lit.: Arctander*

Citrates
→Citric Acid

Citrem
→Citric Acid Esters of Mono/Diglycerides

Citric Acid
Syn.: 2-Hydroxy-1,2,3-propanetricarboxylic Acid
G.: Zitronensäure; F.: acide citrique

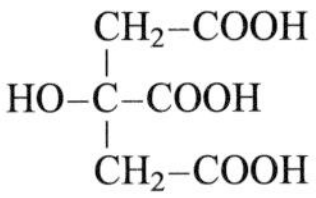

$$CH_2-COOH$$
$$HO-C-COOH$$
$$CH_2-COOH$$

m.w.: 192.12; m.p.: 153 °C

C. is colorless, odorless and soluble in water and in alcohol, slightly in ether and not in hydrocarbons. Below 36.6 °C, it crystallizes as a monohydrate [5949-29-1] from a saturated solution, in anhydrous form at higher temperatures. Upon rapid heating, the crystalline monohydrate melts at 100 °C, solidifies in its anhydrous form and melts again at 153 °C.

C. is a strong acid and forms salts (citrates) and water-soluble complexes, which are responsible for the use of c. to inactivate metal ions. At 175 °C, various dehydration and decarboxylation reactions take place under formation of aconitic, itaconic and citraconic acid and their anhydrides. Only →itaconic acid is commercially important.

C. is an intermediate in the citric acid (Krebs) cycle. A human body forms and transforms ≈2.000 g/day. It is contained in lemon juice (5–9%) and in lower concentrations in many other fruits and plants.

It was produced formerly from lemon juice by precipitation as Ca-salt and subsequent acidification with sulfuric acid. Large-volume production today is carried out by fermentation with certain strains of *Aspergillus niger*, either by the surface fermentation method or the submerged process from →sucrose molasses, →sucrose, →glucose syrups, black liquor from →wood pulping (sulfite process) and other sugar media. Purification is accomplished, among other methods, by precipitation of the Ca-salt and subsequent acidification.

C. and citrates are extensively used in food as an acidulant in drinks, jams, candies, jellies and canned fruit, in baking powder and as flavor enhancer. It has a synergistic effect to antioxidants, especially for fats and oils (→food additives).

Technical uses are metal cleansers, complexing of Fe-ions, as a scale remover, in →electroplating, as →textile auxiliary and as starting material for making ester plasticizers (→plastic additives) that are approved for food packaging. C. is also considered and used as builder in →detergents. Many applications are known in pharmaceutical and cosmetic formulations e.g., triethyl citrate in deodorants (→skin preparations).

Up to 70% of the annual production goes in food applications, the rest in cosmetic, pharmaceutical and technical areas.

The annual production worldwide is ≈ 400 000 mt. The markets for c., especially in food applications, are attacked by cheaper →malic acid and →fumaric acid.

Lit.: M.Roehr,C.P.Kubicek,J.Kominek: "Biotechnology, Microbiology, Products, Biomass and Primary Products" H.Dellweg (ed.) 3, 456−465 (1983) Verlag Chemie Weinheim
Kirk Othmer* (3.) **6,** 150−179
[77-92-9]

Citric Acid Esters of Mono/Diglycerides

Syn.: CMG; Citrem; Mono/diglyceride Citrate
G.: Zitronensäuremono/diglycerid;
F.: esters citriques de mono/diglycérides

CH_2-O-CO-R^1
|
CH-OR^2 OH
| |
CH_2-O-CO-CH_2-C-CH_2-COOH R^1: C_{15-17} alkyl
                              |
                            COOH R^2: H or −CO−R^1

s.v.: 330−350; m.p.: 55−60 °C

There are numerous variations of the above formula possible because the hydroxyl and carboxyl groups of citric acid participate in many esterification and transesterification reactions.

C. are produced by esterification of a mixture of fatty acid, glycerol or of mono/digycerides with citric acid.

C. are waxy and viscous depending on the nature of fatty acid.

C. are able to form fat-soluble metal complexes, which enhances the efficiency of antioxidants.

They are hydrophilic, anionic emulsifiers, which are used in margarine (antispattering), in sausages and in meat products.

Lit.: G.Schuster, "Emulgatoren für Lebensmittel" Springer Verlag Berlin/Heidelberg/NY/Tokyo (1985)
N.J.Krog "Food Emulsions" ed. K.Larsson, S.E. Friberg,Marcel Dekker NY/Basel (1990)

Citronella Grass

Cymbopogon winteranus Jowitt
Cymbopogon nardus
Gramineae/Poaceae
G.: Citronellgras; F.: herbe de citronelle

Cymbopogon winteranus is cultivated in China, Taiwan, Indonesia, Brazil, Columbia and Guatemala and produces →citronella oil of the so-called "Java" type. *C. nardus* originates in Sri Lanka and yields a slightly different oil of lesser importance.
Lit.: Arctander*

Citronellal

Syn.: 3,7-Dimethyl-6-octen-1-al
G.: Citronellal; F.: citronellal

$$
\begin{array}{c}
CH_3 \\
| \\
CH \\
H_2C \overset{\displaystyle|}{} CH_2 \\
H_2C \qquad CHO \\
| \\
CH \\
| \\
C \\
H_3C \qquad CH_3
\end{array}
$$

m.w.: 154.25; m.p.: 77 °C; b.p.: 206 °C

This mono→terpene is isolated from →citronella oil or produced synthetically and is used as such in →fragrances and →flavors as well as being further processed to yield several perfume raw materials of floral qualities, such as citronellol, geraniol and hydroxycitronellal. C. has a green-citrusy, powerful, slightly woody, metallic note, which makes it useful in citrus fragrances and green-herbal types. In flavors it is used in small amounts in citrus, tutti-frutti and ginger ale notes.
Lit.: Arctander*

Citronella Oil

G.: Citronellöl; F.: essence de citronnelle

C. is steam-distilled from fresh and partially dried leaves (grass) of →citronella grass and is a yellow to yellow-brownish liquid with a typical, fresh and sweet, citrus-rosy odor.

Main constituents are 30−45% geraniol and 40−50% →citronellal.

The oil is used by itself as perfume raw material for low-cost citrus-type →fragrances in applications such as soap, soap flakes, →detergents and cleaning products. In addition, c. and its isolates serve as starting materials for the production of several perfumery materials, such as →citronellal and derivatives. Its insect-repelling qualities have lead to some use in repellant sprays, lotions and candles.

Lit.: E.Guenther, "The Essential Oils" Van Norstrand/
Reinhold, NY (1952)
Arctander*

Citrus aurantifolia
→Lime Tree

Citrus aurantium var. *amara*
Citrus bigaradia
→Orange Tree (bitter)

Citrus aurantium var. *dulcis*
→Orange Oil (sweet)

Citrus bergamia
→Bergamot Tree

Citrus limetta
→Lime Tree

Citrus medica
→Lemon Tree

Citrus reticulata
→Mandarin Tree

Civet Absolute
G.: Zibet Absolut; F.: civette absolue

C. is one of the four traditional animal-derived raw materials used in fragrances (the others being musk, ambergris and castoreum, which have lost importance completely) and the only one that can be obtained from farmed animals without killing them.

Despite this fact and its widespread use in high-class perfumes, c. is gradually substituted by re-constitutions based on detailed analytical studies of the natural material due to the increasing number of fragrance users that oppose the use of ani-mal-derived raw materials in fragrances destined for their products.

C. is produced by hydrocarbon extraction via a →concrete from natural crude civet, a glandular secretion from various varieties of the civet cat, *Viverra civetta*, the Ethiopien variety being the most common. Other producing countries are India, Indonesia, China, Somalia and Kenya. The cats are kept in narrow cages, fed raw meat, frequently teased to increase production of the glandular se-cretion, which is then regularly scraped from the glands.

C. is a brown to dark brown, viscous to pasty ma-terial with an extremely intense animal-like, sweet, honey-like note, combined with some fae-cal undertones due to the presence of scatol. Be-cause it is extremely long-lasting, it needs to be carefully dosed and blended to produce its posi-tive effect of giving warmth, depth and radiance to all types of fragrances, even delicate florals, without the animal-faecal note becoming apparent and obnoxious.

Lit.: Arctander*
Gildemeister*

Clary Sage
Salvia sclarea L.
Labiatae
G.: Muskateller Salbei; F.: sauge sclarée

C. is a native of the Mediterranean countries, where it is cultivated in Spain, France, Italy and Morocco, but it has spread to Eastern Europe, especially Southern Russia, Ukraine and the Cau-casian region, where it is cultivated on a large scale. It has also been introduced in the US.

C. is a relatively tall plant with abundant foliage and delicate purple flowering tops. C. is the source for →clary sage oil and absolutes.

Lit.: Arctander*
The H&R Book*

Clary Sage Oil/Absolute
G.: Muskateller Salbeiöl;
F.: Essence de sauge sclarée

C. is steam-distilled from the flowering tops and foliage of →clary sage; the →essential oil is a col-orless to pale yellow or pale olive mobile liquid with a sweet, herbaceous, initially fresh odor, which gradually fades into a soft tobacco-like, bal-samic note also reminiscent of tea and ambergris. These characteristics are even more obvious in the →absolute, produced by hydrocarbon solvent ex-traction via concrète from the same plant.

Main constituents are linalol, linalylacetate and sclareol, which is isolated and used as starting ma-terial in the manufacture of synthetic ambergris substitutes.

Lit.: Arctander*
Gildemeister*
The H&R Book*

Cleaners

Syn.: Cleansers; Cleaning Agents
G.: Reiniger; F.: produits à nettoyer

There are numerous household, institutional and industrial c. available, which have multipurpose or specific functions. The products are formulated from many ingredients, but the most important are →surfactants and sequestering agents. The use of RR-based products is rather limited.

Aside from petrochemical anionic and nonionic surfactants, such as alkylbenzene sulfonate and alkylphenol ethoxylates, there are →fatty alcohol sulfates or →fatty alcohol ether sulfates and →fatty alcohol ethoxylates in use in multi-purpose c., scouring c., dishwashing agents, floor c., carpet c. (Li-salts of sulfates are used), toilet hygiene products, automotive c. and in products used in the food industry. The use of sulfates is limited because of their →hydrolysis in extreme pH ranges. →Acylamino alkane sulfonates and →acyloxy alkane sulfonates are rather stable in these applications. Fatty alcohol ethoxylates are the preferred nonionics because of their good biodegradability.

→Fatty alcohol ethoxylates, whose terminal OH-groups are blocked, are used in products for automatic dishwashing and in the food industry.

→Soaps are used in floor c., metal c. and in industrial food c.

→Fatty acid ethanolamides and →fatty amine oxides are contained in dishwashing agents, the latter also in products for the food industry.

→Quaternary ammonium compounds find use in multipurpose c. and in c. for food processing because they act also as disinfectants. In automotive c., they are used in the final rinse in protective finishes.

Terpene alcohols (→pine oil) are disinfective additives in multipurpose c.

→Stearic acid and →cetyl alcohol are used as waxy base material for toilet hygiene blocks, used as automatic tank dispensers.

→Citric and →lactic acid are sometimes used as sequestering agents in metal c. An important application is the use of →citric acid in liquid household products for automatic dishwashers in the final rinse.

→Enzymes, especially proteases and →amylases, are applied in cleaning operations in the food industry to remove protein and starch contaminats, especially in membrane cleaning (ultrafiltration, reverse osmosis). →Gluconic acid and →gluconic acid lactone are used as sequestering agents to remove beer scale.

In industrial cleaning, e. g., prior to metal surface treatment (phosphatizing, enamelling and electroplating), efficient cleaning and degreasing is necessary.

Chlorinated solvents, which have been used widely in the past, are now substituted by aqueaous systems with low-foaming →surfactants of the nonionic type (→fatty alcohol ethoxylates, which are terminally blocked for some applications). Corrosion inhibitors, based on fatty acids and →fatty amines, are used together with the surfactants.

Lit.: Ullmann* (5.) **A7**, 137

Cleansers

→Cleaners

Clove Bud Oil

G.: Nelkenblütenöl, Gewürznelkenöl;
F.: huile de clous de girofle

C. is distilled from the dried flower buds of *Eugenia caryophyllata*, Myrtacea (Syn.: *Syzygium aromaticum*), a slender, medium-sized tree from Indonesia. Today's main production areas of clove buds also include Madagascar, Tanzania and Sri Lanka, from where the dried plant material is shipped to European and American distilleries. Main constituents are →eugenol (80–90%), eugenyl acetate and caryophyllene, with traces of other eugenyl esters and lower aliphatic ketones, such as methyl n-amyl ketone. These additional substances explain the olfactory differences between →eugenol and c.: the latter has the same typical spicy-pungent, yet sweet note as eugenol, but with an additional fruity freshness in the topnote and an overall richer and rounder body. Due to its extensive use in flavor blends, c. is one of the biggest-volume essential oils in the world; this, combined with a yield in essential oil of around 15% of the dried buds, accounts for a reasonably priced material.

In general, flavor and perfumery uses are similar to those of →eugenol with which it is also often combined, where the essential oil adds a more natural and rich impression.

Lit: Gildemeister*
    The H&R Book*
    Arctander*

Cluster Bean

→Guar

CMC

→Carboxymethyl Cellulose

CMG
→Citric Acid Esters of Mono/Diglycerides

CMS
→Carboxymethyl Starch

Coating Additives
Syn.: Paint Additives
G.: Lackadditive;
F.: additifs pour vernis

In addition to binders, pigments and solvents, →coatings or paints contain additives at concentrations of 0.01–1%. In former times, they were called "auxiliaries" and were often used only to adjust off-spec paint batches. Today, they have an important and specific function in almost all formulations. They are used to prevent negative effects in paint manufacturing and application as well as to impart certain properties to the coating.
C. are classified according to the following functions:

Defoamers:
During manufacturing, air is entrapped in the coating and causes reduced efficiency in production. Foam appears not only during manufacturing but also during application (surface defects).
Additives are used to control the foam in paints as they do in many other applications (→defoamers and foamers). They are not soluble but easily disperable, must penetrate in the foam lamella and spread across the surface.
Silicone and mineral oil as well as finely dispersed particles are used. →Fatty acid ethoxylates and →fatty alcohol ethoxylates are part of the emulsifying system. →Metal soaps are used as hydrophobic particles.

Wetting and Dispersing Additives:
Wetting agents reduce the interfacial tension between binder solution and pigment surface. In the dispersing process these additives stabilize the dispersion of the pigment particles by van der Waal forces. They avoid floating and flooding problems, which occur during drying of the paint film and are caused by local eddies, leading to separation of the pigments and to a nonuniform surface appearance. Antisettling agents have similar functions, which especially avoid settling of pigments and formation of a hard sediment during storage.
A large variety of products, mainly RR based are in use:
→Lecithin, →fatty acid esters and →α-sulfo fatty acid esters, →glycerol ethoxylates and →fatty al-

cohol ethoxylates, →fatty alcohol sulfonates, →alkylpolyglucosides, →fatty amine ethoxylates and →lignosulfonates.

Surface Additives:
Surface defects, such as orange peel, crater formation and poor leveling are the result of differences in interfacial tension. Silicones are frequently used as additives to lower the surface tension.
→Fatty acid amides also give smooth surfaces. Leveling is enhanced by fatty alcohol derivatives. Fatty alcohols and fatty acid esters are good antiblocking agents. Woolwax (→lanolin) and →fatty acid ethoxylates are also helpful surface improving additives.
Emulsions of →fatty alcohols are in use as workability improvers in synthetic resin plasters. They limit water evaporation and prolong the wet film phase for easier application. Cracks in coatings are also inhibited.

Driers:
→Drying oils.

Preservatives:
The only RR-based preservative is phenyl-mercury oleate.

Rheology Additives:
Good leveling, no sagging and storage stability are influenced by addition of thickeners, such as →xanthane gum, →methylcellulose, →carboxymethyl cellulose or →ethyl hydroxypropyl cellulose in emulsion paints. Other additives in this respect are hydrogenated →castor oil, stand oils (→drying oils) and fatty acid →metallic soaps, which are used exclusively in solvent-based formulations.

Light Stabilizers:
No RR-based products are in use.

Corrosion Inhibitors:
→Fatty acid amine salts and monoamides as well as →tannin derivatives are RR-based products used.

Others:
Epoxidized oils (→epoxides) are used as plasticizers, stabilizers and antioxidants for some coatings. Other plasticizers are →fatty acid esters, stand oils (→drying oils) and →castor oil. →Fatty acid amides and →metallic soaps (Zn and Ca) are used as matting or flattening agent and as hydrophobic additive. Fatty acid derivatives, such as →glyceryl monostearate, chlorides and ammonium salts, improve the electric conductivity while →fatty amine ethoxylates, fatty acid amides and ammonium sulfates function as antistatic additives.
Lit.: Ullmann* (5.) **A18,** 465

Coatings

Syn.: Enamels; Lacquers; Paints; Varnishes,

G.: Beschichtung, Überzug; F.: revêtement, enduit

C. represent all materials and methods to cover a surface. In terms of RR, only organic coatings are of interest.

Organic c. normally consist of a binder, pigments or colorants, a carrier liquid and various additives (→coating additives). The binder is a polymer, which may or may not be reactive (cross-linkable by polymerization or oxidation). The liquid may be organic or water, acting either as solvent or diluent and which evaporate during drying and film formation. The liquid component allows the adjustment of the viscosity to achieve easier handling in manufacturing and in application. They may be omitted in case of powder-c., redispersable powders and hot melts. Sometimes, they do not evaporate and react with the c. system (reactive diluent).

All kinds of applications are common: brushing, dipping, (electrostatic) spraying, electrocoating (electrophoresis), rolling, powder coating and the hot melt technique.

C. are applied to give the surface protection (against corrosion, aging, mechanical damage) and decorative appearance.

The term "coatings" encompasses paints, lacquers, enamels and varnishes. The borderlines between these terms are not always clearly defined. Especially the term "paints" is frequently understood as a synonym for c.

Binders of paints are normally vegetable oils, e. g., →linseed, dehydrated →castor, →soybean, →tung and → oiticaca oils as such (→drying oils) or in modified form (→alkyd resins). They are often combined with numerous synthetic resins. Natural resins (→resins, natural) are also applied in special applications where their high price is acceptable. Decorative →casein paints for plaster walls have lost importance. Chlorinated and cyclized →rubber are also in use (swimming pool coatings).

An endless variety of colors and pigments exists. In paints, the film formed after evaporation of the solvent is cured by oxidative polymerization of the double bond in case of alkyd resins and drying oils.

Varnishes are unpigmented paints.

Enamels contain higher-molecular binders than paints with lower solid concentration and a lower pigment-binder ratio, thus giving them more glossy surfaces.

Lacquers have no or low reactivity. They dry only by evaporation of the solvent. Their films remain soluble or swellable and thermoplastic. Binders used in lacquers are →cellulose nitrate, →cellulose acetate, sometimes also →cellulose ethers and synthetic polymers. They are replaced slowely by water-based formulations.

Plastisols are a special form of c. They are dispersions of polymers (mainly PVC) in a plasticizer, which is a nonsolvent at room temperature but a good solvent at elevated temperature. After heating, they produce a gel, resulting in a homogenous film.

For highly special coatings, resins (→resins, natural), such as dammar, copal, mastic, →shellac and →rosin, are in use. Such applications serve as protective coatings on valuable paintings and furniture.

Lit.: Encycl.Polym.Sci.Engng.* (2.) **3**, 615
          Kirk Othmer* (3.) **6**, 427

Cobalt Soap

→Metallic Soaps

Coccos ilices

→Dyes, natural

Cochineal

→ Dyes, natural

Cocoa Butter

Syn.: Cacao Butter; Theobroma Oil

G.: Kakaobutter; F.: beurre de cacao

s.v.: 190–200; i.v.: 35–40; m.p.: 28–30 °C

C. has a pleasant aromatic flavor. It melts within a range of 4–5 °C with an oral cooling effect.

The →glycerides in c. consist of 14% 2-oleo-dipalmitin, 40% 1-palmito-3-stearo-2-olein, 27% 2-oleodistearin, 8% palmitodiolein and 8% stearodiolein.

It is isolated from cocoa beans (*Theobroma cacao*, L. Sterculiaceae) by fermentation, roasting, dehulling, grinding and expelling.

Uses are exclusively in chocolate production and confectionary products. Because of its high price, there are many attempts to design c. substitutes from other fats.

World production of beans is 1.3×10^6 mt/a which contain 55% c.

Lit.: Ullmann* (5.) **A10**, 222

Coconut

Cocos nucifera L.

Palmae

G.: Kokosnuß; F.: noix de coco

C. is the fruit from a tree of the Pacific tropics and the source of copra (the dried meat of the opened nut), which by oil extraction is manufactured into coconut oil and coconut meal. It represents the most important source of lauric oils (→fats and oils).

A c. that is seeded in a nursery will grow its roots during the first 4 months inside the fibrous coir. For optimum productivity, seedlings must be spaced at 9 m. →Intercropping with cocoa, coffee or legumes is possible during the first years. Productivity depends a lot on germplasm (hybrids with dwarf types, such as the Malayan dwarf, are the highest yielding and earliest fruit-bearing crops), fertilization of the ovaries (wind or artificial) pollination is obligatory due to protandric (male flowers will open 10 to 20 days ahead of the females) fruit development, fertilizer application and age of the palms (Phillipine trees seem to be over-aged and have received comparatively low levels of fertilizer).

Average yields are 30–50 nuts per palm with a maximum of 8000 nuts per ha per year. After halving the nuts and drying, the copra is separated from the shells. The copra then needs to be dried in the sun or in local tapahan dryers (see below) and transported to the oil mill. Copra yields 65–70% oil. Average copra yields in the Philippines range between 0.65 and 1.0 mt/ha. High yields are 6 mt copra/ha, which produce about 4 mt oil and 2 mt press cake. Today, oil mills are mostly located in the c. producing countries, whereas up to the late 1960's, substantial amounts of copra were shipped to the consuming European and American countries.

The most important c. oil-exporting country are the Phillipines with a share of 37% of all exported c. oil.

Production in 1994 ($\times 10^6$ mt)	Copra	Coconut oil
Philippines	2.10	1.31
Indonesia	1.28	0.78
India	0.60	0.37
Mexico	0.18	0.10
Papua New Guinea	0.14	0.05
Vietnam	0.13	0.08
World	4.96	3.08

Other products of the c. – mostly of local importance – are:

the coir (the outer fibrous coat of the nut, →cellulose fibers), which is used for floor covers, door mats, insulation and similar material; the shells of the nuts, which are useful for firewood, for the so-called tapahan drying of the copra in simple oven-like devices in the Phillippines, and for high-quality →charcoal with excellent application for →activated carbon, e. g., in gold recovery (→mining chemicals); the leaves, which are useful for any kind of covering, such as roofing; the trunks, which are useful for timber and similar applications.

Lit.: K.Satyabalan "Coconut" in Röbbelen*, 494–504 (1989)

Coconut Oil

G.: Kokosöl; F.: huile de coco, huile de coprah
s.v.: 250–258; i.v.: 7.5–12; m.p.: 20–28 °C

C. is a white, semi-solid fat, soluble in alcohol. Derives from →coconut palm. Fatty acid composition of c.:

C8:0	C10:0	**C12:0**	**C14:0**
8–10%	4–10%	**44–51%**	**13–19%**
C16:0	C18:0	C18:1	
7–11%	1–3%	5–8%	

C. is used for nutrition (margarine), as ointment base in cosmetics and for the manufacture of soap (→saponification).
→Hydrolysis yields coconut oil fatty acids and →glycerol. The medium-chain fatty acids (→fats and oils) can be used directly for many applications but may be further processed by → distillation to obtain a broad cut (C_8–C_{18}) or a low-boiling cut (C_8–C_{10}), a middle section (C_{12}–C_{14}) and a higher-boiling fraction (>C_{16}). →Fractional distillation yields the pure acids (→ caprylic capric acid, → lauric acid, →myristic acid, →palmitic acid, →stearic acid).
→Transesterification yields →fatty acid methyl esters, an important intermediate for the production of →fatty alcohols.
C. is the main source for medium-chain oleochemicals.

World production of c. is in the range of 2.9×10^6 mt/a (1993)

Lit.: Ullmann* (5.) **A10**, 220
[8001-31-8]

Coconut Oil Fatty Acids
→Coconut Oil

Cocos nucifera
→Coconut

Codeine
G.: Codein; F.: codéine

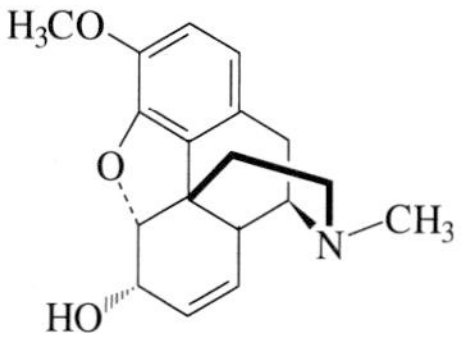

m.w.: 299.39; m.p.: 154–156 °C

C. is present in →opium from 0.7–2.5%, depending on the source (→poppy), but most often it is prepared by methylation of →morphine with phenyltrimethylammonium chloride in methanol/potassium hydroxide under pressure, or from →thebaine. It is used as a narcotic analgesic and antitussive.

Lit.: Ullmann* (5.) **A1**, 377
      [76-57-3]

Coir
→Cellulose Fibers
→Coconut

Cold-water-soluble Starches
→Pregelatinized Starches

Colewort
→Crambe

Collagen
→Proteins

Collodion
→ Cellulose Nitrate

Colophony
→Rosin

Colza
→Turnip Rape

Concrete
G.: Concrete, Konkret; F.: concrète

C. is a perfumery material prepared from clearly defined (by species/origin) plant materials, such as flowers, barks, herbs, leaves and roots, by hydrocarbon solvent extraction. Solvents used are petroleum ether, benzene, butane, etc. The resulting c. consists of all hydrocarbon-soluble substances contained in the plant material, which, apart from the odorous materials, includes →fatty acids, such as →lauric acid and →myristic acid. This explains why concretes (from the French language for solid) are usually solid-waxy masses, which need to be further purified if to be used in →fragrances for alcohol-based →perfumes. The alcohol-soluble matter is known as →absolute, of which most concretes contain about 50% (range from 20–80%).

Lit.: Arctander*

Conjugation
→Isomerism

Construction Materials
Syn.: Building Materials
G.: Baustoffe; F.: matériaux de construction

C. used by humans for several thousand years are rocks, wood and man-made bricks.
Modern **inorganic** construction materials require some RR-based products as auxiliaries and additives:
– Concrete additives, mainly surfactants, are:
    →lignosulfonate used as liquifier, →rosin soaps for making foamed materials, and fatty acid salts (→soaps) used to reduce water penetration of concrete. Sulfonates and hydroxycarboxylic acids reduce the speed of cement binding.
– →Wood-derived products (chips, shavings and saw dust) or →cellulosic fibers (→flax, →sisal, →jute, →hemp, coir (→coconut), →chinagrass, →straw) are sometimes added as extenders and reinforcing materials to cements and mortar in construction itself or in preformed bricks and boards. The necessity to substitute asbestos has enhanced developments and new applications of RR in this area.
– Glass fiber or asbestos are gradually substituted in gypsum compositions, such as boards or plasters, by the a.m. →cellulosic fibers.
– Gypsum plasters, especially when spray-applied, need the addition of water-soluble polymers such as →methylcellulose, to avoid the fast loss of water before the binding reaction of gypsum is finished.
The oldest **organic** construction material was and still is →wood. Modern reconstituted wood materials are plywood and various kinds of chip boards.

Sometimes, RR-based products are used as binders for their production.

Paperboard coated with bitumen is widely used as a roofing material.

Synthetic polymers (polyester, epoxy) need reinforcing fibers to form stable construction panels. As substitutes for the usual glassfibers, again natural fibers (→flax, →jute, →sisal, →hemp and many others) are subject of intensive investigation.

For thermal insulation wool is of increasing interest, either as such or in composites. →Cork is a good noise-control material in floorings and wall covers. Peat and →straw, seaweed and coir (→coconut), mostly in the form of boards or molded articles, are also good insulation materials.

All these RR-based materials, whether used as such or in composites, have the advantage of biodegradability and mostly good physical properties. However, they swell more or less in water and are sensitive to attacks of microorganisms, insects and small mammals. Flammability is an other property that can simultaneously be an of advantage and disadvantage.

Economic figures for the use of RR as c. are not available, but actual developments show encouraging future outlets for these raw materials.

Lit.: Ullmann* (4.) **8**, 312

Copernicia cerifera
Copernicia prunifera
→Carnauba

Copper Soap
→Metallic Soaps

Copra
→Coconut Oil

Corchorus olitoris
Corchorus capsularis
→Jute

Core Binders
→Foundry Industry

Coriander
Coriandrum sativum L.
Umbelliferae
G.: Koriander; F.: coriandre

C. is an annual herb with slender, solid, smooth stems around 0.5 m tall. It is native to the Mediterranean region, where it has been grown as a spice since ancient times. It is extensively cultivated as a crop in India and also in South-East Europe, the Middle East and Brazil.

The dried fruits are an important ingredient of spices, such as curry powder, seasonings, confectionary and flavoring spirits. The fruits contain a volatile oil, which is used for flavoring and also in pharmacy. Besides the →essential oil, c. seeds also contain a fatty oil at about 20% of the →seed weight, where the major fatty acid component is →petroselinic acid at about 80%.

Lit.: Cobley* (1976).

Coriandrum sativum
→Coriander

Cork
G.: Kork; F.: liège

C. has been used commercially for at least 2500 years. It has a closed-cell structure and is produced from the outer bark of the cork oak (*Quercus suber* L.). It is harvested from trees between 25–200 years of age by stripping them every nine years.

The nonfibrous material is highly resistant to water, organic solvents, weak acids and alkali. Its specific gravity is 0.1–0.3, which provides excellent flotation and insulation.

Chemically, c. consists (30–50%) of hydroxy- and dibasic fatty acids, which are linked together by lactone and ester groups to form a polymer of complex structure called suberin. Main acids are phellonic acid (C_{22} mono hydroxy) or phloionic acid (C_{18} dihydroxy) and other more hydroxylated or unsaturated acids as well as dibasic acids (→suberic acid). Other components are →lignin (13–18%), →waxes (5–15%) and →tannins.

C. is used to make stoppers. The scrap from this production and lower grades are used for composites in insulation, floor coverings, wall panelling and industrial applications.

The total world consumption in 1979 was estimated at 300 000 mt.

Due to price increases and the development of alternatives, which however never have been able to copy the properties of c., consumption is declining.

Lit: Kirk Othmer* (3.) **7**, 110
    [61789-98-8]

Corn

Syn.: Maize, Indian Corn
Zea mays L.(ssp. *mays*)
Gramineae, Poaceae
G.: Mais; F.: maïs

C. in its present form is probably the crop most influenced by mankind. This is probably due to the fact that c. is a →C_4 plant. It is native to Central and South America. Due to its different sites of origin and application the following genetic pools of c. are segregated:

- Waxy maize, with its origin probably in South-East Asia, the →starch of which consists of →amylopectin only;
- flint and dent c., two hard c. types important for human nutrition;
- yellow dent c., which is the main corn type for →wet milling;
- flint c. is early-maturing and has a hard aleuron coat;
- soft c. is well adapted for starch production;
- pop or puff c. has a limited production volume and is mostly used for snacks. New developments extend its application to packaging material;
- sweet c. synthesizes low-molecular polymers and sugars (it contains 70% water);
- white c. is preferentially used for → dry milling, e. g., cereal products.

Since the 1940's, c. is the classical example for practical applications of the →hybrid breeding technique that makes use of the different gene pools of c. by combining the best fitted individuals to create more vigorous, cold-tolerant and high-yielding varieties, which made c. the no.1 crop in the US and also the most important feed and starch crop in northern Europe. For food applications, it was necessary to raise the protein content to over 20% and to improve the amount of the essential →amino acids lysine and tryptophane. C. is used for 75% of the world's starch production. For food as well as for technical applications, →amylo maize starch varieties (high-amylose c.) were bred. They contain 50–80% →amylose in the starch as compared to 22% in yellow dent c. starch.

→C. steep liquor, →c. germs (12% of the total seed), →c. oil, →c. germ meal, →c. gluten feed and →c. gluten meal are the by-products of the →wet milling process of →c. starch production.

C. cobs, which is the by-product after removing the kernels, are widely used as feed component and as a technical source for the production of →furfural.

C. is planted in rows and combine-harvested (silo c. will be chopped before reaching full maturity). Sweet c. is used as vegetable.

Production in 1995 ($\times 10^6$ mt)	Corn (seed)
US	200.7
China	102.0
Mexico	18.2
Brazil	33.0
France	13.5
Argentina	11.5
Romania	9.0
Italy	7.8
World	510.4

Lit.: Fehr/Hadley* (1980)

Corn Germ Meal

→Corn Germs

Corn Germ Oil

→Corn Oil

Corn Germs

Syn.: Maize Germs
G.: Maiskeime; F.: germes de maïs

C. are the commercial and industrial term used by the starch industry to designate the embryo of the caryopses of maize and are the by-product of →corn starch production.

The germs are isolated from the maize kernels after removal of the →corn steep liquor.

The dried germs contain:

48–52% →corn oil
13%   protein,
10–12% starch,
2–4%   water
plus fiber and ash.

The residue left after removal of the oil is called corn germ meal.

It is of a higher nutritional value than corn as a whole and consists of:

25%   Crude protein (d.b.),
1.5%   residual oil (d.b.) after solvent extraction,
6–10% residual oil after expelling,
10%   moisture,
plus starch and fiber.

It is used as a special, medium-protein, medium-energy component in feed for hogs and poultry or generally as a component of → corn gluten feed.

Lit.: Blanchard*, 87–96, 298–301, 357–369 (1992)

Corn Gluten Feed

Syn.: Maize Gluten Feed
G.: Maiskleberfutter;
F.: gluten de maïs pour la fourage

The commercial product, obtained in the course of →corn starch production, contains fiber, gluten, starch and a small amount of oil.

C. is the major product of corn →wet milling. It comprises the cellulosic and hemicellulosic materials from the husk, termed fiber, as the main component.

It is separated from the mill-stream after the third grind (fine milling) of the degermed corn kernels (→corn starch production). The yield and composition of c., coming from corn starch production, depend largely on the intended use in animal feed: Generally, 20–25 % of the corn input, sold as c., is a medium-protein (18–25 % d.b. crude protein), medium-energy feed (1 % crude lipids, 10–15 % starch) with 8 % ash and fibers (adding up to 100 %).

Before final drying, other components may be added as desired: germ meal, concentrated →corn steep liquor, →corn gluten meal or residues from the refinery filters . Final drying to 10–12 % water is usually performed in rotary drum dryers, flash dryers with a hot-air stream from direct firing or steam tube dryers. Dry blending of all the components of c. may be employed with the advantage of higher purity and better color.

Lit.: Blanchard*, 96–99, 105–108, 298–301, 528 (1992)

Corn Gluten Meal

Syn.: Maize Gluten, Maize Gluten Meal
G.: Maisgluten; F.: gluten de maïs

C. is the usual name for the insoluble protein complex from maize, an industrial material obtained from →corn starch production as a by-product.

For details of the separation of c. and starch: →corn starch production.

The composition of c. derived from this process is (d.b.):

Protein 69 %, starch 19 %, lipids 3 %, ash 2 %, fibers ≈ 3–5 % and xanthophyll (100–225 mg in 100 g).

For special purposes (→hydrolized vegetable protein), c. may be enriched up to 72–75 % protein content .

As a consequence of the sulfur dioxide treatment and the conditions during steeping of maize corn gluten, proteins are denatured. The gluten proteins and →zein contain →glutamic acid as a major component (17–20 %). C. contains only a small amount of essential amino acids, such as lysine

(1.2 %) and tryptophane (0.4 %). Only one third of all amino acids in maize are essential with respect to human nutrition.

Maize with enhanced protein content and enriched in essential amino acids is an important object of research and development to serve as a food basis for countries in Africa, South and Middle America and Asia.

Utilization of c. includes:

- protein basis in animal and poultry feed; the latter is important due to its high content of xanthophyll and provitamin A;
- substrate for acid hydrolysis to yield seasonings, →vegetable protein, hydrolized →amino acids;
- isolation of →zein as raw material for industrial nonfood applications;
- alkaline heat treatment, resulting in natural glues, is an ancient nonfood application, which may gain industrial interest as RR.

Lit.: Blanchard*, 105–108, 299–301, 528, (1992)

Corn Oil

Syn.: Corn Germ Oil; Maize Oil
G.: Maisöl; F.: huile de maïs
s.v.: 187–196; i.v.: 109–133

C. is obtained from →corn germs. The total maize grain contains 3–7 % of c., up to 87 % of which is in the germ and is called corn germ oil, while the rest is distributed in the other part of the kernel. Both oils have roughly the same fatty acid pattern:

C16:0	C18:0	C18:1	**C18:2**
8–12 %	1–3 %	20–32 %	**40–68 %**

The crude oil is produced by →expelling and/or →extraction of the dry (97 %) germ (45–50 % oil). For further processing of the residual cake and its uses: →corn germ. The oil can be further refined by the usual methods (→fats and oils). It is pale yellow, free of cholesterol and has a high nutritional value due to the high content of unsaturated fatty acids.

C. is used mainly in the food industry (cooking and salad oil, dressings). It is hydrogenated for margarine production. There are also some applications in cosmetics and pharmaceuticals (as oleaginous vehicle).

World production (1984) was 800 000 mt, with an increasing tendency due to the worldwide expansion of corn starch production.

Lit.: Kirk-Othmer* (3.) **4**, 836; **9**, 800
Ullmann* (5.) **A10**, 225
Blanchard* 72, 88–92, 357 (1992)
[8001-30-7]

Corn Starch Production

Syn.: Maize Starch Production; Corn Wet Milling
G.: Maisstärkegewinnung;
F.: amidon de maïs, procédés de fabrication

The wet-milling process leads to the production of food and industrial starches as well as to by-products: →corn steep liquor; →corn germs, containing corn oil and corn germ meal; →corn gluten feed; and →corn gluten meal (→zein).

All components of the original maize are recovered without waste and with low input of fresh water.

Process outlines:

Commercial dent corn (USA Yellow II or Yellow III) or waxy corn is worked up in the shelled state as kernels after removal from the corn cobs (→corn), purification and sifting out broken kernels and odd material. First, the kernels are steeped by soaking for about 40–50 h in process water from the later stages of the process, containing 0.8–1% of maize d.m. at 49–53 °C, with 0.1–0.2% sulfur dioxide being added to obtain a pH of 2.5–3.5. This procedure, running in a series of steeping tanks, begins with the tank in which the corn has steeped longest. A lactic acid fermentation by *Lactobacillus* spp. takes place in which the soluble sugars are used as substrate. This is essential to minimize fouling during steeping and further processing. Steeping softens the kernels for grinding, removes solubles, and loosens the germs as well as the starch-protein bonds in the endosperm. The steep water is drained off when its concentration reaches 5–7% of d.s. For further processing: →corn steep liquor.

The softened kernel is then passed through two stages of degerminating mills (with hollow spacings), which tear the kernels apart without crushing them and free the rubbery germ. The specific weight of the latter is lower than that of the kernel, giving rise to its separation (hydrocyclones, flotation channels). The →corn germ fraction, which is 7.5% of the total maize kernel, is worked up for →corn oil and corn germ meal (→corn germs).

The degermed kernels are subjected to a third grind in fine mills with close spacings, resulting in a finely ground liquid mixture of starch, gluten, fiber and bran. Multi-step screening systems (6–10 steps of screen bands or jet refiner sieves) are employed for removing corn grits, starch and gluten and for thoroughly washing of the fibers, which latter are then dewatered by centrifugal screens or screw-presses. Final drying yields →corn gluten feed.

The mill stream after concentration by centrifuga-tion, is fed to a nozzle-disc centrifuge for primary separation. Here, the highly hydrated gluten is separated from the heavier starch stream (underflow) by high-speed centrifugal machines or by multi-hydrocyclones as specifically light, higly diluted overflow. Enrichment and washing of the latter comprises concentration by high-speed centrifuging, mechanically dewatering by decanter centrifuges or rotary vacuum belt-filters, and finally, drying to 90% d.m. in flash-type dryers or rotary steam-tube dryers. After cooling, grinding and sifting, dry →corn gluten or →corn gluten meal is obtained as yellow powder ready for shipping. One ton of maize (12% water) will yield about 50–66 kg corn gluten.

The starch stream from the primary separator proceeds now to a multi-stage countercurrent washing and concentrating step (separator or multicyclone units), finally resulting in a highly concentrated starch milk (about 450 g/liter), ready for further processing (→modified starch; →starch hydrolysis products). Fresh water is introduced into the last washing steps only. The washings are returned to earlier steps of the process until they provide the process waters for the steeping. No more fresh water is needed to equalize the sum of all the water losses by evaporation and drying processes for starch and by-products (max. 2 mt/1 mt maize).

For dry starch production, the concentrated starch milk is dewatered (vacuum filters or basket centrifuges) and finally dried to storage humidity (12–13%), usually by flash dryers.

The following amounts of starch and by-products result from 100 parts of maize (d.b.):

Corn steep liquor solids	6.5%
Germ (45–50% oil)	7.5%
Bran (fiber)	12.0%
Gluten (72–75% protein)	5.6%
Starch	68.0%
(Losses	0.4%)

Technology similar to c. is applied in the production of sorghum starch, which is the smallest in world starch production (≈50000 mt/a).

Steeping is also applied in the manufacture of starches from other cereal crops, such as rice, barley, oats and wheat grain, as well as from legumes.

Recent innovations:

Process development relates to the steeping as well as to the disintegration techniques[1].

Continuous high-pressure steeping is performed in steeping vessels at 1.5 MPa/15 bar and 70 °C without sulfur dioxide. Within 4 h, the kernels have absorbed sufficient water for further processing by wet milling. This technique may be applied to all starch crops that need primary steeping. The short steeping time, compared with traditional steeping, results in a higher yield of gluten; steep water can be recycled, thus reducing consumption of water and energy.

The disintegration of the degermed grains may be improved by a high-pressure disintegration technique, which is multivalent in its application in starch manufacture.

The annual world production of corn starch is estimated to be 75% of the total starch production of about 37×10^6 mt.

Lit.: van Beynum/Roels* 47–72 (1985)
　　　Blanchard* 69–125, 155–175 (1992)

[1] F.Meuser, F.Althoff in: "Stärke im Nichtnahrungsbereich" **41**, 41–55 Landwirschaftsverlag Münster-Hiltrup (1990)

Corn Steep Liquor

Syn.: Maize Steep Liquor
G.: Maisquellwasser; F.: extrait soluble de maïs

C. is the commercial term for the viscous liquid after low-temperature concentration of the water that has been used for steeping of maize in → corn starch production. C. can be further concentrated to the dry state and is then shipped as dry corn steep liquor.

For process outlines: →corn starch production.

The steepwater, containing 6–7% d.s., is concentrated in multi-stage evaporators. The average composition becomes then :

Dry matter	35–50%
Crude protein (d.b.)	44–48%
Ash (d.b.)	16%
Lactic acid (d.b.)	25%

The crude protein includes free amino acids and ammonia, polypeptides of various chainlength, as well as vitamins, enzymes and other biocatalysts.

Utilization includes:

– Nutrient in industrial fermentation operations for production of →enzymes, →antibiotics, nutritional (feed) yeast and →amino acids, and is usually shipped at 50% d.m.;
– additional component in feed formulations, such as →corn gluten feed; it is shipped at 35–45% d.b. and dried together with the other ingredients.

Lit.: Blanchard*, 73–87, 293–295 (1992)

Corn Sugar Molasses

→Hydrol

Cosmetics

G.: Kosmetische Mittel; F.: cosmétiques

C. are intended for cleaning, care, health-prophylaxis and improvement of appearance of skin, hair and teeth. In article 1 of the EC Cosmetics Directive, c. are filed as follows: A "cosmetic product" shall mean any substance or preparation intended to be placed in contact with the various external parts of the human body (epidermis, hair system, nails, lips and external genital organs) or with the teeth and the mucous membranes of the oral cavity with a view exclusively or mainly to cleaning them, perfuming them, changing their appearance and/or correcting body odors and/or protecting them or keeping them in good condition. Article 2 prescribes that c. must not cause damage to human health when applied under normal or reasonably foreseeable conditions of use. The 6th amendment of the EC Directive 23.06.93 demands product information with listing and labelling of ingredients and product function, information regarding overall safety and manufacturing processes and regulations concerning animal testing (a ban of animal testing for cosmetics is planned for 1998).

The EC regulations, in general, are similar to or patterned after USA regulations. Ingredient labelling on the container has been required there since 1988. Unlike the EC regulation, the Food and Drug Administration (FDA) identifies acne-treating products, antidandruffs, antiperspirants, adstringents, oral care products, skin protectants and sunscreen products as "Over the Counter Drugs" (OTC), which has to follow regulatory requirements for drugs.

Customers expect c. to give them healthy skin, hair and teeth, and an overall fresh and elegant appearance.

The large area of c. is treated under separate keywords in the following three groups:

→**Skin preparations**: They include soaps, bath and shower products, skin creams and lotions, sunscreen products, deodorants, lipsticks, antiperspirants, make-up products and perfumes and are most commonly used c. for cleaning, care and improvement of appearance of skin. Face masks, genital hygiene products and tanning products are less important.

→**Hair preparations**: The most important are shampoos, hair conditioners, hair lotions, setting lotions, hair sprays, permanent wave preparations, hair colorants and shaving products.

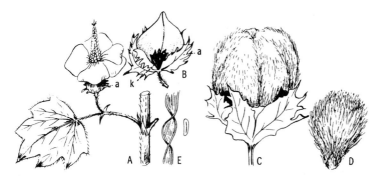

Cotton; A: Stem with inflorescence. B: Immature capsule with outer calyx (a) and calyx leaves (k). C: Opened mature capsule. D: Seed with seed hairs (linters). E: Fiber of seed hair, longitudinal view (left), cross-section (right) (from Franke*, 1981, with permission).

→**Oral hygiene products**: Important c. for teeth are toothpastes and mouth washes.

Nearly all of the c. mentioned may be produced by using more and more RR-based ingredients, such as →surfactants, emulsifiers (→emulsions), →fats and oils, →waxes, thickening agents, and solvents. Additionally, special agents, such as dyes (→dyes, natural), →fragrances, preservatives and humectants, are used, some of which may also be produced or extracted from RR. The raw materials of c. are referred to under the above-mentioned keywords.

Sales of c. in 1993 in Europe were (in 10^9 $):

3.1, women's fragrances; 3.6, color cosmetics; 3.9, facial skin care; 1.6, body care; 0.9, sun care; 3.8, hair care; 2.0, hair styling; 2.5, bathroom products; 1.6, deodorants; 2.1, oral hygiene; 2.6, men's lines, or 27.7 in total.[1]

These figures are relatively vague, because product categories cannot be clearly defined.

Lit.: Kirk Othmer* (4.) **7**, 572–619

Hilda Butler (ed.) "Cosmetics" Poucher's (9.), **3**, Chapman & Hall, London/Glasgow/NY/Tokyo/Melbourne/Madras (1993)

[1] The "Market Report: Cosmetics and Toiletries: European overview 1993: France, Germany, Italy, Spain, UK" (published by European Cosmetic Markets Magazine, SÖFW-Journal, 11/94, 666)

Cotton

Gossypium L. spec
Malvaceae
G: Baumwolle; F: coton

C. is a good example for multiple-use crops: fiber and seed oil and meal (other examples: →linseed and →hemp). New world c. species of commercial importance are *G. barbadense* L. and *G. hirsutum* L., old world species are *G. arboreum* L. and *G. herbaceum.*

Their use as fiber crops (→cotton fiber) has been known from the Inkas as well as from India and Pakistan since 2500 to 3000 B.C. due to archeological findings of precious clothes from ancient cultures.

After pollination, the capsules swell. At maturity they will open so that the five leaves which originally formed the calyx and the fruit, turn to the peduncle, thus displaying a white flake-like ball of cotton fiber with 25 to 50 seeds inside.

The cotton ball consists almost entirely of pure cellulose, which originates from the fibers formed by the upper cell layer of the seed coat. Machine- or hand-harvested cotton consists of fiber and seeds at a proportion of 1:2. Mechanical harvesting is done by stripping and vacuum collecting the flakes.

The harvested mixture of cotton and seed normally will be pressed into bales and carried to a gin where seeds and fiber are mechanically separated. Short, feltlike fibers will remain on the seed and are called linters (→cellulose), which have to be separated in a special delintering process from the seed. All further processing occurs at different places: the fibers are processed in a spinning mill/textile factory, and the seeds in an oil mill. Linters are well suited for chemical derivatization and for specialty paper making, such as cigarette paper.

C. seed contains 33% oil (→cottonseed oil) and 38% protein. Typical for the protein are the two dyeing agents, gossypurin and gossypol. The latter is poisonous for pigs and poultry, which allows their application in mixed feed for cattle only.

The oil may be used for any kind of nutritional and oleochemical applications.

Production in 1994 ($\times 10^6$ mt)	cotton	c. seed	c. seed oil
China	4.33	7.70	0.90
US	4.28	6.90	0.56
India	2.22	4.35	0.47
FSU	2.00	3.66	0.52
Uzbekistan	1.27	2.34	0.33
Pakistan	1.37	2.74	0.23
Turkey	0.63	0.97	0.15
World	18.44	32.62	3.64

Lit.: R.J. Kohel "Cotton" in Röbbelen*, 404, (1989)

Cotton Fiber

G.: Baumwolle; F.: fibre de coton

C. is by far the most dominant natural →cellulose fiber (87% by value). There are three chief types: Egyptian (*Gossypium barbadense*), American Upland (*G. hirsutum*) and Indian (*G. herbaceum*) cotton.

C. is almost (95%) pure cellulose. The remainder consists of fibrous nitrogen compounds (proteins), waxes and pectins, which are found on the surface (i.e., cell wall). The fiber diameter is 12–22 μm and the length of the staple fibers runs in the range of 10–60 mm depending on the type of c. The originally round fiber takes on the shape of a twisted ribbon. For details on internal morphology: →cellulose.

The chief parameters for processing of the fiber are: fiber length, uniformity of length, grade (amount of impurities, called trash), micronaire (maturity and coarseness), strength, elongation (which is approximately 7%) and cohesiveness (cohesion between the fibers). Moisture uptake is 8% at 65% r.h.

The seed hairs, called lint, are removed by saw-like machines (gins). Slots in the machine allow lint to pass and keep the seed back. Modern gins are multifunctional that also dry, blend and clean the lint and the seed. Packing of the long cotton fibers into bales of 220 kg follows. The seed, 65% by weight and 15% by value, is source of →cottonseed oil and cotton linters (→cellulose).

The first step of processing is the selection of the bales and blending operations (sometimes with synthetics). This goes together with a bulk increase and removal of trash. Further cleaning is achieved by carding and blending through drawing. A roving is spun next, which is drawn and twisted to a yarn by a ring and traveller during spinning. For further processing: →textile auxiliaries.

World production is about 19×10^6 mt (1992).

Lit.: Ullmann* (5.) **A5**, 391
　　Encycl.Polym.Sci.Engng.* (2.) **4**, 261
　　Kirk Othmer* (4.) **7**, 620

Cottonseed Oil

G.: Baumwollsaatöl; F.: huile de coton
s.v.: 190–198; i.v.: 100–118; m.p.: 0–10 °C

The purified oil is slightly yellow. It is semi-drying and soluble in ether, benzene and hot hexane. The fatty acid composition is:

C16:0	C18:0	C18:1	**C18:2**
21–27%	2–3%	14–21%	**45–58%**

The oil also contains minor amounts of cyclopropenoid fatty acids. It is gained from the seed after separation of linters (→cellulose), dehulling and extraction of the "meat".

Winterization at 0–8 °C allows the removal of the more saturated triglycerides (→crystallization). →Hydrogenation and isomerization (→isomerism) yields an oil with a m.p. of 30–35 °C. Hydrogenation gives a m.p. of 36–37 °C, which is used as couverture for confectionaries. All above mentioned oils are mainly used for nutrition, e.g., in margarines and shortenings.

Technical uses of c. are as drying oil, in →alkyd resins and in →cosmetics. →Hydrolysis yields fatty acid, which is used in →alkyd resins.

Worldwide production is in the range of 4×10^6 mt (1992).

Lit.: Ullmann* (5.) **A10**, 224
　　[8001-29-4]

Cottonseed Oil Fatty Acids
→Cottonseed Oil

Coupling Sugars
→Oligosylfructoses

Cradle-to-grave Analysis
→Life Cycle Analysis

Crambe

Syn.: Colewort
Crambe abyssinica L.
Cruciferae
G.: Krambe, Meerkohl; F.: chou d'abassinie

C. is a typical herbacious representative of the Cruciferae family, but with only one seed formed per pod. Its origin is in the Mediterranean area.

Since 1932, c. has been evaluated in numerous countries, including Canada, Denmark, Germany, Poland, Russia, Sweden, and Venezuela and the USA. In the 1940's, seed stocks were first introduced to the USA and agronomic evaluation began in 1958. In the USA, c. is considered to have potential because it yields dependably higher levels of erucic acid than rapeseed. Commercial-scale production by the US Department of Agriculture has established processing conditions and proven the efficiency of c. seed meal as a feed supplement for cattle. C. can be a winter crop for climates with temperatures not below $-7\,°C$. In colder climates, it should be grown as a spring crop.

C. yields up to 2000 kg of seed per ha with an oil content of 30−40%. The fatty acid pattern is:

C16:0	C18:0	C18:1	C18:2
2−10%	2−4%	12−18%	8−12%

C18:3	**C22:1**
8−10%	**55−62%**

The high volume-to-weight ratio in the threshed seed, with a husk content of up to 40% (by weight), is considered a limiting factor of c. development. The husk cannot be removed easily from the seed and must therefore be transported to the extraction mill. Husked seed usually contains 30−54% oil and 30−50% protein, whereas oil and protein are reduced significantly if husks are not removed, and fiber content increases to 16%. Commercial production of c. in western Canada, which reached several thousand acres in the mid-to-late-1960's, was severely affected by the disease *Alternaria brassicicola.*

C. oil has good stability at high temperatures. It is therefore well suited as a lubricant for steel processing.

Lit.: K. Lessmann, W.P. Anderson, "Crambe" in "New Sources of Fats and Oils" E.H. Pryde, L.H. Princen, K.D.Mukherjee. (ed.) Am. Oil Chemist Soc. Champaign, Il. (1981)
G.A. White, J.J. Higgins, "Culture of Crambe. A New Industrial Oilseed Crop". U.S. Dept. Agric. Agric. Res. Service Prod. Res. Rep. No. 95, 22. (1966)

Cross-linked Starches

G.: Vernetzte Stärken; F.: amidon (fécule) reticulé

C. are chemically →modified starches with cross-links between the macromolecules by means of bi- or poly-functional reagents.

Cross-linking is carried out preferentially on granular starches in heterogeneous-phase reactions under slightly alkaline conditions. The temperature is held below onset on granule swelling.

Cross-linking agents used include phosphorous oxychloride, sodium trimetaphosphate, mixed adipic/acetic acid anhydride, dicarboxylic acid; (ester bonds); and epichlorhydrine, ethylene or propylene oxide, formaldehyde, di-aldehydes and *bis*-hydroxy-methyl ethylene urea (ether bonds).

Cross-linking establishes a small number of inter- and intra-molecular chemical bridges between the polymeric chains in the starch granule. The proof of cross-linking is in the reinforcement of the molecular structure in the starch granule. It is retained during prolonged cooking, pressure cooking, shearing forces, cooking under acidic or other deteriorative conditions. Only one bridge for every 1000 or 2000 AGU is sufficient for distinct change in functional properties. They may be varied by the kind and concentration of the reagent as well as by the starch (granular →starch ethers or →starch esters may be included), time, temperature and pH of reaction.

Desirable changes include higher onset temperature of starch granule swelling, increase in viscosity and stability, development of viscous pastes under pressure or acidic conditions, nonswelling of the starch granules under pressure cooking.

The greatest benefit is gained in stabilization of such highly water-binding starches as →waxy maize, potato and tapioca starches as well as →starch ethers. Additional ester groups are introduced after the cross-linking reaction. The reagent amounts required are generally low. They do not exceed 0.05−0.1% (on starch d.b.) for epichlohydrine or phosphorous oxychloride, or 0.5−1% for sodium trimetaphosphate.

Applications in nonfood industries include corrugated cardboard, wet-resistant films and paper coatings, thickening of textile printing pastes, sizing and finishing of textiles, wet-resistant glues; higly cross-linked products are used as binders for electrolytes in alkaline batteries and as anti-sticking agents in surgical dusting powders. In food industries, the obtainable high paste stability is utilized for thickening and binding in numerous canned or frozen foods.

Reactions during cross-linking have been shown schematically by Blanchard (1992):

Lit.: Tegge* 189–190 (1988)
     Blanchard*, 311–315 (1992)

Crocus sativus
→Dyes, natural

Crystallization
G.: Kristallisation; F.: cristallisation
C. in general is the formation of crystals from a solution or a melt by cooling. C. is used to separate a mixture of various substances or to purify them.

In context with RR, c. is used to separate saturated, high-melting from unsaturated, lower-melting →fatty acids, which are normally hard to separate by →distillation. The technology is commercially applied on a large scale for the separation of →stearic acid from →oleic acid.

The oldest form of c. is the "panning and pressing process": Molten crude tallow fatty acids are cooled in flat pans, the acid cake is wrapped in filter cloth and the liquid fraction is pressed out hydraulically. This process is labor-intensive and unsatisfactory in terms of the degree of separation and is almost totally abandoned today.

Winterization is another process to separate saturated from unsaturated oils. The oil is slowly cooled in large vessels and the crystals of the more saturated oils are separated by filtration or centrifugation. Originally applied only to →cottonseed oil, it is nowadays broadly used in edible oil manufacturing.

Modern processes use fractional c. in combination with multi-stage or continous technologies. In the Henkel hydrophilization process, the mixture is cooled to 20 °C, and an aqueous solution of a wetting agent (e.g., short chain →fatty alcohol sulfate) containing an electrolyte is added. A suspension is formed by agitation, and the liquid fraction is washed off from the stearic acid crystals. The stearic acid is separated in a centrifuge. The oleic acid phase is cooled to the cloud point, and the process is repeated in a second stage.

Other processes apply fractional c. to solutions of the crude acid mixture in organic solvents, e.g., methanol (Emery process), acetone (Amour-Texaco process), and hexane, leading to good separation. The handling of inflammable solvents is a disadvantage compared to the hydrophilization process.

All these processes yield →stearic acid that contains still up to 15% unsaturated acid, while the →oleic acid reaches a maximum concentration of 75%.

Other applications of c. in oleochemistry are the separation of →stearic from →isostearic acid and of →oleic and →linoleic acid from tall oil fatty acids as starting material.

The method of c. is also frequently used in purification and separation in the sugar processing industry (e.g., →sugar beet processing, →sugar cane processing).

Lit.: Ullmann* (5.) **A10**, 264
Kirk Othmer* (4.) **5**, 172

Cucurbita foetidissima
→Buffalo Gourd

Cuoxam Fiber
→Cellulose Regenerate

Cuphea
Syn.: Cinnobar Flower; Cigar Flower, Firefly
Cuphea spp.: *C. ignea* A.DC., *C. viscosissima* Jacquin, *C. lanceolata* Aiton, *C. procumbens* Ortega, *C. wrightii* A.Gray, *C. tolucana* Peyritsch.
Lythraceae
G.: Zigarettenblümchen, Höckerkelch; F.: parsonie, saliquier

Natural occurrence in North, Central and South America (Michigan, Illinois, Texas, Florida, Mexico, all Central and South American states with tropical and subtropical climates). Some species of c., such as *C. ignea* or *C. procumbens*, are well known ornamentals all over the world. They are often misinterpreted as fuchsias.

Experiments to domesticate c. started in the late 1950's in the USA and the 1970's in Germany. Today there are two working research projects in Oregon and Germany (→agronomy, →plant breeding, →genetic engineering). The interest in c. lies in its

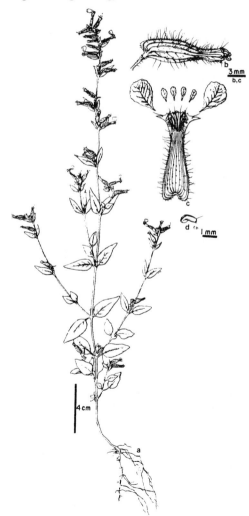

Cuphea wrightii; a: Habit. b: Lateral view of flower. c: Flower opened dorsally. d: Disc [from Graham, Shirley A., 1988, Revision of Cuphea Section Heterodon (Lythraceae), Systematic Botany Monographs, 20, the American Society of Plant Taxonomists, with permission].

high amounts of different medium-chain triglycerides in different species (→fats and oils), hitherto not known in any other plant species.
Fatty acid content of various c. species (in %):

Cuphea

	C8:0	C10:0	C12:0	C14:0	C16:0	C18:0	C18:1	C18:2
painteri	**65.0**	24,0	0.2	0.4	2.8	0.4	3.3	3.9
lanceolata	0.6	**83.2**	2.1	2.0	3.4	–	3.4	4.6
wrightii	–	29.4	**53.9**	5.1	2.3	0.4	3.1	4.6
aequipetala	24.6	1.3	1.8	**56.0**	6.6	–	4.9	3.6

There are more than 300 different described species of c. Domestication work concentrated on those species that appeared to be best adapted for field propagation and those annual species, that seemed to be best adapted for oleochemical applications. The goal was to obtain a possible annual substitute for the traditional lauric oil sources of →coconut and →oil palm.

Agronomical and breeding work was performed with the help of induced mutations for reaching a quick change of wild plant characteristics, such as indehiscence of the fruit ("shattering" of the seeds), formation of sticky glandular hairs, improved germination and emergence in the field (dormancy) and determinate growth and maturity of the crop.

Field experiments with *C. wrightii* in California resulted in yields of 900 kg seed/ha with a multiple harvesting technique with the help of vacuum picking. No commercial production of c. has been reported.

New experiments with genetic engineering in Germany and the US are aiming at the genetic transfer of such genes capable of synthesizing medium chain triglycerides e.g., into rapeseed or other commercial oilcrops. Another successful way to gain high-lauric rapeseed was reached by the gene transfer of the Californian bay laurel to rapeseed.

Lit.: F.Hirsinger, in Röbbelen* 518–532 (1989)

Cuprammonium Fiber
→Cellulose Regenerate

Curdlan
G.: Curdlan; F.: curdlan

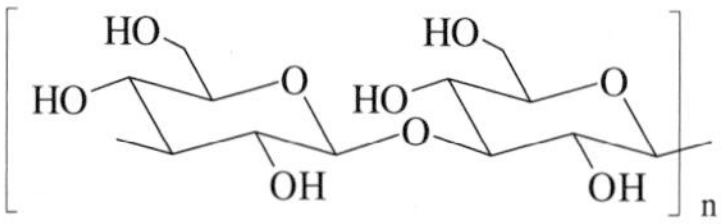

C. is a neutral, linear microbial polysaccharide (→microbial gums) that consists almost completely of β-1,3-joined glucose units. C. is insoluble in water at r.t. and forms thermostable elastic gels at temperatures above 54 °C that are stable within the pH range 3–9.5. C. is not degraded in the gastrointestinal tract. It is produced by fermentation with *Alcaligenes faecalis* var. *myxogenes* or *Agrobacterium radiobacter*. C. is used in foods as a low-energy thickener and gelatinizer but also for the production of water- and air-tight biodegradable films and foils, and for the immobilization of enzymes and cells. C. shows some antitumor activity.

Lit.: Morris, Cheetham in Food Biotechnology I 219–224, Elsevier Applied Sci., London (1987)
[54724-00-4]

Cutting Fluids
→Metalworking Fluids

Cyamopsis tetragonoloba
→Guar Gum

Cycloamyloses
→Cyclodextrins

Cyclodextrins
Syn.: Cycloamyloses; CD; Schardinger dextrins
G.: Cyclodextrine; F.: cyclodextrines

Grouping in toroidal form of at least six α-1,4-linked AGU, resulting from the action of enzymes produced by certain bacteria, such as *Bacillus macerans*, on starch.
The three charactetic types of c. $(C_6H_{10}O_5)_n$, n = 6, 7, 8, are :
α-C.: m.w. 972; β-C.: m.w. 1134; γ-C.:, m.w. 1296.
The β-type with the following structure is the most frequently prepared and technically applied c.:
It is prepared from 5% solutions of →potato starch, which is incubated at 40 °C with the →enzyme cyclomaltodextrin glucanotransferase (CG-tease) (EC 2.4.1.19 nomenclature: α-1,4-D-glucan-4-α-D-(1,4-α-glucano)-transferase).
Reaction results in a mixture of the three types of c. The largest amount, β-c., is separated from the others by crystallization after inactivating the enzyme, removing unreacted starch by hydrolysis, precipitation of the c. by acetone, and purification.
Crystalline β-c. is soluble in water (25 °C) 1.79%, in ethanol 0.1% and easily in dimethylformamide.
All c. possess the outstanding property of having hydrophilic outer and hydrophobic interior surfaces. They are able to form inclusion complexes

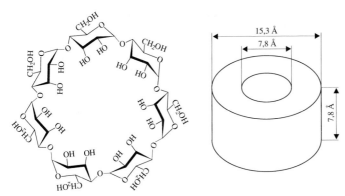

with hydrophobic substances, if they fit into the space, which is formed by the toroidal arranged AGU.

Complexing by c., always in the presence of water, causes:

- formation of storage-resistant dry forms of aliphatic or aromatic compounds, etheric oils, flavorings, →essential oils, biocatalysts, medical agents;
- stabilization, prolonged storage times and reduced losses of highly oxygen-sensitive valuable agents;
- enhanced water solubility and bio-availability of drugs and their controlled release;
- selective absorption from substance mixtures and enrichment of special compounds.

These properties are industrially utilized in food, cosmetics, pharmaceuticals, in plant protecting agents, in fine chemicals and in purification of liquids.

At present, the annual world production amounts to about 1000 mt of mainly β-c.

Despite not being a large scale product, rising production and decreasing prices will enhance numerous new special applications. Developments that will lead to →c. derivatives and →c. polymers are indicative for the extending carbohydrate discipline of "CD-chemistry".

Lit.: Szejtli, J.: Nahrung/Food **29**, 301–310, (1985)
      Eggersdorfer* 197–210, 301–310, (1992)
      Ruttloff* 590–591, 604, 620–622, (1994)

Cyclodextrin Derivatives

G.: Cyclodextrinderivate;
F.: dérivés de cyclodextrine

Aim of derivatization is increased versatility by increasing the hydrophilic character of the outer surface area. This is accomplished with substituents such as sulfuric acid esters and alkyl, hydroxy-

alkyl, carboxymethyl, aminoalkyl ethers; formation of carboxyl groups by glycolic oxydation; formation of branched c. with glucose or oligosaccharide side chains by pyrolytic reactions.

Derivatization is mostly performed with β-cyclodextrin. Solubilities depend on DS as well as on the nature of the substituent groups; levels up to 60% have been reached. In such cases, the viscosity of the solution may become the limiting variable. C. exhibit higher complexing activity for guest molecules, binding capacity and solubility of the inclusion complexes.

Lit.: Lichtenthaler* 302–310. (1990)
      Eggersdorfer* 197–210, 301–310. (1993)

Cyclodextrin Polymers

G.: Cyclodextrinpolymere;
F.: polymères de cyclodextrine

This group of chemically polymerized cyclodextrins is formed by crosslinking and exhibit reduced solubility or total insolubility.

Crosslinking is performed by the action of epichlorhydrin, alkylene oxides or phosphoroxychloride, mainly on β-cyclodextrins, and lead to nonlinear, more or less branched oligomers or polymeric networks. Solubility is decreased when the DP increases.

The capacity for inclusion depends on the structural peculiarities of the guest molecules. A typical inclusion capacity may be increased by interchain linkages.

Based on their insolubility, c. are well suited for fixed-bed operations, such as removing organics from aqueous dispersions, retarded release of active agents according to their rate of diffusion, or as uniform beads for chromatographic column packing. Linear arrangement of cyclodextrin units is accomplished by formation of inclusion compounds with hydrophobic polymers; polyethylene oxide c. or polypropylene oxide c. are insoluble in water. By use of polymers with combined hydrophobic and ionic groups (polyamines reacted with nicotinic acid chloride) in the presence of cyclodextrins water-soluble "polyrotaxanes", consisting of 40 cyclodextrin units, were prepared: the units are permanently stringed on a linear polymer molecule as an ampholytic polymer compound.

Lit.: Szeijtli,J.: Nahrung/Food **29**, 911–924 (1985)
    Eggersdorfer* 301–310 (1993)

Cymbopogon citratus

Cymbopogon flexuosus
→Lemon Grass Oil

Cymbopogon nardus

Cymbopogon winteranus
→Citronella Grass

Cytology
G.: Zytologie; F.: cytologie

C. is the branch of biology dealing with the structure, function, pathology and life history of living cells.

Special emphasis in plant c. is put on the number and structure of the chromosomes of a cell, thus explaining genetic relationships between plant genera, species, and varieties. By counting the number of chromosomes of, e.g., → rapeseed (n = 19), it may be deduced that rapeseed (*Brassica napus*) is an amphidiploid bastard between *B. oleracea* (cabbage n = 9) and *B. campestris* (turnip rape n = 10).

By counting chromosomes, different levels of the ploidy [e.g., n = haploid, 2n = diploid (the normal two sets of chromosomes), 4n = tetraploid, 6n = hexaploid] of a certain plant species may be detected. Also, aberrations of the chromosome count of bastards of related plant species give an indication of their relationship. In wheat, this technology was capable to locate the chromosomes for certain characteristics.

Lit: G.L.Stebbins, "Chromosomal Evolution in Higher Plants", Arnold, London, (1971)

D

Dactylopius coccos
→Dyes, natural

Dalmatian Pyrethrum
→Pyrethrum Plant

Dammara orientalis
→Resins, natural

Dammar Resins
→Resins, natural

Danish Agar
→Furcellaran

Datem
→Diacetyl Tartaric Acid Esters of
Mono/Diglycerides

Datura stramonium
→Atropin

DE
→Dextrose Equivalent

Decanedioic Acid
→Sebacic Acid

Decanoic Acid
→Capric Acid

1-Decanol
→Capric Alcohol

Defoamers and Foamers
Syn.: Antifoaming and Foaming Agents; Foam In-
hibitors and Foam Boosters; Foam Depressants.
G.: Entschäumer und Schäumer;
F.: antimousses et moussants

Foam consists of a gas dispersed in a liquid. Pure
liquids are not able to form a stable foam. However, the presence or addition of →surfactants or a
small amount (impurities) of →polysaccharides or
proteins stimulates foam formation.

In some applications, such as the production of
whipped cream, beer, meringue (→food additives),
foamed plastics, →fire fighting foams, ore-
(→mining chemicals) and waste paper-flotation
(→paper additives), foamed concrete and →deter-
gents, (controlled) foam formation is desired. Products enhancing this property are called foamers
and are →surfactants or →water soluble polymers
or mixtures of both.

In other applications, however, foaming is highly
undesirable, leads to process problems (reduces
tank and reactor volume or the efficiency of pumping, stirring etc.) and causes other troubles. Products counteracting foaming are called defoamers,
antifoaming agents or foam inhibitors. They are
mostly formulated products that are tailored to the
specific application.

D. contain the following RR-based ingredients:
Vegetable →fats and oils, liquid or solid →fatty
acids,→fatty alcohols, →glycerol, waxy products,
→fatty acid amides and sulfonamides.

In some applications nonionic →surfactants
(→fatty alcohol ethoxylates), which show the phenomenon of a cloud point, are used. Above their
cloud point they act as defoamers, but below they
may even act as foam stabilizers. Another special
d. is silica that is made hydrophobic by treating it
with →fatty amines. All these ingredients are used
in solid or liquid (emulsion) form, with emulsifier
to enhance dispersion.

Unwanted foam formation occurs in the following
applications:
- Cleaning processes in the beverage industry
  (tanks, bottles);
- Sugar and yeast manufacturing;
- Potato processing;
- Metal treatment (→metal working fluids, alka-
  line and neutral →cleaners);
- Emulsion →polymerization and applications of
  polymer dispersions; →Coatings and →adhe-
  sives (uneven film formation);
- Textile pretreatment and dying (→textile auxil-
  iaries);
- Pulp and →paper manufacturing at almost any
  stage of the process;
- Phosphoric acid manufacturing; and
- Wastewater treatment.

US consumption in 1990 was estimated to have been in the range of 150 000 – 170 000 mt, and the rest of the world uses the same or even a slightly higher volume.

Lit.: Ullmann* (5.) **A11**, 465
      Kirk Othmer* (4.) **7**, 928

Degradable Polymers

G.: Abbaubare Polymere;
F.: polymères dégradables

D. are materials for packaging, single-use table-ware, fillings, decorations, agricultural mulch and other short-life utilization without the possibility of recycling, to be destructed in a land-fill or by composting within less than 1 year.

Municipal waste management is confronted with the problem of slow or no degradability of most synthetic plastics (10 – 200 years), leading to in-creasing garbage mountains that occupy 17 – 25 vol.% of the landfills. Consequences are rapidly disappearing space in landfills as well as limiting incineration capacity and decreasing fossil re-sources. The magnitude of plastic waste disposal is characterized by the following amounts (10^6 mt/a): USA 13, EC 6.5 (Germany 0.7), Japan 4.5. In addi-tion to utilized strategies of waste management, such as prevention, recycling, chemical and ther-mal utilization and traditional landfill, introduction of degradability into polymers could be a remark-able contribution for reducing the world waste problem. The following possibilities are known for waste disposal in landfills, mostly to be applied in combination:

- Photodegradation by sunlight in the presence of oxygen, catalyzed by the introduction of activa-tors into the plastic matrix. Some photodegrad-able plastic materials are already commercially available: beverage rings, plastic bags;
- Chemodegradation by hydrolytic action in the presence of water and/or oxygen;
- Mechanical degradation by wind, abrasion, pos-sibly after crushing of the material;
- Biological degradation (biodegradation) by ac-tion of living bacteria, yeasts, fungi or low ani-mals after deposition in landfills, soil (compost-ing) or water, provided the plastics contain or are composed of natural RR, such as →hemi-celluloses, →cellulosic materials, →starch and →gluten.

In 1992, about 15% (386 000 mt) of the post con-sumer waste in the USA consisted of "degradable plastics", including retail bags, refuse bags, agri-cultural mulch, beverage containers and disposable polystyrene products.

Starch as a highly degradable material has been processed to →starch- containing plastics as well as to →starch-based plastics. Composting will be the most convenient form of package waste man-agement if recycling or feeding (pure starch speci-mens) is not possible. Inevitable prerequisites to extented use of d. are their price and performance in competition with the traditional fossil-based plastic materials.

Lit.: H. Röper et al.: starch/stärke **42**, 123 – 130 (1990),
      W. M. Doane: Cereal Foods World, **39**, 556 – 563 (1994).

Degree of Substitution

Syn.: DS
G.: Substitutionsgrad; F.: degré de substitution

The average number of hydroxyl groups in →poly-saccharides that have been substituted per mono-meric unit is called d.

The maximum DS will be 3.0 in native polysac-charides. This means that each of the three hydrox-yl groups between the reducing and nonreducing chain end is occupied by a substituent. DS 0.1 means that one OH- group out of 10 monomeric units is substituted.

Degumming

→Fats and Oils

Dehydrated Castor Oil

→Ricinenic Acid

Dehydration

G.: Dehydratisierung, Dehydration;
F.: déshydratation

D. is the physical drying of water-containing mate-rial as well as the elimination of chemically bound water from, e. g., a hydroxyl-containing organic compound at elevated temperature (200 – 300 °C) in the presence of a catalyst (e. g., Al_2O_3):

$$-CH_2-\underset{\underset{OH}{|}}{CH}- \xrightarrow{-H_2O} -CH=CH-$$

The reaction is used for manufacturing →ricinenic acid, →alkyloxazolines, →furfural, →hydroxy-methyl furfural, →levulinic acid, fatty acid anhy-drides, →caramel color and →levoglucosans.

Lit.: Patai-Rapport "The Chemistry of the Hydroxyl Group" Vol 2, 641 Wiley NY (1971)

Deinking

→Paper

Deodorization

→Fats and Oils

Derris elliptica
Derris malaccensis

→Insecticides

Detergents

Syn.: Washing Agents; Washing Powders
G.: Waschmittel; F.: détergents, détersifs

D. are solid (powder) or liquid products used for
textile washing. Prior to the introduction of formulated products at the end of the 19th and the beginning of the 20th century, only →soap and soda ash
were used to assist simple laundering practices.

The term d. is sometimes used more broadly, and
somewhat misleading, for all kinds of cleaning
agents, even when used for other substrates than
textile. Sometimes, it is also misused by calling a
→surfactant a "detergent".

The success of a textile wash is determined by the
following parameters and components:

– water is used as a solvent for water-soluble soil
  and as a transport medium for the dispersed soil;
– the kind and amount of soil that has to be removed (pigments, fat, protein, dyes, water-soluble materials, starch etc.);
– the kind of textile to be washed; synthetic fibers
  behave completely different than cotton or wool;
– the physical input during machine- or manual
  washing (mechanical component, temperature
  and time of treatment); and
– the d. used.

D. have to be formulated or selected respecting all
a.m. factors. Four basic groups of chemicals are
known:

Surfactants:

In general and with respect to RR, →surfactants
are the most interesting ingredients of d. because
aside of LAS (linear alkylbenzene sulfonate), almost all major surfactants are derived from →fats
and oils. Anionic and nonionic surfactants are used
in heavy-duty d. for soil removal, pigment dispersion, desirable foaming, wetting, and solubilization
of all kinds of stains. From the RR-derived products, mainly →fatty alcohol sulfates and →fatty
alcohol ether sulfates are, and →α-sulfo fatty acid
methyl esters may be, used. →Fatty alcohol
ethoxylates are added as nonionic surfactants and
impart detergency at lower temperatures and for

synthetic textiles. They are also used together with
fatty alcohol ether sulfates in light-duty d. (delicate
laundry). In 1982 ≈ 600 000 mt of fatty alcohol
(ether) sulfates and 500 000 mt of ethoxylates were
used in USA, Western Europe and Japan. The demand will increase greatly if sometimes in the future 1.1×10^6 mt of LAS would be substituted because of ecological considerations.

The use of cationic surfactants is restricted to fabric softeners in the rinse cycle because they are incompatible with anionic surfactants. Amphoteric
surfactants are used only in special formulations.

Builders:

Builders impart the function of complexing and sequestering, mainly Ca-ions in the water used (hardness), from soil and from the fiber surface. The
only RR-based sequestering agents are →citric
acid and citrates, but they are of minor importance.

Bleaching agents: None are based on RR.

Optical Brighteners:

With respect to RR only products based on →furan
dicarboxylic acid are reported to be active.

Auxiliary Agents:

Additional ingredients include enzymes, soil anti-
redepositioning agents, foam regulators and fragrances.

• →*Enzymes*, derived from pancreatic extracts of
  slaughtered animals, were introduced already in
  1913 into formulations of d. Only with the invention of proteolytic enzymes, derived from *Bacillus subtilis* or *B. licheniformis,* which are stable
  against alkali and higher temperatures, enzymes
  received a permanent position in d.
  Amylases have been added to the proteases to attack carbohydrate-based soil. Lipases are not
  used because surfactants are able to remove
  greasy soil sufficiently. Cellulases are added in
  modern d. to remove cotton fluffs and smoothen
  the surface of cotton fibers.

• *Soil anti-redeposition agents* are used to avoid
  depositioning of already dispersed soil on the
  fabric again. →Carboxymethyl cellulose and
  →carboxymethyl starch are effective, especially
  on cellulosic fabrics (cotton). To have an effect
  on other fabric-types, combinations of →carb-
  oxymethyl cellulose and →methylcellulose are
  contained in d.

• *Foam regulators* (→defoamers and foamers) are
  important ingredients of d. In the early days of d.
  use, foam was understood to be an indicator for
  washing power and is still seen that way today.
  Foam boosters, based on →fatty acid amides,
  →fatty acid ethanolamides, →betains and

→amine oxides, are frequently used. In house-hold drum-type washing machines, foam is a pro-blem and has to be kept on a low level. →Soaps of long-chain fatty acid (up to C_{22}) are and were used because the active foam-controlling species are their Ca-soaps, which are formed during the washing process. Recent developments have shown a move towards silicone- or paraffin-based foam depressants.

● →*Fragrances* are used to mask unpleasant odors and to give a fresh odor to d., to the wash liquor and to the laundry itself. RR-based perfuming agents are in use together with synthetic sub-stances.

Aftertreatment: Main products are fabric soft-eners (→quaternary ammonium compounds). Starch formulations, used to stiffen the final fabric, are more and more substituted by synthetic poly-mers. Laundry dryer (tumbler) aids are nonwoven materials that are impregnated with softener and temperature-resistant fragrance oils or sprays based on these materials.

World production (1982) of soaps, detergents, scouring agents and other cleaning compounds was 30×10^6 mt, with d. representing more than half (18×10^6 mt). The volume is increasing because the gap between industrialized countries (stagnant at a per-capita consumption of 20–30 kg) and Asia and Africa (at low level of 2–3.5 kg) will be filled.

There is a substantial move towards high density d. in order to minimize the volume of packaging ma-terials and the problem of their disposal. New for-mulations and components (→alkylpolyglucosides and →fatty alcohol sulfate granules) are necessary. This will, among other ecological and economical considerations, enhance the tendency towards RR-based products.

Lit.: G. Jacobi, A. Löhr "Detergents and Textile Wash-ing" VCH Verlagsgesellschaft Weinheim (1987)
W. G.Cutler, R.C.Davis (ed.) "Detergency, Theory and Test Methods" Surfactant Sci. Series Vol. 5 (3 parts) Marcel Dekker Inc NY/Basel (1981)

Dextran

G.: Dextran; F.: dextrane

D. is the longest-known and first commercially used microbial gum polysaccharide and is naturally formed during processing of sugar cane or sugar beet as undesirable mucilage by the action of *Leu-conostoc mesenteroides* on →sucrose. Other d.-forming bacteria are *Streptococcus* and *Lactobacil-lus*. The formation of d. by the mouth flora is con-sidered as the onset of dental caries. Biosynthesis is performed by the action of the enzyme dextran-su-crase (EC 2.4.1.5.), which acts as an exoenzyme:

$$Sucrose \xrightarrow{Dextransucrase} Dextran + n\ Glucose$$

D. is a branched homopolysaccharide:

The relation between the different linkage types as well as the chainlength of the branches depend on the bacterial origin. Structures, consisting of 95% α-1,6-linkages as well as of alternating α-1,6- and α-1,3- linkages have been found. Industrial produc-tion is run preferably by batch processes of culture broth that contains living cells, in cell-free enzyme solutions, or continuously by immobilized en-zymes. Batch conditions are 22–28 °C, pH 5.5–7.2, 36–48 h, sucrose (15%) as C-source. The re-action proceeds nearly quantitatively with high yield. The m.w. of native d. is in the range of $0.5 \times 10^6 - 5.0 \times 10^8$. Further processing is accom-plished by precipitation to isolate crude d. from the culture broth, purification by precipitation, redis-solution and limited hydrolysis, and fractionation into different m.w. ranges. Fine powders are usually available after spray drying.

D. are produced with several degrees of purity, de-pending on the final application. They are tasteless, white to yellowish, fine or coarse powders, which ex-hibit high water-binding and swelling activity and give high-viscosity solutions. Weak gels are formed at concentrations of >20% and m.w. >100 000.

The main present applications require highly puri-fied, well-defined m.w. fractions with narrow dis-tribution:

– Clinical d. as plasma-extenders, m.w. 70 000; for viscosity regulating of blood, m.w 40 000; half of world production;

– d. for gel-permeation chromatography range be-tween m.w. of 1000 and 200 000 and are fre-quently cross-linked and/or derivatized with ionic groups for uses as specific absorbers or ion-exchangers;

– most important anionic derivatives are d.-sul-fates with an average DS of 1–2; they are used for two-phasic ionogenic separation systems; products of m.w 7000–20 000 may be used as

exchangers for heparin as inhibitors for blood clotting;
- d. of m.w. 1000- 3000 is used to stabilize Fe^{3+}-preparations for injections;
- Diethylaminoethylamine-d. is applied as microcarrier in cell-culture breeding and as therapeutic agent against hyperlipaemia.

Production of 2000 mt/a remains constant.

Lit.: Ruttloff* 493–500, (1991)
     Ullmann* (5.) **A8**, 449–454

Dextrins

Syn.: Pyrodextrins
G.: Dextrine; F.: dextrines

D. are one of the oldest types of →modified starches (1821), prepared by heat treatment in the dry state at >80 °C, with or without addition of small quantities of chemical reagents. Commercial d. are classified according to their preparation :
- **White dextrins**, in the presence of acids, lower temperatures;
- **Yellow dextrins** (Canary dextrins), dry heating, lower quantities of acid or catalyst;
- **British gums,** absence of acids, very high temperatures, sometimes small amounts of alkali are added;

All d. have retained the →starch granule shape in the dry state. Their solubility in cold water as well as viscosity in solution are largely depending on the mode of preparation. The viscosity is used for classification. In white d., increasing heat and time of treatment increases solubility and diminishes viscosity. All commercial starches may be applied for production of d., provided the content on crude protein is minimum. In Europe, →potato starch, but in the USA, →corn starch is mostly used.

Commercial production is preferably carried out in continuous equipment, with the following principal steps:
- Addition of the catalyst to the dry starch, hydrochloric or nitric acid to pH 2 for white d. and pH 3 for yellow d., in gaseous or liquid form; catalysts for yellow d. or british gums are $CaCl_2$, $AlCl_3$, Na_3PO_4, oxydants;
- Drying to 1–5 % moisture in the starch;
- Conversion (roasting): temperature and time depend on the type to be produced, varying between 80–220 °C and from a few minutes to 8 h;
- Cooling and neutralization for immediate interruption of the reaction: acidic products are neutralized with ammonium-hydrogen carbonate or gaseous ammonia;

- Remoistening and screening: the reaction products are dusty (moisture content 1–2 %). They will pick up water when exposed to humid air (swelling) and tend to form lumps when suspended in water. To avoid this, humidity is adjusted to 10 % by spraying a fine water mist into the reaction product. Lumps are screened off by passing through 80 mesh sieves. Finally, blending with other d. qualities is performed to ensure uniform quality.

	White d.	Yellow d.	British gums
pH	2	3	>7
Temperature	80–120 °C	150–200 °C	180–220 °C
Reaction type	Acid hydrolysis, Recombination.	Reversion, Transglucosylation, Acid hydrolysis.	Thermal decomposition, Levoglucosan.
Cold water solubility.	30–80 %	95–100 %	100 %
Solution	Clear,colorless, yellowish, thermally reversible gels	Yellowish-brown, tacky	Dark brown, caramel odor
Viscosity 40%,20 °C	Gels	50–300 mPas	–
Properties	Brittle films, similar to →thin-boiling starch	Good adhesion, stable viscosity	

Utilization: Glues (→adhesives) as such or as blends in paper processing, quick-setting adhesives for labels and fine paper; paper coatings (→paper additives), corrugated-paper glues; core binders in the →foundry industry; thickener for textile printing; soluble nutritive carbohydrate source in culture media, coatings of capsules in pharmaceutical industry, spray-drying adjuvant.

Annual world production is difficult to estimate. The quantities are either listed under the term →modified starches or in connection with paper, cardboard and glue production.

D.-production in USA was 45 000 mt in 1985.

Lit.: Van Beynum/Roels* 85–88, 345,(1985)
     Tegge* 173–179, 312, 314 (1988)
     Blanchard* 346–356. (1992)

Dextrose

Syn.: D-Glucose
G.: Dextrose; F.: dextrose

D. is the primary product of plant photosynthesis, with an annual production $\approx 200 \times 10^9$ mt; 95 % is further metabolized to other carbohydrates (sucrose, starch, cellulose and others); 5 % (10×10^9 mt), is converted to all other organic material in nature; only 2.6% (5×10^9 mt) is utilized by humans as food.

The term d. is also used for D-glucose obtained by the complete hydrolysis of starch (or cellulose), followed by purification and crystallization.

Dextrose can be obtained either in the anhydrous form

$[C_6H_{12}O_6]$

or as a monohydrate

$[C_6H_{12}O_6] \cdot H_2O$

the latter being of predominant economical and industrial significance.

(α-D-glucose) (β-D-glucose)
D-glucose anomeric forms (chair conformation)

m.w.: 180; m.p: 83 °C (monohydrate); 146–150 °C (anhydrous).

Solubility in water: 47% (20 °C), 78% (70 °C), only slightly soluble in EtOH or organic solvents; r.s.: 61; DE: 100%;

Specific rotation: α-pyranose +113.4°
β-pyranose + 19.5°
equilibrium + 52.5° (36% α and 64% β)

Dextroses are produced by total hydrolysis (→hydrolysis, glucosidic linkages) of starchy materials via liquefaction of >30% aqueous suspensions, either by mineral acid or by bacterial →α-amylase (DE 14–18%) in continuous reactors, through saccharification by →glucoamylase 48–96 h in holding tanks at ≈60 °C or in continuous flow reactors to a DE of 96–98%. Purification is achieved by desludging, deionization and decolorization of the raw hydrolysates and subsequent concentration. The last step is crystallization of the concentrate (74–76% d.s.) over two or three days with temperatures descending from 40 °C to <20 °C, followed by centrifugation of the massecuite, and washing, drying and sifting of the dextrose monohydrate. Anhydrous dextrose crystallizes only if the concentrated solutions are heated to 80 °C. In both instances, the mother liquor after crystallization (→hydrol) is further worked up for a second crystallization or as a component for different liquid sugar grades.

Hydrolysates with high purity may be worked up by limited crystallization with 68% solids at 20 °C, followed by spray-drying the massecuite at 50 °C. The solidified fine granular mass, termed →"solid glucose" or "total sugar", is a mixture of monohydrate crystals and glassy-state dextrose.

The main share of d. produced is further reacted as a solution with glucose-isomerase (→isomerization) to an equilibrium mixture of glucose and →fructose, leading to the field of →isomerized sugars (iso-sugars, high-fructose starch syrups, fructose).

Another considerable amount is fermented to organic acids, organic solvents, chemical intermediates, yeast and fuel ethanol (→fuel alternatives). Growing interest is focused on its use as C-source in fermentation for the production of "biopolymers", such as →polyhydroxybutyric acid or →polylactic acid.

A substantial amount is converted by →hydrogenation to →sorbitol.

Under special demands of transportation to the users, refined glucose solution of <50% solids is shipped as "liquid dextrose".

Application in food production is based on its unique properties, such as solubility, digestibility and fermentability, slight sweetness (→glucose syrups).

The following applications are representative:
Instant soft drink powders, nonalcoholic beverages, instant deserts, sweet creams, puddings, confectionary fillings, chocolates, chewing gums, energy providing drinks.

Its high purity, as well as its fast absorption into the human blood system determine the applications in pharmaceutical preparations and in medicine:
Sterile infusion solutions, nutritive mixtures and energy providing preparations, carrier and sweetener for application of drugs in tablets and medicated comprimates.

The a.m. nonfood applications and its forseeable expansion as chemical and biotechnological raw material lead to the conclusion that d. is one of the most important and versatile RR. At present, economical sources are only starch or starch-bearing plant material. Future interest is focusing on cellulose (waste) if reasonable enzymic saccharification of this molecule can be accomplished.

Lit: Schenk/Hebeda* 121–176 (1992)
     Ruttloff* 603–617 (1994)

Dextrose Equivalent

Syn.: DE
G.: Dextroseäquivalent; F.: équivalent en dextrose

D. is the reducing power of a sugar or a starch hydrolysis product, expressed as a percentage of D-glucose on d.b. (% DE).

The term is frequently used in the starch hydrolyzing industry and in other fields of application to classify and characterize different intermediates and commercial products, such as →maltodextrins, →maltooligosaccharides and →glucose syrups, as well as the purity of mono- and disaccharides (dextrose = 100% DE).

Lit: ISO 1227, 2nd ed. 5.6.
Ullmann* (5.) **A9**, 391 and **A11**, 502

Diacetin

→Glyceryl Diacetate

Diacetyl Tartaric Acid Esters of Mono/ Diglycerides

Syn.: Datem; Acetylated Tartaric Acid Esters of Mono/Diglycerides
G.: Diacetylweinsäureester von Mono/Diglyceriden;
F.: esters mono/diacétyl tartriques de mono/diglycérides

$$CH_2-O-CO-R^1$$
$$CH-OR^2 \qquad OX \qquad R^1: C_{15-17} \text{ alkyl}$$
$$CH_2-O-CO-CH-CH-COOH \qquad R^2: H \text{ or } -CO-R^1$$
$$\qquad \qquad \qquad OX \qquad X: H \text{ or } CH_3 -CO-$$

s.v.: 435–450; m.p.: 45–47 °C

Due to numerous parameters that influence the composition during and after production, d. are complex products. The formula above should therefore be seen as representative but schematic.

D. are liquid to waxy and smell normally for acetic acid due to →hydrolysis. They are anionic and hydrophilic emulsifiers.

They are manufactured either by reaction of diacetyl tartaric acid anhydride with mono/diglyceride in the presence of acetic acid or by esterification of mono/diglycerides with tartaric/acetic acid in the presence of acetic acid anhydride.

The main use is in yeast-raised baked goods, especially in bread, due to the strong interaction with →gluten, causing increased volume und uniform pore size.

US consumption in 1993 was 2000 mt, in Europe 20 000 mt.

Lit.: G. Schuster "Emulgatoren in Lebensmittel" Springer Verlag Berlin/Heidelberg/NY/Tokyo (1985)
N.J.Krog "Food Emulsions" K.Larsson, S. E. Friberg (ed.) Marcel Dekker N.Y./Basel (1990)

Dialdehyde Starches

→Oxidized Starches

Dicarboxylic Acids

G.: Dicarbonsäuren; F.: acides dicarboxyliques

The following d., based on or related to RR, are important:
→fumaric acid, →succinic acid, →sorbic acid, →sebacic acid, →azelaic acid, →brassylic acid, →suberic acid and →dimer acid.

While the short-chain d. ($\leq C_6$) are obtained easily by petrochemical or biotechnological synthesis (starting from carbohydrates), longer-chain d. are made by oleochemical pathways.

Carbohydrates have high functionality and are also starting materials for (di)hydroxy d. (fruit acids). Oleochemical sources have the disadvantage of contributing only monofunctionality. It is therefore highly desirable to find routes of incorporating difunctionality to make them useful, e. g., as building blocks in polymers (polyamides, polyesters, polyurethanes). →Ozonolyses and →dimerization are processes that lead to d.

There are several attractive methods for incorporating a second carboxylic group at various positions into, e. g., →oleic acid or its methyl ester, yielding C_{19} d.:
a.) →Hydroxyformylation (Oxo reaction)
b.) →Hydroxycarboxylation (Reppe reaction)
c.) →Hydroxycarboxylation (Koch synthesis)

If erucic acid is used as starting material, C_{23} d. result.

D. esters are used as plasticizers in PVC, and the acids as starting material for polymers, epoxy resins, →lubricants and adhesives.

Lit.: Ullmann* (5.) **A8**, 523

Dicots

→Dicotyledonous Plants

Dicotyledonous Plants

Syn.: Dicots; Dicotyledons; Broad-leaved Plants
G.: Zweikeimblättrige Pflanzen; F.: dicotylédonés

D. are the majority ($\approx 170\,000$ species or 72%) of all flowering plants (*Angiospermae*) (28% are →monocotyledonous plants).

Typical is the formation of two cotyledons after germination. The morphology of the cotyledons normally is distinct from the other leaves, which form typical net-type veins. Flowers will part in fours and fives. Cambium is present. There is much variation of seed coats, seed contents, flowers and other morphological characteristics.

Lit.: G.L.Davis,"Systematic Embryology of the Angiosperms" John Wiley Sons, NY (1966)
P.Raven et al. "Biology of Plants" (5.) Worth Publish. NY (1992)

Digitalin
→Digitoxin

Digitalis purpurea
Digitalis lanata
→Digitoxin

Digitoxin
Syn.: Digitalin; Digoxin
G.: Digitoxin; F.: digitoxine

R¹=R²=H Digitoxin
R¹=OH, R²=H Digoxin

m.w.: 765.05; m.p. 256−257 °C

D. is a scondary glycoside (→saponins) from *Digitalis purpurea* or *Digitalis lanata* and is extracted from the dried leaves with 50% alcoholic solution. Acid hydrolysis yields digitoxigenin and digitoxose. Ten kg of leaves gives 6 g d. and an almost equal amount or even more digoxin, which is more active than d.

D. is highly toxic but is used as a carditonic agent.

Lit.: I. M. Jakovljevic "Analytical Profiles of Drug Substances", Vol. 3, 149−172, K. Florey (ed.), Academic Press NY (1974)
[71-63-6]

Diglycerides
→Glycerides

Digoxin
→Digitoxin

2,3-Dihydroxybutanedioic Acid
→Tartaric Acid

Diketenes
→Fatty Diketenes

Dimer Acid
Syn.: Dimer Fatty Acid
G.: Dimerfettsäure; F.: acide dimère

D. is a highly viscous liquid. It does not cristallize, even at low temperature. It is soluble in hydrocarbons.

D. is produced by →dimerization of unsaturated fatty acids.

The resulting C_{36}-dibasic acid is a mixture of many different structures. Because the reaction follows mainly a Diels Alder pathway, one of the dominating species is characterized by the following formula:

Isomers of this basic structure are also formed. There are also other cyclic structures (polycyclic, aromatic).

The monomers are normally stripped off by vacuum distillation. They are partly isomerized and have interesting properties (→iso-stearic acid).

The quality of d. is characterized by the mono:di: tri ratio.

Normal d. is a mixture of 70−80% dimer and 15−20% trimer. For special applications, d. is purified further. Hydrogenation improves the color and molecular distillation (high-vacuum distillation in a wiped-film evaporator) increases the dimer content up to more than 95%.

Dimer acid has surface active properties because it contains hydrophobic and hydrophilic parts in its molecule. This is the reason for its use in →corrosion inhibitors, →detergents, →lubricants and pigment dispersers in →coatings.

The main volume of d. goes into the manufacture of →polyamides. Modification of →alkyd resins is another application.

Dimer diamine (→fatty amines) is produced from the acid.

Annual production (1986) is in the range of 60 000 mt worldwide.

Lit.: Kirk Othmer* (3.) **7**, 769
Encycl.Polym.Sci.Engng.* (2.) **11**, 476
E.C.Leonard "The Dimer Acids" Humko Sheffield Chemical, Memphis TN (1976)
[61788-89-4]

Dimer Fatty Acid
→Dimer Acid

Dimerization
G.: Dimerisierung; F.: dimérisation

Unsaturated →fatty acids are dimerized (or oligomerized) to dimer or oligomer acids at temperatures of 200–250 °C under 0.7–1 MPa/7–10 bar pressure in the presence of 4–7% selected montmorillonite. Water should be present (2–5%) to avoid decarboxylation. Usually, acids are used as starting material; however, unsaturated →fatty acid esters or →fatty alcohols can be dimerized also. For commercial production mainly →tall oil fatty acids are used. Other mono- and poly-unsaturated →fatty acids (→linseed oil, dehydrated →castor oil, →soybean oil) are also suitable starting materials.

Yields are in the range of 55–60% and depend on the nature of the acid used.

The mechanism of the reaction is mainly a Diels-Alder type. A large number of different structures in the final product indicates that also other reactions (isomerization, conjugation, hydrogenation and dehydrogenation, formation of polycyclic and aromatic compounds) take place. Trimers and a small amount of tetramers are formed also. The ratio of dimer : trimer is 80 : 20.

For qualities, properties and uses of the final product: →dimer acid.

Lit: Encycl.Polym.Sci.Engng.* (2.) **11**, 476
R.Paschke, J.Johnson, D.Wheeler Ind. Eng. Chem. **44**, 1113 (1952)

3,7 Dimethyl-6-octen-1-al
→Citronellal

Dioecious
G.: Diözisch, zweihäusig; F.: dioïque, diœcique

D. plants have the male reproductive organs in one individual and the female organs in another. Thus, the sexes are separated in different individuals as is common for most animals. In plants, this separation of the sexes is rare.

Examples are: asparagus, hops, spinach, →hemp, dates, →jojoba.

See also: →monoecious

Lit.: R.Frankel, E.Galun "Pollination Mechanism, Reproduction and Plant Breeding" Springer Verlag New York (1977)

Dioscorea spp.
Syn.: Yam
Dioscoreaceae
G.: Dioscorea Arten, Yam; F.: dioscorée, igname

D. is the principal genus of this →monocotyledonous family, which is close to the Liliaceae. It consists of about 600, mostly tropical, →monoecious species in which the rhizome is modified to produce an annual tuber.

The yam (*D. alata, D. bulbifera*) certainly is the most prominent species out of 11 cultivated d. It is grown and eaten throughout the wetter tropics, especially in Africa, SE Asia and the Caribbean. It is planted from tubers or cut tubers and will produce new tubers 8–10 months later. The tubers generally are difficult to store. The fresh tubers consist of 60–70% water, 30–40% carbohydrates and 4–8% protein. They are considered a good source of vitamin C.

Wild d., such as *D. composita* Hemsl., *D. terpenapensis* and *D. mexicana* Guill. are important to the pharmaceutical industry for producing →diosgenin from the tubers. From the middle 1950's to

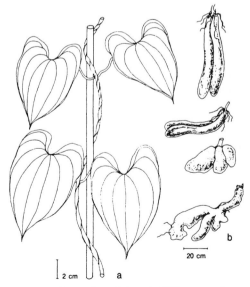

Dioscorea alata; a: Part of shoot; b: Different root forms (from Rehm and Espig*, 1995, with permission).

the early 1960's, well over 50% of all steroids manufactured worldwide originated from Mexican sources. In the early 1970's, the situation changed dramatically when prices for diosgenin jumped from $10 to more than $100 per kg. This was caused by a centralization and nationalization policy of the Mexican authorities to monopolize the collection and marketing of d. Simultaneously, the diosgenin content decreased from 6 to 4% due to overharvesting. In the mid 1970's, other →steroid sources, such as →cholesterol and β-sitosterol, proved also to be suitable for producing steroid hormones for contraceptives by applying newly developed microbiological derivatization techniques. Future world requirements for steroids are likely going to be met by →fermentation and →genetic engineering techniques, and the vegetable diosgenin sources will become less important.

Lit.: D.G.Coursey, "Yams". Longman. London (1967)

Diosgenin

Syn.: Nitogenin
G.: Diosgenin; F.: diosgénine

HO

m.w.: 414.61; m.p.: 204–207 °C

D. is the aglycone from the steroid-saponin dioscin [19057-60-4] and commercially extracted from the tubers of *Dioscorea composita* and *Dioscorea terpenapensis*. The extracts are concentrated by evaporation; the glycoside bond is cleaved by heating with dilute acid, and crude d. is isolated by filtration. Several hundred mt are produced. It is converted to pregnenolone and progesterone and many other →steroids. The crystals of d. are soluble in the usual organic solvents.
D. is the most suitable starting material for the production of →steroids (corticosteroids and contraceptives).

Lit.: Ullmann* (5.) **A13**, 112
Rehm-Reed* 6a, 31
[512–04–9]

Diploid

→Cytology

Diseases

G.: Pilzliche, bakterielle, viröse Krankheiten;
F.: maladies des plantes

D. are caused by fungi, bacteria and viruses as compared to pests.
Plants may easily be infected by fungal d., such as mildews (*Erysiphe* and *Peronospora* spp., such as in →cereals, →Brassicaceae, fruits) and rusts (*Puccinia* and *Ustilago* spp., such as in cereals, vegetables and several fruits). Bacterial infections are common in crops that have comparatively high water contents, such as →potatoes (*Corynebacterium* spp., *Erwinia phytophthora*) or *Brassica* spp. *(Agrobacterium* spp.*)*. They are also causing wilting d. in vegetables and fruits or root rots. Viral d. are common in tobacco (TMV = tobacco mosaic virus), potato, peas and some →cereals. Protecting the crops with different →pesticides, keeping track of good crop rotation, good resistance in the used →germplasm, and proper phytosanitary rules in crop management are important necessities for agronomical production of crops. When producing more than an inspectable number of plants together in large areas, it is almost impossible to trust in a natural balance between plants and d.

Lit.: R.B. Stevens, "Plant Disease". Ronald Press. N.Y. (1974)

Di-starch Phosphates

→Cross-linked Starch

Distillation

G.: Destillation; F.: distillation

This method of purification, widely used in the chemical industry, is also applied to RR-based products, especially in →oleochemistry.
Fatty acids, methyl esters and alcohols, as well as many other oleochemicals, are subjected to d.
Crude fatty acids from fat splitting (→hydrolysis) are normally distilled to produce high-quality products that have excellent color and low levels of impurities. Odor bodies and low-boiling unsaponifiables (hydrocarbons, aldehydes) are removed in the light or head fractions, while higher-boiling impurities (e.g., triglycerides, color bodies, decomposition products and polymerized material remain in the pitch or bottom fraction.
The middle fraction contains a mixture of the fatty acids, which can be used directly in most applications. For some uses however, they have to be fractionated (→fractional distillation) into narrower cuts that consist of a limited mixture of "neighbor-

ing" fatty acids or even into pure fatty acids that consist of up to 99% of a distinct fatty acid.

With increasing temperature, fatty acids are damaged by oxidation, decomposition and polymerization, giving colored and odorous by-products that are detrimental to quality and yield. Therefore, vacuum distillation in the presence of small quantities of steam is the general technolgy used today.

There are three principles of evaporation:

a) Still distillation evaporates the fatty acids from a large volume of liquid; light ends are cut off, the middle section is distilled and the residue remains in the still bottom.

b) In flash distillation, heating of the crude acid takes place outside the distillation unit. The hot material is flashed into the vacuum d. zone where the fatty acids evaporate while using the internal heat of the material. The higher-boiling ends leave the column at the bottom.

c) In film evaporation a thin film of fatty acid is formed and the fatty acids evaporate from the large surface at operating vacuum. With this method, heat transfer is excellent and rapid, the residence time is only a few minutes, and the distillate is stressed at a minimum.

D. units for fatty acids can be divided into batch, semi-batch continous straight or continous fractional d. Batch d. is only used in small plants, and modern technolgy uses continous straight d. with flash or film evaporation. The whole procedure is running continuously by using one or two units. In two units, the first one separates the light ends, while the second distills the main product and the pitch is collected at its bottom. Many different technologies are practiced to obtain a high degree of purification with minimum use of energy (heat exchange) and air pollution.

The equipment has to be made from stainless steel with <2.5% molybdenum. A large variety of complete units is commercially available.

Computers and on-line gas chromotography are frequently used today to control and optimize the d. procedure.

Some of these d. techniques allow a certain degree of fractionation by passing the material through the unit a second time to obtain a lower- and a higher-boiling fraction.

Finer separation, however, is carried out in special units for →fractional distillation.

Another kind of d. is steam distillation: steam is blown into the material with or without vacuum being applied. Some, especially neutral and nonpolar substances (→essential oils, →tocopherol), will distill together with the steam and thus can be separated from the nonvolatile material. The distillate separates into two layers upon cooling or is extracted with proper organic solvents. Because steam d. is a gentle procedure, it is used frequently in the isolation of →fragrance raw materials and in oil refining (→fats and oils).

Lit.: Ullmann* (5.) **A10,** 260 and **B3,** 4 – 1
Kirk Othmer* (4.) **5,** 175

cis,cis-5,13-Docosadienoic Acid
→Meadowfoam

Docosanoic Acid
→Behenic Acid

1-Docosanol
→Behenyl Alcohol

cis-13-Docosenoic Acid
→Erucic Acid

trans-13-Docosenoic Acid
→Brassidic Acid

Dodecanoic Acid
→Lauric Acid

1-Dodecanol
→Lauryl Alcohol

1-Dodecylamine
→Lauryl Amine

Douglas Fir Wax
→Waxes

Dried Glucose Syrup
→Solid Glucose

Dry Cleaning
G.: Chemische Reinigung; F.: nettoyage à sec

D. is a process to remove soils and stains from fabrics and garments (including leather) with organic solvents, e.g., hydrocarbons, perchloroethylene or trichloroethane. A little water is added to enhance the cleaning efficiency against hydrophilic soils.

→Surfactants and emulsifiers (0.5 – 1.25%) of mostly formulated anionic detergents, such as →soap/→fatty acid mixtures, →fatty alcohol ether sulfates, →fatty alcohol phosphate and their →ethoxylates, →sulfosuccinates, →sorbitan esters of fatty acids and →fatty acid ethanolamide ethoxylates, are added.

Lit.: A.R.Martin, Technical Bulletin T-403 National Inst.
of Drycleaning (1964)
Ullmann* (5.) **A9**, 49
Kirk Othmer* (3.) **8**, 50

Drydown
→Odor Description

Dryers
→Drying Oils

Drying Oils
G.: Trocknende Öle; F.: huiles siccatives

Drying oils are mostly vegetable oils that dry to a solid film when being exposed to air/oxygen as a thin film. They are triglycerides of polyunsaturated fatty acids. The type and content of these acids determine the drying characteristics: higher unsaturation, more conjugation and more *trans* isomers (→isomerism) enhance the drying capacity. They are divided in three groups:

a.) Nonconjugated oils, such as →linseed oil, →soybean oil, →sun-flower oil, →safflower oil, grape seed oil, →cottonseed oil, →Chinese tallow oil (stillingia oil) and various types of →fish oil have at least two single bonds between two double bonds. They have excellent drying characteristics but also a tendency to get brittle and yellow if they have a high content of linolenic acid.

b) Conjugated oils, such as →tung oil, →oiticica oil, dehydrated →castor oil and non conjugated oils after isomerization (→isomerism), have alternating single and double bonds. They are drying fast and give excellent finishes, which are flexible and nonyellowing and have good resistance to weathering.

c.) Other oils, e.g., →castor oil after dehydration or →tall oil have to be built into an oil- or resinbody to function as an oily component.

D. have to by purified prior to use:

Degumming is accomplished by treatment with 0.1 % phosphoric acid at 80–90 °C for 30 min.

Deacidifying is done by either adding caustic to remove a soap stock or better by →distillation.

Bleaching is done by treatment with activated fuller's earth.

Vacuum steam distillation deodorizes the oil, while cooling (4–10 °C) and filtering is a method to dewax it.

The mechanism of drying is not yet fully understood. Hydroperoxides, cyclic peroxides and dimerization lead to crosslinking through the formation of C–C, C–O–O–C and C–O–C bonds. The reaction with oxygen is catalyzed by dryers, which are also called siccatives. They are oil-soluble salts of Co, Fe, Mn, Ce, Pb, Zr (order of decreasing activity) with mainly →linoleic, naphthenic or isooctanoic acids. Secondary or auxiliary dryers are the salts of Ca, Ba and Zn. They do not catalyze the autoxidation but enhance the efficiency and stability of primary dryers.

Sometimes, d. are modified and treated:

Boiled oils are the result of treating d. with siccatives at elevated temperature to get higher viscosities. The same effect can be reached with →factice varnish (reaction with sulfur). **Stand oils** are produced by heating (280–300 °C) in the absence of oxygen. A Diels Alder reaction takes place and increases viscosity. Mixtures of stand oils and alkyd resins are used. **Blown (stand) oils** are made by blowing air into the 90–120 °C hot oil. Hydroperoxide formation results in a higher molecular weight. →Isomerization is used to increase the amount of conjugation and *trans* isomers thus changing the drying characteristic of the oil. Reaction with monomers, such as styrene and cyclopentadiene yields copolymers or graft polymers (but also Diels Alder adducts), which dry faster to form clear, glossy and nonyellowing films with higher water and chemical resistance than unmodified alkyd resins. Similar effects are the result of maleinizing: Maleic acid anhydride is reacted with conjugated oil at 80–120 °C to form cyclohexane dicarboxylic acid anhydrides. Nonconjugated oils need higher temperatures. These adducts can be neutralized with ammonia and give water-thinnable binders. All these modifications involve reactions with the alkyl chain of the trigyceride.

→Transesterification is the first step in making (modified) →alkyd resins.

D. are widely used in the manufacture of oil based paints (→coatings), →printing inks, →linoleum, →putty and core binder (→foundry industry).

The worldwide consumption in 1978 of drying oils was in the range of 187 000 mt.

Lit.: Ullmann* (5.) **A9**, 55
Kirk Othmer* (3.) **8**, 130

Dry Milling
G.:Trockenvermahlung; F.: mouture sèche

D. is a process to cut up cereal grains or legume seeds into fractions like grits, meal, flour, germ and bran with different contents of starch, proteins, lipids, fibers and biocatalysts. D. is performed by grinding operations (stone mills, attrition mills, refiner mills, impact mills, pin mills) combined with different sieving operations (screens, sieves) and blending after removal of the hulls and the larger part of the germ.

D. techniques are the most widespread and important processing operations for agriculturally based products. Technical interest in nonfood applications is to be focused on the preparation of special flour for →wheat starch production. Corn, in the form of corn grits, is the main raw material for the production of beverage alcohol, fuel ethanol (→fuel alternatives) and other →fermentations.

Shelled and dry-milled peas as well as beans will be used as new starting materials for legume starch (→pea starches, →broad bean starch) production.

DS
→Degree of Substitution

Dyeing Agents
→Dyes, natural

Dyer's Saffron
→Safflower

Dyes, natural
Syn.: Dyestuffs, natural; Dyeing Agents
G.: Farbstoffe, natürliche; F.: colorants naturels

In addition to inorganic pigments, mankind has used many natural organic dyestuffs based on plants, insects and other animal sources for dyeing textiles (→textile auxiliaries) and other materials, e. g., leather and food, hair and skin. In the second half of the 19th century pioneering industrial chemistry opened synthetic routes to low-cost organic dyes in quality and variety that exceeded the spectrum of natural d. within a few years. Their use declined dramatically and was almost at a zero level when growing consumer awareness and legislative regulations again favored natural dyestuffs, especially in food and cosmetic applications.

The following survey (grouped by chemical structure) lists several d. of historical importance and those that are still in use today, mostly as food d.:

Anthraquinones:
● Madder (alizarin) is derived from more than 35 species of Rubiaceae roots (mainly *Rubia cordifolia* L.). Once the main cotton dye in Europe (cultivated in France and Holland) and the Mid-

dle East, it has been substituted totally by synthetic dyes. Turkey red was a common dyeing process with a combination of turkey red oil (→sulfated fats and oils), alum and madder.
● Kermes is the dye of a shield louse (*Coccos ilices*) growing on *Quercus ilex* and *Q. coccifera* and was used as a substitute for the famous "royal purple". It was substituted later by cochineal. Today, the name kermes is used for a berry (*Phytolacca americana* L.), the red dye of which is used for coloring alcoholic beverages.
● Cochineal is gained from dried female shield lice (*Dactylopius coccos* Costa) that live on a cactus (*Opuntia coccinellifera*), in Central America, the Mediterranean Area and the Canary Islands. Chemically, the dye of kermes and cochineal are similar.
● Carmin is the alumina lake of cochineal and is approved as food dye and in cosmetic and pharmaceutical applications.

Naphthoquinones:
● Henna is the dye gained from *Lawsonia inermis* L. (Lythraceae), grown in India and Egypt and isolated by extraction with hot water. Once a textile d., it is still in use as hair dye and in other cosmetic applications.

Flavones:
● Flavones are natural yellows of a phenyl benzopyrone structure, present in plants as glycosides or tannic acid esters. Quercetin and its derivative quercitrin and some other derivatives are approved in food applications. Yellow wood extracts were used as leather d. (→leather auxiliaries).

Dihydroprans:
● Haematin and haematoxylin are the coloring matter of logwood and have lost importance as a natural leather dye (→leather auxiliaries).

Anthocyanidins (flavylium salts):
● Cranberry and grape extracts yield two approved food d. with this basic structure.
● Betanin is obtained from the red beet (*Beta vulgaris*) and is approved for use in food and cosmetics.

Indigoid dyes:
● Tyrian Purple the most expensive dye (main component is dibromo indigo), historically also called royal purple, is obtained from purpura shellfish (*Murex brandaris*) and is substituted by synthetic dye today (12 000 mollusks were necessary to obtain 1.4 g of the dye).
● Indigo (*Indigofera tinctoria* L.) has been completely substituted as the main blue cotton dye (blue jeans) by synthesis.

Carotenoids:
- Carotene, obtained from β-ionone, is a mixture of three isomeric, long-chain and conjugated compounds and used extensively as food dye.
- Anetto, derived from *Bixa orellana* (India, South America), is used in food coloring and contains Bixin as main compound.
- Saffron, also containing many conjugated double bonds, is gained from *Crocus sativus* and used not only as food dyestuff but also as spice and fragrance.
- →Safflower (*Carthamus tinctorius* L.), mainly grown as an oil crop, is also a source of a dye.

Chlorophyll:
The compound, extracted from green leaves, is not seen normally as a dye because it has no affinity to fibers. However, it is widely used for coloring soaps and edible fats and as an additive for cosmetics.

Lit: Kirk Othmer* (3.) **8**, 351

Dyestuffs, natural
→Dyes, natural

E

East India Resin
→Resins, natural

EC
→Ethylcellulose

Ecklonia
→Marine Algae

Eicosanoic Acid
→Arachidic Acid

5,8,11,14,17-Eicosapentaenoic Acid
Syn.: EPA; C20:5 (Δ 5,8,11,14,17)
G.: Eicosapentaensäure;
F.: acide eicosapentaenoïque

$$CH_3-CH_2-(CH=CH-CH_2)_5-CH_2-CH_2-COOH$$

m.w.: 302.46

The colorless oil is an important unsaturated (all double bonds are *cis*), essential (→fats and oils) fatty acid, which can be gained from →fish oil. It is a predecessor of →prostaglandin and →thromboxane, which show many important pharmacological properties. E.-containing medical preparations are of ecomomic significance, especially in Japan.

Lit.: Am.J.Cardiol. **59**, 155 (1986)
        [1553-41-9]

5,8,11,14-Eicosatetraenoic Acid
→Arachidonic Acid

cis-5-Eicosenoic Acid
Syn.: C20:1 (Δ 5)
G.: Eicosensäure; F.: acide eicosenoïque

$$CH_3-(CH_2)_{13}-CH=CH-(CH_2)_3-COOH$$

m.w.: 310.52

The unsaturated acid derives in high percentage from →meadowfoam.
Lit.: [7050-07-9]

Elaeis guineensis
→Oil Palm

Elaidic Acid
Syn.: Elaidinic Acid; *trans*-9-Octadecenoic Acid; C18:1 (Δ 9)
G.: Elaidinsäure; F.: acide élaïdique

$$CH_3-(CH_2)_7-CH=CH-(CH_2)_7-COOH$$

m.w.: 282.47 m.p.: 45 °C b.p.: 226–228 °C (1.33 kPa/10mm)

E. is obtained from →oleic acid by treatment (→isomerism) with nitrous acid.
Lit.: Ullmann* (5.) **A10**, 269
        [112-79-8]

α-Eleostearic Acid
Syn.: *cis,trans,trans*-9,11,13-Octadecatrienoic Acid; C18:3 (Δ 9,11,13)
G.: α-Eläostearinsäure; F.: α-acide éléostéarique

$$CH_3-(CH_2)_3-(CH=CH)_3-(CH_2)_7-COOH$$

m.w.: 278.44 m.p.: 45–51 °C

The highly conjugated fatty acid is contained in →tung and other wood oils. It is also contained (65%) in the oil of *Centranthus macrosiphon* (Compositae), an →annual plant.
Only the urea adduct is stable enough to be stored for a long period of time.
The undistilled acid is used in →alkyd resins.
Lit.: Ullmann* (5.) **A10**, 232
        [4337-71-7]
        [13296-76-9]

β-Eleostearic Acid
Syn.: *trans,trans,trans*-9,11,13-Octadecatrienoic Acid; C18:3 (Δ 9,11,13)
G.: β-Eläostearinsäure; F.: β-acide éléostéarique

$$CH_3-(CH_2)_3-(CH=CH)_3-(CH_2)_7-COOH$$

m.w.: 278.44 m.p.: 71 °C

The all-*trans* e. is obtained by isomerization (→isomerism) of the α-acid by UV radiation.
Lit.: [544-73-0]

Elephant Grass
→China Grass

Emulsions

G.: Emulsionen; F.: émulsions

E. are thermodynamically instable systems consisting of at least two immiscible liquids, one being dispersed in the other in the form of fine droplets. The phases described are normally referred to as "oil" and "water" or nonpolar lipophilic ("fat-loving") and polar hydrophilic ("water-loving") materials. Because every emulsion simplistically can be described as having an oil phase and a water phase two types of emulsion are possible. The first type is one in which discreet droplets of oil are dispersed in water, referred to as an oil-in-water (O/W) emulsion, while the second contains discrete droplets of water in a continuous oil phase and is known as a water-in-oil (O/W) emulsion. Irrespective of type, the discrete phase of an emulsion is known as the internal or dispersed phase, while the continuum liquid is referred to as the external or continuous phase. E. can be formed by simply applying external mechanical energy (e. g., rapid stirring). Once the stirring is stopped, the e., in absence of any stabilizing force, will break down into its oil and water phases. The reasons for the collapse can be explained in terms of the free surface energy of the system. When e. are formed by stirring, the interfacial surface energy of the system will rise enormously, due to the large increase in interfacial surface area. However, fundamental laws of thermodynamics state that any system will, in the absence of external influences, occupy the lowest energy state. Removal of the mechanical energy will therefore result in collapse of the emulsion, as it reverts to the lowest energy state by minimizing the interfacial area. Therefore, to produce a stable emulsion system, it is necessary to reduce the surface energy at the oil-water interface, thus preventing collapse of the emulsion into the two component phases when the mechanical energy is removed (stirring is stopped).

This can be achieved by the addition of →surfactants. Suitable surfactants have an affinity for both nonpolar (oil) and polar material (water) in the same molecule and are referred to as emulsifiers. The part of the emulsifier molecule that exhibits an affinity for oily materials is referred to as lipophylic, while the part that has an affinity for the water is referred to as hydrophilic. When the emulsifier is added to the system, it migrates to the oil-water interface, the lipophilic part of the emulsifier being orientated into the oil phase and the hydrophilic part being orientated into the water phase. This lowers the energy of the large interfacial area created by forming a huge number of small droplets from a single large drop.

A big number of ionic and nonionic emulsifiers are used in food (→food additives), drugs and →cosmetics. For water-in-oil e. more lipophylic, and for oil-in-water e. more hydrophylic emulsifiers are prefered. The selection of emulsifiers and other components for stabilizing e. is complicated and still largely empirical. Nevertheless, the hydrophile/lipophile balance (→HLB value) is the most useful approach for the selection of emulsifiers, especially for nonionic emulsifiers. A specific procedure, based on laboratory tests, should be established before preparing an e. on a plant scale. Order and rate of addition may affect the acceptibility of the product. An apt illustration is the preparation of an O/W emulsion by the inversion technique. Oil phase and emulsifier are blended in a tank. Water is then slowly added to the oil-phase with stirring. The initially hazy oil and emulsifier blend usually clears at first and then becomes cloudy again. As more water is added, the emulsion assumes a milky cast while the viscosity increases. At some point, called the inversion point, the viscosity suddenly decreases. At this point, the emulsion has changed from W/O to O/W. Further additions of water may then be made rapidly. If the oil is initially added to the water, a poor emulsion results unless enough of the right emulsifier is employed.

Lit.: Ullmann* (5.) **A9**, 297–339
    Kirk Othmer* (3.) **8**, 393–413
    Hilda Butler (ed.) "Cosmetics" Poucher's (9.), **3**, 534–555 Chapman & Hall, London/Glasgow/NY/Tokyo/Melbourne/Madras (1993)

Enamels

→Coatings

Enanthic Acid

Syn.: Oenanthic Acid; n-Heptanoic Acid; C7:0
G.: Önanthsäure; F: acide œnantique

$CH_3-(CH_2)_5-COOH$

m.w.: 130.18; m.p.: –10.5 °C; b.p.: 223 °C

Decomposition of →ricinoleic acid esters at 500–600 °C yields oenanthal (heptanal) which can be oxidized to enanthic acid.
The main product of this reaction is →undecylenic acid.

The aldehyde made from the acid is used as →fragrance raw material.

Lit.: Ullmann* (5.) **A10**, 271
        [111-14-8]

Endosperm

G: Endosperm; F: endosperme

E. is the tissue that surrounds the developing embryo of a seed and provides food for its growth. In →monocotyledonous plants, the e. is often the major part of the kernels.

Lit.: P.Raven et al. "Biology of Plants (5.) Worth Publish. NY (1992)

Enfleurage

→Pomade

Enzymes

G.: Enzyme; F.: enzymes

E. are the catalysts of microbiological reactions that take place in the broad range from the individual living cell to the large-volume fermenters of modern biotechnology. They are →proteins with a specific chemical structure, and they are often specific in their activity.

In living plant cells, they are the workhorses for producing plant ingredients, which make crops so interesting as RR. Modern →plant breeding and →genetic engineering (recombinant DNA techniques) provide the chance of improving yields and composition of industrial chemicals, tailored to specific needs by means of e.

The following classification and nomenclature ("International Union of Biochemistry") is used for e.:

1. Oxidoreductases: $AH_2 + B \rightleftharpoons A + BH_2$
2. Transferases: $AB + C \rightleftharpoons A + BC$
3. Hydrolases: $AB + H_2O \rightleftharpoons AOH + BH$
4. Lyases: $AB \rightleftharpoons A + B$
5. Isomerases: $ABC \rightleftharpoons BAC$
6. Ligases: $A + B + ATP \rightleftharpoons AB + ADP + P_i$

Each of these classes contains subgroups, the names of which are frequently identical with the trivial names. The systematic name of each enzyme consists of two parts: the first one characterizes the substrate (coenzyme), the second indicates the type of reaction with the suffix "ase".
Example: β-glucosidase:
Reaction: β-D-glucoside + H_2O $\rightleftharpoons$ D-glucose + aglucon
Name of e. : β-D-Glucoside-glucosidase (EC-number 3.2.1.21.)

Isolated from living cells, hundreds of different e. are in use for biotechnological processes today, either added directly to the broth or -in modern fermentation technology- as immobilized e. (fixed to a carrier). The latter have the advantage of higher versatility, less e. consumption, better yields and lower costs due to the possibility of continuous processing. The development in the last few decades in identification, isolation, modification and engineering of e. changed the image of biological processes from curiosity to large-volume reality. E. are among the most powerful tools in RR processing.
E. are used in the following three RR-related product groups:

Carbohydrates:
→Amylases split the glucosidic bond in →starch and →dextrins, amyloglucosidases degrade the starch chain from the nonreducing end, cellulases are able to hydrolize the β-glucosidic bond of →cellulose, while hemicellulases are able to degrade →hemicellulose, which are heteropolymers of xylans, arabans, mannans and galactomannans, but also →glucans. Others are dextranases (splitting →dextrans) and inulases (transferring →inulin to →fructose). Pectinases attack →pectins, either by depolymerization or by splitting the methyl ester link to yield pectic acid. Isomerases catalyze the transfer of glucose to fructose and are used in large-scale →isomerization for the production of high-fructose syrup. Glucose oxidases transfer glucose to →gluconic acid. Debranching e. are pullulanase and isoamylase. Lactases hydrolyze lactose to glucose and galactose. Some e. are used in the food industry (→food additives) to improve bread quality, in beet sugar refining and in fruit juice clarification.

Proteins:
→proteins are split by proteases. There are animal-derived proteolytic e., such as pancreatic proteinases, which are able to split collagen and other proteins, while pepsin and rennet are used in the food industry (cheese manufacture). Papain, one of the large-volume e., derives from the papaya fruit (*Carica papaya* L.) and is used in making cookies, in breweries and as meat tenderizer. Another interesting group of proteases are the alkaline bacterial proteases that are used in →detergent formulations to attack proteinic stains. Neutral bacterial e. find use in the leather industry, in breweries and in the production of hydrolysates.

Fats and oils:
Lipases are able to hydrolyze the ester bond in →fats and oils.

However, they are not used on a large-scale for fat →splitting but are able to split ester bonds on distinct triglyceride positions to yield specific →mono- or diglycerides. In the food industry, lipases are used to give flavor to butterfat-containing dairy and confectionary products.

Production figures are not easily available, but the European market (1992) was split into the following segments (based on value): 54% glucosidases; 22% proteases; 17% amylases; 7% lipases.

Lit.: Ullmann* (5.) **A9**, 389
      Ruttloff* (1994)

EPA
→Eicosapentenoic Acid

Epoxidation
G.: Epoxidierung; F.: époxidation

Epoxidation is the reaction of a double bond with oxygen to form an epoxy or oxirane group:

$$-CH=CH- + R-\underset{O}{\overset{\|}{C}}-OOH \longrightarrow -CH-CH- + R-\underset{O}{\overset{\|}{C}}-OH$$

R: H or CH_3

The oxygen is transferred -especially in e. of oleochemicals- by performic or peracetic acid. There are two principal methods:

- The peracid is formed in situ: the unsaturated compound to be epoxidized is dissolved in the carboxylic acid and hydrogen peroxide is added. Reaction takes place at 50–65 °C within 30–40 minutes. Formic acid is preferred because of its better reactivity. Acetic acid needs stronger acidic catalysts.
- The peracid is preformed: Acetic acid is reacted with hydrogen peroxide (90%) in the presence of 2% sulfuric acid. This equilibrium peracetic acid is relatively safe in handling and is used for reaction with the unsaturated compound.

E. is used for the manufacturung of epoxidized vegetable oils (soybean and linseed oil) and of epoxidized alkyl esters of oleic acid and tall oil fatty acid (→epoxides).

Lit.: Kirk Othmer* (3.) **9**, 251
      Ullmann* (5.) **A9**, 534

Epoxides
Syn.: Oxiranes
G.: Epoxide; F.: époxydes

E. are made by →epoxidation of unsaturated compounds. In context with RR, the following epoxides are of interest:

- epoxidized vegetable oils, mainly →soybean oil and, in smaller volume, linseed oil. They are colorless or pale oils with oxirane oxygen contents of 6–7% in soybean and 8–9% linseed oil.
- epoxidized alkyl esters obtained by epoxidation of alkyl esters (mainly butyl) of →oleic or →tall oil fatty acid. Oxirane oxygen contents of 3–4.5% are achievable.
- there are also some plants (→euphorbia and →vernonia) the oils of which contains epoxy fatty acids (→vernolic acid) in significant (>80%) quantities.

All these e. are used as stabilizers and plasticizers in PVC (→plastics additives). In this application, solid e. are sometimes desired. These can be made by →esterification of oleic acid from the modern high-oleic →sunflower with glycol, yielding a dioleate, which is then epoxidized.

E. are interesting intermediates for introducing functional groups by reaction with nucleophilic reagents[1]:

$$-CH-CH-\underset{O}{\diagdown\diagup} + HX \longrightarrow -CH-CH-\\ OH \quad X$$

X: –OH, –OR, –OOCR, –NH$_2$, –NR$_2$, –CN, –Cl

This reaction is a tool to introduce multifunctionality into an oleochemical. The methyl ester of oleic acid can be epoxidized and reacted in the above-mentioned manner. A large variety of polyols (e. g., by reacting epoxidized soybean oil with alcohols) are available for making polyurethanes with interesting properties.

E. can be rearranged to form keto-compounds (→keto fatty acids).

Lit.: Ullmann* (5.) **A9**, 531
      Kirk Othmer* (3.) **9**, 251
[1] Angew.Chem. Int Ed. (in Engl.) **27**, 41–62 (1988)

Epoxidized Oils
→Epoxides

13-Epoxy-*cis-cis*-9,12-Octadecenoic Acid
→Vernolic Acid

Erucic Acid
Syn: *cis*-13-Docosenoic Acid; C22:1 (Δ 13)
G.: Erucasäure; F.: acide érucique

$$CH_3-(CH_2)_7-CH=CH-(CH_2)_{11}-COOH$$

m.w.: 338.56; m.p.: 34.7 °C; b.p.: 255 °C
(1.33 kPa/10 mm)

E. derives from →mustard or →rapeseed oil by →saponifaction. Other sources are →crambe and →fish oil. E. is available in up to 94% purity.

Treatment of e. with nitrous acid yields the *trans* isomer → brassidic acid.

→Behenic acid and →behenyl alcohol are produced by →hydrogenation.

→Ozonolysis yields →pelargonic acid and →brassylic acid. The amide (→fatty acid amides) is used as antiblocking agent for polyolefin films (→plastic additives).

The world production is 25 000 mt/a.

Lit.: Ullmann* (5.) **A10**, 245
       Kirk Othmer* (4.) **5**, 147
       [112-86-7]

Erucic Acid Amide
→Fatty Acid Amides

Erwinia Gum
→Microbial Gums

Essential Amino Acids
→Amino Acids

Essential Fatty Acids
→Fats and Oils

Essential Oils
G.: Etherische Öle; F.: huiles essentielles

E. are a volatile materials (→terpenes) derived by either distillation or expression from odorous plant materials of a single botanical form and species with which they agree in name and odor. Plant materials used include fruit peels, flowers, seeds, fruits, leaves, twigs, grasses, rhizomes, roots and woods. The most currently used process for production is indirect steam distillation.

Lit.: Arctander*

Esterification
G.: Veresterung; F.: estérification

E. is the reaction of an organic or inorganic acid or anhydride with an alcohol according to the general formula:

$$R^1-COOH + R^2OH \leftrightarrow R^1-CO-OR^2 + H_2O$$

If water is eliminated from the reaction mixture, the equilibrium tends to the right side, and almost all acid or alcohol is transferred to the ester. In context with RR, the following esters should be mentioned: →Cellulose esters, →starch esters, →sucrose fatty acid esters, →sorbitan esters of fatty acids, →sulfosuccinates, → fatty acid esters, →waxes, →fats and oils and all other glycerides,

→alkyd resins, esters of inorganic acids with fatty alcohols, e.g., →fatty alcohol sulfates, →fatty alcohol ether sulfates, →fatty alcohol phosphates, phtalates, azelates and sebacates (→plastic additives), as well as the esters of →citric acid, →tartaric acid and →lactic acid.

Lit.: Kirk Othmer* (4.) **9**, 755

Ester Oils
→Lubricants
→Plastic Additives

Estersulfonates
→α-Sulfo Fatty Acid Esters

Ethanol
Syn.: Ethyl Alcohol
G.: Ethanol, Ethylalkohol; F.: éthanol

$$CH_3-CH_2-OH$$

m.w.: 46.07; m.p.: −114 °C; b.p.: 78.32 °C

E. for industrial purposes (solvent, chemical intermediate) is mainly made nowadays by petrochemical synthesis from ethylene.

The oldest fermentation process known uses modern technology today and produces e. from all kinds of carbohydrates (hexoses, pentoses, hydrolyzed starch, sucrose and cellulose and waste liquor from →paper pulping) in aqueous solution with the aid of living yeast (Species of *Saccharomyces, Candida* and *Kluyveromyces*):

$$C_6H_{12}O_6 \rightarrow 2 CH_3-CH_2-OH + 2 CO_2$$

Fermentation ends at ≈18% e. content. Further concentration is accomplished by →distillation (98%) or azeotropic distillation (100%).

For food applications, alcoholic beverages, cosmetics, fragrances and pharmaceutical formulations, fermentation alcohol is preferred. In food applications, such as wine or beer, the fermentation broth is used without any further concentration. In most countries, the price and volume of e. produced by fermentation is government-controlled. E. is used as a →fuel alternative or additive in some countries.

Manufacturing costs of petrochemical e. versus fermentation e. at a comparable plant size (both 50 000 mt/a), based (1986) on an ethylene price of 350$/mt and a molasses price of 45$/mt, are said to be roughly the same. Because →sucrose molasses is not always available in sufficient quantity, other more expensive sugar sources have to be used, which makes the fermentation e. price actually higher than the alternative.[1] Thus, if for any reason (e. g. as fuel additive) e. is desired, synthetic e. will be normally preferred because of its lower price.

There are 28 manufacturing plants (1994) for e. synthesis, which produce 2282×10^6 liter/a, and there are 723 fermentation plants with an output of 5895 liter/a worldwide. The fermentation capacity is, however, only partly used due to raw material campaigns. Sixty % of the world capacity is shared by Brazil, USA and India, whose governments are (were) supporting these activities heavily.

During the oil crisis of the 1970's and encouraged by government support, it was even considered to dehydrate fermentation-e. to ethylene and base the present petrochemistry on this starting material. Such ideas have been abandoned.

Lit.: Kirk Othmer* (4.) **9**, 812

[64-17-5]

[1] Ullmann* (5.) **A9**, 587

Etherification

G.: Veretherung; F.: étherification

E. is the formation of an ether, a compound with an oxygen between two similar or different alkyls, which may contain other functional groups. In context with RR, the following ethers are of importance: →cellulose ethers; →starch ethers; →guar gum ethers and →fatty alcohol ethers.

There are in principle three ways of etherification:

–the reaction of an alcoholic group with an alkyl halide:

$$R^1–OH + R^2X \rightarrow R^1–O–R^2 + HX$$

R^1 and R^2: alkyl   X: halogen

The reaction is carried out under alkaline conditions.

Especially for cellulose ethers, very strong alkali is used (alkali cellulose formation).

– the reaction (→alkoxylation) of the alcoholic group with epoxides, e. g., ethylene oxide and propylene oxide, with alkali as a catalyst

$$R^1–OH + CH_2–CH–R^2 \longrightarrow R^1–O–CH_2–CH–R^2$$
$$\underset{O}{\diagdown\diagup} \qquad\qquad\qquad \underset{OH}{|}$$

R^1: alkyl; R^2: alkyl or H

This reaction may continue because the new OH groups can react with further alkoxide, which is important in manufacturing of →fatty alcohol ethoxylates or cellulose ethers (e. g., →hydroxyethyl cellulose).

– less frequently used industrially is the alkali-catalyzed addition of activated double bonds to the OH group:

$$R–OH + CH_2=CH–Y \rightarrow R–O–CH_2–CH_2–Y$$

R = alkyl   Y = –CN; –CONH$_2$

Lit.: Ullmann* (5.) **A1**, 215

Ethoxylated Fatty Acid Ethanolamides
→Fatty Acid Ethanolamides

Ethoxylated Fatty Acids
→Fatty Acid Ethoxylates

Ethoxylated Fatty Amines
→Fatty Amines

Ethoxylated Sorbitan Esters of Fatty Acids
→Sorbitan Esters of Fatty Acids

Ethoxylated Starches
→Starch Ethers

Ethyl Alcohol
→Ethanol

Ethylcellulose

Syn.: EC

G.: Ethylcellulose, F.: éthylcellulose

E. is water-soluble in a →DS range of 0.7–1.7. Higher DS yields organic-soluble types.

For production, ethyl chloride is used as alkylating agent. The technology is similar to that used in manufacturing →methylcellulose. However, higher temperatures (110 °C) and longer reaction times (8–16 h) are necessary. Resulting salts are washed out with water.

Mixed ethers of ethyl and hydroxyethyl cellulose (EHEC) with an ethyl DS of 0.9 and a hydroxyethyl DS of 0.8–2 have properties like hydroxyethylated or hydroxypropylated MC and are used in similar applications. They have a gelling temperature of 65–70 °C in aqueous solutions. Due to the high degree of etherification, they are resistant to biodegradation.

E. and mixed ethers improve workability and moisture retention of cements and plasters in the construction industry. They show good performance as thickeners in latex paints (→coatings) and in organic-based →paint removers. Pure e. is used as an organic-soluble and compatable film-forming material in paints and →coatings, as base material for molded goods, →hot melt adhesives, and tablet binders and coating agents in pharmaceutical applications. Mixed ethers (EHEC) function as binders for →tobacco sheets.

Lit.: Ullmann* (5.) **A5**, 473
    Encycl.Polym.Sci.Engng.* (2.) **3**, 254
    [9004-57-3]

Eucalyptol

Syn.: Cineole; Cajeputol; 1,8-Oxido-p-menthane
G.: Eucalyptol, Cineol; F.: eucalyptol, cinéole

m.w.: 154.24; m.p.: 1–1.5 °C; b.p.: 176–177 °C

E. is a colorless, mobile liquid which solidifies in the cold. E. is one of the most widely distributed chemicals occurring in →essential oils, many of which belong to the eucalyptus species and are available in great quantities and at low cost. Therefore, synthesis of e. is uneconomical, and the material is produced by isolation from eucalyptus oils, mainly from → *Eucalyptus globulus* and *Eucalyptus smithii*. By themselves they are used in low-cost applications, such as masking odors for industrial purposes.

E. has a pungent, fresh camphoraceous and cool odor and a sweet and fresh, camphoraceous taste with a cool mouthfeel, which accounts for it being extensively used in →flavor compositions for oral hygiene preparations (dental and mouthwash preparations) as well as in candies and chewing gums.

In perfumery, its freshness makes it useful for a wide range of fragrance types, ranging from herbal to medicinal, where it adds lift and an impression of cleanliness.

Lit.: Arctander*
    The H&R Book*
    [470-82-6]

Eucalyptus globulus

Eucalyptus smithii

→Eucalyptol

Eucalyptus Globulus Oil

G.: Eucalyptus Globulus Öl;
F.: essence d' eucalyptus globulus

E. is steam-distilled from the fresh or partly dried leaves of *Eucalyptus globulus* (Myrtaceae). This oil finds widespread use in pharmaceutical preparations (e. g., chest rubs, inhalers etc.) and at the low-cost end of flavors and fragrances, where the use of →eucalyptol is prohibited by cost.

Odor and taste are similar to →eucalyptol with some herbaceous, aldehydic notes appearing in the topnote and a resinous note at the →drydown stage, especially with non rectified oils.

The tree originates in Australia. Main production countries now include Australia, Brazil, Spain, Congo and Portugal.

Lit.: Arctander*

Eugenia caryophyllata

→Clove Bud Oil

Eugenol

Syn.: 2-Methoxy-4-allyl-phenol
G.: Eugenol; F.: eugénol

m.w.: 164.2; m.p.: –9 °C; b.p.: 225 °C

Eugenol is one of the few important fragrance and flavor chemicals that is actually an isolate from an →essential oil. It is extracted or distilled from →clove bud oil and clove leaf oil. It is also a main constituent of cinnamon leaf oil, pimento oil and bay oil as well as a main contributor to the smell of carnation flowers. This phenol is widely used in flavors for spice blends, seasonings, pickles as well as in pharmaceutical and alimentary preparations such as mouthwashes, toothpastes, candies and chewing gums. Another application is in tobacco flavorings, particulary in Indonesia, where spiced cigarets are highly appreciated.

In perfumery, the material is used in a multitude of floral fragrances to give warmth and richness and to complement the heavy sensuous notes of oriental and tobacco fragrances.

E. shows also some pesticidal activity against bacteria and fungi.

Lit.: Ullmann* (5.) **A8**, 558
    [197-53-0]

Euphorbia

Syn.: Caper Spurge; Mole Plant
Euphorbia spp.
Euphorbiaceae
G: Wolfsmilch; F: euphorbe

Two representatives of the e. family are *E. lathyris* L. (caper spurge) and *E. lagascae* Spreng. Both have been suggested as potential new crops for the production of industrial hydrocarbons and oleochemically useful sources of fatty acids.

Other related crops of commercial importance in the family are →castor (*Rizinus communis* L.), →manioc (*Manihot esculenta* Crantz), →candelilla (*E. antisyphilitica, E. cerifera*) and the →rubber tree (*Hevea brasiliensis* H.B.K.)

E. originated in the Mediterranean area, from where it spread throughout Europe. During the Middle Ages, the oil of the seed was widely used as lamp oil. The whole seed or extracts of the seed meal had some importance as an officinal herb. Later, e. was introduced to the New World. Typical for all e. species is the milky sap that is excreted from the plant wherever its stem, leaves, flowers or any other organ have been damaged. This milky sap has been considered for the production of industrial hydrocarbons by using the dried whole plant as a source of renewable energy (so-called gasoline plants).

Besides its application as lamp oil, the use of *E. lathyris* as an oil crop is comparatively new and has not been developed for commercial production. This interest resulted from its high oil content (50%), of which 80–90% is oleic acid. Because of this high-oleic acid content, the oil has a higher value than today's most common sources of oleic acid, such as →lard and →tallow. In this sense, e. is a substitute or competitor of other high-oleic acid sources, such as →sunflower or →rapeseed. *E. lathyris* may be grown either as a winter or spring crop. Due to its comparatively long growing period, the development of winter-hardy forms could be beneficial. At full maturity in many e., seed pods of the wild species open with an audible snap, shattering the seed. Since 1985, the first non-shattering cultivar of *E. lathyris* was given cultivar protection in Germany. Also canopy establishment and winter hardiness will have to be improved, as well as early maturity and determinant growth. In *E. lathyris* the milky sap of all green parts of the plant has cocarcinogenic effects, and it can also cause acrid injury to eyes and skin. Because other species of the family do not produce such irritants,

elimination of these chemicals should be possible. Yields of 20 dt/ha are reported, suggesting that 1 mt of oil per ha might be attainable. For applications of the oil, see also →sunflower.

E. lagascae Spreng. was suggested as an interesting oleochemical raw material because of its high content (60–70%) of epoxy fatty acids (→vernolic acid) in the seed oil.

Oil extraction of the seeds of E. is possible by conventional methods that are applied for soft seeds, such as cold →pressing or solvent →extraction.

Lit.: W. Hondelmann, "Domestication of *Euphorbia lathyris* as a potential new oil crop" in: "New Crops for Temperate Regions", K.R.M. Anthony et al. (ed.), 206–221 (1993) Chapman and Hall, London

Euphorbia antisyphylitica
Euphorbia cerifera

→Candelilla Wax

Evening Primrose

Oenothera biennis L.
Onagraceae
G.: Nachtkerze; F.: onagre

E. is a herbaceous plant with →biennual mode of growth. E. was introduced in 1614 from Virginia, USA to European botanical gardens and has since then become a wild plant. During the reproductive second year, e. forms tall stems, each bearing numerous yellow blooms, which as the common name implies, open in the evening and are pollinated by moths. After pollination, a long seed pod is formed, which is divided longitudinally into 4 compartments, each containing many tiny seeds (→seed weight, 0.5 g).

Commercial production of e. has only started on a small scale by pharmaceutical companies in UK in the late 1960's, when the potential of e. was discovered as a reliable supplier of gamma →linolenic acid, similar to →black currant and →borage.

Agronomic production still has problems due to many remaining wild plant characteristics, such as seed dormancy, indeterminate flowering, high variability in yield characteristics and lacking disease resistance. It is reportedly also grown on a contract basis in countries of the southern hemisphere. The oil content of the seeds is around 22% with about 72% C18:3 and 9% of gamma linolenic acid.

Lit.: A.Uzzam,"Borage Oil and other Plant Oils Rich in γ-Linolenic Acid". Rev.Fr.Corps Gras **35**, 501–504 (1988)

Evernia prunastri
→Oakmoss Absolute/Resinoid

Expelling

Syn.: Pressing; Expressing
G.: Pressung; F.: pression

Modern processing of oilseeds is normally carried out by a combination of expelling (pressing) and →extracting. Oil content of the cake after e. is normally ≈20%. The following extraction brings it down to 1–2%.

Some seeds have to be dehulled to reduce fiber content of the meal. Most seeds are heated or broken in rolls to facilitate e. Various pretreatment techniques (e. g., heat) are known to adjust the process to the specific seed material. The subsequent expelling can be done either batchwise or continuous by using levers, wedges or screws. Continuous screw presses are widely used today.

The term expression is used more in the fragrance industry, where citrus peels are expressed to obtain →essential oils.

Lit.: Ullmann* (5.) **A10,** 192

Explosives/Propellants

G.: Sprengstoffe; F.: explosifs

There are primary explosives (detonator, initiator) of high brisance and secondary explosives. Only two of the latter are based on RR: →Glyceryl trinitrate and →cellulose nitrate.

Glyceryl trinitrate, commonly called nitroglycerol, gelatinized with nitrocellose, is primarily used as an explosive in dynamites. Its importance has somewhat diminished due to substitution by or combination with nitroglycol. G. is used as a plasticizer for →cellulose-nitrate in double and multibase propellants. It is only applied when desensitized by coating with waxes (formerly →bees wax but now paraffin) or by addition of lubricating or antistatic agents, such as →gyceryl triacetate (triacetin), or when compounded with cellulose nitrate. Otherwise it is very sensitive to shock, impact and friction.

The sensitivity of nitroglycerol decreases with temperature. However, frozen it is sensitive to friction due to intercrystalline contact. It decomposes above 60 °C. It exhibits a low-velocity propagation regime with a detonation rate of about 2000 m/s compared with its high-energy detonation rate of about 7800 m/s.

Glyceryl trinitrate is sometimes used in extinguishing the fire of burning oil wells.

Cellulose nitrate, also called nitrocellulose, has exceptional explosive properties due to its polymeric and fibrous structure. It combines mechanical strength with readily available energy for explosives and rocket propellants. The degree of nitration is important for its explosive properties. A maximum of 13.5% N is reached with mixed acid, which is sufficient for uses as explosives and propellants. Nitrocellulose can be adjusted to distinct applications by selecting the DS and by compounding.

Cellulose nitrate decomposes at 125 °C. Addition of week alkaline compounds increases stability.

Other kinds of explosives are **ammoniumnitrate based blasting agents** in mining and construction. The largest volume consists of ammoniumnitrate and 5% fuel oil (ANFO). The only use of RR-based products is in aqueous explosives, such as ammoniumnitrate/aluminum powder slurry or gel blasting agents. Thickeners, such as →guar gum and →carboxymethyl cellulose, sometimes together with a gelling agent, are added.

Explosive emulsions are a more modern form of aqueous e. and may contain RR-based emulsifiers (e. g., →sorbitan ester).

Lit.: R.Meyer:"Explosives" 2nd ed. Verlag Chemie Weinheim NY (1981)
Kirk Othmer* (4.), **10,** 21

Expression
→Expelling

Extraction
G.: Extraktion; F.: extraction

E. is the eluation of a (valuable) ingredient from a complex material by a suitable solvent. There are two types of e.: liquid/solid and liquid/liquid. There are many different technologies and equipment applied (batch and continuous). A recent method is e. with supercritical fluids (mainly CO_2). There are already large-scale uses, e. g., in the e. of hops and of caffeine from coffee.

In context with RR, there are several applications of extraction technology:

Fats and oils are extracted from seeds with a content of less than 20% or from cakes remaining after →expelling with 15–25% oil. For fats and oils, hexane is the normal extractant. If higher-temperatures are needed higher boiling solvents, such as heptane are used (e. g., extraction of castor oil). Steam is used for either direct or indirect heating. In many cases, the counter current principle is applied, i. e., fresh solvent is contacted with the already most extracted material. Oil concentrations in the solution are finally 35 % or more. The solvent is distilled off in vacuum and recycled. The remaining meal is used in most cases as animal feed.

Extraction is also used in oleochemical technology to regain absorbed vegetable oils from spent bleaching earth.

Another large application is the e. of **sugar from beets** (→sugar beet processing).

In the **flavor and fragrance**[1] industry, e. is also an important technology, next to →distillation, to gain desired materials from natural sources. E. is mainly used when distillation and the involved heat would affect the odor or taste profile too much (e. g., flower petals such as rose or jasmine) or when extracts cannot be obtained otherwise (→oleoresins from trees). For some natural sources, both methods can be used, and the resulting distillation product is used for its economic advantage and higher portion in volatile ingredients, while the e. product is used for its better reflection of the natural smell. Fragrance and flavor materials obtained by e. are →absolutes, →concretes, →resinoids.

Lit.: Ullmann* (5.) **A10**, 194
Kirk Othmer* (3.) **9**, 721

[1] Arctander*
The H&R Book*

Extrusion-cooked Starches

G.: Extrudierte Stärke; F.: amidon (fécule) extrudé

E. are physically and chemically →modified starches, obtained by extrusion of granular starch with or without addition of chemical reagents.

They swell and solubilize on contact with cold or hot water.

Extrusion cooking is a technique for production of →pregelatinized starches, of →starch esters, of slightly hydrolized starches, of biodegradable →starch plastics and of →thermoplastic starches.

Production of pregelatinized starches is performed frequently with starches of 12–22 % moisture content in single-screw or twin-screw extruders between 90 and 200 °C. During the residence time of 30–45 seconds the starch may undergo several changes: Disappearance of the granular and crystalline structures, chemically caused depolymerization and melting. The pressure, developed behind the discharge die, finally forces the material through the die. Due to the pressure drop, the melt is more or less expanded and solidifies.

Functional properties depend on the kind of →starch, pH, water content, temperature zones and residence time in the extruder. Additional effects are possible by added reagents, which cause chemical modification or hydrolytic splitting.

Compared with pregelatinized starches from roller dryers, e. show slow dispersibility and lower paste viscosity in cold aqueous systems but rapid dispersion and higher solubility in hot water without lumping. The paste stability is improved, but →retrogradation and setting to gels are largely suppressed. Energy costs for extrusion cooking are lower.

Applications are similar to those listed for →pregelatinized starches with the exception of food uses.

Lit.: T.Galliard (ed.) "Starch: Properties and Potential" 79–114 John Wiley & Sons, Chichester/NY/Brisbane/Toronto/Singapore, (1987)
W.Wiedemann, E.Strobel starch/stärke **43**, 138–145 (1991)

F

Factice

Syn.: Vulcanized Oil
G.: Faktis; F.: factice

F. is a rubberlike colorless or yellowish to brown product made by reaction of vegetable oils, such as →rapeseed oil, →tall oil, →castor oil and →linseed oil, with sulfur at 150–160 °C. Sulfur dichloride reacts at room temperature (sulfur-f.) and yields less colored products used in erasers. Different types and qualities of f. are commercially available. The reaction of the above-mentioned oils with oxygen or peroxides at elevated temperatures yields oxygen-f., and with isocyanates they produce isocyanate-f.

All these are crosslinking reactions.

Once f. was developed as a rubber substitute, extender or filler, but today it is used more as a processing aid, which increases the tolerance for liquid plasticizers. Many rubber goods (e. g., erasers) contain f.(→rubber chemicals).

Lit: D.Swern, "Bailey's Industrial Oils and Fat Products" (3.) Interscience Publ. Co. Inc (1964)
Encycl.Polym.Sci.Engng.* (2.) **12**, 679

False Flax

→Camelina

Family

→Taxonomy

Fats and Oils

G.: Fette und Öle;
F.: corps gras et huiles

Fats and oils are triglycerides, i.e., esters of →glycerol

$$CH_2-O-CO-R^1$$
$$|$$
$$CH-O-CO-R^2$$
$$|$$
$$CH_2-O-CO-R^3$$

with three molecules (R^1-R^3) of mostly even-numbered →fatty acids of various chainlength (8–22 C-atoms) and unsaturation (one or more double bonds). Some of these fatty acids may have also functional groups (e. g., hydroxy or epoxy groups). F. contain 87–90% fatty acids and 10–13% glycerol, depending on the type of fatty acids.

Natural f. are the depot of energy for the first stage of plant growth from seed and for animals in times of malnutrition. As nature-derived products they contain various by-products in small amounts, some of which have significant biological functions:

Fatty acids: due to enzymatic activity, a considerable amount (5% and more) of free fatty acid (ffa) may be present. Producers and suppliers try to keep the level as low as possible because they have a negative influence on quality and price.

Phospholipids: depending on their origin f. contain small amounts of →lecithin and other phospholipids, which are removed and regained during →degumming. They are valuable ingredients and used in edible fats as oil soluble emulsifiers. Soybean oil is rich in phospholipids.

Antioxidants: →Tocopherols (Vitamin E) are natural antioxidants that protect vegetable-based f. from autoxidation and rancidity. It is isolated from various oils during →deodorization and is used for various applications. Main sources are the deodorizer distillates of →soybean oil.

Pigments: Major pigments are carotenoids, which are removed during alkaline refining.

Vitamins: Aside of the tocopherols, some vitamin A (palm oil) and Vitamin D (fish oil) are found.

Sterols: Soybean oil (deodorizer distillates) contains a considerable amount of →phytosterols (sito- and →stigmasterol, →steroids), which are isolated and used in pharmaceuticals. Animal fats contain →cholesterol, which can be recovered from the distillation residues of fatty acids.

Properties and origin:

Customarily, fats are solid at room temperature while oils are liquid. The borderline of this definition is not distinct.

They are derived from vegetable, animal and marine sources. Detailed composition and physical properties can be found under the respective key words.

→babassu oil, →butter fat, →castor oil, →chinese tallow, →cocoabutter, →coconut oil, →corn oil, →cottonseed oil, →fish oil, →linseed oil, →oiticica oil, →olive oil, →palm oil, →palmkernel oil, →peanut oil, →rapeseed oil, →rice bran oil, →safflower oil, →sesame oil, →shea butter, →soybean oil, →sunflower oil, →tallow, →tung oil.

Some f. are grouped together, depending on their origin or chemical nature, especially in trade terminology. The following are of importance:

- **Milk fats**: (→butterfat).
- **Lauric oils** are also called medium-chain triglycerides (MTC). Such oils are →coconut, →palmkernel, →babassu oil and some →cuphea species. Fatty acid chainlength is in the range of C_8 and C_{14} with a peak at C_{12}. They have low unsaturation. They are used in →food and in →soap manufacture. Derivatives find outlets in →cosmetics.
- **Vegetable butter**: (→cocoabutter, →shea butter).
- **Animal fats** [→tallow (beef and mutton) and →lard] are used for edible purposes and are a cheap source of C_{16} and C_{18} saturated and unsaturated fatty acids.
- **Oleic/linoleic oils** are →cottonseed, →peanut, →olive, →corn, →safflower, →tall and →sunflower oils. They are mainly used for edible purposes. Their fatty acids are C_{18} mono- and di-unsaturated.
- **Erucic oils**: →rapeseed and →crambe oil.
- **Linolenic oils** are from →soybean and →linseed. They contain large amounts of fatty acids with more than two double bonds. Their main use is in technical applications, especially →drying oils.
- **Conjugated oils** are →tung and →oiticica oil, which are used as →drying oils almost exclusively because of their high content of conjugated double bonds and their high reactivity.
- **Marine oils**: →fish oil. →Whale oil is another marine oil which has lost importance in the market due to international restrictions on killing whales.
- **Essential Fatty acid containing oils** are triglycerides with many (multiple) unsaturated fatty acids, so-called essential fatty acids, which are reducing artherosclerotic and other pathological problems via proper human nutrition (→linolenic acid). Important triglycerides are linoleic acid containing oils and omega-3 fatty acids contained in →fish oil (→eicosapentaenoic acid).

Manufacture, processing and refining:

→Triglycerides are normally not produced synthetically except when defined structures are desired (→glycerides). An other exception was the production of fats during the 2nd world war in Germany, where synthetic fatty acids (paraffin oxidation) and glycerol (fermentation) were esterified to yield fat for edible purposes.

Fats and oils are available in a large variety and in huge quantities from animals and vegetables. Therefore, manufacturing is exclusively concentrated on RR. Biotechnological research has shown that f. can be made from starch and sugars by certain yeasts.

The following processes are used for isolation:

Oil from seeds are normally gained by pressing (→expelling). Some seeds have a high hull content (e. g., cottonseed) and are dehulled prior to pressing. →Extraction with hexane follows. Three products result: the meal, the hull (both are valuable animal feeds) and the oil. Many different processes and types of equipment are used in these operations.

Oil from fruit pulps (e. g., palm) are processed in the same way. To obtain good qualities (low color and ffa), it is necessary to organize the whole procedure well, starting with the harvest, fast fruit handling, prompt sterilization and finally, work up procedures. Today, producers are able to offer high quality reliably.

Animal fats are recovered by rendering. This is a melt process, which separates the fat from the tissues. Dry rendering is done by heating the crude in vessels while wet rendering is the direct blowing of steam into the material. Many variations in technology and machinery are known. To separate residual cake from water and oil, pressing and centrifuging are common.

Marine oils are processed the same way, with wet rendering being preferred.

Several methods of purification are necessary to improve the quality of raw oils:

Degumming is a treatment with water (steam), which is mainly carried out with oils of high phospholipid content, e. g., soybean oil.(→lecithin).

Alkali-refining is used to eliminate ffa. Almost all edible oils are processed in continous plants today. The added NaOH forms salts, and the soapstock ("foots") is separated in a centrifuge from the oil, which is then washed and dried. Soapstock is a low-cost and valuable source for fatty acids.

Steam-refining is done to remove the ffa by steam→distillation.

Bleaching is accomplished by treating colored oils with acid-activated earth (bentonite clays). Hydrogen peroxide bleach is used only in special cases.

Winterization is a process of cooling an oil, followed by filtration or centrifugation, to separate solids from the liquid oil (→crystallization).

Deodorization is applied to most f. (exception: olive oil) to improve color and taste. This is accomplished by high-temperature (210–275 °C) steam vacuum (1–6 mm)→distillation. Odorous and bad-tasting substances, such as hydrocarbons, ketons and aldehydes are removed together with ffa, sterols and some of the tocopherols. The distillate is a valuable source for sterols and vitamin E. Modern equipment and advanced technology for batch and continuous deodorization are available.

Chemical modification:
→Oleochemistry, which is the modification and derivatization of f., is a wide field of industrial activity in the area of RR. The following "next steps" are of importance:
– Hardening (→hydrogenation)
– Splitting (→hydrolysis,→transesterification,→ saponification)
– Reaction with H_2O_2 (→epoxidation)
– Reaction with atmospheric oxygen (→drying oils)
– Sulfonation (→sulfated f.)

Economic aspects:

Production in 1994 ($\times 10^6$ mt)	total oilseeds	total oils
USA	80.85	8.44
China	42.37	5.94
India	23.50	5.44
Brazil	26.59	3.93
Argentina	18.30	3.39
FSU	8.92	1.89
Canada	9.60	1.19
World	259.57	67.75

While over 30% of the US production of f. is used in nonfood applications, only 14% of the total world production is used industrially. The reminder is used for food applications.

Of all f. 68% are of vegetable origin, and 30% are derived from animals and 2% from fish.

Lit.: Kirk Othmer* (3.) **9**, 796
      Ullmann* (5.) **A10**, 173

Fatty Acid Amides
G.: Fettsäureamide; F.: amides d'acides gras

Unsubstituted (primary) f. have the following structure

R–CO–NH$_2$ R: C$_{7-23}$-alk(en)yl

With the exception of →fatty acid ethanolamides and bisamides, mono- and di-substituted (secondary and tertiary) amides (R–CO–NHR and R–CONR$_2$) are of low commercial interest.

Bisamides have the following structure:

R–CO–NH–(CH$_2$)$_x$–NH–CO–R
R: C$_{15-17}$-alk(en)yl; x: 2 or more

Primary f. are produced by reacting fatty acids with ammonia (200 °C, 345–690 kPa/3.5–7 bar, 12 h). It is also possible to use methyl esters as starting material (220 °C, 12.4 MPa/125 bar, 1 h). F. are (not always isolated) intermediates in →fatty acid nitrile manufacture.

Bisamides are made by reacting a diamine (mostly ethylenediamine) with the fatty acid (mostly →stearic acid) or its methyl ester (180 °C, 6 h).

F. and the bisamides are used primarily for their lubricating properties, their nonreactivity, incompatibility and their surface-active nature due to strong hydrogen bonds. They provide slip and antiblocking properties to polymer surfaces, especially polyolefin films, and are used (amides of →oleic acid and →erucic acid) in this application in large volume. F. are also used as water repellents for textiles, e. g., quaternary salts made by reaction of f., formaldehyde and pyridine are applied to the fabric and heated to form water repellant-finishes (→textile auxiliaries). They are also used as →release agents and are added to formulations of →lubricants and as modifiers in coatings (→coating additives), which change the surface properties.

Bisamides are generally used in the same applications as primary f. and are even more efficient in terms of lubricity and antblocking properties.

In 1986, approximately 55 000 mt of f. were produced worldwide, with the bisamides having the largest share.

Lit.: Kirk Othmer* (4.) **2**, 346

Fatty Acid Amino Amides
G.: Fettsäureaminoamide;
F.: amino amides d'acides gras

F. are formed by reaction of fatty acids and diamines and have the following structure:

R^1–CO–NH–(CH$_2$)$_2$–NH–R^2
R^1: C$_{7-21}$ alkyl; R^2: R^1 or H;

→Dehydration yields →imidazoline derivatives
Lit.: Falbe* p. 113 (1987)

Fatty Acid Anhydrides
→Fatty Diketenes

Fatty Acid Bisamides
→Fatty Acid Amides

Fatty Acid Cyanamides

Syn.: Amide Soaps

G.: Amidseifen;

F.: cyanamides d'acides gras

$$R-CO-CH_3 + NaHN-CN \xrightarrow{100-250\,°C}$$
$$R-CO-N^- -CN + Na^+ + CH_3OH$$

R: C_{7-21} alkyl

F. have properties similar to →soaps. They show better solubility and are less sensitive to water hardness. They were proposed as soap substitutes for many years, but due to expensive manufacturing procedures (from acid chlorides), they never gained market importance. However, recent process developments starting from methyl ester, provide new opportunities to this interesting class of →surfactants.

Increasing importance is likely.

Lit.: Falbe* 81-82 (1987)

Fatty Acid Esters

G.: Fettsäureester; F.: esters d'acides gras

Aside of the important →fatty acid methyl esters and the various →glycerides, there are wax esters (→waxes), →sorbitol esters, →sucrose fatty acid esters, ester oils (→lubricants) and various esters of fruit acids with mono/diglycerides. Some other f. of significance include:

Isopropyl myristate (IPM) is a colorless and odorless liquid, (b.p. 137-175 °C, 0.66kPa/5 mm), an ester of →myristic acid and isopropanol, which is produced by →esterification or →transesterification. It has oxydation stablity and is mild to the skin and therefore widely used as a solubilizer in →skin preparations.

Butyl stearate (colorless liquid, m.p. 19.5 °C, b.p. 220-225 °C) is used as a solvent in many applications and as plasticizer and lubricant in →plastic additives.

Vinyl laurate or stearate produced by →transesterification or by reaction of lauric or stearic acid with acetylene, are sometimes used as plasticizing monomers in vinyl polymers (→polymerization additives).

Etyleneglycol mono-[111-60-4] and di-[627-83-8] stearate are used as opacifiers in shampoos (→hair preparations).

Lit.: W.C.Eisenhard in "Fatty Acids in Industry" R.W. Johnson, E.Fritz (eds.), p. 139 Marcel Dekker Inc. NY/Basel (1989)
Kirk Othmer* (4.) **9**, 781

Fatty Acid Ethanolamides

G.: Fettsäureethanolamide;

F.: éthanolamides d'acides gras

If fatty acids or their methyl esters are reacted with ethanolamine or diethanolamine, ethanolamides are resulting, which can be ethoxylated (→alkoxylation) further.

$$R-CO-N \begin{array}{l} (CH_2-CH_2-O)_nH \\ \\ (CH_2-CH_2-O)_mH \end{array}$$

R: C_{11-17} alkyl
n: 1 or 2 or more
m: 0 or 1

F. are widely used nonionic →surfactants, mostly in combination with others.

Monoethanolamides are only slightly soluble in water and have excellent alkaline resistance. They are used in heavy-duty powder →detergents as foam stabilizers and rinse improvers.

Diethanolamides are used in light-duty liquid detergents, where higher alkalinity is not desired but better solubilities. Large volumes are also consumed in personal care products, such as shampoos and toothpastes.

They find also use as emulsifiers and corrosion inhibitors in →metalworking fluids, in →dry cleaning and textile processing.

Main products are the diethanolamides of the C_{10-14} range.

77 000 mt of f. were produced in 1986 worldwide.

Lit.: Kirk Othmer* (4.) **2**, 350

Fatty Acid Ethoxylates

Syn.: Fatty Acid Polyglycolesters

G.: Fettsäureethoxylate; F.: acides gras éthoxylés

$$R-CO-(O-CH_2-CH_2)_x-OH \quad R: C_{7-23} \text{ alkyl}$$

F. are made either by alkaline-catalyzed ethoxylation(→alkoxylation) of fatty acids or by their →esterification of preformed polyglycols. They contain, depending on the reaction conditions, some level of diesters and polyglycols. Pure monoesters are obtained via borate esters. F. are sensitive to →hydrolysis.

They are used as emulsifiers in →cosmetic products and find wide application in the textile industry as emulsifiers for processing oils, antistatic agents, softeners and fiber lubricants (→textile auxiliaries).

Lit.: Kirk Othmer* (3.), **22**, 368

Fatty Acid Imidazolines

→Imidazoline Derivatives

Fatty Acid Ketenes
→Fatty Diketenes

Fatty Acid Methylesters
G.: Fettsäuremethylester;
F.: esters méthyliques d'acides gras

By far the most important →fatty acid esters are the methyl esters.

R–COO–CH$_3$ R: C$_{7-21}$ alkyl

There is almost no specific, pure methyl ester of commercial importance. Most are used as broad cuts that still bear the fatty acid pattern of the original →fats or oils.

Up to a range of C$_{16}$ they are liquid at r.t. They boil 20–40 °C lower than the corresponding fatty acids and are easier to distill and fractionate and do not need stainless-steel equipment. F. are therefore frequently used as a "masked" form of fatty acid.

F. are made on a large scale either by →esterification or by →transesterification.

Esterification is carried out at 200–250 °C under slight pressure with acidic or alkaline catalysts. An excess of methanol and an elimination of water during reaction promotes ester formation. Batch and continous processes are in practical use. Esterification is used especially if fractionated or special fatty acids are the starting material to make defined methyl esters (e. g., methyl oleate from oleic acid from high-oleic →sunflower).

Transesterification is the predominant process for manufacturing methyl esters. Triglycerides with no free fatty acids react under mild conditions (50–70 °C, atmospheric pressure). Because most fats and oils have considerable contents of ffa, it is necessary to either eliminate them or chose transesterification conditions that allow also esterification of ffa. In technical processes, the raction is carried out at 220–250 °C and 9–10 MPa (90–100 bar). Under these conditions, fats and oils that contain up to 20 % ffa can be transferred into methyl esters. The resulting esters can be purified by →distillation.

F. are primarily used as intermediates for the commercial production of →fatty alcohols. They are also an important starting material to make →fatty acid ethanolamides and are the base for making a relatively new class of →surfactants: →α-sulfo fatty acid methyl esters. F. react with ethyleneoxide in the presence of hydrotalcid to yield nonionic surfactants by inserting EO in the ester function.

Especially the palm oil and rapeseed fatty acid methyl esters are considered -and there are many large-scale tests running- as diesel →fuel alternatives.

They are considered and already partly used as →lubricants in engines running in sensitive environments. In agriculture f. are used as shoot inhibitors for →tobacco. There are also attempts for using f. as a substitute for kerosene as a solvent (e. g., in →insecticides, household products and coatings). All these applications focus on their good biodegradability and, therefore, higher ecological acceptance.

World production in 1985 was ≈ 400 000 mt with strong annual increases. Volumes may increase extremely if one or more of the modern applications, mentioned above, materialize.

Lit.: Falbe* 34–36 (1987)
        Ullmann* (5.) **A10,** 281

Fatty Acid Nitriles
Syn.: Fatty Nitriles
G.: Fettsäurenitrile; F.: nitriles d'acides gras

F. are important intermediates in the manufacture of fatty amines and amides. They are produced from fatty acids and ammonia in the presence of dehydrating catalysts (Al$_2$O$_3$, ZnO) in batch liquid-phase or continuous vapor-phase reactors above 280 °C. Ammonium soaps and amides, formed as intermediates, are readily dehydrated:

R–COOH + NH$_3$ → R–COO–NH$_4^+$ →
$$R–CO–NH_2 + H_2O$$
R–CO–NH$_2$ → R–CN + H$_2$O

R: C$_{7-21}$ alkyl

F. are used directly or after fractionation for →hydrogenation to amines. There is little use of f. other than as intermediates, but they may find some utility as surfactants and corrosion inhibitors.

US production capacity in 1980 was estimated to be in the range of 140 000 mt.

Lit.: Ullmann* (5.) **A17,** 363

Fatty Acid Polyglycol Esters
→Fatty Acid Ethoxylates

Fatty Acid Propylene Glycol Esters
→Propylene Glycol Esters of Fatty Acids

Fatty Acid/Protein Condensates
Syn.: Protein/Fatty Acid Condensates
G.: Fettsäure/Eiweiß-Kondensate;
F.: produits condensés de protéines et d'acides gras

Partially hydrolyzed collagen (→proteins) from leather scraps or shavings with a mole mass of 2000–5000 is reacted with fatty acids or their chlorides. The salts of the reaction products, which contain 10–16% protein, are highly water-soluble and show good lime dispersion and foaming.

Their outstanding use is as extremely mild surfactants in →cosmetic preparations for skin cleansing, shampoos and children's bathing products. They are marketed in 30–40% aqueous solutions.

Lit.: Ullmann* (5.) **A22,** 329 + **A25,** 747

Fatty Acids

G.: Fettsäuren; F.: acides gras

Older definitions say that fatty acids are derived from →fats and oils and have the following structure:

$$CH_3–(CH_2)_n–COOH \qquad n = 6–20$$

With respect to RR this definition is still valid. Today, however, the term "fatty acid" is more broadly understood as saturated and unsaturated aliphatic carboxylic acids of a chainlength of $C_8–C_{22}$ and include also synthetic acids made from petrochemical sources. The latter are, however, of minor commercial importance and not considered further here.

Thus "fatty acids" in context with RR are straigt-chain, even-numbered, saturated or unsaturated carboxylic acids in the range of $C_8–C_{22}$, derived mainly from →fats and oils. Sometimes, they contain functional groups in addition to the carboxylic group. Due to chemical modification of nature-derived acids, some branched and odd-numbered acids and derivatives are also covered under RR-related fatty acids.

Some general remarks to the physical data of f. are pertinent:

The m.p. is an important characeristic. It increases with chainlength and degree of saturation. Odd-numbered acids are lower in m.p. than the acid with one C atom less. In unsaturated acids, the *cis* isomers are lower in m.p. than the *trans*, while acids with conjugated double bonds are higher in their m.p. than nonconjugated acids. Specific gravity of all f. is below 1.0 g/cm^3.

B.p. are increasing with chainlength and are less sensitive to saturation and unsaturation. This makes separation by →distillation difficult.

Solubility in water increases with temperature and decreases with chainlength. The solubility of water in f. is higher than that of f. in water. Solubility in organic solvents increases with chainlength and is good, especially at temperatures above the m.p. of the acid.

Fatty acids are made from →fats and oils (triglycerides) by steam →hydrolysis and subsequent neutralization, or by →saponification. An interesting source is the soap stock that results from refining of crude →fats and oils for human nutrition (foots). Other sources are →tall oil, →waxes and sterol esters.

Normally, the first step of further processing is →distillation as a measure of purification. Cuts (mixtures of "neighboring" acids) are made by further fractional distillation. If desired pure acids (up to 99%) are obtainable by →fractional distillation. However, for most applications, simple distilled f. or cuts are fully satisfactory.

The table lists important f. and gives their most common names and short symbols (number of C atoms, number and position of double bonds and functional groups). For details, see specific keyword:

saturated f.:

C6:0	→caproic acid
C7:0	→enanthic acid
C8:0	→caprylic acid
C9:0	→pelargonic acid
C10:0	→capric acid
C12:0	→lauric acid
C14:0	→myristic acid
C16:0	→palmitic acid
C18:0	→stearic acid
C20:0	→arachidic acid
C22:0	→behenic acid
C26:0	→cerotic acid
C28:0	→montanic acid
C30:0	→melissic acid

unsaturated f.:

C11:1 (Δ 10)	→undecylenic acid
C16:1 (Δ 9)	→palmitoleic acid
C18:1 (Δ)	→petroselenic acid
(Δ 9)	→oleic acid
(Δ 9)	→elaidic acid
C18:2 (Δ 9,12)	→linoleic acid
(mixture)	→ricinenic acid
C18:3 (Δ 6,9,12)	→γ-linolenic acid
(Δ 8,10,12)	→calendulic acid
(Δ 9,11,13)	→eleosteric (α + β)
(Δ 9,12,15)	→α-linolenic acid
C20:1 (Δ 5)	→eicosenoic acid
C20:4 (Δ 5,8,11,14)	→arachidonic acid
C20:5 (Δ 5,8,11,14,17)	→eicosapentaenoic acid
C22:1 (Δ 13)	→erucic acid
(Δ 13)	→brassidic acid
C22:2 (Δ 5,13)	docosadienoic acid
	(→meadowfoam)

f. with functional groups:

C18:0 + OH (12-)	→hydroxystearic acid
C18:1 (Δ 9) + OH (12-)	→ricinoleic acid
C18:1 (Δ 9) + epoxy (12,13-)	→vernolic acid

C18:3 (Δ 9,11,13) + keto (4-)   →licanic acid
C20:1 (Δ 11) + OH (14-)   →lesquerolic acid

Aside from →distillation, other methods of separation are applied.

→Crystallization is the method of separating unsaturated and saturated acids, especially of oleic and stearic acid. For the same purpose, extraction with supercritical gases (e. g., CO_2) has been investigated. Adsorption to crystalline silica or cross-linked polystyrene is used to separate different f. (oleic/linoleic acid; lauric/myristic acid; palmitic/stearic acid). The property of urea to form addition products with preferably saturated fatty acids when crystallizing from methanol solutions can be used to obtain 97–99% oleic acid from →olive oil fatty acids.

Modifications of unsaturated f. include →hydrogenation and isomerization (*cis-trans*) (→isomerism) of the double bonds. Addition reactions, →ozonolysis and →metathesis are other possibilities for modifying unsaturated f. More than one double bond enables conjugation (→isomerism) and oligomerization (→oligomers).

Derivatives of commercial importance are: →fatty acid esters, →alkyd resins, →fatty acid amides, →fatty acid nitriles, →fatty amines, →fatty alcohols and →soaps.

Aside from being such important intermediates, f. are used as such for →candles and as →rubber chemicals.

In 1980, the applications of f. in the West European market were structured as follows:

Fatty alcohols, amines, esters, soaps and plastics	35–40%
Detergents, soaps, cosmetics	30–40%
Alkyd resins, paints	10–15%
Rubber, tires	3– 5%
Textile, leather, paper	3– 5%
Lubricants, greases	2– 3%
Others (incl. candles)	2– 5%

World production is estimated to be 2.6×10^6 mt/a (1988)

Lit.: Ullmann* (5.) **A10**, 245
      Kirk Othmer* (4.) **5**, 147
      R.W.Johnson, E.Fritz (ed.) "Fatty Acids in Industry"
      Marcel Dekker Inc., NY (1989)

Fatty Acid Salts
→Metallic Soaps
→Soaps
→Saponification

Fatty Acid Sucrose Esters
→Sucrose Fatty Acid Esters

Fatty Alcohol Alkoxylates
→Fatty Alcohol Ethoxylates
→Fatty Alcohol Propoxylates

Fatty Alcohol Ether Phosphates
→Fatty Alcohol Phosphates

Fatty Alcohol Ether Sulfates
Syn.: Alkyl Ether Sulfates
G.: Fettalkoholethersulfate;
F.: sulfates d'alcools gras éthoxylés

F. are the most important derivatives of fatty alcohols. They are the salts of the sulfuric acid half-esters of →fatty alcohol ethoxylates:

$R-O-(CH_2-CH_2-O-)_n-SO_3^- Na^+$

R: alkyl C_{8-18}, mainly C_{12-16}; n: mainly 2–4;

The properties are determined by the chainlength of the alcohol, number of ethylene oxide molecules and the cation, which may be NH_4^+ or $HO-CH_2-CH_2-NH_3^+$ (alkanolamine salt) instead of Na^+. The medium-chain alcohol may also be branched and petrochemically based.

Compared to →fatty alcohol sulfates f. have better solubility in water and greater stability against water hardness due to the glycol segment in the molecule. F. are stable in alkaline media, but they are sensitive to acid →hydrolysis. Their use in cosmetics is based on their good dermatological compatibility. They have excellent foaming properties and show synergistic effects with other →surfactants, such as →fatty alcohol sulfosuccinates, →amphoterics and →amineoxides, in terms of improved skin compatibilty. Combinations with →fatty acid alkanolamides have increased foaming power and oil-dispersing ability. A remarkable property, especially in shampoos and bubble baths is the thickening effect of f. by addition of electrolytes (salts).

F. are produced by →sulfation of →fatty alcohol ethoxylates with chlorosulfonic acid and to a large extent with SO_3, followed by neutralization, which has to be carried out fast to avoid degradation of the acidic intermediate. For specialties, amidosulfonic acid is used as sulfation agent.

The most suitable product for cosmetic applications is a f. with C_{12-14} and 2–3 EO units. Large volumes of shampoos and bath formulations are based on this structure. Liquid dishwashing (→cleaners) and light-duty →detergents are other areas of application. Technical uses comprise emulsifiers for polymerization (→polymerization additives), →fire-fighting foams, many applications where foaming is desired (e. g., foamed gypsum plaster, latex and resin foaming), and as a dispersant and emulsifier in hundreds of special applications. By varying the chainlength of the alcohol and the number of glycol units, f. can be tailored easily to specific needs.

Production in 1990 was estimated at 400 000 mt worldwide.

F. represent a stable market with some tendency towards natural products due to price and ecological considerations.

Lit.: Falbe* 67–69 (1987)
     Ullmann* (5.) **A25,** 747

Fatty Alcohol Ether Sulfonates

G.: Fettalkoholethersulfonate;
F.: sulfonates d'alcools gras éthoxylés

Because fatty alcohol ether sulfates are sensitive to →hydrolysis it is attractive to use f., where the sulfur is directly linked to the carbon. There are several possibilities for synthesis but none of them gained large-scale importance.

A new reaction scheme circumvents earlier difficulties by starting from unsaturated fatty alcohol ethoxylates, which are terminally blocked (→ fatty alcohol ethoxylates) and finally sulfonated at the double bond(→sulfation/sulfonation).

Such products are used in enhanced oil recovery (→oilfield chemicals) and in special textile applications (→antistatic agents).

Lit.: Falbe* p. 84 (1987)

Fatty Alcohol Ethoxylates

Syn.: Alkyl Polyglycol Ethers; Fatty Alcohol Polyglycol Ethers
G.: Fettalkoholethoxylate;
F.: alcools gras éthoxylés

Polyglycol ethers of fatty alcohols are an important class of nonionic →surfactants of the following structure:

$$R^1-O-(CH_2-CH-O)_nH$$
$$\underset{R^2}{|}$$

R^1: C_8-C_{22} alkyl R^2: H or CH_3- n: 3–15

Some of the ethylene oxide may be substituted by propylene oxide. Such a modification increases the hydrophobic character of the alkyl group with some influence on the foaming performance and other properties of the nonionic surfactant.

For propoxylated alcohol see: →fatty alcohol propoxylates. Propoxylation reduces biodegradability.

Ethoxylates and mixed ethers are produced by →alkoxylation.

Properties of f. depend on the chainlength of the alcohol, the number of ethylene oxide (or propylene oxide) units attached to the alcohol molecule and their homologue distribution.(→alkoxylation). Variations of these parameters give f. either emulsifying, dispersing or wetting properties.

Aqueous solutions of f. show unusual behavior: depending on the number of ethylene oxide units for a given alkyl chain, they get turbid at elevated temperatures and have a definite cloud point, which can be used for analytical characterization. They have their optimum surface activity close to this cloud point.

F. are low-foaming and have excellent wetting and emulsifying properties.

F. have become the major nonionic →surfactant in →detergent formulations. They are highly effective at low washing temperatures and at low concentrations and show a strong soil antiredeposition effect, especially on synthetic fibers. All these properties follow modern trends. F. find application in formulations for liquid, concentrated light-duty →detergents and in products for manual and machine dishwashing. They are used as emulsifiers in cosmetics, in →polymerization, for formulating →insecticides, and with mineral oil in →metal working fluids (drilling and cutting oils). There is a broad use in →textile auxiliaries as a washing and wetting agent, in industrial →cleaners and in oil drilling (→oilfield chemicals). These are only a few examples because there are hundreds of small technical applications that cannot all be mentioned here.

F. function as intermediates to make →fatty alcohol ether sulfates, →fatty alcohol ether phosphates, →fatty alcohol ether carboxylates and →fatty alcohol ether sulfosuccinates.

They are also the starting materials for making nonionic detergents with a terminally blocked OH-group (e. g., the OH group is reacted with butylchloride to form a butoxy-group at the end of the glycol chain). Such products are used preferably in low-foaming, strongly alkaline formulations in bottle →cleaners in the beverage industry, in special textile processes (mercerization) or in →metal working applications. They are used also as low-foaming emulsifiers in polymerization (→polymerization additives).

It is estimated that more than half of all fatty alcohols produced are ethoxylated prior to any further use. The dominance of f. as nonionic surfactants is increasing due to the fact that petrochemical alkyl phenol ethers are less biodegradable and have higher toxicity to fish.

Lit.: Falbe* 88–90 (1987)
Ullmann* (5.) **A25,** 747

Fatty Alcohol Phosphates

G.: Alkylphosphate; F.: phosphates d'alcools gras

The salts of mono- and diesters of phosphoric acid with fatty alcohol or their ethoxylates are anionic →surfactants. The water solubility and acid strength of such partial esters decrease with increasing chainlength. However, the sodium salts have good solubility. They are resistant to water hardness and to →hydrolysis, especially in alkaline media. They have good wetting and emulsifying properties.

$$R-O-\overset{OH}{\underset{OH}{\overset{|}{P}}}{=}O \quad \text{and} \quad H-O-\overset{OR}{\underset{OR}{\overset{|}{P}}}{=}O \qquad R: C_{8-18} \text{ alkyl}$$

F. are manufactured by reaction of →fatty alcohols and their ethoxylates with P_2O_5, $POCl_3$ or PCl_3, followed by neutralization.

The monoesters inhibit foaming of other anionic and nonionic →surfactants. Phosphates are mild to the skin, and therefore they are used in alkaline cleaners and in cosmetics.

In technical application, f. are specialties that are used in emulsion polymerization (→polymerization additives), →dry cleaning detergents, →textile auxiliaries, industrial →cleaners of high alkalinity and electrolyte concentration, corrosion inhibitors, pesticide formulations, papermaking and drilling muds (→oilfield chemicals).

US consumption was 17000 mt in 1980.

Lit.: Kirk Othmer* (3.) **22,** 359

Fatty Alcohol Polyglycol Ethers

→Fatty Alcohol Ethoxylates
→Fatty Alcohol Propoxylates

Fatty Alcohol Propoxylates

G.: Fettalkoholpropoxylate;
F.: alcools gras propoxylés

Fatty alcohols that are only propoxylated (→alkoxylation) are water-insoluble, have low surface activity and are therefore only useful in a few special applications, such as →defoamers and as foam inhibitors in technical cleaning formulations. They can be used also as spreading agents (cosmetic).

For applications were propylene oxide is used to modify ethoxylates, see: →fatty alcohol ethoxylates.

Lit.: Manufact. Chemist Aerosol News **48,** No.5, 23 (1977)

Fatty Alcohols

G.: Fettalkohole; F.: alcools gras

F. are aliphatic alcohols with chainlengths between C_8 and C_{22} according to the formula

$$CH_3-(CH_2)_n-CH_2OH \qquad n = 6-20$$

F. derived from RR have straight chains and are even-numbered, primary alcohols. F. containing less than 10 C atoms are called plasticizer alcohols, while those between 12–18 C atoms are detergent-range alcohols. F. with more than 22 C-atoms are called wax alcohols.

They may be saturated or unsaturated. When f. are derived from fats, oils and waxes, they are called natural alcohols. Alcohols of similar structure made by petrochemical processes (Ziegler and Oxo reaction) are called synthetic alcohols. They are not discussed here.

As usual in oleochemistry, broad cuts of f. with a mixture of neighboring chainlengths are satisfactory for many applications. They are called after their origin, e. g., coconut alcohol, tallow alcohol, etc., and are still bearing most of the fatty acid pattern of the original fat or oil. For special purposes, pure f. are made by →distillation and →fractional distillation.

Additional information can be found for a selection of f. under the following keywords:
→caprylic alcohol, →capric alcohol, →lauryl alcohol, →myristyl alcohol, palmityl alcohol (→cetyl alcohol), →stearyl alcohol, →behenyl alcohol and the unsaturated →oleyl alcohol, →linoleyl alcohol and →linolenoyl alcohol.

Physical properties:

Properties of pure alcohols are described under their respective keywords. In general, the b.p. increases by about 20 °C and the m.p. by about 10 °C for each additional $-CH_2-$ unit. Water solubility is appreciable for hexanol and decanol but drops rapidly with increasing chainlength. In general, f. are soluble in alcohols, ethers and hydrocarbons. Lower alcohols have a fruity odor. Below C_{12}, f. are clear liquids at r.t., while higher alcohols are pasty and the longer-chain f. are crystalline waxy materials.

Chemical properties/derivatives:

Chemical reactivity of f. is determined by the OH group.:

- Esters can by made by →esterification with carboxylic acids. (→fatty alcohol esters);
- Sulfates result from →sulfation with SO_3 or chlorosulfonic acid (→fatty alcohol sulfates);
- →Ethoxylation and →propoxylation lead to →f. ethoxylates or →f. propoxylates;
- →Alkylhalides are the result of reaction with, e. g., PCl_3;
- α-Olefins are gained by →dehydration. They are, however, more easily obtained by petrochemical processes (e. g., Shell Shop Process);
- Aldehydes and fatty acids are the result of →oxidation;
- Fatty amines can be manufactured from f. (→fatty amines);
- Branched alcohols may be gained by dimerization (→Guerbet alcohol);
- Dimers (→dimerization) comparable to fatty acid dimers are available in a similar reaction.

Manufacture:

In the first quarter of the 20th century, small volumes of f. were produced by →saponification or by sodium reduction of wax esters, such as spermaceti. Both alternatives are not used any more due to inavailabilty of the raw material and difficult technology.

→Hydrogenation is the best method to obtain f. from RR today.

Fatty acids, their methyl and butyl esters and also triglycerides are transferred to f. by high pressure hydrogenation.

Primary feedstock is coconut or palm kernel oil for the C_{12-14} range, and tallow, palm oil and many other sources are used for the C_{16-18} range. Rapeseed is the source for behenyl alcohol.

$$R^1\text{-COO-}R^2 + 2\,H_2O \xrightarrow[\text{copper chromite}]{200\text{--}300\,°C/\approx 30\,MPa} R^1\text{-CH}_2OH + R^2OH$$

R^1: C_{17-21} alkyl

R^2: $H-$, CH_3-, C_4H_9- or $CH_2-CH-CH_2-OR^1$
$\qquad\qquad\qquad\qquad\qquad\quad |$
$\qquad\qquad\qquad\qquad\qquad OR^1$

At a first glance, the →triglycerides seem to be the most appropriate starting material. The rough reaction conditions, however, transfer most of the valuable by-product glycerol into propanediol or propanol. More hydrogen and catalyst is consumed than would be necessary for the methyl ester. Improvements of economy and technology are necessary to make this process attractive for a large-scale application.

→Fatty acids are a better starting material. The reaction temperature has to be higher than for hydrogenation of methyl esters, and the equipment has to be made of corrosion-resistant material. The catalyst has to be acid-resistant. Hydrocarbons are formed as by-product. These problems are partly resolved by the Lurgi process in which preformed f. is recycled and forms wax esters with the acid, which are hydrogenated easier.

By far the most used process for the production of f. from RR is the hydrogenation of →fatty acid methyl or butyl ester. They are made first by →transesterification of triglycerides or by →esterification of fatty acids.

There are three principle types of hydrogenation processes:

The suspension process: $1-2\%$ copper chromite catalyst is suspended in the ester (or the acid), and hydrogenation is carried out at $250-300°$ C and 30 MPa. After the reaction, the catalyst is separated in a centrifuge.

The gas phase, fixed bed (Henkel) process: the copper chromite catalyst is pelletized and forms a fixed bed in a reactor in which the ester is mixed with a large excess of hydrogen. Pressure is $20-25$ MPa and the temperature is $230-250\,°C$. The liquid and gaseous phase are separated. Hydrogen is recycled, and the liquid phases releases the methanol during depressurizing. Highly pure f. is recovered.

The trickle-bed process can be used for hydrogenation of liquid products, such as wax esters and fatty acids. The copper chromite is supported by SiO_2. The reaction is carried out at $20-30$ MPa and $250\,°C$.

The catalyst copper chromite, used in all three processes, also hydrogenates all double bonds in the chain, and completely saturated alcohols are the result. With use of a zinc-containing mixed catalyst, the double bonds in unsaturated starting materials are retained to an extent of 98%. Highly unsaturated alcohols are recovered.

Narrow cuts may be made by →distillation and →fractionation of the final product or of the starting material, e. g., the methyl ester or the fatty acid prior to hydrogenation.

Uses:

F. are important intermediates; 95% of the total production flows into the manufacture of derivatives:

→Fatty alcohol esters, →fatty alcohol ethers, →fatty alcohol phosphates, →sulfosuccinates, →fatty alcohol sulfates, →fatty alcohol ether sulfates, →Guerbet alcohols and alkyl halides. Their uses are described under the respective keywords.

Only 5 % of f. production is used as such. These direct uses include:

The shorter-chain alcohols function as (co)solvents in many technical applications. F. of various chain-lengths are used as lubricants in plastic processing (→plastic additives), →defoamers and foamers in →detergent formulations (increase of foam stability and washing power), →lubricants in cold rolling of aluminum, and as agents to give consistency to creams, ointments and lotions in cosmetic formulations.

Unsaturated f. find use, due to their more liquid consistency, as oily component in →cosmetics and as fatting agent in washing lotions, foam baths and hair lotions (→oleyl alcohol).

Economic aspects:
The worldwide capacity for f. based on RR in l985 was estimated to be at 520 000 mt worldwide.
A complete survey of the market is given in[1].
In the last years a strong increase in f. production based on RR could be realized due to the favorable raw material market, the improvements in the manufacturing processes and the high priorty given to ecological considerations today.

Lit.: Ullmann* (5.) **A10**, 277
      Kirk Othmer* (4.) **1**, 865

[1] J.Knaut, H.J.Richtler J.Amer.Oil Chem.Soc., **61**, 317 (1984)

Fatty Alcohol Sulfates

Syn.: Alkyl Sulfates; FAS
G.: Fettalkoholsulfate; F.: alcool gras sulfatés

F. represent a class of important anionic →surfactants. They are the salts of the halfesters of sulfuric acid and fatty alcohol:

$R-O-SO_3Na$

R: C_{8-18} alkyl

The properties are determined by the chainlength. They are foam-intensive and wash-effective, and they have a high interfacial activity. Water solubility decreases with increasing chainlength, while sensitivity to water hardness increases. Unsaturated alkyl chains impart better solubility.

F. are produced commercially by →sulfation of fatty alcohols with chlorosulfonic acid or mainly with SO_3, followed by neutralization with sodium hydroxide. They are traded as pastes (mostly 30 %) or spray- or roller-dried powder.

The uses of f. depend on the chainlength of the alcohol:

Short-chain (C_8-C_{10}) alcohol sulfates are technical wetting agents (insensitive to water hardness and electrolytes, hydrotropic properties) and are preferably used in stronger alkaline media, e. g., mercerization baths in →textile auxiliaries, in cleaning and degreasing, electro-plating baths.

Other applications are as wetting agents in stearic acid-oleic acid separation (→crystallization), as a hydrotrope in liquid detergents, and in emulsion polymerization.

Medium-chain ($C_{10}-C_{16}$) alcohol sulfates are used in some parts of the world for liquid shampoos and bubble baths, especially as the alkanolamine- instead of the Na-salt. The salts are also used in dish-washing formulations, →fire fighting foams, hand lotions and household →cleaners. Because of their limited water solubility, the sodium salts are used in shampoos, hand cleaners, syndet soaps and foamers in tooth paste.

Long-chain ($C_{16}-C_{18}$) alcohol sulfates find broad application in many technical applications, such as →textile and →leather auxiliaries, frothers in →mining chemicals, emulsifiers in polymerization and latex dispersions (→polymerization additives), and in paints and foams. F. of the tallow range find increasing use not only in specialty household products but also in heavy-duty →detergents as a RR-based supplement or substitute for linear alkyl benzene sulfonate (LAS). Modern high-density powder →detergents make use of recently developed f. in granulated form.

In 1990, 300 000 mt were produced worldwide. Volume may increase in the future if the trend to function as an alternative to petrochemical surfactants continues.

Lit.: Kirk Othmer* (3.) **22**, 355
       Ullmann* (5.) **A25**, 747

Fatty Amine Ethoxylates

G.: Fettaminethoxylate;
F.: amines grasses oxygenées

F. are prepared by →alkoxylation (150 °C) of primary fatty amines in the presence of alkaline catalysts. Only one of the hydrogen on the N-atom reacts, and further ethoxylation proceeds on the OH group. If both hydrogens need to be ethoxylated one has first to react the amine without a catalyst at 100 °C at mild EO pressure. Further alkoxylation needs an alkaline catalyst again. F. are used as finishing agents and antistatics in textile treatment (→textile auxiliaries), as emulsifier in drilling muds (→oilfield chemicals) and in bitumen processing. They show synergism in the herbicidal activity of various pesticides.

Lit.: J.M.Richmond (ed.) "Cationic Surfactants" Marcel Dekker Inc. NY/Basel (1990)

Fatty Amine Oxides

G.: Aminoxide; F.: amines grasses oxygenées

F. are products that are on the borderline between nonionic and cationic →surfactants. They are produced by oxidation of tertiary amines such as fatty alkyl dimethyl amine, with H_2O_2:

$$\underset{\underset{CH_3}{|}}{\overset{\overset{CH_3}{|}}{R-N}} + H_2O_2 \xrightarrow{60-70\,°C} \underset{\underset{CH_3}{|}}{\overset{\overset{CH_3}{|}}{R-N}}{\rightarrow}O + H_2O$$

R: C_{12-18} alkyl

(In some commercial products, the two methyl groups are substituted by hydroxyethyl groups).

A. are frequently used in →detergents and personal care products.

They are insensitive to water hardness. They are cationic in acid solutions and nonionic in neutral or alkaline media.

They foam satisfactorily, are mild to the skin, and are therefore mainly used in manual dishwashing formulations (→cleaners), shampoos and light-duty →detergents. They are replacements for fatty acid alkanolamides as foam builders. Their higher price is compensated by better performance. They are also used as phase transfer catalysts and in the textile industry as dyeing auxiliaries and antistats.

A. are marketed in solution or paste form (30–55% d.s.): $\approx$45 000 mt were consumed in the United States with 80% going into household products.

Lit.: Kirk Othmer* (4.) **2**, 357
      Ullmann* (5.) **A25**, 747

Fatty Amines

Syn.: Alkyl Amines
G: Fettamine; F.: amines grasses

The chemical structures of f. are as follows:
$R-NH_2$: primary amines or monoalkylamines
R_2-NH: secondary amines or dialkylamines
R_3-N: tertiary amines or trialkylamines

To be called f., at least one of the R's has to be an alkyl of a chainlength of C_8-C_{22}. Most f. are derived from RR. However, there are long-chain amines made by petrochemical methods, which are called f. also.

In →oleochemistry, pure fractions are used commercially only in rare cases. Broad cuts that still bear the fatty acid pattern of the original fat or oil are fully satisfactory for most uses. This is also true for f. Broad cuts that are used commercially are based on the following →fats and oils:
→tallow (also hydrogenated), →coconut, →soybean and →palm.

Fractionated amines of some importance are →lauryl amine, →palmityl amine, →stearyl amine and →oleyl amine.

Properties of f. depend on structure, chainlength and purity. The m.p. increases with chainlength. Secondary f. melt higher than primary amines with the same alkyl, and tertiary melt lower than secondary f. Unsaturation in the chain lowers the m.p. as usual. The b.p. increases with chainlength by approximately 10 °C per C-atom. Thus pure f. can be obtained by →fractional distillation. F. are insoluble in water, but they are taking up water to form hydrates. Alkalinity of f. decrease in the following order:

secondary amines > primary amines > tertiary amines > ammonia.

The property of f. to be local irritants of eyes and skin is reduced by ethoxylation. Manufacturing from RR is predominantly done via the →fatty nitriles, which are made from fatty acids by reaction with ammonia. The nitriles are reduced by →hydrogenation (batch or continous) into amines. The reaction conditions determine whether primary, secondary or tertiary amines result. Saturated f. are formed when nitriles are hydrogenated at 80–140 °C and 1–4 MPa/10–40 bar over Ni catalyst. Unsaturated f. can be obtained with Raney Co or Cu-chromite as catalyst. Ammonia has to be present to suppress the formation of secondary amines. Saturated and unsaturated secondary amines are gained in yields of more than 90% if ammonia is vented from the reactor, the reaction temperature is 160–210 °C, and pressure is maintained at 5–10 MPa/50–100 bar. Trialkylamines are produced via the imine R–CH=NH and Schiff base R–CH=N–CH$_3$ with Ni catalyst at 230 °C and 0.7 MPa/7 bar hydrogen pressure. The most important amine types, which are the base for making dimethyl dialkyl ammonium compounds, are made by the following reaction:

$$R_2NH + CH_2O + HCOOH \longrightarrow R_2NCH_3 + CO_2 + H_2O$$

These compounds can also be made by reductive methylation with formaldehyde, hydrogen and a Ni catalyst.

Other starting materials to make f., aside from nitriles, are fatty alcohols. F. are obtained by reductive alkylation of ammonia or methylamine with fatty alcohols at 90–190 °C under low hydrogen pressure in the presence of Raney nickel. If water is removed continously from the reaction mixture, only secondary or tertiary amines are obtained. This reaction is mainly used when starting from synthetic fatty alcohols and focusses on the manufacture of textile softeners. It is, however, also used for making medium-chain (C_8–C_{10}) trialkylamines from RR used in uranium recovery (→mining chemicals).

There are other possibilities to make amines in a one-step hydrogenation reaction from the methyl ester or fatty acid (2.5×10^7 Pa/250 bar), which are not used to a large extent.

Narrow cuts can be made by fractional distillation of either the starting acid, the nitrile or the finished amine.

F. are able to undergo a series of chemical reactions. The most important are:

– salt formation, e.g., with HCl or acetic acid, resulting in improved solubility in water;
– alkylation with methyl chloride or dimethyl sulfate, forming quaternary ammonium compounds. Reaction is carried out in aqueous alcohol in the presence of alkali;
– →alkoxylation (ethoxylation) of primary f., carried out at 150–200 °C without a catalyst, yields the mono-adduct, while alkaline catalysis at 100 °C gives the polyalkoxy amine;
– Michael addition of acrylonitrile to primary f., followed by hydrogenation, is the best method to make the so-called diamines:

$$R-NH_2 + CH_2=CH-CN \longrightarrow R-NH-CH_2-CH_2-CN$$
$$R-NH-CH_2-CH_2-CN + 2H_2 \longrightarrow$$
$$R-NH-CH_2-CH_2-CH_2-NH_2;$$

– with halogenated carboxylic acids or lactones, betaines (→amphoteric surfactants) are obtained;
– oxydation with hydrogene peroxide yields →amine oxides;
– isocyanates result by reaction of f. with phosgene.

By far the largest (46% of the total amine market) use of f. is in fabric softeners (→quaternary ammonium compounds). Predominantly used are dimethyldialkylammonium chloride and the corresponding methyl sulfate. The alkyl is hydrogenated tallow. Other quaternaries contain hydroxyethyl- or imidazolinium structures. Dispersions containing 5–7% f. are used in rinse cycles of household and commercial laundry machines, and 0.1–0.2% remains on the fabric. In drying cycles, the softeners are applied to the substrate (paper, wovens and nonwovens as well as foam pads).

F. and especially their salts are used in mining operations (→mining industry, →flotation) in concentration operations in potash and phosphate mines but also in many others. Fatty amine salts are used in products such as →fertilizers to prevent caking; 7% of the total market goes in this application.

Primary amines, diamines, →fatty amine ethoxylates and →quaternary ammonium compounds are used as emulsifiers in bituminous emulsions used in road coverings (7% of total amine market): 0.5–1% are added to form the asphalt emulsion, which is broken in contact with the negatively charged aggregate and simultanously improves adhesion to it.

Primary amines and diamines, neutralized with, e.g., oleic acid, are used as corrosion inhibitors. The main product is N-tallow-1,3-propanediamine dioleate. The cationic amino function sticks to the metal surface and prevents corrosive attack.

Quaternary ammonium compounds are used in sugar refining, as bactericides, sanitizers and desinfactants, and together with ethoxylated f. in pesticide formulations and in agricultural adjuvants. Guanidins, derived from dodecylamine, are fungicides for fruit trees.

F., also ethoxylated, are used as antistatic agents as →plastic additives and in the →textile industry. Ethoxylated f. are fiber finishes (→textile auxiliaries).

Pigment grinding aids and pigment dispersants in →coatings and in magnetic tape manufacturing are based on f., diamines and ethoxylated amines. The amine covers the pigment surface, making it more compatible with the hydrophobic formulation.

A large outlet is the manufacture of organo-clays. Methyl- and benzyl- quaternary ammonium compounds, based on di-hydrogenated tallow amine, are mainly used to modify the surface of the clay, which is used as thickening agent in drilling muds (→oilfield chemicals), →lubricants and oil based →coatings. In oil fields, f. function as corrosion inhibitors, emulsifying/deemulsifying agents and surfactants (→oil field chemicals).

F. function as epoxy curing agents, as additives for gasoline (oleyl amine) and fuel oil.

The total world market for f. in 1983 was estimated at 270 000 mt. This is a strong increase from 120 000 mt in 1971. The growth rate remains strong: in 1990, the US market alone was estimated to be at 280 000 mt.

Lit.: Kirk Othmer* (4.) **2**, 405
 Ullmann* **A2**, 1

Fatty Diamines
→Fatty Amines

Fatty Diketenes
Syn.: Diketenes
G.: Fettdiketene; F.: dikétènes d'acides gras

$$CH_3-(CH_2)_{15}-CH=\underset{\underset{\displaystyle O-C=O}{|\quad\quad|}}{C-C}-(CH_2)_{15}-CH_3$$

Long-chain fatty acid ketenes are formed by removal of HCl from the acid chlorides and dimerize easily to the lactone dimer.

Thus, the acid chloride of →stearic acid forms octadecylketene dimer (formula above); m.p. 80 °C [24430-01-1].

→Palmitic acid yields the hexadecylketene dimer; m.p. 64 °C [10126-68-8] and →myristic acid forms the tetradecylketene dimer; m.p. 57–58 °C [42272-70-8].

These f. are used as effective paper sizes (→paper additives), forming covalent bonds with the OH groups of →cellulose and resulting in a hydrophobic paper surface.

Lit.: Kirk Othmer* (4.) **14**, 963
 Ullmann* (5.) **A15**, 72

Fatty Ethers
There is not much use of f., with the exception of di-n-octyl ether as a solubilizer (oil component) in cosmetic applications. F. spread well, are mild to the skin and biodegradable.

Fatty Nitriles
→Fatty Acid Nitriles

Fennel
Foeniculum vulgare Mill.; *F. officinale* Gaertn.
Umbelliferae
G: Fenchel; F: fenouil

F. is native to the Mediterranean region, and it has become naturalized in many temperate countries. F. is a stout, erect glaucous →perennial with greatly divided leaves with hair-like segments. Flowers are 1–2 mm in diameter, yellow, in typical compound terminal umbels. The fruits are ovoid, 5–6 mm long, greenish-brown, with typical ridges.

All parts of the plant are aromatic. The leaves are used as a pot-herb. The petioles are used as such for salads or as vegetable. The bulbs of the whole plant as well as the stems are used as vegetable also. The seeds contain, beside the →essential oil, also around 20% of fatty oil, which is composed of about 80% of →petroselenic acid, which may be used in oleochemistry in the same way as the fatty oil of →coriandrum.

Lit: Purseglove* (1974).

Fermentation
G.: Fermentation; F.: fermentation

In biology, f. is defined as the process of extracting energy from organic compounds without the involvement of oxygen. In industrial terms, f. is the processing or derivatization of organic materials to new products by means of →enzymes, such as in the manufacture of traditional food products (beer, wine and cheese). Important new applications are the production of pharmaceuticals, such as →antibiotics, →fragrances and dyeing agents. F. is carried out on a large scale in industrial plants of several mt.

The most common feedstocks (culture medium) are carbohydrates, such as sugar, molasses, starch and many other RR-based materials.

The advantages of f. are in the comparatively low energy costs because biological processes are being copied, where temperatures above the denaturalization of protein (>40 °C) may not be engaged. The disadvantages lie in the low concentrations of the desired products, which require efficient cleaning and concentration procedures. Also, f. processes need to operate under sterile conditions, so that other enzymes or microorganisms do not interfere with the desired processing scheme. This is why f. processes have gained importance only in those situations where higher production costs pay off because of noncompeting traditional chemical processes or because of high-value products that result from f. processing.

Lit.: H.G. Schwartzberg, M.A. Rao, (ed.) "Biotechnology and Food Process Engineering", Marcel Decker, NY, (1992)

Fertilizer
G.: Dünger; F.: engrais

Substances that are used to improve the growth of plants.

Organic f.: manure, compost, peat, →straw, stalks, toppings, etc. (as green manure). The oldest way to return nutrients and organic matter (which once were harvested from the field) back to the soil via animal residues and straw or simply by biologigal →respiration.

Mineral f.: inorganic substances (mostly water-soluble salts) applied to the soil. May also be of organic origin, such as urea. For proper growth, plants need the following:

macro nutrients: N, P, S, K, Ca, Mg; and micro (trace) nutrients: Cl, B, Mo, Fe, Mn, Zn, Cu.

Typical mineral f. are: $(NH_4)_2SO_4$, NH_4NO_3, NH_3, $CO(NH_2)_2$, $CaCN_2$, $Ca(H_2PO_4)_2$, $Ca_3(PO_4)_2$, KCl, K_2SO_4, $CaCO_3$, CaO.

Micro nutrients often are contained in some of these f. due to their origin from mining processes. Some nutrients (such as Fe or Zn) are plentiful in soils. They will, however, need to be added when substrates other than soil are used, such as aqua culture (e. g., rockwool).

Formulated mineral f. contains →fatty amines as anticaking agent.

Lit.: Martin/Leonard* (1967).

Ferula galbaniflua
Ferula rubricaulus
→Galbanum Resinoid/Oil

Fibers

G.: Fasern; F.: fibres

Many natural fibers have been used by mankind for generations. Others are based on mainly cellulose and modified by processes developed during the last 150 years.

For details: →wool, →silk, →cotton and other →cellulose fibers (→viscose, →cellulose regenerate, →cellulose acetate).

Field Mustard
→Turnip Rape

Films

Syn.: Foils

G.: Filme, Folien; F.: films, feuilles

There are only two types of free films or foils based on RR:

→Cellophane and →cellulose acetate (various DS).

Lit.: Encycl.Polym.Sci.Engng.* (2.) **7**, 73

Fire Fighting

Syn.: Fire Extinguishing Agents

G.: Feuerlöschmittel;

F.: agents d'extinction des incendies, extincteurs

There are not many techniques and materials in which RR-based products are used for f.

Keratin hydrolysates (→proteins) and →surfactants such as →fatty alcohol ether sulfates, →alkylpolyglucosides and fluorinated surfactants (→perfluoro alkyl compounds) are used as basic protein foaming agents in the formulation of fire extinguishing foams.

→Carboxymethyl cellulose, →methylcellulose, →alginates and especially →guar gum and its derivatives are applied as thickeners in foam-based agents and as gellants for water used in forest firefighting by airplane.

Dry chemical fire extinguishers, based on sodium bicarbonate, contain sometimes Ca- or Mg-soaps (→metallic soaps).

→Glyceryl trinitrate (nitroglycerin) is, among others, used in techniques to extinguish burning oil-wells by explosion.

Lit.: Ullmann* (4.) **A11**, 567
        Kirk Othmer* (3.) **Suppl.** 443

Firefly
→*Cuphea* spp.

Fish

G.: Fisch; F.: poisson

F. is a versatile source of RR. Aside of human nutrition, fishmeal is extensively used as protein-rich animal feed, and the →fish oil (marine oil) is used in →oleochemistry.

The total fish catch in 1993 was $\approx 100 \times 10^6$ mt.

Lit.: Vital Signs 1994, Worldwatch Inst. 32 WW. Norton NY

Fish Oil

G.: Fischöl; F.: huile de poisson

s.v.: 180–197; i.v.: 150–175

Depending on origin and season, the properties of f. varies in a broad range. The fatty acid composition fluctuates also in a broad range:

C14:0	C16:0	C16:1	C18:1	C18:x
6–9%	11–20%	7–11%	2–15%	1–3%

C20:1	C20:x	C22:1	C22:x
2–16%	6–19%	1–20%	5–14%

x= more than one double bond

The oil has weak drying properties.

Sources are →fish processing plants, mainly in Japan, North and South America and Scandinavia, where the oil is recovered by cooking, pressing (→expelling) or →extraction. The co-product is fish meal, which is used for animal feed and human (surimi) food. Raw materials are the whole fish bodies or the wastes from fish processing.

A special f. is **menhaden oil**, which is derived from an nonedible fish caught on the US east coast. It contains a high percentage of →linoleic and poly-unsaturated C_{20}/C_{22} fatty acids (PUFA). It is used in →coatings, in fat liquors (→leather auxiliaries) and in oil drilling (→oilfield chemicals) but also in edible fats after →hydrogenation.

Another oil to be mentioned is **herring oil.** It is used mainly in fat liquors (→leather auxiliaries). A slow drying oil is obtained when the oil is blown (→drying oils).

→Hydrogenation yields hardened fish fat (m.p.: 31–45 °C). Due to the content of sulfur and poly-unsaturated fatty acids, there is a large consumption of catalyst and hydrogen in this operation.

Uses of especially the hardened fat are mainly in the food industry for margarine and shortening blends.

Cod and halibut liver oils are of medicinal interest because of their high content in vitamins A and D.

F. contains a large amount of polyunsaturated fatty acids (up to 6 double bonds), which are of high nutritional and dietetic value (essential fatty acids, →fats and oils). Especially C20:5 (→eicosapentaenoic acid) and C22:6, called omega-3 fatty acids, are used as either nutrional supplements or therapeutic agents to inhibit a variety of pathological impacts in man.

Technical applications of f. are as extenders in drying oils in combination with dryers, as starting materials for making →textile and →leather auxiliaries and as sources of fish oil fatty acid, which are made by →hydrolysis of the oil.

Production of fish oil in 1994	($\times 10^6$ mt)
Japan	0.10
Peru	0.25
Chile	0.28
USA	0.09
Denmark	0.11
FSU	0.10
Island	0.08
Norway	0.10
World	1.24

Lit.: Ullmann* (5.) **A10**, 237
[8016-13-5]

Fish Oil Fatty Acids

→Fish Oil

Flavons

→Dyes, natural

Flavors

G.: Aromen; F.: arômes

F. are mixtures of single chemicals (mostly nature identical, depending on each country's legislation) and natural extraction products, such as →essential oils, →absolutes and →resinoids. They also include super-concentrated juices or fruit extracts, used to enhance the taste of any industrially made food product. Applications range from candies, chewing gum, baked goods, dairy products, beverages, snack foods to canned goods, powdered soups, ready-made meals (frozen or canned) to f. for tobacco, oral hygiene products and orally taken pharmaceutical products such as cough syrup.

F. come in different forms:

- liquid f., based on alcohol, propylene glycol, water or edible oils as carriers for the flavoring principles;
- powdered spice blends, simple mixtures of crude, ground spices and dried herbs;
- powdered f., either dry dispersions of flavoring materials into a powder, such as sugar or micro-encapsulated products that are mainly produced by spray-drying.

Lit: Ullmann* **A11**, 141

Flax

Syn.: Linseed
Linum usitatissimum L.
Linaceae
G: Flachs, Faserlein; F: lin

F. describes special fiber-producing varieties of →linseed, which is mainly grown for seed oil production. For the purpose of fiber production, planting of the seeds have to be at narrower spacings, and higher seed rates (1800 plants/m^2) than for linseed production (1000 plants/m^2). This will prevent formation of branches and help develop a uniform stand. The f. crop will have to be harvested when 30–50% of the seed bolls are brown or yellow with fully developed brown seeds. The stem then will have turned from green to yellow. In former times, f. was harvested by hand-pulling of the whole plant with roots. Today, most f. is machine harvested. Yields range between 40–60 dt of straw per ha.

The fibers have to be dried and processed in a mill, where they are threshed or deseeded. In the next step -retting or partial rotting- the fibers are separated from other organic products of the stem with the help of bacteria. This may be done in the field with dew retting or in water of nearby rivers, which will take 1–3 weeks. As f. production is receiving more attraction, new physical and chemical retting processes (retting in tanks or in the field before harvest) are being developed. Because production and extraction methods have not been changed for more than 30 years, this is a major goal of development. In the f. mill, the fibers (length 4–60mm) go through breaking and scrutching before they are spun into linen yarns, which are used in threads and twines of various kinds.

Wet spinning, where water acts as a lubricant, yields a fine and strong but less elastic yarn.

Fiber yields of 10–15 dt are reported per ha. The fiber consists of 62% →cellulose, 16.7% →hemicellulose, 1.8% →pectin and 2% →lignin. F. is difficult to blend with other fibers and is used for apparel, table and bed linen.

In USA f. is the best and sole source for cigarette paper. After farmers harvest the seed heads, the companies go in with their own machines to harvest and take the straw.

Main producers (75% of world production) are the former "East block" countries as well as Belgium and France.

Lit.: Fehr/Hadley* (1980)
        Ullmann* (5.) **A5,** 399

Flocculants

G.: Flockungsmittel; F.: floculants

The large-volume f. are synthetic (polyacrylamide). F. derived from natural products include →starch, →starch derivatives, plant gums (→guar gum), seaweed extracts (→alginic acid), →cellulose ethers, →proteins and →tannins. They are relatively cheap but require high dosages.

Among this group of f., starch is the most used, either as such or in the form of derivatives (→ cationic starch; →anionic starch). Guar gum and derivatives, as well as animal →glue (gelatin), are also of some importance. Albumin is used for clarifying wine.

General uses are in water/wastewater treatment and in the →paper and sugar industries. Starch is used in bauxite extraction (aluminium production). Guar is used in gold and uranium recovery (→mining chemicals).

In 1984, 40 000 mt f. were used in USA and 15 000 mt in Western Europe.

Lit.: J. Bratby, "Coagulation and Flocculation", Upland Press Inc. Croydon UK (1980)
        Kirk Othmer* (4.) **11,** 64

Flooring Materials

G.: Bodenbeläge; F.: revêtements du sol

Out of the many f. in use, only those based on plastics, wood and textiles are of interest in relation to RR.

By far the largest volume is plastic, especially PVC-based. Most types of →plastics additives are applied in their formulation. Backings are often textile, such as →cellulose fibers.

→Linoleum is a tradional f. and is almost entirely based on RR (→linseed oil).

→Wood is also an old f., which is still in use in different kinds and shapes. Hardwoods are especially used as parquet.

The following fibers are in use as f. (carpets): →wool, →cotton, rice straw (tatami), →sisal, →hemp, coir (→coconut) and →jute (for backing).

Lit.: Encycl.Polym.Sci.Engng.* (2.) **7,** 233

Flotation

G.: Flotation; F.: flottation

F. is a process for the selective separation of solids dispersed in an aqueous medium. Finely ground and milled material is agitated in water in a flotation cell, while finely dispersed air is being introduced. Chemicals (flotation reagents) are added to modify the surface of the particles, the consistency of the pulp and the stability of the air bubbles.

The solid particles, which are naturally hydrophilic, are rendered hydrophobic by the collector reagents. This allows the attachment of the particles to the rising air bubbles. The resultant mineral-enriched froth is floated off and collected.

The particles with a hydrophilic surface remain in the pulp and are discharged at the bottom of the flotation cell. The addition of depressants before flotation increases the hydrophilic properties of these particles.

Flotation is a continuous progressive process. The particles are gradually upgraded by arranging the cells into lines or banks and by refloating the flotation concentrates.

F. is used extensively by the →mining- and →paper industries. The latter uses this technology for deinking of recycled paper. This versatile process is being developed and refined for use in many other applications.

Lit.: Kirk Othmer* (4.) **11**, 81–107

Fluoroalkyl Compounds
→Perfluoro Alkyl Compounds

Foam Depressants
→Defoamers and Foamers

Foamers
→Defoamers and Foamers

Foeniculum vulgare
Foeniculum officinale
→Fennel

Foils
→Films

Food

G.: Ernährung, Nahrungsmittel; F.: produits alimentaires

Most plants are grown, animals bred and natural products collected for use in human nutrition. This area is, however, not the subject of this dictionary, which deals with the use of the same sources for functional, technical and industrial applications. There are some major food crops that are almost exclusively being used for nutrition today, but they have a high probability of gaining importance as RR, either as such or after some plant breeding work. The use of RR as →food additives in the food processing industry is within the scope of this book, because food additives are more a processing aid than of direct nutrional importance.

Food Additives

G.: Nahrungsmittel-Zusatzstoffe;
F.: additifs alimentaires

Food additives are substances that are added to food products to increase or decrease the nutritional value of food, to improve stability and shelf life, to affect sensory properties and to facilitate certain technological processes. There are various definitions for f., depending on the legal situation of a country. Due to the fact that natural products are considered to be nontoxic, easily biodegradable and derived from sources that are reproducable, f. are mainly and with increasing tendency based on RR.

Vitamins:

Vitamins are sometimes added for revitaminization (losses during processing), standardization and enrichment. Mostly vitamins A, D and B are added. Vitamins and provitamins, such as →tocopherol, →ascorbinic acid and carotenoids (→dyes, natural), have antioxidative or coloring functions.

→Amino acids:

Deficiencies and losses of essential amino acids, such as lysine, can by adjusted by addition during food production. →Glutamic acid salts are flavor enhancers, especially in oriental foods. Amino acid mixtures are added in special diets.

Bulking agents:

To give food a bulky, rich appearance without increasing the nutritional value, air (foam) or water are added, mostly with the support by an emulsifier (see below). Other possibilties are thickening and bulking agents, such as cellulose (→cellulose, microcrystalline), →methylcellulose, →carboxymethyl cellulose, →guar gum, →locust gum, →xanthan gum and →polydextrose, a polycondensate of →sucrose with →sorbitol or →isomalt (disaccharide alcohol).

Preservatives:

Only a few RR-based products are able to avoid or at least delay the spoilage of food by microorganisms. →Sorbic acid (and sorbates) act against molds that form carcinogenic aflatoxins.

→Lactic acid and →acetic acid lower the pH and can inhibit the growth of bacteria especially. Full efficiency is mostly given by a combination of several preservatives and physical methods.

Antioxidants:

Food is frequently subject to oxidative degradation and spoilage. Especially, fats with unsaturated fatty acids can get rancid. Radical scavengers are the most effective a. →Ascorbic acid and its salts, isoascorbic acid, gallates (gallic acid) and →tocopherols are used as well as some petrochemical-based products.

Sequestrants form complexes with metal ions, which accelerate oxidative degradation, and show synergistic effects with the a.m. antioxidants. →Phytic acid, calcium gluconate (→gluconic acid), →lactic acid and lactates, →tartaric acid and tartrates, →citric acid and citrates and →lecithin are such RR-based products.

Emulsifiers[1]:

E. are often necessary in combination with thickeners (see below) to form stable →emulsions but also to improve the consistency, texture and mouthfeel of food. They improve also the structure of food that contains small amounts of fat, such as baked goods. There is a broad variety of emulsifiers available that are tailored to the specific needs. Proper selection and combination of emulsifiers is often the key to a successful food product.

The following food emulsifiers are mainly used: →Lecithin, sodium and potassium salts of fatty acids (→soaps), calcium stearate (→metallic soaps), mono- and diglycerides of fatty acids (→glycerides), →acetic acid esters of mono/diglycerides, →lactic acid esters of mono/diglycerides, →citric acid esters of mono/diglycerides, →diacetyl tartaric acid esters of mono/diglycerides, →succinic acid esters of mono/diglycerides, →mono/diglyceride phosphates, →ethoxylated mono/diglycerides, →sucrose fatty acid esters, →polyglycerol esters of fatty acids, →fatty acid propyleneglycol esters, →lactylic esters of fatty acids and their sodium or calcium salts, →stearyl tartrate and →sorbitan fatty acid esters. E. are widely used in food products. They enable the production of W/O emulsions, such as margarine, and prevent their spattering during frying. For mayonnaise and salad dressings, lecithin is used. They are, in combination with thickeners, responsible for stability and melting behavior of ice cream. Mono/diglycerides and their esters are often used with fruit acids in baked, yeast-leavened goods to increase elasticity of the dough and to improve the uniformity of pores through interaction with the proteins and →starch. Instant foods that contain fat, shortenings and pastries need emulsifiers. They lower the viscosity in chocolate.

Thickeners:

In contrast to emulsifiers, thickeners are not surface-active but are →water-soluble polymers that increase the viscosity of the aqueous phase of a food product, thus changing their flow characteristic. They have also stabilizing effects on (fat) emulsions and influence the size and uniformity of ice and sugar crystals (ice cream). Products used are: alginic acid, →sodium alginate, →agar, →carrageenan, carob gum (→locust bean gum), →guar gum, →tragacanth, →gum arabic, →xanthan gum, →pectin, →starch, →modified starch, →starch derivatives and cellulose ethers, such as →methylcellulose and →carboxymethyl cellulose.

Gelling agents:

To impart not only thickening but also gelling in deserts, jams, preserves and aspic for fish and meat, g. are used such as →gelatin, →sodium alginate →alginic acid, →agar, →carrageenan and →pectins.

Foam stabilizers:

Hydrocolloids, such as →methylcellulose, →hy-droxypropyl cellulose or →tragacanth, are used to stabilize foamy products in baked goods, beverages and confectioneries. Emulsifiers can be used also, especially in W/O emulsions.

Humectants:

A certain moisture level of a food product may be maintained by h. They sometimes prevent the crystallization of sugar and give chewiness to confectionery. →Sorbitol and →glycerol are the most widely used products.

Anticaking agents:

To ensure the free flow of powdered products, such as salt, spices, vegetable- fruit- and beverage-powders, powdered soups and sauces, confectionaries and leavening agents, a. are added. Their characteristic property is that they are insoluble in water. The main products based on RR are →starch, flour and Na- and K-stearates (→soaps) and Ca-stearate (→metallic soaps).

Coatings:

Food products, such as citrus fruit, meat or cheese, are sometimes coated for protection and to improve appearance. →Waxes (e. g., →carnauba wax), resins and cellulose esters (→hydroxypropyl cellulose) or acetylated monoglycerides are used.

Coloring:

Coloring of food can be accomplished by synthetic or natural dyes. The laws pertaining to food colorings are different from country to country. However, natural products (→dyes, natural) are normally accepted. Examples are cochineal, chlorophyll, carotenes, bixin and betanin as well as →caramel color.

Modifiers of taste and flavor

Sugar (→sucrose), the normal sweetener in food, may be substituted by peptides, e. g., →aspartame (→amino acids) for dietary and diabetic reasons (intensive sweetener) or by sugar alcohols, e. g., →sorbitol, →xylitol, →mannitol for diabetics (nutritive sweetener).

To impart a sour taste to foods such as soft drinks, fruit products, pickles, salad dressings, mayonnaise and some fish products, acidulants are added. Most common is vinegar and →acetic acid. →Citric acid, →tartaric acid, →lactic acid, →malic acid and →fumaric acid are also used, some of them contributing a taste of their own (citric acid). →Quinine creates a bitter taste (tonic water). Flavor enhancers (umami) are the salts of →glutamic acid and the more modern →inosine 5′ monophosphate and disodium guanylate, which are effective in lower concentrations than sodium glutamate.

Spices and flavors are widely used natural resources. The extracts consist of complex mixtures of numerous partly still unknown substances.

Processing aids:

Many additives are used during food manufacturing; however, only a few RR-based materials should be mentioned here: →ethanol as an extractant, mold →release agents, such as →lecithin, Ca-stearate (→metallic soaps) and waxes, antifoaming agents (e.g. →sorbitan esters of fatty acids), →yeast as leavening agent, cysteine hydrochloride and ascorbinic acid) as dough conditioners, and various enzymes.

Lit.: T.Branen, "Food Additives", Marcel Dekker NY (1990)

R.J.Lewis Sr. "Food Additives Handbook" Van Nostrand Reinhold NY (1989)

Y.Pomeranz," Advances in Cereal Science and Technology" Vol. VI Am. Ass. of Cereal chemista ISBNO-913250-33-3 (1984)

Kirk Othmer* (4.) **11**, 805

[1]) G. Schuster,"Emulgatoren in Lebensmitteln" Springer Verlag, Berlin/Heidelberg/NY/Tokyo (1985)

Monographs for Emulsifiers for Foods E.F.E.M.A. 250 Avenue Louise (Bte 64) B-1050 Bruxelles(1985)

Foots

→Fats and Oils

→Fatty Acids

Fougère

→Odor Description

Foundry

G.: Gießerei; F.: fonderie

For casting metals, molds are used. There are permanent molds (metal, graphite and ceramic) or sand molds. Sand cores for hollow castings are used for both mold types. Sand molds and cores need binders. Usually, 1–2 parts are added to 100 parts of sand. There is a large variety of inorganic and organic binders. Some foundry resins are based on RR:

Furan resin based binders are prepared by the acid-catalyzed (co)polymerization of →furfuryl alcohol, which is obtained by hydrogenation of →furfural, derived of fibrous by-products of agricultural production, such as →bagasse, →corn, corn cobs, →oats, wood (→furan derivatives).

Furan "no-bake" resins are two-component, r.t.-curing, acid-catalyzed systems of furfuryl alcohol prepolymer with additional monomer. "Hot and warm box" resins are mainly based on urea-modi-fied furfuryl alcohol-formaldehyde condensates that are cured at 100–170 °C. Also phenol-formaldehyde resins, modified with furfuryl alcohol, are used. In "cold box" binder systems, the sand is mixed with a low-viscosity furan resin and a peroxide. The core is formed, and SO_2 is blown into or generated in situ in the sand to cure the resin rapidly.

Drying oils: Core oil binders are used in oven-dried or no-bake systems and are formulated with →drying oils, such as →linseed, →tung, →oiticica or →soybean oil. They are mixed with the sand, shaped and polymerized by oxygen induction at 200–260 °C. →Driers may be added. Sometimes, wood →rosins, →alkyd resins, →sucrose molasses, →dextrins, starch and sulfite waste liquor (→lignosulfonate) from pulping are added.

Alkyd binders: →Alkyd resins are also used as core binders. Isocyanate-modified alkyd resins have excellent core making and molding properties in no-bake systems. They are cross-linked by multifunctional isocyanates as well as by oxydation and polymerization of the unsaturated alkyd resins.

Carbohydrates: The mother liquor of dextrose crystallization (→hydrol) is also used as a core binder.

About 110 000 mt of organic binders are used in this application in USA (1985).

Di- and tri-glycerides are used to cure silicate binders (→glyceryl triacetate).

Lit.: Encycl.Polym.Sci.Engng.* (2.),**7**, 290

Fractional Distillation

G.: Fraktionierung; F.: fractionnement

F. is precise isolation of a given chemical by →distillation. It is used in oleochemistry, e.g., for fatty acids, methyl esters and alcohols.

F. is applied to distilled fatty acid mixtures of →coconut or →palmkernel oil (C_6–C_{18}) and separate unsaturated fatty acids from →rape-seed (C_{18}–C_{22}) oil, as well as for the separation of tall oil fatty acids. The fractionation of tallow, soybean, cottonseed and linseed fatty acids is practiced only in rare cases. High degrees of purity (up to 99%) are reached. F. is carried out in vacuum columns that are filled with internal packings and/or trays. These necessary devices cause a pressure increase (vacuum loss) which has to be compensated by unwanted higher temperatures. The pressure loss has to be kept as low as possible to avoid decomposition. With normal trays, the drop per theoretical plate is 4 mm, but in modern packings only 0.4 mm. Multi-stage jet pumps generate a vacuum of less than 4 mm.

With this technology, such difficult f. as the separation of →erucic acid from similar fatty acids is possible.

With increasing numbers of fatty acids in the feed, the mixture has to pass the unit several times to get satisfactory separation. In modern units, four or six columns are linked together in series. The fatty acid mixture is preheated (heat exchange is used to safe energy), dried and fed into a system of falling film evaporators, fractionation columns and coolers. Computer-aided control, together with on-line gas chromatography, is used in large modern units. A variety of complete units is offered by equipment manufacturers.

Lit.: Ullmann* (5.) **A10**, 261

Fragrances

G.: Duft; F.: fragrance

The term f. is used to describe a mixture of single chemicals, →essential oils, →absolutes, →resinoids and other extraction products from natural sources. Such mixtures are then incorporated in all kinds of cosmetic and household products, such as →soaps, →detergents, air fresheners, toiletries and shampoos, or they are diluted with alcohol and water to become →perfumes. What raw materials are combined in which proportion depends on the base in which f. are going to be used, the desired odor, persistance and the cost of the finished product.

Originally entirely composed of natural raw materials, f. today are always mixtures of synthetic and natural raw materials. Apart from economic reasons, such as constant and generally lower prices, constant quality, almost unlimited supplies and more predictable stability, synthetic raw materials allow the creation of novel notes and provide, if they are captive/patented developments, a competitive advantage for a particular f.- supplier.

Natural products, on the other hand, impart an inimitable complexity and richness to any f. and are therefore used in all types of application. Only the end-product's price affects the choice of what natural raw material to use but not necessarily the quantity. In general, 1/3–1/4 of any f. will be constituted of natural raw materials. The progress in analytical methods has lead to the development of excellent reconstitutions for many of the more expensive →essential oils and →absolutes. This, combined with additional increases in labor costs,

will definitely further reduce the amount of natural raw materials used.

Lit.: Ullmann* (5.) **A11**, 141

Fructans

Syn.: Fructose Polymers; Polyfructoses
G.: Polyfructosen; F.: polyfructosanes

Natural polymers of fructose are called f. and occur in the storage organs of certain higher plants as well as in the extracellular secretions of microorganisms.

F. are chemically classified into two groups: Inulins with β-1,2-, and →levans with β-2,6 linkages. Branched chains will be found in both groups.

Oligomeric and low-polymeric inulins and levans are typical for higher plants, whereas high-molecular f. are produced by microorganisms. F. obtain increased significance as RR due to the availability of high-fructose solutions by acidic or enzymic splitting of polyfructose chains.

Inulin

Levan

Promising higher-plant sources of inulins are chicory (*Cichorium* sp.) and Jerusalem artichoke (→topinambur).

Chicory is native in Europe and Asia. The roots contain 16–20% of inulin, which is composed of 80% fructose and 20% glucose, the latter occupying the terminals of the oligomeric and polymeric chains. Cultivation of chicory is located in Belgium, the Netherlands and France on 14000 ha of land; chicory yield is 44 mt/ha, with 17% inulin. At present, the annual output of inulin based fructose syrup is estimated to 84000 mt, and an additional 60000 mt are planned.

The Jerusalem artichoke is grown in the Mississippi valley and in South-East Russia. It will likely also be cultivated as an interesting RR for fructans in central Europe. It contains 13–18% carbohydrates of which nearly 80% are fructans. Further products of hydrolysis are 15–25% glucose and 3–7% difructose anhydrides.

The isolation of f. is performed on the basis of their high solubility in hot water, and similar techniques as in the beet sugar diffusion step are used. The polysaccharides will precipitate on cooling. An advanced process, immediately leading to fructose syrups, consists of in-situ hydrolysis of the sliced bulbs by acids or fructanases.

Bacterial f. of the high-molecular levan type are extracellular products of *Bacillus polymyxa* when grown on 4–16% sucrose solutions or on →sucrose molasses. The m.w. is $\approx 2 \times 10^6$, the fructose: glucose ratio is 12000:1, and 72% of the fructose has a β-1,2 backbone with 12% branch points of β-1,2-linkages and 13% terminal groups. The water solubility of f. decreases with increasing m.w.

F. may be used as low-cost sources of rather pure fructose. Their viscosity makes them useful as thickeners, binders and as protective encapsulating agent for flavors, colors and pharmaceuticals. Inulin is used for the production of dietary food. I. is not degraded by the enzymes of the mucosa and can be used in diagnostics to determine the filtration rate of the kidneys.

Lit.: Salunkhe/Desai*, 122–129 (1988)
Lichtenthaler* 172,,179 (1991)
Ullmann* **A12**, 49 + 471

Fructo-oligosaccharides

G.: Fructooligosaccharide;
F.: fructooligosaccharides

F. are storage oligomers that consist of β-1,2-linked anhydro-fructose units with terminal AGU. They are widespread in the plant kingdom, in at least 36000 species of 10 higher plant families.

→Fructans are the source for preparation of f. by partial hydrolysis with the enzyme endo-inulinase (pH 5.4, 56 °C). The resulting f. consist of 1–6 monosaccharide units.

Total hydrolysis is performed either indirectly after extraction of inulin from the plants or directly on the macerized and pasteurized plant tissues. After purification and concentration, monosaccharides and sucrose may be excluded by ion-exchange chromatography.

The synthesis of f. is performed by action of immobilized fructosyltransferase on sucrose (pH 5.5, 50–60 °C), the latter acting as an acceptor for the transferred fructosyl groups. Final product composition: DP 3–5 56.7%; DP >5 5.6% ; mono- and disaccharides 39.7% (the latter may be removed by ion-exchange chromatography).

Important characteristics are sweetness and sweetness enhancement, nondigestibility by human and animal amylases, fermentable by intestinal (*Bifidus* bacteria) and other microorganisms.

These properties are used in special health-diet formulations as soluble dietary fiber and for suppressing production of intestinal putrefacient substances.

Lit.: Lichtenthaler* 73–78 (1991)

D-Fructose

Syn.: Levulose, Fruit Sugar
G.: Fructose, Fruchtzucker; F.: fructose, lévulose

F. is one of the most widely distributed monosaccharides (ketohexoses) of the plant kingdom and occurs in numerous sweet fruits, in vegetables and in flowers. It is also the major component of bees' honey. Industrial crystallized products or aqueous solutions are isolated preferentially from mixtures with glucose.

F. exists in crystalline form as β-D-fructopyranose, which turns in aqueous solution, by mutarotation, into a temperature-sensitive equilibrium of five tautomers:

β-D-fructopyranose

α-D-fructopyranose

β-D-fructofuranose

α-D-fructofuranose

M.w.: 180.16 $(C_6H_{12}O_6)$; m.p.: 100–104 °C; $[\alpha]_{20}^D$: −91–93.5°; solubility in water at 20 °C about 80% (w/v).

F. is a constituent of some polymers (→fructans). Steady increases of the cultivation of chicory in

Europe is indicative of the tendency to use its polymers for the production of →fructose syrups.

Most f. is found as a component of →sucrose, from which it is obtained by inversion (→invert sugars). There are many other f.-containing di- and trisaccharides, e. g., →isomaltulose, →leucrose.

The highest potential source for f. is the →isomerization of high-glucose syrups to equilibrium mixtures of glucose and f.(→high-fructose syrups, →"isomerized sugars").

Production of crystallized fructose is rather problematic because of the high solubility of f. Starting from high-fructose sirups or from →invert sugar, enrichment of f. is performed by cyclical or continuous chromatographic separation from glucose. F. contents of 85–96% and 90% of d.s. are necessary for crystallization. The process can be either batch or continuous.

One technique involves cooling the feed material and seeding with f. crystals from 60–80 °C to 25–35 °C. Another possibility is advanced crystallization by addition of methanol (for nonfood uses) or ethanol.

Functional properties are: high sweetness (r.s. 120–180), synergistic sweetening effects, flavor enhancement, moisture management, crystal growth controll, browning reaction and freezing point depression. F. and its syrups are mainly used in food and pharmaceutical industries, in the latter for infusion fluids, for parenteral nutrition forms, and for health and sports dietetics.

F. is easily metabolized in the human body by following the liver pathway, and physiological energy content is 17.2 kJ/g. Postprandial glucose and insulin levels are low, so that f. can be utilized by diabetic patients.

Crystalline f. world production approaches 90 000 mt/a with a rising tendency because of improved enrichment and crystallization techniques. The largest capacity is reported from the USA, ≈ 45 000 mt/a. Other important producers are Finland, Germany, Israel and Japan.

Future interest as RR is focussing on f. being starting materials for the production of →5-hydroxymethyl furfural and →levulinic acid. Hydrogenation leads to →mannitol.

Lit.: Rymon-Lipinski/Schiweck* 183–211 (1991)
      Schenk/Hebeda* 177–231 (1992)

Fructose Polymers

→Fructans
→Levans

Fructose Syrups

→D-Fructose

Fructosyl Sucroses

Syn.: Neo Sugars
G.: Fructosylsaccharosen; F.: fructosylsaccharoses

F. are mixtures of, preferentially, nonreducing tri-, tetra-, and penta-saccharides from two, three or four →fructose moieties that are enzymatically transferred to one →sucrose molecule. F. are linked together by β-1,2-disaccharide bonds under the action of fructosyl-transferase (FTase):

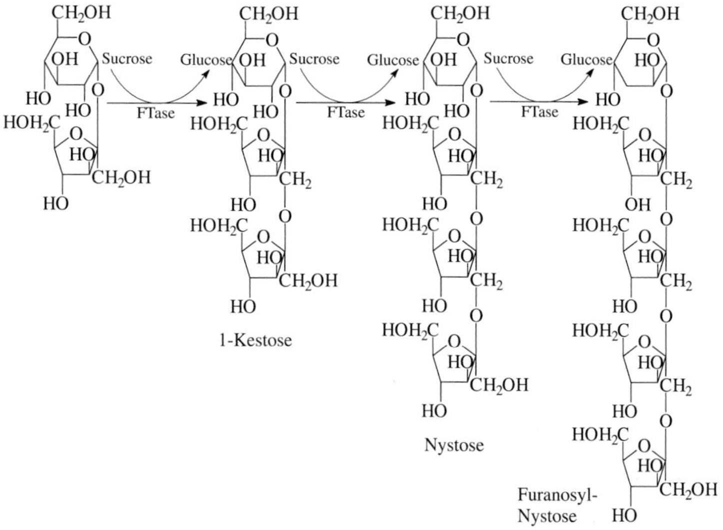

1-Kestose

Nystose

Furanosyl-Nystose

The reaction mixture contains 55% of these compounds and 45% →invert sugar and sucrose. Chromatographic enrichment leads to f. of 95% purity.

They are marketed for food application as 70% syrups or as spray-dried powders, mainly in Belgium and Japan.

Products of similar structure and composition are available by hydrolizing inulin(→fructans)-containing plant materials by action of the endo-inulase enzyme. The refined concentrated products with 85% solids consist of fructo-oligomers with 7 mono-saccharide groups.

The linkages cannot be entirely split by the human intestinal enzymes; the physiological energy value is 1.9 kcal/g, the cariogenity seems to be low, r.s. 30–60.

They are approved for production of candy, baking goods, ice cream and beverages.

Lit.: Rymon-Lipinski/Schiweck* 254–256 (1990)

Fruit Sugar
→Fructose

Fuel Alternatives
Syn.: Biodiesel; Bioethanol; Biofuel; Bioalcohol
G.: Alternative Treibstoffe;
F.: biocarburants, biocombustibles

There may be four reasons to consider f.:
1. Limited availability and high price of crude oil (e.g., oil crisis in the 1970's);
2. Lack of hard currency in countries without their own oil production and with a "closed" economy (e.g.,m Brazil in the 1970's);
3. Ecological considerations (e.g., global warming, smog formation);
4. Large agricultural surpluses.

At present, the high manufacturing costs of all alternatives to petroleum-based fuels give f. only a chance if at least one of the a.m. reasons gains significant importance. Presently, only the ecological reasons keep f. under considerations. In some countries, the agricultural production surpluses have to be cut back. Farmers are interested in using the fallow land for the production of f.

The actual f. are: Natural gas, liquid gas (propane/butane), hydrogen, methanol, ethanol (in Brazil 12×10^9 liter/1989) as well as vegetable oils and their methyl esters. Methane made by anaerobic fermentation from feed lot wastes (biogas) may gain local interest but will not be a bulk fuel. Most of these are still linked to petroleum and other fossil sources. Only the following four are important and based on RR and are therefore considered:
1. →methanol (gasification of wood);
2. →ethanol from carbohydrates by fermentation;
3. vegetable oils (e.g.,→rapeseed or →palm oil); and
4. their →fatty acid methyl esters as diesel substitutes (biodiesel).

Methanol can be excluded from further consideration because it can be made from liquid natural gas (LNG) or coal much cheaper than from biomass if demand arises e.g. as a fuel additive.

The pro and con arguments of each of the remaining three and the conclusions that can be drawn are as follows:

Ethanol:

Pro:
- Easily accessible from various carbohydrate sources (sugar, starch, wood);
- Can be produced almost everywhere;
- Well-established production technology but needs improvement;
- Less global warming due to partial recycling of CO_2;
- Practical experience of use (Brazil/USA) at substitution rates of 25% and 100%;
- Improves octane value (allows higher compression), and can be seen as gasoline additive (higher price level);
- Comes closest to gasoline in handling and use.

Con:
- Needs high acreage (2.2 m²/liter) even if the best crop management (→sugar beet or →sugar cane) is applied;
- High production cost (→ethanol) demands tax exemption or subsidies or tax increase on gasoline to be competitive;
- Investment for large fermentation capacity necessary;
- Many by-products are useless or of low value.
- One liter of ethanol produced creates ten liters of waste water (vinasse);
- No continuous production due to campaign harvesting (beet sugar).

Rapeseed oil:

Pro:
- Easily accessible (local alternatives, e.g., palm oil in Malaysia);
- Can be used as such after simple purification;
- Not much investment in processing equipment;
- Theoretically closed CO_2 cycle;
- Better energy efficiency than ethanol;

Con:

- Present diesel engines must be adjusted;
- Production costs $7 \times$ higher than diesel (present prices);
- Large acreage (8 m^2/liter) necessary;
- Crop rotation allows only one harvest in 3 years at same area;
- 30–70% of energy output is needed as (fossil) energy input,
- Poor flow behavior at low temperatures;
- Other and additional emissions compared to fossil diesel (e. g., CH_4, N_2O, CO, NO_x released during growing and combustion).

Methyl esters:

Pro:

- Availabile from rapeseed crop (arguments see above);
- Theoretically closed CO_2 cycle;
- Normal diesel engines can be used and existing small technical problems may be resolved.

Con:

- Higher costs than the oil as such due to additional processing step (transesterification);
- Additional investment necessary;
- High output of by-product glycerol (collapse of present price level) may burden balance of calculation for this f.;
- 30–85% of energy output is needed as (fossil) energy input;
- Other and additional emissions compared to fossil diesel (e. g., CH_4, N_2O, CO, NO_x released during growing and combustion).

Conclusion:

The discussion of f. today is focussed mainly on ecological advantages. These are not as high as originally expected, because growing and processing needs fossil energy input, which consumes a part of the energy output and makes the CO_2 argument less convincing. Figures derived from →life cycle analysis differ widely due to different co-product allocations, involvement of local tax advantages or given subsidies and many other parameters.

Some, still arising, general technical difficulties seem to be resolvable sooner or later.

The key problems are the noncompetitive production costs.

Convincing and provable ecological facts may, however,. compensate for this disadvantage in the future.

Lit.: Kirk Othmer* (4.) **1**, 826
      Ullmann* (5.) **A16**, 744

Fumaric Acid

Syn.: Butenedioic Acid
G.: Fumarsäure; F.: acide fumarique

HOOC–CH
‖
CH–COOH

m.w.: 116.07; m.p.: 286–287 °C

F. forms monoclinic, prismatic needles and is found as a metabolite in Iceland moss, fungi, lichens and some plants, e. g., *Fumaria officinalis*. It is produced industrially by fermentation of glucose with fungi (e. g., *Rhizopus nigricans* or *Brevibacterium ammoniagenes*) and in large scale by catalytic isomerization of maleic acid.

F. is the starting material to make →malic acid and is used in polyesters instead of maleic acid because of its innocuous nature. It is also an acidulant and flavoring agent in the food and pharmaceutical industries as a substitute for →tartaric or →citric acid.

Annual production in United States (1978) was $\approx 12\,000$ mt.

Lit.: Kirk-Othmer (3.)F **14**, 770–793
      Ullmann (5.) **A16**, 59 + **A21**, 217
      [110-17-8]

Furan

Syn.: Furfuran; Oxole
G.: Furan; F.: furan, furfurane

m.w.: 68.07; m.p.: −85.6 °C; b.p.: 31.4 °C

F. is a colorless, highly flammable liquid, which is miscible with most organic solvents but not with water.

It is produced commercially by decarbonylation of →furfural.

Furfural   Furan

Commercial grade is 99%.

The most important reaction of f. is the →hydrogenation to →tetrahydrofuran. Reaction at 400–450 °C over Al_2O_3 with ammonia yields pyrrole and with H_2S thiophene. F. can react with maleic anhydride to form a Diels Alder adduct. Both can copolymerize radically to form an alternating 1 : 1 copolymer, which is used as a copper-complexing agent.

Uses of f. are as intermediate in the synthesis of pharmaceuticals, agricultural chemicals, stabilizers and fine chemicals. Largest outlet is the production of →tetrahydrofuran, thiophene and pyrrole.

Lit.: Ullmann* (5.) **A12**, 119
    Kirk Othmer* (3.) **11**, 516
    [110-00-9]

Furan-2,5-Dicarboxylic Acid

G.: Furan-2,5-dicarbonsäure;
F.: acide 2,5-furan-dicarboxylique

F. is the end product of the one-step oxidation of →hydroxymethyl furfural.

$$HOOC-\underset{\underset{O}{\diagdown\diagup}}{C}\overset{HC-CH}{\underset{}{C}}-COOH$$

m.w.: 156.08

F. is a highly reactive but stable compound. It is structurally similar to terephthalic and isophthalic acid, which are monomeric units in the synthesis of petroleum-based, large-volume polymers. Thermally stable polyesters, polyamides and polyaramides based on f. represent present and future potential replacements for fossil raw materials in polymer synthesis.
Hydrogenated f. can be further reacted to adipic acid.
Derivatives are reported as components of optical brighteners (→detergents) as well as intermediates for the manufacture of pharmaceuticals and cosmetic ingredients.

Lit.: Eggersdorfer* 184–196, (1993)

Furanol

→Furfuryl Alcohol

Furan Resins

G.: Furanharze; F.: résines furaniques

F. are based on RR and derived from →furfural via →furfuryl alcohol, which undergoes an exothermic selfcondensation under strong acidic conditions. Prepolymers of the following structure result:

$$\underset{O}{\underset{\diagdown\diagup}{HC}}\overset{HC-CH}{\underset{}{C}}-CH_2 \left[\underset{O}{\underset{\diagdown\diagup}{HC}}\overset{HC-CH}{\underset{}{C}}-CH_2 \right]_n \underset{O}{\underset{\diagdown\diagup}{HC}}\overset{HC-CH}{\underset{}{C}}-CH_2OH$$

Condensation is stopped by adjusting pH to 5–8. The prepolymer is stable for several months (>38 °C). Further condensation with additional furfuryl alcohol or with aldehydes (furfural or formaldehyde) and with phenol or urea is again activated by strong acids.
F. are mainly used in core binders (→foundry industry) but also to make corrosion-restistant mor-

tars and concrete (installation of acid-proof bricks) and in laminating for corrosion- and heat- resistant fiberglass-reinforced equipment.

Lit.: Encycl.Polym.Sci.Engng.* (2.) **7**, 454

Furcellaran

Syn.: Danish Agar
G.: Furcellaran; F.: furcellaran

F. is contained in a special kind of red seaweed extract (→marine algae) from *Furcellaria fastigiata*, Rhodophyceae, constituted from α-1,3-linked D-galactose units and of 3,6-anhydrogalactose as repeating units; xylose and glucose are minor constituents.
The structure is related to that of →carrageenan with lower DS (11% sulfuric acid groups).
F. is found in depths of 4–8 m, mainly at the northern and eastern coasts of the Baltic Sea and the Kattegat, where it is harvested by means of drag nets from the bottom of the sea by trawling during the cold months. About 200–250 mt/a of dry f. are produced in Denmark. Amounts in the order of 40000 mt/a are available from the Russian coasts of the Baltic Sea.
The structural relationship with carrageenan results in similar functional properties. Gels are somewhat less brittle and more smoother than carrageenan gels, and the protein reactivity is lower.
In addition to →carrageenan food applications f., is applied in the cosmetic industry for stabilizing aqueous pastes.

Lit.: Levring* 336–338, (1969)
    J.M.V. Blanchard, J.R.Mitchell, "Polysaccharides in Food" 188–189, Butterworths, London/Boston/ Sydney/Wellington/Durban/Toronto(1979),

Furfural

Syn.: Furfurol
G.: Furfural; F.: furfural

m.w.: 96.08; m.p.: –36.5 °C; b.p.: 161.6 °C

F. is the key chemical for the manufacture of furan and its derivatives made from RR. Its precursors are the pentosans xylan and arabinan, which are, next to cellulose, among the most widely distributed chemicals in nature.
It is a colorless liquid (freshly distilled), which darkens rapidly if exposed to air. It has good solvent properties and is miscible with most organic solvents, saturated hydrocarbons excluded. It is used as selective solvent in some industrial applications.
F. is produced from fibrous residues of food crops. Raw material (→hemicellulose) include →corn,

corn cobs, →oats and →rice bran and →bagasse. The pentosan content of these materials must be in the range of 25–40% for effective production. Less frequently used are wood, wood products and sulfite waste liquor (→paper pulp). Important factors for selecting raw materials are the pentosan content, price, availability and cost for collection, transportation and handling. Production is carried out in batch or continuous digesters with strong mineral acid as catalyst. High-pressure steam is blown into the digester, and furfural is recovered by steam distillation, followed by separation and purification procedures. Normally, 18–23% of f. can be obtained.

Pentosan   Pentose

Furfural

The transfer of f. to →furfuryl alcohol and →furan are the two most important reactions. →Methylfuran is obtained by →hydrogenation and careful nitration or halogenation yields the nitro or halogen derivatives. With phenol, →furfuryl alcohol and urea, polymers are obtained. Catalytic vapor-phase oxidation yields maleic acid.

Uses are the manufacture of →furan (→tetrahydrofuran), →furfuryl alcohol, →methylfuran and nitrofurans, which are intermediates for making antimicrobial reagents. It is also used as a selective solvent in the separation of saturated and unsaturated compounds in lubricating oils, gas oil and diesel fuels, as well as in extractive distillation of C_4/C_5 hydrocarbons for manufacturing synthetic rubbers. Other uses are as reactive solvent in corrosion-resistant →furfuryl alcohol resins and in high-carbon phenol resins (abrasive wheels and brake linings). Furfuryl acrylate is used as monomer in vinyl-polymerization.

In 1985, 133 000 mt of f. were produced while using only a part of the capacity of 200 000 mt world wide.

Lit.: Ullmann* (5.) **A12**, 122
      Kirk Othmer* (3.) **11**, 501
      [98-01-1]

Furfural, 5-Hydroxymethyl-
→5-Hydroxymethylfurfural

Furfurol
→Furfural

Furfuryl Alcohol
Syn.: Furanol
G.: Furfurylalkohol; F.: alcool furfurylique
m.w.: 98.1; m.p.: −31 °C; b.p.: 170–171 °C

F. is a colorless liquid, which darkens on exposure to air. It is miscible with water and most organic solvents, saturated hydrocarbons excluded.

F. is produced by liquid- or vapor-phase →hydrogenation of →furfural, preferably with selective copper catalysts that do not promote hydrogenation of the ring structure:

Furfural   Furfuryl Alcohol

The most important chemical property of f. is the formation of →furan resins under strong acidic conditions. It can be hydrogenated to →tetrahydrofurfuryl alcohol and undergoes all reactions of a primary alcohol, e.g., (trans)→esterification. Under mild acid conditions, levulinic acid (or, in presence of alcohol its ester) is formed.

The main outlets for f. are →furan resins. Other uses are as solvent, diluents and modifiers for epoxy, phenolic and urea resins and for the production of →tetrahydrofurfuryl alcohol.

The demand of f. in 1985 was in the range of 80 000 mt.

Lit.: Ullmann* (5.) **A 12**, 125
      Kirk Othmer* (3.) **11**, 510
      [98-00-0]

G

Galactans

→Hemicelluloses

Galactomannans

→Guar Gum
→Locust Bean Gum

Galalith

G.: Galalith; F.: galalithe

G is the trademark of a horn-like plastic, which is made from milk →casein and formaldehyde. It was used for making molded goods and was the only protein-based commercial plastic.

Production was given up in the 1970's for economic reasons.

Commercial revitalization of manufacturing has occasionally been consicered in times of large milk surpluses.

Lit.: Ullmann* (4.) **19**, 559

Galbanum Resinoid/Oil

G.: Galbanum Resinoid/Öl;
F.: resinoïde/essence de galbanum

Crude g. is an exudation collected from the roots of *Ferula galbaniflua* (Boiss.) and *F. rubricaulus* (Umbelliferae/Apiaceae), which are wild-growing in Iran, Syria, Turkey and Lebanon. Initially, a milky latex-like substance, it dries on contact with air into a gum resin, which is then either steam-distilled to produce galbanum oil or extracted with hydrocarbon solvents to yield galbanum resinoid.

Galbanum oil is a colorless to pale-yellow liquid with a typical, intense and diffusive leafy green odor that is reminiscent of pea-pods or green peppers with slight pine-like woody undertones. The resinoid is a semi-liquid, amber-colored material that possesses the same green vegetable-like notes as the oil but with a more rich, balsamic, woody-resinous quality.

Both products are frequently used in small amounts in all kinds of fragrance types to impart a natural, foliage-like impression. Larger amounts are used in green-floral notes such as hyacinth or violet, or in herbal or coniferous fresh notes.

Lit.: Arctander*
      The H&R Book*

Gamma-Linolenic Acid

→Linolenic Acid (γ-)

Gelatin

G.: Gelatine; F.: gélatine
m.w.: 65000–300000

Partial →hydrolysis of collagen (→proteins)- containing animal products (skin, white connective tissues and bones) leads to g. Acid-catalyzed hydrolysis yields type A g., while alkali treatment gives B g. Edible g., used in pharmaceutical, food and photographic applications is made from carefully selected raw materials, while technical g. (→glue, animal) is similar in structure but starts from waste protein material.

While collagen is an insoluble, fibrous protein, g. is soluble in hot water and gives heat-reversible gels upon cooling (a 0.5% solution solidifies at 35–40 °C), which is the most important property of g. Commercial g. is a vitreous, taste- and odorless solid with 9–13% water. G. is amphoteric. It contains high percentages of glycine and unusually high amounts of proline and hydroxyproline (→amino acids).

Chemical composition and physical properties vary somewhat and depend on starting material and conditions of hydrolysis.

Raw material for edible qualities (A type) is mainly pigskin, while crushed bones and dehaired cattle hides are used for making B types. Pigskins are carefully cleaned, soaked in dilute mineral acid for several hours and extracted several times with hot (50–60 °C) water. Final steps are evaporation, purificaton, drying, milling and blending. Pure B type g. is made by alkaline (lime) soaking of bones and hides for several weeks. Final work-up procedures are the same as with A type g.

Production of technical qualities starts from collagen-containing waste material but is carried out also in the a.m. manner, with the difference that glue is more completely hydrolyzed. For details of properties and uses: →glue, animal.

The main outlet (65%) for pure g. is the food market: confectionary, desserts, dairy products and canned meats are only a few products where g. is used. In the pharmaceutical industry (10%), the most important application is for the manufacture of capsules. Another use is as tablet binder. An old but sensitive application is photographic emulsions (20%). No synthetic polymer has succeeded to attack the unusually strong position of g. in this area. There are some small industrial uses of pure g.

The total world capacity is estimated to be 175 000–200 000 mt/a, with USA producing 30 000–32 000 mt.

Lit.: A.G.Ward, A.Courts (ed.) "The Science and Technology of Gelatin", Academic Press Inc. NY (1977) Ullmann* (5.) **A12**, 307
[9000-70-8]

Gelidium spp.
→Agar

Gellan
→Microbial Gums

Gelose
→Agar

Genetic Engineering
Syn.: Recombinant DNA
G.: Gentechnologie; F.: manipulation génétique

The branch of biology dealing with the splicing and recombining of specific genetic units from the DNA of living organisms is used to modify the existing genetic codes to produce new or improved species, valuable biochemicals, etc. With g., it is now possible to introduce new genetic material beyond the natural barriers from one plant, animal or microbial species to another.

Typical methods include tissue culture, protoplast fusion and specific gene transfer methods. Tissue culture techniques are adapted from methods normally used for growing microorganisms by adding specific phytohormones and selected culture media to reach callus growth of cells. These may be modified to suitable growth of branches, leaves, roots and, finally, →inflorescences. By adding, e. g., herbicide to the tissue culture medium, cells with resistance to these herbicides may be selected and regenerated to productive plants. In protoplast fusion techniques, the cell walls are degraded by →enzymes (cellulases, hemicellulases), and the resulting protoplasts of different plant species may be fused. If this is successful, and the combination of the nuclei as well as the plastides and the mitochondria is successful, new →hybrid varieties may be regenerated.

Specific gene transfer methods use the naturally occurring transformation techniques, such as gene vectors (Ti plasmids from *Agrobacterium tumefaciens*) or plant-pathogenic DNA viruses, for obtaining new characteristics in other plants.

Typical goals of g. are the transformation of →pesticide resistance into crops, the incorporation of →fatty acid patterns of →coconut into the pattern of →rapeseed to open another source for lauric oil (→fats and oils), the improvement of crop and vegetable species by adding specific flavor genes.

Lit.: D.J.Murphy (ed.) "Designer Oil Crops" VCH, Weinheim (1994)

Gentiobiose
G.: Gentiobiose; F.: gentiobiose

m.w.: 342.2 ($C_{12}H_{22}O_{11}$)

G. is formed as one of the reversion products during the acid →hydrolysis of starch. It is found enriched in the mother liquor →hydrol of dextrose crystallization, together with →isomaltose, →panose and other oligosaccharides. G. cannot be split by α-glucosidases (→amylases); it develops a strong bitter taste. It lowers the crystallization yield of dextrose in starch hydrolysates after acid hydrolysis.

Lit.: Tegge* 66–68, 273 81 988)

Genus
→Taxonomy

Geranial
→Citral

Geranium Oil
G.: Geraniumöl; F.: essence de géranium

G. is produced from various →hybrids and subspecies of *Pelargonium graveolens* and other Geraniaceae and yields distinctly different products, depending on the country of origin. The oil is steam-distilled from the leaves and stalks of the plants; an →absolute is also commercially available but of lesser importance.

G. is usually yellowish, dark-yellow or greenish-olive in color with a powerful rosy-leafy character, accompanied by minty-herbaceous, sulfuric and sometimes earthy qualities.

Main production areas are Reunion and Madagascar, Egypt, Algeria and Morocco, China, Russia and France.

Main constituents are citronellol, geraniol and menthone.

G. is rarely used in flavors but is one of the most frequently used →essential oils in perfumery, where its rosy-fresh qualities make it useful in rose and other floral types. The minty-herbaceous aspects are main constituents of the fantasy fougère note and other herbaceous-floral types. It blends well with lavender and citrus notes in men's colognes.

Lit.: Arctander*

Ghee Fat

Syn.: Ghi

G.: Ghi Butter

G. derives from butter (→butterfat), which is heated to 122 °C for about 30 min. Water evaporates and lactose, salts and albumins sink to the bottom. The melt is poured off and solidifies to a yellow, finely grained fat with a m.p. of 30 °C and a water content of 0.7%. G. can be stored much longer than butter. G. is used as multipurpose fat, mainly in India and is made from buffalo butter.

Lit.: D.Swern "Bailey's Industrial Oil and Fat Products" Wiley Interscience Publ. Co. Inc. (1979)

Gibberellins

G.: Gibberelline; F.: gibberellines

ent-gibberellan as basic structure of gibberelins

G. represents a group of ca. 70 tetracyclic, diterpenoic (→terpenes) phytohormones that were first isolated in 1938. They are found in many higher plants, mosses, fungi, algae and bacteria. G. regulate the development of plants and are produced by fermentation in cultures of *Gibberella fujikuroi*. They are used as plant growth regulators for the promotion of growth of plants, especially of seedlings, bacteria and tissue cultures (→genetic engineering).

Of special interest is g. A_3, which is produced biotechnologically from the fungus *Fusarium moniliforne* and it is used especially in the USA to produce seedless dessert grapes and to increase fruit and vegetable yields of celery and artichokes when applied in concentrations of 1–1000 ppm. A mixture of g. A_3, A_4 and A_7 is used also as a food additive in the malting of barley.

Lit.: Ullmann* (5.) **A20**, 415
      [77-06-05]

Gingelly Oil

→Sesame Oil

D-Glucit/Glucitol

→D-Sorbitol

Glucomannan

→Hemicelluloses

D-Gluconic Acid

Syn.: Dextronic Acid

G.: D-Gluconsäure; F.: acide D-gluconique

m.w.: 196.16; mp.: 131 °C

G. is prepared by the oxidation of D-glucose (by hypochlorite, hypobromite, electrochemically in alkaline medium, by microbial oxidation with bacteria (e.g., *Acetobacter aceti*) or fungi (e.g., *Penicillium* sp. or *Aspergillus* sp.), or by enzymatic oxidation with glucose oxidase (from *Aspergillus niger*). G. is used as a metal-etching agent, →textile auxiliary, in bottle washing liquids and as an additive in beverages. Ferrous gluconate is used for the therapy of iron insufficiency; sodium gluconate finds an application as a sequestrant in →cleaners; the magnesium salt is antispasmotic, and the ammonium salt is used as a latent acid catalyst in textile printing. In aqueous solutions, the acid is partially transformed into an equilibrium mixture with the corresponding gamma- and delta-gluconolactones.

The combination of sodium gluconate with sodium carbonate or sodium hydroxide solutions are useful for removing grease and corrossion from aluminium and rust from steel. Gluconate is also useful in the pretreatment of certain surfaces, e.g., in the galvanic deposition of nickel-cobalt brazing surfaces onto aluminium. Concrete manufacturers have found sodium gluconate to be a highly effective agent for retarding the curing process.

Current annual worldwide production capacity of g. is estimated to be 60 000 mt. The bulk of production (85 %) is in the form of sodium gluconate and other alkali gluconate salts.

Lit.: Ullmann* (5.) **A12**, 449
      [526-95-4]

Gluconic Acid-5-lactone

Syn.: Gluconolactone
G.: Gluconsäure-5-lacton;
F.: 5-lactone d'acide D-gluconique

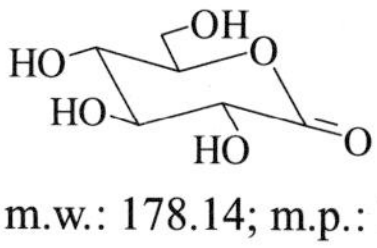

m.w.: 178.14; m.p.: 153 °C (d.)

G. is obtained by treatment of →D-gluconic acid with water; an equilibrium between g. and its corresponding gamma-lactone is observed. G. is used in many →cleaners. In the dairy industry, it is used for the protection against the formation of milk incrustation because of the sequestring ability of the gluconate radical, which remains active in alkaline solutions, and in breweries as a beer-scale destroying agent. In the food processing industry, g. is used as an acid component in baking powder and as an ingredient of curing solutions.

Lit.: Ullmann* (5.) **A12**, 450
      [90-80-2]

Gluconolactone

→Gluconic Acid-5-lactone

D-Glucose

→Dextrose

D-Glucose, anhydrous

→Dextrose

D-Glucose, monohydrate

→Dextrose

Glucose, solid

→Solid Glucose

Glucose Syrups

G: Glucosesirup, Stärkesirup; F: sirop de glucose
G. are purified concentrated aqueous solutions of nutritive saccharides, obtained by →hydrolysis of glucosidic linkages of →starch.
Commercial g. are produced in various degrees of hydrolysis (→dextrose equivalent, DE), ranging from >20 % DE to <80 % DE; the latter is termed "starch hydrolyzate" and consists largely of →dextrose. G. of DE <30 % are available as spray-dried, hygroscopic, white powders: "dried glucose syrup". Hydrolysis may be performed in different ways:

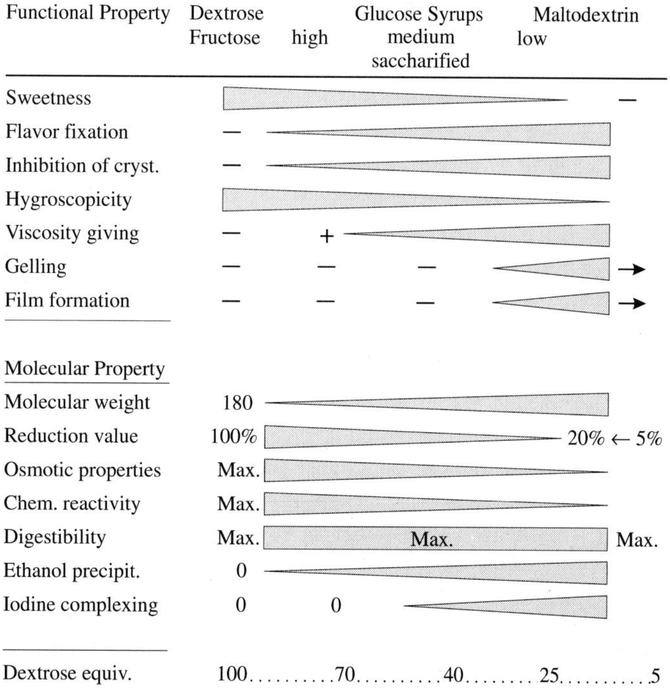

Functional Property	Dextrose Fructose	high	Glucose Syrups medium saccharified	Maltodextrin low
Sweetness				—
Flavor fixation	—			
Inhibition of cryst.	—			
Hygroscopicity				
Viscosity giving	—	+		
Gelling	—	—	—	→
Film formation	—	—	—	→
Molecular Property				
Molecular weight	180			
Reduction value	100%			20% ← 5%
Osmotic properties	Max.			
Chem. reactivity	Max.			
Digestibility	Max.		Max.	Max.
Ethanol precipit.	0			
Iodine complexing	0	0		
Dextrose equiv.	100 70	....	40 25	5

- Single-step acid hydrolysis under pressure (the original process);
- Single-step hydrolysis with thermoresistent bacterial →α-amylase until 110 °C;
- A (first) starch liquefaction step by either acid or bacterial α-amylase, followed by a (second) saccharification step with a fungal α-amylase or/and glucoamylase.

Neutralization (pH adjustment), desludging, de-colorizing, deionization and concentration of the raw hydrolyzates lead to the commercial colorless and clear syrups with about 75–80% d.s. Turbid, white syrups are those of low conversion (<30% DE).

Composition and properties depend on the catalysts used as well as on the conditions of hydrolysis. Commercial g. are:

low, 20–38%DE; normal, 38–48%DE; medium, 48–58% DE; high, 58–68%DE; extra high, >68% DE.

Differences in composition of saccharides (equal DE 42%) depend on the kind of hydrolyzing agent:

Hydro-lyzing Agent	Mono	Di	Tri	Tetra	Higher
		Saccharides (all in %)			
A	18.5	14.5	12.0	10.0	5.0
B	5.9	44.4	12.7	3.3	33.7
C	2.5	58.0	18.0	2.0	19.5

A: Acid Hydrolysis; B: Acid liquefaction + fungal α-amylase; C: Bactertial α-amylase + β-amylase

The most important properties of g. with respect to their applications include:

viscosity increase, stabilization of moisture content, foam stabilization, inhibition of crystallization (other sugars, coarse ice crystals), freezing point depression, color formation, enhanced solubility, sweetness, mouth feel and body, digestibility and fermentability.

These properties and their relations to DE value, m.w. and other molecular properties are shown on p. 121.

Food industry applications are confectionary, baking, fruit processing, composed foods, alcoholic beverages, soft drinks.

Pharmaceutical industry applications are medicated confectionaries, carriers and soft sweeteners for liquid medicines, coatings and binders for pills. The world production of g. in $\times 10^6$ mt (1992) was 5.6 (USA 2.8, EC 1.7, Japan 0.5). There is generally a rising trend, caused by nonfood uses, mainly for antibiotics and chemicals via biotechnology.

Lit: Schenk/Hebeda* 227–317 (1992)
    Ruttloff* 597–601, 610–617 (1994)

Glucosidases
→Amylases

Glucoside Hydrolysis
→Hydrolysis Glucosidic Linkages

Glucurone
→D-Glucurono-6,3-lactone

D-Glucuronic Acid
G.: D-Glucuronsäure; F.: acide glucuronique

m.w.: 194.14; m.p.: 167 °C

G. forms white needles and is obtained by oxidation of the primary hydroxyl group of →D-glucose. G. is a constituent of the heteropolysaccharide →xanthan; it usually occurs as a glycosidic combination with phenols, alcohols, etc., as glucuronides. By lactonization, →D-glucurono-6,3-lactone is formed.

Lit.: G. J. Dutton (Ed.), "Glucuronic Acid, Free and Combined", Academic Press, NY (1966)
[6556-12-3]

D-Glucurono-6,3-lactone
Syn.: Glucurone; Dicurone
G.: Glucuronsäure-γ-lacton;
F.: glucurono-γ-lacton

m.w.: 176.12 m.p.: 176–180 °C

G. is found in many plant gums in polymeric combination with other carbohydrates. It is an important structural constituent of practically all fibrous and connecting tissues of animal organisms but also of many mucilages. It is prepared from many polysaccharides by oxidation or by lactonization of →D-glucuronic acid. It is used as a detoxifying therapeutic agent for treatment of hepatitis, arthrosis and ischemia. G. decreases the toxicity of sulfonamides.

Lit.: Carbohydr. Res. **92**, 51 (1981)
    The Merck Index* (11.) 4362
[32449-96-6]

Glue, Animal

Syn.: Animal Glue; Technical →Gelatin
G.: Tierischer Leim; F.: colle forte

G. derives from the protein →collagen, extracted from hides and bones of animals. Animal glue and gelatin have the same structure (→gelatin). G. can be made from bone or skin wastes. Waste fish proteins are sometimes also used.

For production details: →gelatin.

G. is marketed as a dry solid with a color ranging from yellow to brown. It is a hydrolysis product of collagen with an average m.w. between 20 000 and 90 000. Hot water (50–60 °C) dissolves g. readily and boiling water slowly hydrolyzes it. Polyvalent metal ions form insoluble products. A g. solution gels on cooling and can be liquified by heating.

G. once was the only wood →adhesive but is substituted today almost completely in this application by synthetic products.

The manufacture of abrasives (sand paper), mainly in combination with formaldehyde, and gummed tapes are the main markets today. Another still strong outlet for g. is as a binder in match heads. Laminating →adhesives and stabilizers in latex formulations are other uses as well as in copper plating and extraction, flotation aids in uranium mining, paper sizes, especially for making banknotes and as protective colloid in various applications.

World demand is in the range of 30 000–50 000 mt/a.

Lit.: Kirk Othmer* (3.),**11**, 911

Glues

→Adhesives

Glutamic Acid

Syn.: 2-Amino-glutaric Acid; Glutaminic Acid
G.: Glutaminsäure; F.: acide glutamique

$$HOOC-CH-CH_2-CH_2-COOH$$
$$\underset{NH_2}{|}$$

m.w.: 147.13; m.p.: 247–249 °C (d.)

G. is a proteinogenic, nonessential →amino acid, mainly occurring in →casein (23.6%), →wheat gluten (31.4%), →maize gluten (18.4%), →soybean (18.5%) and →sucrose molasses. A part of g. is obtained from its natural sources by acid pressure-hydrolysis of →proteins, such as gluten, casein or soybean and is now produced on large scale either by fermentation (with *Corynebacter glutamicum*, *Brevibacter flavum* or *Micrococcus glutamicus*) or by enantioselective enzymatic hydrolysis with hydantoinase from *Bacillus brevis*. In addition,

g. is obtained by the microbial conversion of α-keto glutaric acid using strains of *Aeromonas*, *Bacterium megatherium-cereus* or *Bacterium pumilu* (28–30 °C, neutral pH, within 36–48 h). Most of g. (90%) is applied as its mono sodium-salt as a flavor enhancer (→food additives) in precooked meals and in flavoring essences. G. is the active ingredient of →hydrolyzed vegetable protein from →corn gluten meal and →wheat gluten. It is of limited pharmaceutical use in the therapy of hyperammonaemia and neuroses due to its identification as an excitatory neurotransmitter; the hydrochloride is used as a gastric acidifier, the magnesium salt hydrobromide as an anxiolytic. The hydrochloride has been used to improve the taste of beer.

Lit.: Ullmann* (5.) **A4**, 149; **A16**, 711; **A19**, 2
      [6899-05-4]

Gluten-Cereal

G.: Getreide-Gluten; F.: gluten de céréales

G. is the water-insoluble protein complex in the cereal grain endosperm. It is the commercial term for industrial products that are extracted in the course of starch production by →wet milling (→corn gluten, →corn gluten meal, →zein, →wheat gluten).

Gluten proteins are composed of two main classes:

Glutelins	Prolamins
Avenin (Oats)	Gliadins (Wheat, Rye)
Orycenin (Rice)	Hordein (Barley)
Glutenin (Wheat)	Zein (Corn, Sorghum).

Glutelins are insoluble in water but soluble in diluted acids and alkaline solutions. Prolamins are soluble in 50–80% EtOH.

M.w. are between 30 000 and 40 000.

Varying small amounts of albumins (leucosins) and globulins make up the water-soluble share of the cereal endosperm and aleurone proteins.

The g. proteins are somewhat different with respect to their amino acid compositions (%):

Protein	Arginine	Leucine Isoleucine	Valine	Glutamic Acid	Proline
Gliadine	3.2	6.0	3.0	46.0	13.2
Glutenine	4.7	6.0	1.0	27.2	4.4
Zein	1.6	3.0	3.0	35.6	9.0

About 20–25% of the crude-protein nitrogen is bound as amide nitrogen in glutamine and asparagine. In all cereals, the amino acids are not sufficiently balanced with respect to human nutritional requirements.

The high content of →glutamic acid makes wheat as well as maize gluten well suited for the production of →hydrolyzed vegetable protein.

Only corn gluten and wheat gluten are of industrial significance. Fields of utilization are:

- Feed component in →corn gluten feed and wheat feed;
- Human nutrition: gluten upgrading in yeast-raised bakery products; special gluten-enriched, low-polysaccharide bakery products; pasta product enrichment (wheat gluten meal);
- Substrate for hydrolysis to →hydrolized vegetable protein;
- Nonfood industrial applications are advancing with →wheat gluten as binders, adhesives, coatings and bulking adjuvants as a consequences of largely enhanced →wheat starch production as well as development of →wheat gluten derivatives.[1]

Lit.: J.B.S.Braverman "Introduction to the Biochemistry of Food", 126–129, Elsevier Publ.Co. Amsterdam/ London/NY (1963)

[1] "Stärke im Nichtnahrungsbereich" Reihe A/Heft 388, 189–196, 230–238, Schriftreihe des Bundesministers für Ernährung, Landwirtschft, Forsten, Landwirtschaftsverlag GmbH, Münster-Hiltrup (1990)

Glutineous Starches

→Waxy Starches

Glycerides

G.: Glyceride; F.: glycérides

G. are the esters of →glycerol with →fatty acids

CH_2-O-R^1 (position 1)
$CH-O-R^2$ (position 2)
CH_2-O-R^3 (position 3)

R^1, R^2, R^3: $H-CO-C_{7-21}$ alkyl

Nomenclature and definitions are not always clear in this group of frequently used products.

Structure:

There are various possibilities to substitute the three OH-groups of glycerol either by the number of fatty acid acyls (R^1, R^2, R^3) and by the position of different acyl groups (1,2 and 3).

In monoglycerides, only one of the three OH groups is acylated, the other two are H. There are two isomers possible: one where R_2 is the acyl and another where R_1 or R_3 contains the acyl group (which are identical).

Diglycerides carry two acyls. There are also two possible isomers: one in which the acyls are on the 1 and 2 positions, and another where the 1 and 3 positions are acylated.

There is only one possible structure in triglycerides because all three OH-groups are acylated.

There is another option to increase the number of possibilities and vary the properties of g.: the nature of the fatty acids used (chainlength, saturation/unsaturation). Rather impure, broad cuts (→fatty acids) can be used to yield a very complex mixture. If fatty acids of high purity are used, the resulting g. are rather well defined.

Properties:

Triglycerides, including those made by synthesis, are hydrophobic fatty or oily materials. For their properties: →fats and oils.

The mono- and the diglycerides and their mixtures are called "mono/diglycerides" (MDG) and show surface activity due to their combination of hydrophilic and hydrophobic molecule parts. Depending on the nature of the fatty acid used, they are solids or liquids at r.t. Their m.p. generally increases from tri- to di- to monoglycerides

Odor, taste and color depend also on the nature of the fatty acid and the degree of purity. Monoglycerides show various polymorphic structures with different m.p., and they are able to form hydrates with a gel structure.

Intra- and intermolecular acyl migration before and during use, especially enhanced by higher temperatures and hydrolytic conditions, may lead to structures with other properties.

So-called self-emulgating mono/diglycerides are made by addition of small amounts of →soap. Thus, their lipophilic character (W/O emulsifier) is changed to an O/W system. The use of soaps in g. used in food applications is restricted by legal regulations in different countries.

Production:

Normally, g. are made by acid-catalyzed →esterification of free fatty acid and glycerol or by →transesterification of glycerol and →fatty acid methyl esters.

Triglycerides are mostly derived from nature (→fats and oils) or are made by →esterification or →transesterification, if special structures and properties are desired. Triolein (→olein) or tristearin (→stearin) are made this way. Purity depends on the quality of the fatty acid or methyl ester used.

The widely used mono/diglycerides (MDG) are frequently produced by →transesterification of a trigyceride (→fat and oils) with glycerol to yield an equilibrium mixture that contains 35–60% monoglyceride and 35–50% diglyceride, and the rest is 1–10% of trigyceride, glycerol and fatty acid, depending on the ratio of triglyceride and glycerol in the reaction mixture. The reaction is carried out with alkaline catalysts at 200–250 °C for less than an hour. The fatty acid composition remains that of the starting triglyceride. Fats and oils or fatty acids used are mostly of the tallow range (C_{16}–C_{18}). Such mixtures are called "glyceryl monostearate" (GMS) in the market, a term, which does not reflect its inhomogenous character.

Monoglycerides of 90–96% purity can be made by molecular distillation (high vacuum, thin film). 90% are substituted at the 1-position and only 10% in the 2-position.

Very specific structures of g. can be made by use of enzymatic (lipase) methods.[1]

Uses:

For triglyceride uses: →fats and oils.

Mono-g. (MG) and the mixtures of mono/di-g. (MDG) are widely used as nonionic food emulsifiers (→food additives). They find also application in →cosmetic preparations and in the plastic industry (→plastic additives). All three areas use large volumes of these materials.

As food emulsifiers, they interact with lipids, proteins and carbohydrates and are used in baked goods, margarines, confections, icings, toppings and in the manufacture of peanut butter.

In USA (1981), 100 000 mt were used in food applications (≈66% of the total food emulsifier market), with 57 000 mt going into bread production.

Dervatives and their uses:

Mono/diglycerides are the starting material for ethoxylates yielding ethoxylated MDG (EMG). By addition of 20 moles of EO, food emulsifiers (W/O) are obtained (m.p.:28–32 °C) that are rather stable against temperature and →hydrolysis. In 1981, the USA used 6500 mt, mainly in yeast-raised baked goods (bread).

They act also as intermediates for →ethoxylation and/or →sulfatation, yielding nonionic or anionic surfactants. They are also the starting material for further →esterification with acetic, lactic, citric, succinic and acetylated tartaric acid, as well as with the mono-sodium phosphate esters. All are well-known →food emulsifiers.

For specific monoglycerides see: →glyceryl monooleate, → glyceryl monostearate and →glyceryl monoricinoleate.

Lit.: G.Schuster,"Emulgatoren in Lebensmittel", Springer Verlag Berlin/Heidelberg/NY/Tokyo (1985)
N.J.Krog "Food Emulsions" K.Larsson,S.E. Friberg (ed.)
Marcel Dekker NY/Basel (1990)

[1] J.Falbe et al.; Angew.Chemie International Edition **27**, 56 (1988)

Glycerin(e)

→Glycerol

Glycerol

Syn.: Glycerin(e); 1,2,3-Propanetriol
G.: Glycerin; F.: glycérol, glycérine

CH_2–OH
|
CH–OH
|
CH_2–OH

m.w.: 92.09; m.p.: 18.2 °C; b.p.: 290 °C (d.); 182 °C (2.92kPa/22mm)

The simplest triol (trivalent alcohol) is a colorless, clear, viscous, hygroscopic, odorless, sweet-tasting liquid. G. is miscible with water and alcohol in any ratio but insoluble in ether, benzene and chloroform. G. lowers the freezing point of water with a maximum at −46.5 °C (66.7% g. and 33.3% water). This property was the basis for a major market as antifreeze for automotives for many years, which today has been completely substituted by ethylene glycol.

G. is always the by-product of the →splitting of →fats and oils, which contain -depending on the fatty acid composition- between 8 and 14% g.

High-pressure →hydrolysis of →fats and oils is the main source and yields fatty acids and g. as a 15% solution in water. Their →transesterification, mainly with methanol, gives → fatty acid methyl esters and g. in a concentration of 90–92%. →Saponification was the traditional method to obtain g. but has been abandoned almost totally.

To purify and concentrate aqueous g. from the above mentioned processes, treatment with active carbon, →distillation in thin film evaporators and ion exchange techniques are used.

Various concentrations (86–88%, 90–95%, 99.8%) and purities are marketed.

The largest market for g. as such is for pharmaceuticals and especially cosmetics (35% of the total consumption). It functions as moisture retainer, solvent, substance carrier and plasticizer for hydrophilic polymers. It is easy to perfume, nontoxic and has good skin care properties. It is used in many formulations, such as ointments, shaving creams, creams and lotions, toothpastes, ear and nose drops, and many more.

Another large market is the →tabacco industry for moisture retention with a positive effect on burning behavior (10% of the market).

Food applications are numerous, where g. acts as preservative, plasticizer, flavor enhancer, improver of consistency and freshness (→food additives) and as intermediates for food emulsifiers (10% of the total market).

There are thousands of different applications for g.: in manufacturing of parchment paper for wrapping papers in the food industry, production of →cellophane, →rubber, →plastic and → textile. In the metal industry, it is used as lubricant and as hydraulic fluid. Adhesives, paints, coatings, ceramics and inks are other outlets.

G. is used as an intermediate for making →alkyd resins (second largest outlet with 12%), →polyesters and →polyurethanes.

As a triol, g. is able to form mono, di-, and tri-esters of inorganic and organic acids. The OH- group in the 2-position is the least reactive. The most important inorganic ester is →glyceryl-trinitrate (nitrogycerine), while the most common organic esters are those of fatty acids in various chainlengths and degrees of substition (→glycerides). The ester of acetic acid is also used in considerable volume (→glyceryl triacetate, →glyceryl diacetate and →glyceryl monoacetate).

G. can be ethoxylated and easily undergoes self condensation to yield →polyglycerol.

Future markets may be the use as animal feed and as carbon source in →fermentation.

G. production worldwide is estimated to be in the range of 700 000 mt/a (1995), which are almost entirely made from natural →triglycerides (RR). In the 1970's some production units for synthetic g. were installed that are shut down today, with the exception of one plant.

Lit.: A.A.Newmann, "Glycerol", Morgan Grampian London (1968)
  Kirk Othmer* (3.),**11**, 921
  Ullmann* (5.) **A12**, 477
  [56-81-5]

Glyceryl Diacetate

Syn.: Diacetin; Glycerol Diacetate
G.: Glycerindiacetat; F.: diacétate de glycérine

G., is the diester of glycerol and acetic acid. There are two isomers: the 1,2- [b.p.: 140–142 °C (1.6 kPa/12 mm)] and the 1,3-diacetoxy-2-propanol (m.p: 40 °C; b.p.: 260 °C). Normal diacetin is a

mixture of both and has similar properties as →glyceryl triacetate with a slightly different polarity. It is used alone or in mixture, e. g., in the food, cosmetic and pharmaceutical industries, as plasticizer and solvent for cellulose derivatives, →alkyd resins and →shellac.

Lit.: Kirk Othmer* (3.) **11**, 931

Glyceryl Monoacetate

Syn.: Acetin; Glycerol Monoacetate
G.: Glycerinmonoacetat;
F.: mono-acétate de glycérine

b.p.: 158 °C (22kPa/165 mm)

Only one OH group of glycerol is esterified. There are two isomers possible. The monoacetate is used in the manufacture of tanning agents and dynamite. It acts as a plasticizer for →cellulose acetate and as a solvent in the food industry.

Lit.: Kirk Othmer* (3.) **11**, 931

Glyceryl Monooleate

Syn.: GMO; Glycerol Monooleate
G.: Glycerinmonooleat;
F.: mono-oléate de glycérine
For structure and production: →glycerides.

m.w.: 356.55; m.p.: 36 °C

G. is insoluble in water but forms gels.

Commercial products are mixtures of mono and diglycerides that contain also other fatty acids than oleic acid.

Distilled products have a monoglyceride content of more than 90%.

Very pure g. can be made by direct →transesterification of high-oleic →sunflower oil (HOSO).

G. is used as emulsifier for W/O emulsions in cosmetic and pharmaceutical formulations and as →food additives.

Lit.: [111-03-5]

Glyceryl Monoricinoleate

Syn.: Glycerol Monoricinoleate
G.: Glycerinmonoricinoleate;
F.: mono-ricinoléate de glycérine
For structure and production: →glycerides.

m.w.: 372.55

G. is a colorless paste, which is water-dispersable. A mixture of mono- and diglycerides is commercially available as well as distilled versions (>90% mono).

G. is used as emulsifier in in →cosmetic, →textile, →paper and →leather industries.

Lit: [1323-38-2]

Glyceryl Monostearate

Syn: GMS; Glycerol Monostearate
G.: Glycerinmonostearat;
F.: mono-stéarate de glycérine
For structure and production: →glycerides

m.w.: 358.55

Colorless wax-like mass. Insoluble in water.
Commercial products are mixtures of mono- and diglycerides of various fatty acids, mainly stearic acid. Products with a mono content of more than 90% are available.
G. is not self-emulsifying but this property can be enhanced by addition of fatty acid alkali →soaps.
Uses are mainly in the cosmetic and food industries (→food additives) as emulsifier and stabilizer, but also in the plastic industry as plasticizer and lubricant (→plastic additives) and in rubber (→rubber chemicals).

Lit.: [31566-31-1]

Glyceryl Triacetate

Syn.: Triacetin; Glycerol Triacetate
G.: Glycerintriacetat; F.: triacétate de glycérine

$$CH_2-O-CO-CH_3$$
$$CH-O-CO-CH_3$$
$$CH_2-O-CO-CH_3$$

m.w.: 218.2; m.p.: 3 °C; b.p.: 258–260 °C

G. is a colorless, almost odorless, nonviscous liquid. It is miscible with most solvents with the exception of hydrocarbons and →fats and oils.
It is manufactured by →esterification of →glycerol with acetic acid or its anhydride, followed by vacuum distillation.
Uses are mainly as plasticizer and stabilizer in the production of → cellulose acetate-based cigarette filters (→tobacco industry). Other outlets are binder for solid rocket fuels, desensitizer for →explosives and plasticizer for →cellulose nitrate.
Mixtures of diacetin and triacetin are used as curing agent for silicate-based core binders in the →foundry industry.

Lit.: Kirk Othmer* (3.) **11**, 931
        [102-76-1]

Glyceryl Trinitrate

Syn.: Nitroglycerine (incorrect terminology)
G.: Glycerintrinitrat, Nitroglycerin;
F.: nitroglycérine

$$CH_2-O-NO_2$$
$$CH-O-NO_2$$
$$CH_2-O-NO_2$$

m.w.: 227.09; m.p.: 13.5 °C; b.p.: 256 °C (d.)

Yellowish, oily odorless liquid. Highly explosive when heated or exposed to shock, impact and friction.
G. is manufactured by dropping glycerol into cooled mixed acid (50:50 mixture of nitric acid 90% and oleum with 25–39% SO_3) under stirring. Temperature is kept below 25 °C. Nitration is complete after one hour. The oily phase is washed several times with water. There are old batch and modern continous processes. Especially the two modern continous processes (Biazzi p. and Nitro Nobel Injector p.) guarantee maximum safety because they transfer the nitrate immediately into an aqueous, rather insensitive emulsion.
G. is primarily used as →explosive (dynamite), in →fire fighting in oil wells and in rocket propellants. It is also used in medicine as a vasodilator in coronary spasm.

Lit.: Kirk Othmer* (3.) **9**, 572
        [55-63-0]

Glycine spp.
→Soybean

GMO
→Glyceryl Monooleate

GMS
→Glyceryl Monostearate

Gold of Pleasure
→Camelina

Gossypium spp.
→Cotton

Gracilaria sp.
→Agar

Grape Seed

Vitis vinifera L.
Vitaceae
G: Traubenkerne; F: pépins de raisins

Grapeseeds are a by-product of the wine industry. Seeds need to be separated from the wine pomace, freed of adhering pulp and dried. The oil (pomace oil) is mechanically pressed out and/or extracted with solvents.

Major suppliers in Europe are France, Spain and Italy, with an annual production of about 6000 mt of grapeseed oil each.

Argentina produces ca. 2000 mt/a.

The oil content varies between 6% (dark and red grapes) and 20% (sweet white grapes). The fatty acid pattern is:

C16:0	C18:0	C18:1	**C18:2**
4–6%	2–4%	13–31%	**50–76%**

Best oil grades are used for edible purposes. The rest is applied for soap and paint making and for oleochemical applications.

Lit.: Gunstone* (1994)

Greases

→Lubricating Greases

Groundnut

→Peanut

Guar

Syn.: Cluster Bean
Cyamopsis tetragonoloba (L.) Taub.
Leguminosae
G.: Guarbohne; F.: guar

The g. bean is probably indigenous to India and Pakistan and has a long history of cultivation throughout Asia. It has been introduced to Africa and the United States (since 1953).

The g. plant is a bushy annual, 1–3 m tall, with a long taproot and lateral roots on which grow many large, lobed nodules. Small flowers are formed in dense clusters on axillary racemes. Ca. 100 days after planting, the crop will bear fruits. At maturity, stiff, erect pods are formed in clusters, each 5–11 cm long. They contain 5–12 oval seeds of the size of a pea, which are white, grey or black. The green fodder yield is around 10 dt/ha, and the seed yield around 300 kg/ha, depending on the water regime of the crop. Under irrigation, such yields may easily be doubled.

The g. bean is hardy and drought-resistant. It is grown commonly in mixed cultivation with other crops. The seeds and young pods are a source of human food, and the whole plant is used as forage. G. seeds contain about 33% protein and 40% carbohydrates. G. meal is ground from the endosperm. It contains 4–5% protein and 80% →galactomannans. From the mucilaginous seeds a gum is obtained after several cleaning procedures.

Lit.: Cobley* (1976).

Guar Gum

G.: Guarmehl; F.: farine de guar

G. is a yellowish powder that is readily soluble in hot water. Solutions are normally turbid. They show pseudoelasticity and begin to flow as soon as shear is applied.

G. has two hydroxy groups in the *cis*-position and therefore forms gels with borax in aqueous, alkaline solution. It liquifies again when pH is dropped below 7. Similar gels are obtained by addition of polyvalent cations (e. g., Ca^{++}). Aqueous solutions of g. are degraded rapidly by acids and enzymes. Stability in alkaline media is rather good.

Chemically, g. is a polysaccharide of the galactomannan type.

A β-(1,4)-glycosidically linked mannose backbone carries galactose side groups, statistically on every second mannose unit.

```
-man-man-man-man-man-man-man-man-man-
        |       |       |       |
       gal     gal     gal     gal
```

Guar gum is obtained by feeding the purified seed (→guar) to an attrition mill to yield two endosperm halves. The germs are sifted off. For many technical applications, the endosperm is ground and used as such. For more qualified use, the endosperm is dehulled, milled and sifted to give various grades of a powder with a galactomannan content of 85–95%.

Commercially interesting derivatives of g. are gained by →etherification. With chloroacetic acid carboxymethyl guar, an anionic polymer, is obtained. With ethylene oxide or propylene oxide, hydroxyethyl guar or hydroxypropyl guar are produced via →alkoxylation in alkaline media. Cationic →quaternary ammonium compounds result by reaction of g. with 2-hydroxy-3-chloro-propyl-trimethylammonium chloride or the corresponding epoxide.

Derivatization changes the properties of g. substantially as to clearity, rheological behavior, reactivity, compatibilty and interaction with surfaces of minerals, etc.

Straight g. is used as is in →food additives, mostly as thickener and water-binding agent. Derivatives are not approved for food uses. G. is used in ice creams, salad dressings, cheese, jams and jellies, milk products and soups. There are also some applications in canned pet food and in cattle feed.

The technical applications of g. and its derivatives are numerous, and some of them consume large volumes. Hydroxypropyl guar (HPG) is used in hydraulic fracturing in oil and gas wells (→oilfield chemicals). G. and derivatives find use in slurry →explosives and in →fire fighting. Carboxymethyl and hydroxyethyl guar are used in the textile industry in printing and dyeing (carpets) and as sizes (→textile auxiliaries). The cable industry uses g. to immobilize penetrating water. The largest outlet for the gum is in paper as a wet-end additive and in dry strength improvement (→paper additives). Another large application for g. and derivatives is their use as →flocculant and depressant in ore flotation in the mining industry (→mining chemicals).

The worldwide annual consumption of g. is estimated to be in the range of 125000 mt. Annual trend (1988–1995) in food use is +3.2%.

Lit.: J.K.Seaman "Handbook of Water-Soluble Gums and Resins" R.L.Davidson (ed.), McGraw-Hill NY (1980)
J.E.Fox "Seed Gums" in A.Imeson (ed.) "Thickening and Gelling Agents for Food" 153–170 Blackie Academic and Professional, London (1992)
P.Harris "Food Gels" Elsevier Appl.Sci.Publ., London (1990)
[9000-30-0]

Guayule
→Rubber, natural

Guerbet Alcohols
G.: Guerbetalkohole; F.: alcools guerbet
The condensation of two moles of fatty alcohols (mostly C_8–C_{12} or C_{16}–C_{18}) at 200–300 °C in the presence of an alkaline catalyst, e. g., KOH leads to branched alcohols (2-alkyl alcohols):

$$2\,R_x\text{–}CH_2\text{–}CH_2\text{–}OH \longrightarrow$$
$$R_x\text{–}CH_2\text{–}CH_2\text{–}CH\text{–}CH_2\text{–}OH + H_2O$$
$$\underset{R_x}{|}$$

x: C_{6-16} alkyl

Modern catalyst combinations allow to work at lower temperture, thus reducing the amount of by-products.

G. are liquid like unsaturated alcohols but are not sensitive to oxydation and rancidity. They are therefore used in a variety of cosmetic products.

Other uses are lubricants for →metal processing, plasticizers for stencil manufacture (nitrocellulose), solvents for printing inks. G. are intermediates to make the corresponding branched carboxylic acids.

About 2000–3000 mt/a are produced.

Lit.: Ullmann* (5.) **A10**, 288
Kirk Othmer* (3.) **1**, 909

Guerbet Reaction
→Guerbet Alcohols

Guizotia abyssinica
→Niger Seed

Gum Arabic
Syn.: Gum Acacia
G.: Gummiarabicum, Akaziengummi;
F.: gomme arabique

G. is obtained from trees of the genus →*Acacia*, which grow in the Sudan and the Senegal. The Kordofan-grade gum from *Acacia senegal* L. Willd. (*A. verek* Guill. et Perrot.) is considered to be the most desirable. The dried exudate is processed by removal of the bark followed by grading, sizing and blending to various specifications. Types differ in mesh size, water-insoluble residue, solution color and clarity. G. is available in powdered (also spray-dried) or granular form, in thin flakes or spheroidal tears. It is white to yellowish and odorless andconsists of arabinic acid, a branched polysaccharide of L-arabinose, D-galactose, L-rhamnose and →D-glucuronic acid (3 : 3 : 3 : 1)

In pharmaceutical formulations and technology, g. is applied as an emulsifying or suspending agent and as a tablet binder, it raises viscosity and can be used as a protective colloid. It is also used as a mucilage (in candy and other food), as a colloid stabilizer, in the manufacture of spray-dried fixed flavors, and for powdered flavors used in packed dry-mix products (puddings, desserts, cake mixtures), where flavor stability and long shelf life are important. It is used for the production of D-galactose.

Annual production is 50000–60000 mt.

Lit.: Ullmann* (5.) **A25**, 46
Kirk-Othmer* (3.) **12**, 55
[9000-01-5]

Gum Dragon
→Tragacanth

Gum Elemi
→Resins, natural

Gum Lac
→Shellac

Gum Rosin
→Rosin

Gums, Industrial
→Water-Soluble Polymers

Gum Tragacanth
→Tragacanth

Gun Cotton
→Cellulose Nitrate

H

Haematin
→Dyes, natural

Haematoxylin
→Dyes, natural

Hair Preparations
G.: Haarbehandlungsmittel;
F.: produits capillaires

The most important hair care products are shampoos, hair conditioners, hair lotions, setting lotions, hair sprays, permanent wave preparations, hair colorants and shaving products.

Shampoos:
Shampoos did not assume importance as a product category for cleaning hair until new →surfactants were developed in the 1930's as the active material instead of →soap. These new →surfactants are relatively insensitive to water hardness, thus they do not leave scum-like residues (calcium soaps) on the hair. Now, shampoos are the most used haircare products. The mainly used →surfactants for shampoos are →fatty alcohol ether sulfates, →fatty alcohol sulfates, →sulfosuccinates, →fatty acid/protein condensates, and →amphoteric surfactants. Lately, nonionic →surfactants, such as →alkylpolyglycosides are increasingly used. Foam builders, such as →fatty acid ethanolamides, are used to improve qualities desired by the consumer, such as creaminess, quantity of lather, foam stability and to reduce the defatting of hair. Conditioning additives, e. g., quaternary hydroxyalkyl celluloses (→cellulose ethers), prevent tangling during combing and brushing of wet and dry hair, impart a smooth feel to the hair after drying, and reduce static charges, which causes "flyaway hair". Thickeners, e. g., salts such as sodium chloride, which have a special effect on the viscosity of →fatty alcohol ether sulfates or polymers, such as →hydroxy ethyl cellulose, must be added to prevent the shampoo from running down the face and into the eyes. Opacifiers give shampoos a rich, pleasing appearance. Relatively insoluble, higher-melting waxlike substances, e. g., ethylene glycol monostearate and distearate (→fatty acid esters) mixed with →fatty acid ethanolamide, impart opacity or pearlescence to the shampoo, depending on their crystallization properties. Coloring agents and fragrances are added to improve appearance and smell of hair after shampooing.

Different types of shampoos with specific properties, e. g., shampoos for frequent use, conditioning shampoos especially for damaged hair, anti-dandruff and special shampoos for fatty hair are offered in the retail and professional markets.

Hair conditioners:
Hair conditioners are applied to hair after shampooing and are intended to promote the following properties: smooth easy combing of both wet and dry hair, reduction of "flyaway-hair" caused by combing and brushing, enhancement of the gloss or luster of hair. The active agents of hair conditioners are first of all oily substances, i. e., mineral oils, →fatty alcohols, →quaternary ammonium compounds, such as cetyltrimethylammonium chloride and cetylpyridinium chloride, and organic acids, e. g., →citric acid and →lactic acid. The ingredients mentioned are dissolved or emulsified in water. Hair conditioners are sold as hair rinses, packs and balms. Other less important products for similar treatment, but which are additionally used for special styling, are W/O and O/W hair dressings, brilliantines and pomades. These products contain mainly →oils and →waxes.

Hair lotions:
Hair lotions are for use of a daily application to keep the hair tidy and the scalp in good condition. All claims that have been made to arrest baldness and to restore hair have not been supported by clinical evidence. The lotions are mixtures of water and →ethanol, which contain small amounts of active agents for hair and scalp. Such agents are →glycerol, →quaternary ammonium compounds, e. g., →cetyltrimethylammonium chloride, and anti-dandruff agents, e. g., piroctone olamine [the ethanolamine salt of 1-hydroxy-4-methyl-6-(2,4,4-trimetylpenthyl)-2-(1H) pyridinone]. Users of hair lotions usually are men, because the big amounts of water used as well as shampooing destroy the style.

Setting lotions:
Setting (styling-) lotions are applied to towel-dried hair after shampooing or conditioning and spread by combing. The hair then is put into curl-

ing rollers and dried in a current of warm air. Finally, the curlers are removed, and the hair is gently combed. Nowadays, the hair is dried and combed mostly in one working process with a special hair dryer. Purpose of setting lotions is to prolong the life of the water wave and to add body or volume to the hair by increasing interfiber friction. Thus, their aim is opposite to the aim of conditioners, i. e., conditioners intend to make hair smooth and easy combing and must improve the quality of single hairs, while setting lotions must improve the hair collective, the hair-style. They are essentially solutions of polymers, such as copolymers of vinyl pyrrolidone and vinyl acetate in water and ethanol. Setting products are more often used as styling mousses than as lotions and gels.

Hair sprays:
Hair sprays are applied to dry hair after styling has been finished. They are intended to give styling a hold and to prevent change of style by wind and humidity. Like setting lotions hair sprays contain polymers, such as copolymers of vinylpyrrolidone and vinyl acetate, copolymers of vinyl acetate and crotonic acid neutralized with alkanolamines, e. g., 2-amino-2-methyl-propandiol, small amounts of plasticizers, →cetyl alcohol, silicones, disolved in ethanol or isopropanol without or with only a small amount of water, because more than 10–20 % of water destroy the hair style. The polymer solution is filled up with propellants, such as propane and butane, in an aerosol can. In the hair care market, hair sprays rank second in sales only to shampoos.

Permanent wave preparations:
Permanent wave preparations are applied every two/three months to condition hair for longer-lasting styling. The hairs are weakened during the first step of the treatment by partial reduction of the disulfide bonds of the hair keratins. The new configuration of the wave is stabilized during the second step by oxidation (neutralization) of the disulfide bonds that were reduced before. The reduction agent mainly used is the ammonium salt of thioglycolic acid dissolved in water. Small amounts of anionic, cationic or neutral →surfactants and →fats and oils, e. g., →lanolin, are added. The oxidation (neutralization) agent is hydrogen peroxide diluted in water.

Hair colorants:
Hair colorants are applied every 4–6 weeks because growth is about 1 cm per month. They are intended to cover grey hair, to change natural hair

color or to bleach hair. Permanent hair colorants contain intermediates, e. g., p-phenylenediamine or resorcinol, dissolved in → emulsions (creams or lotions) or gels. The intermediates are oxidized by addition of hydrogen peroxide. It depends on kind, mixture and percentage of the intermediates whether the hair, even with more than 30 % white hair, become blond, brown or black. So-called semipermanent colorants are applied to receive fashionable hair shades. They contain dyeing agents similar as used for coloring textiles. Lately, natural colors, such as henna (→dyes, natural), which contains lawsone (2-hydroxy-1,4-naphtoquinone), a red-orange dye, are again used for coloring hair. Furthermore, other natural →dyes, e. g., walnut shells, containing juglone (5-hydroxy-1,4-naphthoquinone), which gives a yellow-brown color, indigo (→dyes, natural), which gives a blue color, and chamomile extract, containing apigenin (4′,5,7 trihydroxy-flavone), which gives a yellow color, have been rediscovered for coloring hair. The use of these products is limited, however, because of poor selection of shades, uneven coloring and often unpractical methods of application. Today, as before, the most important hair colors are permanent hair colorants. The market for natural dyes is less than 5 % of the total hair colorant market.

Shaving products:
Shaving products are mainly lather shaving creams, sticks, and aerosols. These S. contain sodium and potassium salts of →stearic acid (→soaps), water and small amounts of →oils, such as →coconut oil, or fatty alcohols, such as →lauryl alcohol or →stearyl alcohol and →glycerol. Dry shaving preparations (pre-electric shave lotions) are solutions of, e. g., diisopropyl adipate in →ethanol. After-shave products are used after wet and dry shaving and are intended to refresh and cool the skin and exert a mild adstringent effect. These products contain often propylene glycol and →menthol in ethanol and water.

Lit.: Ullmann* (5.) **A12**, 571–601
Kirk-Othmer* (3.) **10**, 881–918
Hilda Butler (ed.) "Cosmetics" Poucher's (9.), **3**, 130–212, 259–287, Chapman & Hall, London/Glasgow/NY/Tokyo/Melbourne/Madras (1993)
Clarence R. Robbins, "Chemical and Physical Behavior of Human Hair" (3.), Springer Verlag, Berlin/Heidelberg/NY (1994)

Hardening
→Hydrogenation

Harvesting

G.: Ernte; F.: récolte

Combine harvesting:

The traditional method of harvesting field crops consists of the following three steps:

1. cutting the crop (with mechanical knives, sickels, scythes, cradle, etc.);
2. windrowing the straw plus mature fruits with seeds in a swath;
3. drying the cut crop in bundles or shocks;
4. threshing the swath or the bundles so that seeds, straw and fruits are separated.

Since about 1880, these 4 processes have been combined in a tractor-drawn or self-propelled machine, thus named "combine harvester".

Forage harvesting:

Normally, this is done by cutting and chopping green hay or other plant material (alfalfa, green rye, etc.) for artificial drying or for silage production.

Vacuum harvesting:

This is applied when fruits and other plant parts are not to be destroyed during harvesting, such as for multiple harvesting (mechanical picking of fruits and vegetables, or for →cotton or →cuphea seed). The use of all parts of a plant is called →whole crop harvesting.

Lit.: S.R.Chapman, L.P.Carter,"Crop Production, Principles and Practices" Freeman and Co, San Francisco (1976)

Hashish

→Hemp

HEAR

→Rapeseed

HEC

→Hydroxyethyl Cellulose

Helianthus annuus

→Sunflower

Helianthus tuberosus

→Topinambur

Hemicelluloses

Syn.: Polyoses

G.: Hemicellulosen, Holzpolyosen;

F.: hémicelluloses

H. are, next to →cellulose, the second most abundant natural organic chemicals in the biosphere, present in all parts of the plant, concentrated in the primary and secondary layers of the cell wall and closely associated with cellulose and →lignin. Their biological function is as connecting or integrating units.

Some annual as well as hardwood plant species may contain up to 35 % h. (xylan). H. are heteropolysaccharides. Composition and structure differ widely based on the plant source.

It can be expected that new pulping techniques (e. g., organosolv process) will open new sources for pure h.

The main constituting monomeric units are:

→Pentoses:

β-D-xylose, α-L-arabinopyranose,

α-L-arabinofuranose,

Hexoses:

β-D-glucose, β-D-mannose, α-D-galactose,

Hexuronic acids:

β-D-glucuronic acid, α-D-4-O-methylglucuronic acid, α-D-galacturonic acid,

Desoxyhexoses:

α-L-rhamnose, α-L-fucose.

Naming the polysaccharides is based, like that of the corresponding enzymes, upon the dominating sugar residues that form the main chain. Hard- and softwoods are the most important carriers of h.; they can be separated by their different relationships of xylan and mannan:

| Polyoses | Angiosperms | | Gymnosperms | |
	(%)	Constituents	(%)	Constituents
Xylans	25–30	Xylose, 4-O-methyl-glucuronic acid, Acetylic groups, DP 70–150;	5–10	Xylose, 4-O-methyl-glucuronic acid, Arabinose, DP 150–200;
Mannans	3–5	Mannose, Glucose, DP 60–70;	20–25	Mannose, Glucose, Galactose, Acetylic groups, DP 60–70;
Galactans	0.5–2	Galactose, Arabinose, Rhamnose,	0.5–3	Galactose, Arabinose

Important producers of h. are:

Sugar beet pulp
Soybean, Coffee bean, Larchwood
Lupinus albus seed
Guar gum, Locustbean gum
Konjak-mannan
Triticaceae flour
Grasses
Gymnosperms
Angiosperms

L-arabinan (Homopolysaccharides)
L-arabino-D-galactan
D-galactan (Homopolysaccharides)
D-galacto-D-mannan
D-gluco-D-mannan
L-arabino-D-xylan
L-arabino-D-glucurono-D-xylan
D-galacto-D-gluco-D-mannan
D-glucurono-D-xylan.

Simplified structures of the repeating units of the most important polysaccharides of h. include:

Xylan from angiosperms:

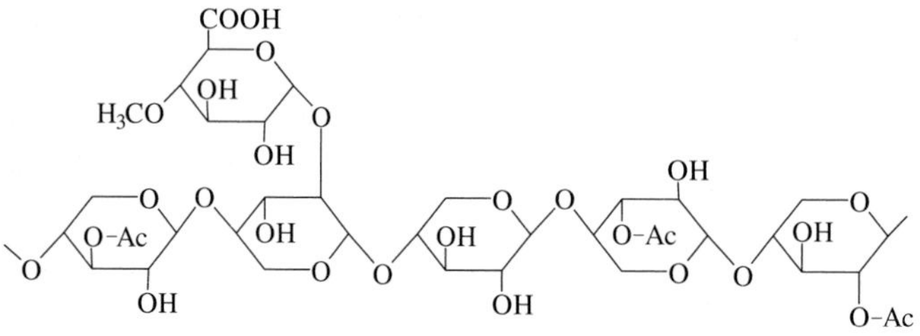

Xylan from gymnosperms:

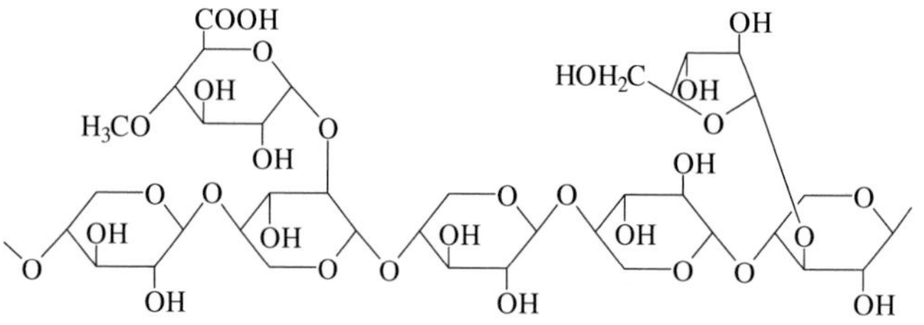

Glucomannan from gymnosperms:

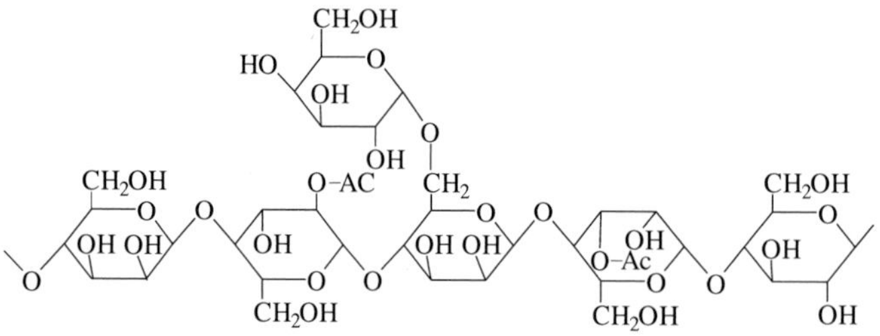

Arabinogalactan from larch wood:

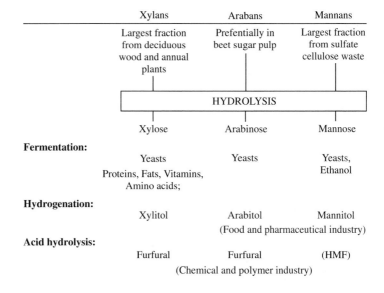

Isolation:

The plant material is extracted by:

ethanol-benzene or chloroform-methanol to remove crude lipids; cold/warm water to remove proteins and soluble carbohydrates; ammonium oxalate-oxalic acid solution or EDTA to remove →pectins; acidic sodium chlorite to remove lignins.

The residue, holocellulose, consisting of cellulose and h., is extracted by alkaline solutions: xylans are dissolved in 5–10% aqueous alkali solution; mannans are soluble in 17–20% alkali or in alkali-borate solutions (24% : 4%). Herbage and wood h. are precipitated by acid (pH 4.5–5) to

yield the rich-in-xylose, higher-molecular, water-insoluble "Hemicellulose A".

The polysaccharides that remain in solution are precipitated with excess ethanol to provide "Hemicellulose B". It contains highly-branched, water-soluble polysaccharides of lower molecular weight, of glucuronic acid and arabinose. The resulting preparations can be further subfractionated by applying techniques, such as iodine complexing, ethanol and copper precipitation, ion exchange and molecular exclusion chromatography.

Utilization:

H. in vegetables, fruits, legumes, cereals, bread, together with lignin and cellulose, constitutes the "dietary fiber": plant polysaccharides that cannot be digested by the enzymes of the human small intestine. But they may be utilized by the microorganisms of the human large intestine and impart a slightly acidic reaction in the colon, especially the glucomannans, arabinogalactans and guar gum. They exhibit important functions in reducing dis-

	Xylans	Arabans	Mannans
	Largest fraction from deciduous wood and annual plants	Prefentially in beet sugar pulp	Largest fraction from sulfate cellulose waste
	HYDROLYSIS		
	Xylose	Arabinose	Mannose
Fermentation:			
	Yeasts	Yeasts	Yeasts, Ethanol
	Proteins, Fats, Vitamins, Amino acids;		
Hydrogenation:			
	Xylitol	Arabitol	Mannitol
		(Food and pharmaceutical industry)	
Acid hydrolysis:			
	Furfural	Furfural	(HMF)
		(Chemical and polymer industry)	

eases by lowering serum, liver and blood cholesterol and triglyceride levels.

Mannans of legume seeds (guar, carob galactomannans, konjak mannans) are constituents of native food in India, Japan and Pakistan. →Guar gum or →locustbean gum are widely used as stabilizers and thickeners in modern food industries with an annual growth of 3.2 % (guar) and 0.6 % (locust).

Most prominent nonfood utilization of the wood polyoses is in the papermaking industry as natural glues for fiber-fiber adhesion. Most of the available h., however, are found as more or less degraded, soluble polysaccharide contents of the waste fluids of prehydrolysis exposure to cellulose isolating and bleaching chemicals. Therefore, their binding functionality is decreased and utilization by hydrolysis to the monomeric sugars is possible (see table, p. 135).

The most important raw material for large-scale industrial use is →furfural.

Lit.: J.M.V.Blanchard,J.R.Mitchell (ed.) "Polysaccharides in Food" 93–108, Butterworths, London/Boston/Sydney/Wellington/Durban/Toronto(1979),
W.Burchard (ed.) "Polysaccharide, Eigenschaften und Nutzung" 43–53, 294–301 Springer Verlag-Berlin/Heidelberg/NY/Tokyo (1985)

Hemp

Cannabis sativa L.
Cannabaceae
G.: Hanf; F.: chanvre

H. is one of the few examples of →dioecious crops. Up to the 19th century, h. was one of the most important fiber raw materials in Europe (→Cellulose fibers). At that time, →jute, →sisal and →cotton were beginning to be imported in large quantities from the commonwealth countries. These crops were less expensive, and h. began to become less competitive. Certain markets for h., such as ropes, twine and thread, were lost later to synthetic fibers. Nevertheless, production of h. in Europe may become promising again in France, the Netherlands, Great Britain and Germany. This is due to the fact that it is one of the highest yielding and least intensive crops in these countries. It is highly selfcompatible so that there is no need for crop rotation, and it has also good possibilities for a sound crop rotation. It oppresses weeds well and is so fast growing that it does not require any kind of herbicide treatment (→pesticide). H. is used in breeding nurseries for preventing outcrossing of other crops because of its tall stature of up to 5 m,

thus having the good properties as an isolation crop, e. g., for sugar beet seed production.

For fiber production, h. is cut when the male plants are in full flower and shedding pollen. The subsequent retting and scutching processes are similar to those for →linseed and →flax. H. seed contains 32 % oil, 25 % protein and 20 % starch. Seed of h. is also useful as food. The oil (rich in C18:2 and C18:3) was used for centuries as burning oil for lamps, and its leaves and flowers were used in pharmaceutical applications. After solvent extraction, the meal may be applied as animal feed. H. fibers are also being used for the manufacturing of paper and technical filters and for the pads of brakes and clutches as well as insulation, e. g., in house construction or in automobiles.

Due to its high d.s. yield, it is being discussed for →whole crop harvesting as a feed for straw furnaces or decentralized power stations.

H. produces the drug tetrahydro cannabinol (THC), also known as marijuana or hashish, in all green parts. Because of this, commercial production of fiber-hemp is forbidden in several countries, although the breeding of species with low amounts of THC (<0.3 %) has been successful.

Best suited for producing the THC-drug is Indian h. [*C. sativa* L. ssp. *indica* (Lam.) Small & Cronq]. Worldwide, about 200 000 mt of h. fibers are produced. Main production takes place in FSU, China, Korea and Southeast Europe.

Lit.: Martin/Leonhard* (1967)

Henna

→Dyes, natural

Heptanoic Acid

→Enanthic Acid

Heritability

G.: Heritabilität, Erblichkeit; F.: hérédité

Measure for the heritable proportion of a plant characteristic or its genetic components. H. is measured by estimating the genetic and the environmental proportion of the variation of a certain trait of one genetic line by means of field plot experiments at several locations. It is calculated as the proportion of genetic variation to environmental variation:

$$h^2 = V_g/V_e$$

The higher its value, the greater is the chance of a trait to be inherited by the next generation without being influenced by environmental effects (→plant breeding).

Lit: D.S.Falconer, "Introduction to Quantitative Genetics". Ronald Press. N.Y. (1960)

Herring Oil
→Fish Oil

Heteropolysaccharides
→Polysaccharides

Hevea brasiliensis
→Rubber Tree

Hexacosanoic Acid
→Cerotic Acid

Hexadecanoic Acid
→Palmitic Acid

1-Hexadecanol
→Cetyl Alcohol

cis-9-Hexadecenoic Acid
→Palmitoleic Acid

Hexadecyl Alcohol
→Cetyl Alcohol

1-Hexadecyl Amine
→Palmityl Amine

Hexadienoic Acid
→Sorbic Acid

Hexanoic Acid
→Caproic Acid

HFCS
→Isomerized Sugars

Hibiscus cannabinus
→Kenaf

High-amylose Starches
G.: Hoch-amylosehaltige Stärken;
F.: amidons riches en amylose

Starches, particularily rich in →amylose, obtained from certain plant cultivars, mainly from →corn and pea, are called h.

Normally, amylose contents of the industrially important starches are between 18% and 26% of the polysaccharide d.s. In contrast to →waxy starches from corn, sorghum, rice and barley, which consist of 99–100% amylopectin, starches of a few varieties contain higher amounts of amylose:

→Amylomaize starches (high-amylose starches), from natural occurring varieties of *Zea mays* L. with 50–55% amylose content. Plant breeders have succeeded in producing hybrids with up to 85% amylose content in the starch. Such hybrids have gained technical significance in the USA.

– Wrinkled →pea starches: The starches of most legumes generally show somewhat enhanced amylose contents of >28–35% (→broad bean starch, →pea starches). Naturally occurring wrinkled pea (*Pisum sativum,* convar. *medullae* ALEF) contains starches with ≈ 50% amylose. Plant breeders intend to generate a wrinkled pea with up to 90% amylose.

Special efforts in plant breeding and cultivation research are aiming on genetically stabilizing high amylose content in combination with high starch yield and suitable conversion properties.

Lit.: Tegge*, 31, 158–160, (1988)
    "Stärke im Nichtnahrungsbereich" Reihe A/Heft 388, 95–99, 171–173, 239–247, Schriftreihe des Bundesministers für Ernährung, Landwirtschft, Forsten, Landwirtschaftsverlag GmbH, Münster-Hiltrup (1990)

High-Fructose Corn Syrup
→Isomerized Sugars

High-Oleic Sunflower
→Sunflower

HLB Value
G.: HLB-Wert; F.: équilibre hydrophile-lipophile

The HLB value is an empirical helpful tool in making →emulsions and to characterize nonionic →surfactants. On a value scale ranging from 1 to 20, it tells in which ratio the **H**ydrophilic and the **L**ipophilic parts are **B**alanced in a surfactant molecule. It is defined as the ratio of the molecular mass of the hydrophilic part of the surfactant (M_h) to the total mass of the molecule (M) multiplied by 20. In other words: the higher the value the more hydrophilic is the surfactant.

Lit.: P.Becher (ed.) Encycl. of Emulsion Technology, vol. 2 Applications, Marcel Dekker NY. (1985)

HMF
→5-Hydroxymethylfurfural

HM-Pectins
→Pectins

Hog
→Pig

Homopolysaccharides
→Polysaccharides

Honesty
→*Lunaria annua*

Hordeum vulgare
→Barley

Horse Bean Starch
→Broad Bean Starch

HOSO
→Sunflower

Hot Melts
→Adhesives
→Polyamides

HPC
→Hydoxypropyl Cellulose

HPG
→ Guar Gum

HVP
→Hydrolyzed Vegetable Protein

Hybrid
G.: Hybrid; F.: hybride

A h. is an offspring produced by plants or animals of unlike genetic constitution. It occurs or is induced naturally in so-called interspecific h., originating from different species, which normally result into sterile offsprings. Such an interspecific h. is marked in →taxonomy by a "x" (→peppermint oil, rose). Sterility may be overcome by doubling of the chromosome number (by, e.g., treatment with colchicine). →Rapeseed (*Brassica napus*) is an example of an interspecific h. between cabbage and →turnip rape (*Brassica campestris*).

In →plant breeding, h. varieties are designed after crossing of especially produced inbred lines of genetically distinct pools or races of one species of plants. →Corn varieties were among the first com-

mercial h. crops, with production starting around 1940.

Whereas the parental lines are low yielding after several inbreeding generations, the h. shows superior vigor and yield, which is often much higher and better than the mean value of the parents before inbreeding. This superiority of the h., however, will only show up in a few out of several hundred crosses, depending on the ability of the plant breeder to realize, select, and cross "fitted" progenies. The h. effect occurs only in the first offspring of the parental cross. In the 2nd and following generations after the cross, there will be little uniformity and a high rate of outcrossing. Because of this the seed producer has a good market position, as the farmers are heavily depending on the special h. seed.

Lit.: F.N.Briggs, P.F.Knowles "Introduction to Plant Breeding" Reinhold Publ. Corp. (1967)

Hydraulic Fluids
G.: Hydraulikflüssigkeiten;
F.: fluides hydrauliques

H. are liquids that transfer energy in fluid-filled systems, either by pressure (hydrostatic drives) or by flow (hydrokinetic drives).

The majority (85%) is based on mineral oil, but 15% are synthetic products, some of them RR-based and used in special applications, e.g., the diesters of →sebacic acid and →azelaic acid with branched alcohols or the esters of trimethylolpropane and pentaerythritol with saturated and unsaturated →fatty acids.

They are gaining increasing interest because of their biodegradability.

There are also fire-resistant h., which are emulsions of oil in water or vice versa. Emulsifiers and additives used may be based on RR.

Auxiliaries are similar to those used for →lubricants and are added to protect the metal surface from damage, such as extreme pressure additives (EP) for heavy loads and fatty acid derivatives for lighter loads (antiwear additives). Fatty acid salts and sulfonates are corrosion inhibitors, while sulfonates and amines act as dispersants to prevent deposits. Viscosity index improvers based on polymethacrylates of →fatty alcohols reduce the temperature sensitivity of viscosity and lower the pour point.

The main applications for h. are construction machines, machine tools, agricultural equipment and automotives.

Automatic transmission fluids may consist of up to 15 components to meet the complex specifications. Not many RR-based products are involved.

Brake fluids are based on glycol or glycol ether or esters, with blown castor oil or its reaction product with polyglycolethers being added to improve viscosity and antiwear.

World consumption of h. was 2×10^6 mt in 1978.

Lit.: J.G. Wills, "Lubrication Fundamentals", Marcel Dekker Inc NY (1980)
Kirk Othmer* (3.) **12**, 712
Ullmann* (5.) **A13**, 165.

Hydrocarboxylation

G.: Hydrocarboxylierung; F.: hydrocarboxylation

There are two possibilities to introduce a second carboxylic group into an unsaturated fatty acid via h.: the Reppe and the Koch synthesis:

$$-CH=CH- + CO + ROH \longrightarrow \begin{array}{c} -CH-CH_2- \\ | \\ COOR \end{array}$$

In the Reppe reaction, $Co_2(CO)_8$ or $Ni(CO)_4$ is used as a catalyst to yield various positional isomers. If cobalt/4-picoline is used as a co-catalyst, yields of 50–60% of linear alpha-omega dicarboxylic acids are obtained.

The Koch synthesis uses sulfuric acid as a catalyst and leads to a strong isomerization of the double bond to yield a mixture of isomers with a large amount of α-branched dicarboxylic acids.

A good starting material is oleic acid from high-oleic →sunflower.

Lit.: H.Baumann et al. Angewandte Chemie (english ed.) **27**, 53 (1988)

Hydrocolloid

→Water-Soluble Polymers

Hydroformylation

Syn.: Oxosynthesis
G.: Hydroformylierung; F.: synthèse oxo

H. is the reaction of an olefin (in case of RR-based products, unsaturated fatty acid, e.g., →oleic acid) with synthesis gas under pressure:

$$-CH=CH- + CO + H_2 \xrightarrow{Co_2(CO)_8} \begin{array}{c} -CH-CH_2- \\ | \\ CHO \end{array}$$

The aldehyde obtained may be transferred to a hydroxy group by →hydrogenation (in case of a fatty acid as starting material, a branched hydroxy fatty acid results) or oxydized by air or oxygen at 20–25 °C with calcium acetate or manganese naphthenate as a catalyst to the desired dicarboxylic acid. In protonic solvents (water or methanol), h. yields the acid or their ester directly.

Because the Co-carbonyl catalysts favor isomerization of the double bond, a mixture of positional isomers is obtained. If rhodium/triphenylphosphane is added as a co-catalyst, the 9- and 10- position is favored in h. of oleic acid.

High yields are obtained if 80–90% oleic acid (derived from high-oleic →sunflower) is used.

Lit.: J.Falbe, "New Synthesis with Carbon Monoxide", Springer Verlag Berlin (1980)
H.Baumann et al. Angewandte Chemie (english ed.) **27**, 53 (1988)

Hydrogenation

G: Hydrierung; F: hydrogénation

H. is the addition of hydrogen to an organic compound that contains double bonds or certain functional groups (e.g., carboxyl or nitrile). Technically important examples are:

A: $-CH=CH- \longrightarrow -CH_2-CH_2-$ = Hardening
B: $R-CO-OR \longrightarrow -CH_2-OH$ + ROH = Fatty alcohols
C: $R-C\equiv N \longrightarrow R-CH_2-NH_2$ = Fatty amines

A. Hardening:

The oldest h. in →oleochemistry is the hardening of fatty oils. In 1901, Norman discovered that liquid →oleic acid can be transferred into solid →stearic acid by treating it with hydrogen in the presence of a nickel catalyst. Shortly after this fundamental reaction was discovered, it was applied to unsaturated trigycerides (→fats and oils), which were transferred to solid fats ("hardening"). This was the beginning of a large-scale margarine manufacturing. Reaction conditions: 2.5–3.5 MPa H_2 are pressed into the 180 °C hot oil in which Ni-silica or Ni-formiate is finely dispersed. Batch and continous processes as well as fixed-bed reactors are in use today. Sometimes partial h. is practiced. Melting point and characteristics are adjusted, bad odors disappear and cheap oils are made useful for nutrition.

The reaction is also broadly used in industrial oleochemistry for h. of unsaturated →fatty acids. Starting materials are predominantly acid oils from the refining of oils for nutrition ("foots"), split (→hydrolysis) fatty acids, e.g., from tallow and the stearin fraction from the olein/stearin separation (→crystallization). H. is also applied for making →hydroxystearic acid from castor oil fatty acid, to eliminate unsaturation in coconut oil prior to →fractionation, to get →arachidic and →behenic acid from →fish and →rapeseed oil, to improve color in →dimer acid and to eliminate unsaturation in the starting material prior to separation by →crystallization of→stearic from →isostearic acid .

Reaction conditions are roughly the same in h. of fats and oils. The plant and equipment may also be the same, but it has to be stainless steel to avoid corrosion. The catalyst reacts easily with fatty acids to form inactive Ni-soaps, which leads to a loss of expensive catalyst. The fatty acid must be dried and purified before reaction by treatment with clay. A large variety of catalysts are available in the market. A typical example consists of 20–25% nickel and 10–15% silica coated by 60–65% hydrogenated fat, which makes handling and transportation safer because activated catalyst is pyrophoric. After the h. the catalyst is filtered off, and the acid is distilled to remove salts and minimize odor.

If there are more than one double bond in the fatty acid, h. takes place in several steps with different velocities (higher unsaturated fatty acids react faster), which provides the opportunity for selective and partial h. Isomerization (isomerism) can also be observed.

B. Production of →fatty alcohols:

→Fatty acids, their methyl and butyl esters and also triglycerides are transferred to fatty alcohols by high-pressure hydrogenation. Primary feedstock is →coconut or →palm kernel oil for the C_{12-14} range and →tallow, →palm oil and many others for the C_{16-18} range.

Rapeseed is the source for behenyl alcohol.

$$R^1-COO-R^2 + 2\,H_2O \xrightarrow[\text{catalyst}]{>T/>p} R^1-CH_2OH + R^2OH$$

R^1: C_{7-21} alkyl

R^2: $H-$, CH_3-, C_4H_9- or $R^1-O-CH_2-\overset{OR^1}{\underset{|}{CH}}-CH_2-$

For further details: →fatty alcohols

C. Hydrogenation of nitriles:

→Fatty amines are produced by h. of →fatty nitriles, which are derived from →fatty acid and ammonia via the amide. Dehydration of the amide is the first step to yield the nitrile, which is hydrogenated in the presence of a catalyst. Aside from Ni and Co, noble metals and iron are used as catalysts in batch processes with suspended catalyst and in continous processes with a fixed-bed catalyst.

For further details: →fatty acid nitriles and →fatty amines.

D. Others

H. is applied in context with RR in the following other synthesis:

- →furan to →tetrahydrofuran
- →furfural to →furfuryl alcohol
- →rosin
- →maltose to →maltitol
- →glucose and →fructose to →sorbitol and mannitol
- →isomaltose to →isomalt

Lit.: Ullmann* (5.) **A10**, 267
Kirk Othmer* (4.) **5**, 161

Hydrol

Syn.: Corn sugar molasses
G.: Hydrol; F.: mélasse de glucose

H. is the noncrystallizing mother liquor, remaining after the final low-grade massecuite crystallization of →dextrose from starch total →hydrolysis by acids. The dark, brown viscous syrups have a bitter taste, which prevents utilization in human nutrition. They are preferentially used as nutritive agents in fermentations and as core binders in the →foundry industry.

Two reactions accompany the hydrolysis:

- decomposition of the glucose molecule, formation of →hydroxy methylfurfural and →levulinic acid, further reaction with amino acids to maillard products and the dark melanoidins;
- reversion of glucose to oligosaccharides →isomaltose, →gentiobiose, →maltose, →panose, →6-isomaltosido-glucose and other 1,6-glucosidic oligosaccharides.

They make up 5–6% of the original hydrolysate d.s. and are accumulated during the crystallization steps up to 30% of the total hydrolysate. Crystallization of the twofold of dextrose is prevented by this amount of reversion saccharides. Acid hydrolysis for dextrose production has been largely displaced by enzymic hydrolysis, which avoids decomposition and reversion. Significance of h. as a by-product of dextrose production is decreasing, but decomposition and reversion products may be found in other products of acid hydrolysis of α-glucans.

Lit.: Tegge* 65–67, 221–225. (1988)

Hydrolysis

G.: Hydrolyse; F.: hydrolyse

Hydrolysis is a reaction of water with another substance to yield one or more reaction products. In context with RR, mainly ester and ether groups are being hydrolyzed. Because h. of esters in →oleochemistry and h. of glucosidic bonds in saccharide chemistry are different as to starting materials, reaction conditions and end products, they are covered under separate keywords: →hydrolysis in oleochemistry, →hydrolysis, glucosidic linkages.

Hydrolysis in Oleochemistry

G.: Hydrolyse in der Fettchemie;

F.: hydrolyse des corps gras

→Fats and oils are split by h. in →fatty acids and →glycerol

$$R^1-CO-O-CH_2 \qquad\qquad R^1-COOH \quad HO-CH_2$$
$$R^2-CO-O-CH \;+3\,H_2O \;\longleftrightarrow\; R^2-COOH \;+\; HO-CH$$
$$R^3-CO-O-CH_2 \qquad\qquad R^3-COOH \quad HO-CH_2$$

R^{1-3}: C_{7-21} alkyl

Crude →fats and oils have to be purified first to remove minerals, gums, soaps, proteins and dissolved packaging material (poly-olefins) by filtering at room or elevated temperature with or without filter aids or by treatment with 1% sulfuric acid. The water used for splitting must be demineralized.

The oldest catalyst for h., the "Twitchell reagent", is a mixture of aromatic hydrocarbons and oleic acid reacted with concentrated sulfuric acid. This aromatic/aliphatic sulfoacid gives within 12–24 hours at 100 °C and normal pressure a degree of hydrolysis of 80–85 % without changing the water phase. This can be increased to 98 % by exchanging the glycerol/water phase by fresh water to force the equilibrium reaction to the right. This process is used today only in small-scale operations.

Acid catalysts accelerate h. effectively but have the disadvantage of material corrosion at higher temperatures and pressures.

Dibasic metal oxides, e. g., ZnO are also known as good catalysts, without corrosion problems. Modern (continous) processes work without any catalyst but need temperatures above 200 °C.

Pressure autoclaves (1–3.5 MPa) in batch or multi-stage batteries are in use to hydrolyze in cycles more or less continuously. In advanced processes, counter-current extraction technologies are applied to get a high degree of h. Many variations are in use, and complete process units are available.

The most modern technology works in columns and utilizes the difference in density of the two

phases to combine high-pressure, no-catalyst technolgy with the counter-current principle. Reaction conditions are 250–260 °C and 5–6 MPa. Fat is pumped in at the bottom of the splitting column, water at the top. The reaction takes place in the middle of the column where the heat exchangers are arranged. High-pressure steam is pressed in there (250 °C). A pressure regulator controls the flow of the fatty acid at the top, while an interface control device commands the flow of the glycerol-water (15 %) at the bottom.

Further steps are →distillation of the acid, followed eventually by →fractional distillation and →crystallization. The aqueous phase is concentrated and purified to get →glycerol. Hydrolysis at low temperature can be carried out with lipases (→fermentation). Despite the possibility to safe energy, this technology is still limited to special cases, e. g., the partial h. at certain positions of the triglyceride (→glycerides).

Lit: N.O.V.Sonntag et al.,"Fatty Acids in Industry", Marcel Dekker Inc. NY (1989)
      Ullmann* (5.) **A10**, 257
      Kirk Othmer* (4.) **5**, 170

Hydrolysis of Glucosidic Linkages

G.: Hydrolyse, Glucosidbindungen;

F.: hydrolyse, rupture de liaisons glucosidiques

H. is the splitting of, at least, one glucosidic linkage between two adjacent groups by introducing one molecule of water in the presence of a catalyst, acid or carbohydrase:

$$R^1-O-R^2 + HOH \xrightarrow[\text{Catalyst}]{} R^1OH + R^2OH$$

This general expression, which depicts beginning and end states of a multistep process, is valid for the following hydrolytic systems:

- R^1: sugar residue; R^2: aglucon, natural glucosides, such as amygdalin, →saponins, anthocyanin;
- R^1: sugar residue; R^2: sugar residue, disaccharides, such as →maltose, →lactose, →sucrose;
- R^1: sugar residue; R^2: oligomer, polymer residue →polysaccharide, (splitting from the nonreducing end, such as glucose or maltose);
- R^1: oligomer, poly-; R^2: oligomer, polymer residue →polysaccharide, mer residue (statistical splitting, oligo and polysaccharides).

For **hydrolysis by acids** (hydrochloric, sulfuric, phosphoric, oxalic, acetic, citric acid), the function of the proton is to activate the glucosidic oxygen.

This is followed by splitting off the glucosidic group and, finally, there is heterolysis of one water molecule to stabilize both molecule residues, the proton is set free and ready for the next step of transmission catalysis. This process, though thermodynamically allowed, needs high energy of activation. Therefore, acid h. are run at elevated temperatures or/and under conditions of pressure cooking. As a consequence, accompanying and succeeding consecuting reactions must be regarded, which lead to

– deterioration of the hydrolysis product: →hydroxymethylfurfural, →levulinic acid, levoglucosans, erythronic acid, formic acid; further Maillard reactions with amino acids to soluble and, finally, insoluble coloring products (melanoidins, humins);

– synthesis of new, unusual glucosidic linkages by reversion of reaction products, which may negatively influence yield, purity, properties of technical hydrolysis products, →hydrol.

Hydrolysis by enzymes (free or immobilized carbohydrases) is characterized by rapidly decreasing the energy of activation of the (thermodynamically possible) reaction. So, processing under mild energetic and environmentally suitable conditions becomes possible.

Necessary conditions for processing are: specificity for the type of glucosidic linkage and substrate, concentration, temperature, time and pH of reaction, activating or protecting substances, and absence of interfering or inhibiting agents. Enzymatic production of numerous final or intermediate hydrolysates is performed in high-scale and high-tech equipment for the production of →dextrose, →glucose syrups, →solid glucose, →maltose, →maltose syrups, maltooligosaccharides, →branched oligosaccharide syrups, →maltodextrins from starch, →inverted sugars, →isomaltulose, →leucrose, →oligosyl-fructoses from →sucrose, →fructose and →fructose oligomers from →fructans or →levans.

Lit.: Schenk/Hebeda* p. 650 (1992)
Ruttloff*, 39–50 and chapters 11–15.(1994)

Hydrolyzed Vegetable Protein
Syn.: HVP
G.: Glutenhydrolysat; F.: gluten hydrolysé

H. is the complex mixture of →amino acids resulting from the acid hydrolysis of →cereal gluten.

→Wheat gluten as well as →corn gluten meal are the prefered starting materials due to their high content of →L(+)-glutamic acid (and glutamine), which is the most active component for flavor enhancement.

Production: Fresh native gluten, as well as dried or devitalized gluten, with at least 70% protein are the raw materials. Hydrolysis is performed, in rubber- or plastic-lined, stirred autoclaves with steam jacket, on 1:1:1 -mixtures of dry gluten, water and concentrated hydrochloric acid at $\approx 113-116\,°C$ for 8–12 hours.

After cooling to 100 °C, disodium phosphate is added to precipitate traces of iron, and sodium carbonate is then added for neutralization to pH 5.5. The hot slurry is first filtered to remove insoluble humins and then stored for about two months for maturation. Then it is filtered again. H. results as a red-brown clear liquid, which exhibits the basic flavor of well-browned beef. This is the result of the complex action of →mono-sodium glutamate as a flavor enhancer and of several →amino acids that contribute to the boullion flavor, such as glycine, alanine, proline, leucine, serine, phenylalanine, asparagine, valine and tyrosine.

Application: H. is used in industry and in the kitchen for flavoring, flavor enhancing and seasoning in meat and poultry, sausages, seafoods, cured products, vegetable dishes, canned, frozen or dehydrated food products, sauces, dressings and bouillion cubes.

In USA, h. is mainly prepared from →corn gluten meal, whereas in Europe and Asia, wheat gluten is the preferred material. Its content of →glutamic acid (1/3 of the entire protein-N) made wheat gluten hydrolysate a raw material for the production of →glutamic acid, which is now mainly produced by fermentation.[1]

Hydrolysates from wheat gluten, maize gluten or soy protein are starting materials for the technical preparation of other →L-amino acids (production was 1810 mt in 1980): cysteine/cystine, histidine, arginine, leucine, serine, threonine and tyrosine.

Lit.: J.W.Knight "The Starch Industry" 124–128, Pergamon Press Oxford/London/Edinburgh/NY/Toronto/Sydney/Paris/Braunschweig (1969)

[1] Ruttloff* 465–484 (1991)

Hydrophilization
→Crystallization

Hydrophobic Starches

Syn.: Molding Starches, Nonwetting Starches
G.: Hydrophobierte Stärken;
F.: amidon transformé hydrophobique

Granular starch can be chemically modified by esterification of the hydroxyl groups with hydrophobic substituents in the usual way. The substituent acid anhydrides are containing usually alkyl groups with 8 or 10 C-atoms:

Starch

C$_8$H$_{17}$—CH—C$\overset{O}{\diagdown}$ NaOH
     H$_2$C—C$\diagup$$_O$

Octenyl succinic anhydride   Octenyl succinate
(OSA)   ester

The DS determines the more or less hydrophobic character of these esters. Charactberic properties include: affinity to fats and oils, surface activity and water repellency. In food and pharmaceutical applications, the →starch granules are nonwetting but may absorb water until the humidity equilibrium is reached. Besides application as emulsifiers, they are used as molding materials in confectionary and drug molding, as encapsulating materials and as carriers for lipophobic agents. The hydrophilic-hydrophobic balance enables compounding of normally incompatible polymers in →starch-plastic compounds.

Lit.: Van Beynum/Roels* 92–93, (1985)
     Blanchard* p. 321 (1992)

Hydroxyalkylated Starches

→Starch Ethers

14-Hydroxy-, *cis*-11-Eicosenoic Acid

→Lesquerolic Acid

Hydroxyethyl Cellulose

Syn.: HEC
G.: Hydroxyethylcellulose;
F.: hydroxyéthylcellulose

The nonionic h. is water soluble above →MS of 1.3. Commercial types are within a range of 1.7–3. These types have 1.5–2.5 ethylene oxide units in the sidechain. Further ethoxylation increases the side chain and make the product also soluble in alcohol.

H. is produced by →alkoxylation of alkali-cellulose suspended in solvents, such as acetone, isopropanol or *tert*.butanol; 0.8–1.5 moles of alkali per AGU are necessary. To decrease viscosity, the alkali-cellulose is degraded by aging (→cellulose) before reaction or by adding hydrogen peroxide to the alkaline reaction mixture. For better effiency, the addition of ethylene oxide is carried out in two stages. After the first reaction step, only catalytic amounts of alkali are necessary. Reaction takes place in 1–4 h at 30–80 °C and is stopped by neutralization with hydrochloric or acetic acid. Salts are removed by washing with alcohol/water mixtures. If retarded dissolution in water is desired, the wet product is treated with glyoxal.

Solutions of h. in water are compatible with other water-soluble polymers, they are less sensitive to electrolytes than MC, have pseudoplastic properties and are not very stable against enzymatic degradation. H. forms clear films, which can be plasticized by →glycerol or polyglycols. They are water-soluble unless treated with dialdehydes.

H. is generally used as a thickening agent, binder, stabilizer, film-forming material and protective colloid.

Because of its water retention, h. and the mixed ether (EHEC) are used as →oil field chemicals (cements and fracturing fluids) and in cements and other applications in construction. Both are used as thickener for paints (→coatings). H. is a sizing agent and a finish in the textile industry (→textile auxiliaries), also as a thickener in textile printing, carpet dyeing and as a binder for nonwovens. H. is used in agriculture as a thickener for pesticide spray emulsions and in seed coatings. Like →methylcellulose and →hydroxypropyl cellulose, it is used as a suspension stabilizer in PVC manufacturing (→polymerization additives). Mixed cellulose ethers, such as EHEC and HEMC are approved for pharmaceutical and food applications. They function as thickening and gelling agents, as coatings and stabilzers for suspensions and emulsions.

US consumption ranges at 30 000 mt/a (1984) and 54 000 mt/a in Japan, while Western Europe production amounts to 16 000 mt/a (1979)

Lit.: Ullmann* (5.) **A5**, 474
   Encycl.Polym.Sci Engng.* (2.) **3**, 242
   [9004-62-0]

Hydroxyethyl Guar
→Guar Gum

Hydroxyethl Starch
→ Starch Ethers

5-Hydroxymethylfurfural
Syn.: HMF
G.: 5-Hydroxymethylfurfural;
F.: 5-(hydroxymethyl)-furfural

Degradation of all hexoses and hexuloses by acid-heat-catalyzed elimination of three moles of water yields h.:

* possible overreaction

m.w.: 126 ($C_6H_6O_3$); m.p.: 28 °C.

100 parts of monosaccharide theoretically yield 70 parts h.

White to yellowish crystal mass, easily soluble in water, ethanol, soluble in ether and dichloromethane.

The best raw materials are high-purity →fructoses, like crystalline fructose, enriched fructose syrups, →invert sugars, →isomerized sugars, →glucose, hydrolysates of →inulin or →levan. The reaction is run mostly in aqueous solution with sulfuric acid (pH 1.8) as catalyst. Reaction time is 2 h at 150 °C in stirred pressure tanks. After completing the reaction, cooling and precipitating the sulfate ions, h. is isolated by cation-exchange chromatography from the solubles (salts and sugars). H. is crystallized from its concentrated solution by cooling.

H. is one of the most reactive and versatile compounds for further processing as an RR in the non-food area. It can become a key substance for substituting fossil raw materials by developing an "HMF-chemistry". The properties are related to the chemical functions of the aldehyde, alcohol, aromatic diene components of the difunctional furan derivative. Typical reactions are:

– Completion of the hydrolysis reaction will lead to an equilibrium mixture of →levulinic acid and formic acid;
– H.-ethers may be formed in situ; they are more stable in anhydrous systems than h. itself;
– Heating with HCl leads to halomethylfurfural, useful for further organic syntheses such as for preparation of surface-active compounds;
– Hydrogenation leads to →tetrahydrofuran derivatives, which are useful intermediates to →furan derivatives as resins and fuel extenders;
– Oxidation yields the interesting →2,5-furan dicarboxylic acid.

Substitution of some fossil commodity chemicals by RR, such as h., will become realistic under convenient conditions of economy, ecology and product properties.

Ref.: Kuster.B.M.F.:starch/stärke 42–(1990),314–321,
   Eggersdorfer* 184–196 (1993)

2-Hydroxymethyl Tetrahydrofuran
→Tetrahydrofurfuryl Alcohol

12-Hydroxyoctadecanoic Acid
→Hydroxystearic Acid

12-Hydroxy-*cis*-9-Octadecenoic Acid
→Ricinoleic Acid

2-Hydroxy 1,2,3-Propanetricarboxylic Acid
→Citric Acid

Hydroxypropyl Cellulose
Syn.: HPC
G.: Hydroxypropylcellulose;
F.: hydroxypropylcellulose

This nonionic →cellulose ether is soluble in cold water and undergoes thermal gelation at 40–45 °C. H. is manufactured (→alkoxylation) by a slurry process similar to that used for →hydroxyethyl cellulose.

Temperatures are higher and reaction times longer. The crude product is purified like →methylcellulose.

H. is soluble in water but also in polar solvents, such as methanol, ethylene glycol and chloroform but insoluble in aliphatic hydrocarbons.

It is thermoplastic and can be extruded at 160–180 °C without a plasticizer to molded goods. It is insensitive to humidity and does not get tacky, although it is still water-soluble.

H. is nontoxic and approved as a →food additive. It is used as stabilizer for whipped food and in →cosmetic emulsions, as a thickening and gelling agent, for coatings in the pharmaceutical and food industry. Stabilization of frozen food is another use in the food area.

Further uses are in the construction industry in plasters and cements to improve workability, water retention and as a retarder (mixed ether with MC and HEC). In the paint industry it is used in solvent-based →paint removers . As mixed ether with MC, it is applied as a binder in extrusion-moulded →ceramics. HPC is a suspension stabilizer and protective colloid in PVC manufacture (→polymerization additives). In the leather industry, it is used as pasting adhesive (→leather auxiliaries). Applications in →cosmetics are in protective creams against irritants and in alcohol-based formulations.

Lit: Ullmann* (5.) **A5**, 476
　　　Encycl.Polym.Sci.Engng.* (2.) **3**, 248
　　　[9004-64-2]

Hydroxypropyl Guar
→Guar Gum

12-Hydroxystearic Acid
Syn.: 12-Hydroxyoctadecanoic Acid
G.: Hydroxystearinsäure;
F.: acide 12-hydroxystéarique

$$CH_3-(CH_2)_5-CHOH-(CH_2)_{10}\,COOH$$

m.w.: 300.49; m.p.: 81 °C

H. is insoluble in water, soluble in alcohol and forms colorless crystals.

It derives from →hydrogenation (125–135 °C, 1.5–2 MPa/15–20 bar, 0.2% Ni-catalyst) of →castor oil and subsequent →hydrolysis or of by hydrogenation of →ricinoleic acid.

H. is used as lubricant in PVC (→plastics additives).

The →metallic soaps are used as costabilizer in PVC (→plastic additives), and the Li-soaps as lithium grease in →lubricating greases.

H. production is 40 000- 45 000 mt from hydrogenated castor oil (10% of total castor oil production).

Lit.: [106-14-9]

Hydroxysuccinic Acid
→Malic Acid

D,L-Hyoscyamine
→Atropine

I

Imidazoline Derivatives

Syn.: Fatty Acid Imidazolines
G.: Imidazolinderivate;
F.: derivés de l'imidazoline

The condensation (180–200 °C) of 1 mole fatty acid with aminoethyl ethanolamine yields imidazolines:

$$R-COOH + H_2N-CH_2-CH_2-NH-CH_2-CH_2-OH$$

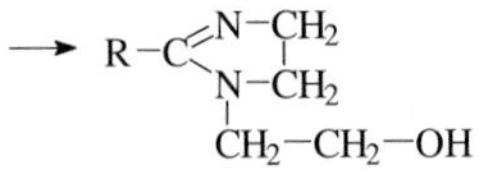

$$\longrightarrow R-C\underset{N-CH_2}{\overset{N-CH_2}{<}}$$

R: C_{15-17}-alkyl CH_2-CH_2-OH

These compounds are cationic →surfactants. However, in acidic solutions they undergo gradual →hydrolysis. Stronger cationic and more stable compounds are obtained by quaternization (→quaternary ammonium compounds).

They are mild to the skin, aseptic ingredients for wetting agents, →cleaners and cosmetics.

By reacting 1 mole of diethylene triamine with two moles of fatty acid and subsequent quaternization with dimethylsulfate (only the double-bonded N can be quaternized), an imidazolinium salt with two fatty acyls is obtained, which is an important fabric softener.

$$R-COOH + H_2N-CH_2-CH_2-NH-CH_2-CH_2-OH$$

$$\longrightarrow R-C\underset{N-CH_2}{\overset{N-CH_2}{<}}$$
$$CH_2-CH_2-NH-CO-R$$
$$+ (CH_3-O-)_2-SO_2$$

$$\left[R-C\underset{N-CH_2}{\overset{\overset{CH_3}{|}}{\underset{N^{+\cdots}CH_2}{<}}} \right] CH_3OSO_3$$
$$CH_2-CH_2-NH-CO-R$$

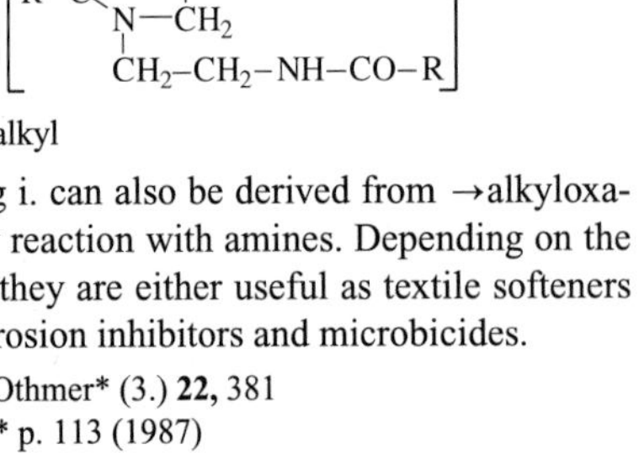

R: C_{15-17} alkyl

Interesting i. can also be derived from →alkyloxazolines by reaction with amines. Depending on the structure, they are either useful as textile softeners or Cu-corrosion inhibitors and microbicides.

Lit.: Kirk Othmer* (3.) **22**, 381
Falbe* p. 113 (1987)

IMP

→Inosine-5-monophosphate

Inclusion Compounds

→Cyclodextrins

Indian Corn

→Corn

Indian Ironweed

→Vernonia

Indian Tragacanth

→Karaya Gum

Indigo

→Dyes, natural

Indigofera tinctoria

→Dyes, natural

Inflorescence

Syn.: Flower
G.: Blüte, Blütenstand; F.: fleur, inflorescence

A flower or a cluster of flowers on a common axis is called an i. (see figure below). The most common i. in crop plants is the cyme and the raceme. A cyme is a determinate i. consisting of primary, secondary, tertiary and often additional floral branches. Opening of flowers will start with the flowers of the primary branch. An example of a cyme is the →potato.

In a raceme, flowering will begin at the base, proceeding towards the top. A raceme consists of a primary stalk (peduncle) with equal flowers along its length on short stalks (pedicels): A typical raceme is that of →soybean.

In most →cereals, such as barley or wheat, and many other grasses the i. is a raceme whose flowers are sessile, which means that the pedicels seem to be attached directly to the primary stalk. Such i. are called spikes.

The panicle of oat or →rice is that of a branched raceme.

Lit.: N.W.Simmonds, "Evolution of Crop Plants" Longman, London/NY (1976).
P.Raven et al. "Biology of Plants" (5.) Worth Publish. NY (1992)

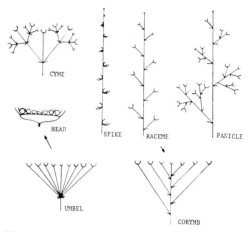

The most common types of inflorescences

Inks

G.: Druckfarben, Tinten; F.: encres d'imprimerie

Inks are fluids or pastes that are used to print or write on substrates like paper, plastics, metal surfaces, etc.

Industrial printing:

I. consist of a vehicle (resin and solvent), pigment and other additives (dryers, lubricants, defoamers etc.). The formulation is adjusted to the printing technology/equipment and to the nature of the substrate. These factors also influence the selection of the resin and the drying characteristic of the i. Drying can be accomplished by absorption or penetration into the substrate, evaporation of the solvent, oxidation by air, precipitation, e. g., by moisture set or polymerization, including UV curing and fusion (plastisol).

Commercial printing in machines is accomplished by one of the following principle techniques:

1. Letterpress: a metal relief wetted by i. is pressed to the substrate (fading out technology);
2. Flexography: like letterpress, but the relief consists of a rubber plate;
3. Lithography: the printing surface consists of the oleophilic printing area that accepts i. and hydrophilic nonprinting areas that do not accept i. (most widely used technology);
4. Gravure: a copper cylinder bears the image engraved by etching;
5. Screen: a squeegee forces the ink through a screen that is stenciled with the image.

I. are adjusted to these technologies and to numerous other different situations and problems, resulting in a host of ink formulations.

The following resins or binders based on RR are used: →drying oils, →oleoresins, (modified)→rosin, rosin soaps, →alkyd resins, →shellac, chlorinated and cyclized →rubber, →polyamides based on dimer acid and, →cellulose nitrate and →ethylcellulose.

US consumption in thousand mt in 1983[1]) of RR-based resins:

Resin	Letter	Litho	Gravure	Flexo	Screen
Alkyd	–	15.4	–	–	1.36
cellulose ethers	–	0.77	–	–	–
nitrocellulose	–	–	3.62	2.04	–
polyamides	–	–	0.45	4.53	–
drying oils	0.45	–	–	–	0.09
rosin esters	1.6	16.3	6.8	4.76	0.18
rosin soaps	–	–	22.68	–	–

The total world production of printing i. (1991) was 1.66×10^6mt, (North America 590000 mt, Europe 565000 mt and South-East Asia 385000 mt).

In 1983 resin consumption to produce 590000 mt of printing i. was 138000 mt, with ≈80000 mt based on RR.

Office and household inks:

A common writing instrument is still the pencil[2]).

A pencil is made from wood with a homogeneous graphite or pigment lead in the center. The finished leads are dipped in a →fat and →wax mixture to improve consistency and sliding ability. Colored leads are made from pigments with the help of binders, e.g., →methylcellulose, →ethylcellulose or →cellulose nitrate, and are also dipped in a fat/wax mixture. Wooden pencils are coated with →cellulose acetate. In 1985 4×10^9 pencils were produced worldwide.

The following inks are used in household and office:

1. Drawing ink: A suspension of carbon black in a colloidal solution of →shellac soap, stabilized with animal glue (→glue, animal, →methylcellulose, →hydroxyethyl cellulose or →carboxymethyl cellulose; →surfactants are also used;
2. Fountain pen i.: Blue-black i. is a solution of iron salts and tannic and gallic acid and other natural tanning agents (→tannins) with many other ingredients, one of which is →gum arabic. Colored i. are made from organic dyes. →Glycerol and/or →sorbitol are used as moisturizer. Surfactants are added also;
3. Ball point i.: Strong dye solutions or pigment dispersions (40%) in →oleic acid (10%) or (blown) →castor oil or acid resin vehicle. In ball

point ink pens, the reservoir is made of bundles of →cellulose acetate fibers. The ink is more liquid than that of ball-points with →surfactants added to regulate the surface tension,

4. Fiber(felt)-tip pens.: The only RR-based products used in i. are surfactants to improve flow-behavior;

5 Chalks: Binding agents, such as →tragacanth, →carboxymethyl cellulose, alginates (→sodium alginate) and →gum arabic, are used;

6. Crayons: A combination of →stearic acid (up to 40%) and ester →wax (up to 25%) is the base material of cast or extruded crayons;

7. Ink jet : A technique used for computer printing needs very fluid and fast-drying dye-colored inks. Natural resins (→resins, natural) and solvents have to be selected carefully;

8. Electrostatic i.: Photocopying machines need toners based on resins (→resins, natural)/pigment mixtures, which have the property to deposit electrostatically on an image and form a homogenous film by heat or solvent exposure.

Lit.: Encycl.Polym.Sci.Engng.* (2.) **13**, 368
 Kirk Othmer* (3.) **13**, 374
 Ullmann* (5.) **A9**, 37; A22, 143

[1] Printing Ink, Hull and Co., (July 1984)
[2] Ullmannn* (5.) **A9**, 38

Inosine-5′-monophosphate

Syn.: IMP, Inosinic Acid, Inosine Phosphoric Acid
G.: Inosin-5′-monophosphat
F.: inosine 5′-monophosphate

m.w.: 348.22

I. is found in meat and meat extracts and in sugar beet. It is produced by enzymatic hydrolysis of yeast ribonucleic acid by means of phosphodiesterases from *Penicillium citrinum* or *Aspergillus* sp., followed by treatment with a desaminase from *Aspergillus oryzae*. I. is mainly produced in Japan ("Umami") and used as a flavor enhancer (→food additives), most often in combination with mono sodium glutamate, which shows synergistic effects.

Lit.: Ullmann* (5.) **A4**, 149 + **A22**, 384
 The Merck Index* (11.), 4882
 [131-99-7]

Inositolhexaphosphate
→Phytic Acid

Insecticides
G.: Insektizide; F.: insecticides

Active ingredients:
There are only a few active insecticides derived from natural resources. The most important is →pyrethrum.
An old i. is **nicotine** [54-11-5], β-pyridyl-α-methyl pyrrolidine:

which is contained (2–14%) in the leaves of *Nicotiana tabacum* and *N. rustica*, (Solanaceae) and is applied in the form of its sulfate. Similar compounds (nicotinoids) are nornicotine [494-97-3] and anabasine [494-52-0]. All are active against aphids (plant lice).

Another class of i. are rotenoids derived from the roots and seeds of *Derris (Deguelia) elliptica* and *Derris malaccensis* (Leguminosae) and over 60 other plant species. A most active compound is **rotenone** [83-79-4], a polycyclic compound that can be applied to food crops shortly before use, due to its rapid degradation after exposure to sunlight. Rotenone-containing i. are applied as dust, powder or emulsions.

Known for centuries as a "louse powder" is sabadilla (Syn.: cevadilla, caustic barley) derived from the ground seeds of *Schoenocaulon officinale* (Liliaceae). The active compounds are two alkoloids: **cevadine** [62-59-9] and **veratridine** [71-62-5]. They are used in thrips (*Thysanoptera*) control in sugar beet.

Ryanodine [15662-33-6] is the active ingredient of the roots of *Ryania speciosa*, (Flacourtiaceae). It is more stable against sunlight and air than pyrethrum and rotenone and is used for codling moth control.

An interesting and promising insecticide is derived from the seeds of the →**neem** tree (*Azadirachta indica* A.Juss)[1]. With water or alcohol,. a mixture of several active ingredients (five of them are powerful i., while ≈20 are less effective) are extracted (→extraction). The most effective compound is azadirachtin (90% of the effect on most pests), while meliantriol, salannin, nimbin and nimbidin contribute more specific activities. All are steroidal compounds with a limonoid structure. Like

→pyrethrum, the active ingredients are degraded when exposed to rain and sunlight.

The i. activity of neem is based on a combination of repellancy and disruption of growth and reproduction and less on a fast knock-down effect. It is known that more than 200 insect species are being repelled. I. effectiveness is comparable to strong synthetic i., such as malathion, DDT and dieldrin. It is believed that there is no chance of resistance due to the multicomponent composition. Neem extracts perform not only as an i. but are also effective against nematodes, some fungi and plant viruses. They act also as systemic agents in certain plant species. Neem products appear to have little or no toxicity to warm-blooded animals. This new generation of natural i. is on its long way of gaining official approval.

I. based on RR lost importance during the last decades and were substituted by synthetic and more powerful i. There is, however, some tendency for a backswing to natural products due to environmental considerations (e. g., neem project).

There are also development efforts and first results to use the biological activity of bacterias and viruses for insect control directly, without the isolation of the active compound.

Auxiliaries:

Some surfactants (→lignosulfonate) are used to emulsify active ingredients in formulated i.

Water-soluble polymers (cellulose ethers, such as →hydroxyethyl cellulose and →carboxymethyl cellulose) promote the adhesion of the active i. on plant leaves and are therefore sometimes added to compounded i.

→Fatty acid methyl ester have been considered and used as substitutes for kerosene as solvents in formulated i. They are biodegradable and therefore less harmful to the environment.

Lit.: M.J.Jacobsen, D.G. Crosby (ed.) "Naturally Occurring Insecticides" Marcel Decker NY (1973) (5.) **A14**, 271

[1] National Research Council (1992) "Neem: A Tree for Solving Global Problems" National Academy Press, Washington, D.C.

Insulin

G.: Insulin; F.: insuline

I. is a peptide/protein hormone that occurs in the pancreatic islets of Langerhans. It regulates the blood glucose content and must be supplemented in case of diabetes mellitus.

It is still gained mainly from animal pancreas (→cattle, →pig). The yield is 70–100 mg from 1 kg of pancreas.

Modern biotechnology is used increasingly to make recombinant i. from modified *Escherichia coli*.

Lit.: Cuatrecas, Jacobs "Insulin" in Handbook of Experimental Pharmacology Vol. 92 Springer Verlag, Berlin (1990)

Ullmann* (5.) **A19**, 122–125

Intercropping

G.: Zwischenfruchtanbau;

F.: culture dérobée

Growing a crop with another crop in the same field, as in alternate rows, is called i. This may be done to cover the soil, for a better feed value, for phytosanitary purposes or to isolate different crops.

Examples for covering the soil are: growing legumes between young →oil palms, or planting cocoa, coffee or bananas in →coconut plantations. Clover or alfalfa are mixed with cereals, such as rye or barley, for fodder crops harvested at green maturity for grazing ruminants or bailed for feed. Rows of →hemp are grown between different lines of →sugarbeets for seed multiplication of the sugarbeets to prevent outcrossing.

Lit.: S.R.Chapman, L.P.Carter, "Crop Production, Principles and Practices", Freeman and Co. San Francisco (1976)

Inulin

→Fructans

Invertases

G.: Invertasen; F.: invertases

Generic term for enzymes (e. g., α-glucosidases or β-fructosidases) that are capable of cleaving →saccharose into glucose and fructose. I. of the human intestine, mammalians and of honey bee are α-glucosidases, whereas i. of higher plants belong to the β-fructosidases. They are produced primarily from baker's or brewer's yeast. The yeast cells are broken up (mechanical, chemical or enzymatic), and the enzyme is purified. They are used in food technology for the production of →invert sugar and as analytical reagents for sucrose. The addition of i. makes the production of sweets with a solid coating and a soft center possible.

Lit.: Ullmann* (5.) **A9**, 423 + **A11**, 578

K. Myrbäck "Invertases", in "The Enzymes", P.D. Boyer et al. (ed.) Academic Press, NY **4**, 379–396 (1960)

[9001-57-4]

Invert Sugars

G.: Invertzucker; F.: sucres invertis

I. are liquid mixtures or syrups of →fructose and →glucose in a ratio between $50:50$ and $95:5$ (d.b.).

They are produced from →sucrose solutions by →hydrolysis of the glucosidic linkage. The reaction is catalyzed either by protons (free acids or strong acidic cation exchangers) or enzymatically by free or immobilized →invertases.

The degree of inversion may be varied by time and temperature of the reaction, which is limited by the formation of →5-hydroxymethylfurfural under forced conditions. Enrichment to high fructose contents is performed by chromatographic separation on columns filled with calcium polystyrene sulfonate ion exchange resins.

Commercial products show the following degrees of inversion and % solids: 0.49 (75%), 0.66 (72.2%) and 0.95 (65%).

A standard quality is: Inversion 0.66; viscosity (20 °C) 450 mPas; physiological energy value 17.2 kJ/g.

Application of all i. is highly specific for the food industry and to a limited extent for the pharmaceutical industry. Nonfood applications could be increased by fructose enrichment to fructose syrups and crystalline fructose.

Lit.: Rymon-Lipinski/Schiweck* 138–142 (1991)

IPM

→Fatty Acid Esters

Iris pallida

→Orris

Iron Soap

→Metallic Soaps

Isethionates

→Acyloxy Alkane Sulfonates

Isoglucose

→Isomerized Sugars

Isomalt

→Isomaltulose

Isomaltose

G.: Isomaltose; F: isomaltose

m.w.: 342 ($C_{12}H_{22}O_{11}$)

I. is 1,6-α-D-glucosido glucose, which is formed during starch hydrolysis (→hydrolysis, glucosidic linkages) by setting free the branching points of the →amylopectin. It is further formed as a reversion product from glucose at the end of the total hydrolysis of starch. Together with →gentiobiose, it hinders →dextrose crystallization in acid hydrolysates of starch. I. is formed as an enzymatic resynthesis product when glucoamylase acts on concentrated glucose solutions.

Lit. Tegge* 66–68, 260–263 (1988)

Isomaltotriose

G.: Isomaltotriose; F.: isomaltotriose

m.w.: 504 ($C_{18}H_{32}O_{16}$)

I. is α-D-glucopyranosyl-1,4-D-glucopyranosyl-1,6-α-D-glucopyranose, which is formed as a component in →branched oligosaccharide syrups by action of transglucosidase (α-glucosidase) on →maltose. It may also be found as a reversion product of the acid hydrolysis of starch, which is enriched in the mother liquor of glucose crystallization (→hydrol).

Lit.: Schenk/Hebeda* 350–351, 361 (1992)

Isomaltulose

G.: Isomaltulose; F.: isomaltulose

I. is the reducing keto-disaccharide that is prepared by action of special bacteria on →sucrose.

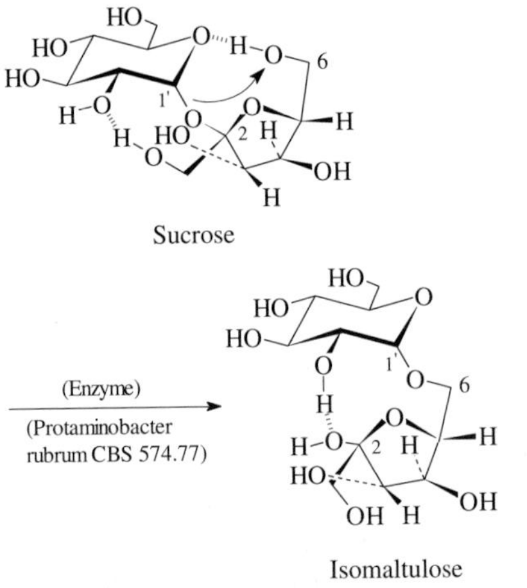

Sucrose

(Enzyme)

(Protaminobacter rubrum CBS 574.77)

Isomaltulose

m.w.: 360 ($C_{12}H_{22}O_{11} \times H_2O$); m.p.: 123–124 °C; $[\alpha]_D^{20}$: 103–104°; DE: 52%; r.s.: 42

Rhomboid, white crystals with a solubility (20 °C) of 49 % (w/w) in water. It is noncariogenic and has a physiological energy value of 17.2 kJ/g.

It is produced by running 40–60 % solutions of pure sucrose through a reactor equipped with immobilized cells of *Protoaminobacter rubrum*. Other important microorganisms that are capable of this transformation are: *Leuconostoc mesenteroides, Serratia plymuthica, Serratia marcescens, Erwinia carotovora, Erwinia rhapontici*.

The yield of transformation is 85–95 %, and the composition of the raw product: 79–84.5 i., 9–11 % trehalose, →sucrose, →inverted sugar, →isomaltose, higher homologues.

After concentration, i. is crystallized in high yield and purity.

Application in the food sector as dietetic sweetener utilizes the noncariogenity and the slow intestinal absorption rate, mainly in Belgium, Germany and Japan. Most i. is further reacted by →hydrogenation (Raney-nickel, neutral to alkaline) to yield isomalt, an equimolecular mixture of α-D-glucopyranosyl-D-sorbitol (GPS) and α-D-glucopyranosyl-D-mannitol (GPM), as a promising low-caloric sweetener.

Annual production is 20 000 mt/a.

Nonfood uses as bulk chemical have been demonstrated.

Lit.: Lichtenthaler* 58–64 (1991)
G.Descotes (ed.) "Carbohydrates as Organic Raw Material II" VCH Weinheim (1993)

Isomerism

G.: Isomerie; F.: isomérie

There are two types of i.: Structural isomerism and stereoisomerism.

Structural I.:

Compounds are identical as to their molecular formula, but the order in which the individual atoms are connected differs. A simple example is ethanol and dimethylether

CH_3-CH_2-OH and CH_3-O-CH_3

RR-related compounds that show structural i. are glucose and fructose and many other pentoses and hexoses (→isomerization). In →oleochemistry, structural i. is rather rare because nature prefers to produce straight-chain fatty acids. In the position of double bonds, structural i. is more common. For example polyunsaturated fatty acids may occur in nature with nonconjugated (e. g., →linolenic acid) or conjugated (e. g.→eleostearic acid) arrangements of their double bonds.

conjugated:
$-CH_2-CH=CH-CH=CH-CH=CH-CH_2-$

nonconjugated:
$-CH_2-CH=CH-CH_2-CH=CH-CH_2-CH=CH-$

Because conjugated fatty acids are more reactive, such as in drying →alkyd resins, or are able to undergo Diels Alder reactions, (→dimerization) nonconjugated fatty acids from sunflower or soybean oil are technically isomerized to get conjugation. This is done by →saponification of the oils with excess alkali and heating the soap paste to 200–300 °C. After acidification, fatty acids with a degree of conjugation of 95 % are obtained. This process goes together with a partial *cis-trans* i. to produce mixtures of many possible combinations.

Stereoisomerism:

Compounds are again identical as to their molecular formula, and the order in which the atoms are connected is also the same, but the isomers differ in their spatial arrangement. A RR-related example is the *cis-trans* isomerism in oleochemistry. Unsaturated fatty acids are able to form two (or more) isomers. Good examples are →oleic and →elaidic acid:

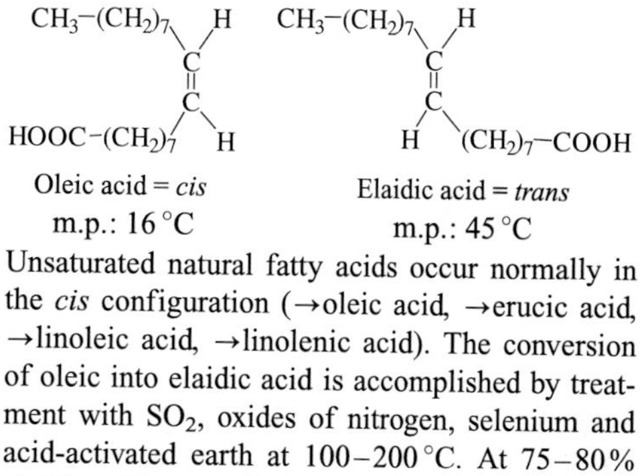

Oleic acid = *cis*
m.p.: 16 °C

Elaidic acid = *trans*
m.p.: 45 °C

Unsaturated natural fatty acids occur normally in the *cis* configuration (→oleic acid, →erucic acid, →linoleic acid, →linolenic acid). The conversion of oleic into elaidic acid is accomplished by treatment with SO_2, oxides of nitrogen, selenium and acid-activated earth at 100–200 °C. At 75–80 % elaidic acid, an equlibrium is reached. The effect is used to increase the melting point of an oil or a fatty acid. Partial →hydrogenation of polyunsaturated fatty acids also increases the proportion of *trans* isomer.

Molecules that contain a C-atom (called an asymmetric atom) that is connected to four different substituent groups are chiral. They may have two structures that are not superimposable mirror images and therefore not identical. They are comparable with the human left and right hands and are called enantiomers. Many physical properties are identical but they can be identified by the ability of their aqueous solutions to rotate polarized light clockwise or counterclockwise [dextrorotary (D or +) and levorotary (L or -) enantiomers]. The

1:1 mixture of the two optical isomers is called the racemate.

This optical activity can be observed with many RR-related compounds. Thus, →amino acids (with the exeption of glycine) can exist in two optically active isomers. Many hydroxy acids (→lactic acid, →tartaric acid, →malic acid) are also active, and optical isomerism is frequently observed in saccharide chemistry.

Lit.: Ullmann* (5.) **A10,** 269

Isomerization

G.: Isomerisierung; F.: isomérisation

The action of "xylose isomerase" ("glucose isomerase", "sweet-zyme"), *D-xylose-ketol-isomerase,* (EC 5.3.1.5.) isomerizes the aldoses D-xylose or D-glucose into the corresponding ketoses D-xylulose or →D-fructose by passing the 1,2 enediol intermediate. The thermodynamically controlled process leads to equilibria of 84% D-xylose and 16% D-xylulose or 49% D-glucose and 51% D-fructose

O ‖ CH │ HC−OH │ HO−CH │ HC−OH │ HC−OH │ CH₂OH	OH │ CH ‖ C−OH │ HO−CH │ HC−OH │ HC−OH │ CH₂OH	CH₂OH │ C=O │ HO−CH │ HC−OH │ HCOH │ CH₂OH
Dextrose	1,2 Enediol	Fructose

Isomerization of glucose is an economically important process, leading to →isomerized sugars (high-fructose syrups, HFCS).

It is run with immobilized enzymes in continuous flow. Isomerase enzymes are produced by the microorganisms *Streptomyces, Actinoplanes, Arthrobacter, Bacillus coagulans* and *Aerobacter levanicum.* After isolation and concentration, the enzyme is fixed physically in high concentration on porous silicate material of well-defined particle size and pore structure. The reaction conditions with →glucose syrups of 30−40% d.b. and 94−96% dextrose purity are: 45−65 °C, pH 6−8.5, Mg-and/or Co-salts as activators, and protective agents for the enzyme. The primary reaction product contains 42% fructose in equilibrium. It can be further enriched to 55% or 90% fructose by column chromatography. The process of enzymatic i. of glucose has opened a broad field of fructose utilization to replace →sucrose in food

utilization but, moreover for fructose chemistry in nonfood utilization.

Lit.: Schenk/Hebeda* 127−199 (1992)
Ruttloff* 591−592, 617−620 (1994)

Isomerized Sugars

Syn.: High-Fructose Corn Syrup; HFCS; Isomerose

G.: Isoglucose, Isomeratzucker;

F.: sirop de glucose isomerisée

I. are liquid mixtures (syrups) that contain →fructose, →glucose and minor amounts of →malto-oligosaccharides.

I. are produced from highly purified →glucose syrups (>94% glucose) by continuous →isomerization with immobilized →glucose isomerase. The enzyme converts the glucose into an equilibrium mixture of 42% fructose, 55% glucose and 5% malto-oligosaccharides. The highly purified product may be further enriched in fructose by ion-exchange chromatography to high-fructose syrups of 90−95%. In practice, they are blended with 42% fructose syrup to a liquid sweetener of 55% fructose, which is adjusted to the sweetness of →sucrose. Both are predominant types of commercial products. I. with 80%, 90%, 95% fructose are commercially available if extraordinary sweetness, energy reduction or pure fructose functionality are desired.

The two main types are sold at 71% and 77% solids, pH 4.0, viscosity of 160 mPas or 800 mPas, m.w.: 190−200.

I. are produced worldwide as caloric sweetener if starch is more available and cheaper as a starting material than sucrose.

Present annual world production (1994/95) amounts to more than 9.55×10^6 mt, 78% of which is consumed in USA and 16% in Asia. In EC, production is limited by legislation to only 281 000 mt/a in the interest of the beet sugar industry.

In the food industry, nonalcoholic beverages take the main portion; other fields of application are bakery products, breakfast cereals, canned products, confectionary, dairy products, ice cream, alcoholic beverages.

Nonfood applications are increasing: in fermentation processes, in pharmacy as a coating component, and in special dietetic formulations. Syrups of >90% fructose may be used instead of solutions of pure fructose or as the starting material for crystalline fructose. →Hydrogenation is performed to yield mixtures of →mannitol and →sorbitol in varied ratios.

Lit.: Schenk/Hebeda* 177–199 (1992)
      Ruttloff* 205–209 (1994)

Isomerose
→Isomerized Sugars

Isopalmitic Acid
G.: Isopalmitinsäure; F.: acide isopalmitique

I. consists mainly of 2-hexyl decanoic acid. It is produced by alkali fusion of the corresponding 2-hexyldecan-1-ol (→Guerbet alcohol).
The acid is used where saturation and a low m.p. is desired, especially in cosmetic.

Lit.: Ullmann* (5.) **A10**, 271

Isopropyl Myristate
→Fatty Acid Esters

Isostearic Acid
G.: Isostearinsäure; F.: acide isostéarique

m.w.: 284; m.p.: ≈ 7 °C

I. is a coined name for a saturated C_{18} acid that is a complex mixture of isomers (primarily methyl-branched, but also cyclized and aromatized), which are mutually soluble and virtually inseparable.
I. derives from the manufacture of →dimer acid. The stripped-off monomers are →hydrogenated and yield i. →Crystallization is used to separate stearic from isostearic acid.
The acid is used especially in cosmetic applications as a substitute for →oleic acid, where a low

m.p. is desired but saturation guarantees color and degradation stability.

Lit.: Kirk Othmer* (3.) **7**, 770
      [2724-58-5]

Itaconic Acid
Syn.: Methylene Butanedioic Acid
G.: Itaconsäure; F.: acide itaconique

$$CH_2=C-CH_2-COOH$$
$$\quad\quad |$$
$$\quad\quad COOH$$

m.w.: 130.1; m.p.: 167–168 °C

Colorless, hygroscopic crystals, soluble in water and ethanol.
I. is a decomposition product of →citric acid with aconitic acid as an intermediate. It is, however, commercially produced by submerged aerobic fermentation of molasses and other glucose sources with certain strains of *Aspergillus terreus* or *A. itaconicus*. Isolation and purification is accomplished by acidification, clarification, concentration and crystallization.
I. is used in small amounts (<10%) as a comonomer in vinyl polymerization (→polymerization additives). Thus, styrene-butadiene latices containing i. are widely used as →carpet backings and →paper coatings. Acrylic latex that contains i. is used as a nonwoven fabric binder with improved adhesion. Polyacrylonitrile fiber gains improved dyability when small amounts of i. are incorporated as a comonomer.
I. is used as an antioxidant, as an additive for drying oils and in the production of plasticizers (→plastics additives). I. is also used in the separation of triorganophosphanes.

Lit.: Kirk Othmer* (4.) **14**, 952
      Ullmann* (5.) **A8**, 535
      [97-65-4]

J

Japanese Lacquer
→Resins, natural

Japan Wax
→Waxes

Jasmine
Jasminum grandiflorum L.
Oleaceae
G.: Jasmin; F.: jasmin

Originating from East India, the j. shrub is now also cultivated in the South of France, Algeria, Morocco, Spain, Italy and Egypt.

Its small white flowers have to be handpicked in the early morning hours to produce the best quality of →jasmine absolute.

Lit.: Arctander

Jasmine Absolute
G.: Jasmin Absolut; F.: absolue de jasmin

J. is produced via →concrete by volatile solvent →extraction from →jasmine flowers with a low yield (0.2% for concrete, 50% of which is absolute) and a high labor cost. This absolute is one of the most valuable flavor and fragrance raw materials.

J. is a dark orange (upon aging, reddish-brown), slightly viscous liquid that possesses a rich, diffusive floral odor with honeywax-like, fruity jam-like, herbaceous and slightly animalic undertones. It is a classic material in high-class perfumes and finds a multitude of applications in all floral combinations. Main constituents are benzyl acetate, linalol, methyl anthranilate and indole.

Lit.: Arctander*
      The H&R Book*
      Gildemeister*

Jasminum grandiflorum
→Jasmin

Jerusalem Artichoke
→Topinambur

Jojoba
G.: Jojoba; F.: jojoba
Simmondsia chinensis (Link) Schneid.; *S. californica* (Link) Nutt
Simmondsiaceae (Buxaceae)

J. is a perennial desert crop (shrub) with hazelnut-like nuts, the oil of which serves as a substitute for the oil of the sperm whale (→sperm oil).

J. is native to Mexico, Arizona, and California. It is a typical example for a recently domesticated crop where the interest in the chemical composition of the oil was the main driving force for cultivation. There are, however, reports about early applications of j. oil by native Mexican people. The name is of Indian origin and is pronounced ho-ho'-ba.

Jojoba; A: Branch with fruits. B: Fruit in longitudinal section (from Franke*, 1981, with permission).

Main producing countries are Mexico, Arizona, California, and Israel. Production experiments have been reported from Argentina, Paraguay, South Africa, India, Australia, and Southern Europe. The j. shrub grows comparatively slow, due to its 10 m long root, its special climatic adaptation to low rainfall (120–200 mm/a) and high temperatures. It will take 3–5 years to bear →inflorescences in a →dioecious mode. A plantation will need to have a ratio of male to female plants of 1:6 to produce good seed set. The seeds contain oblong tripartite nutlets with a typical size of 1×2 cm and have a →seed weight of 800–1000 g per 1000 seeds. Seed yields per plant are reported as 400–500 g after 5 years, 2–4 kg after 12 years and 13.5 kg after 25 years. In the natural desert habitat, a j. plant may become as old as 100–200 years.

Harvesting machines have been developed but handpicking still dominates.

The oil is obtained from the crushed seeds by →expelling and/or →extraction. For chemical composition: →jojoba oil.

The by-product of j.-oil extraction, the meal, may not be fed to animals due to toxic substances in the protein.

With the banning of whale products for food and technical applications, the importance of j. plantations rose tremendously in the late 1970's, especially when market prices of $ 40 to 50 per gallon of oil were reported. A tremendous increase of plantations resulted in the 1980's in oil prices of under $ 10.

Today about 500- 1000 mt of j. oil are produced annually.

Lit.: A.R.Baldwin (ed.) Proceedings of the 7th International Conference on Jojoba and its Uses; American Oil Chem. Soc. Champaign, Illinois (1988)
B.F.Haumann, Jojoba, J.Amer.Oil Chem.Soc. **60,** 44A (1983)
E.H.Pryde, L.H.Princen, K.D.Mukherjee (ed.) "New Sources of Fats and Oils" Am.Oil Chemist's Soc., Champaign Il. (1981)

Jojoba Oil

G.: Jojobaöl; F.: huile de jojoba

s.v.: 92–95; i.v.: 82–88; m.p.: 12 °C

J. is a waxy, yellowish oil, which is slow to turn rancid and is undamaged by repeated heating to high temperatures. It is one of the few naturally occurring →wax esters of unsaturated →fatty acids and unsaturated →fatty alcohols, mainly of a chainlength of $C_{20}-C_{22}$. The oil is obtained from the seeds of the →jojoba plant.

It is used as such in cosmetic preparations (lotions, ointments, shampoos, hair conditioners). This is the main outlet.

The sulfurized oil is a high-performance →lubricant.

Hydrogenated j. is similar to →carnauba wax and can be used as a substitute.

Because j. is similar to the oil of the sperm whale (→sperm oil), which is practically out of the market due to whale catching restrictions, it could be a substitute. However, other cheaper alternatives have been developed in recent years. For this reason, the present price level of j. hampers the use in lower-price applications where spermoil was formerly used.

Lit.: A.R.Baldwin (ed) Proceedings of the 7th International Conference on Jojoba and Its Uses American Oil Chem. Soc. Champaign, Illinois (1988)
B.F.Haumann, Jojoba, JAOCS **60,** 44A (1983)

Juniperus mexicana
Juniperus virginia
→Cedarwood Oil

Jute

Corchorus olitorius L.
Corchorus capsularis L.
Tiliaceae
G: Jute; F.: jute

Among the bast fibers j. is the most important, and it is the second most important vegetable fiber after →cotton. It exceeds →wool in volume.

J. was grown almost exclusively for centuries in the Ganges delta. Since the 1950's, China has also started to cultivate this plant. There are two varieties: *Colchorus capsularis*, which grows in the little delta farms on periodically flooded areas to yield white j., while *C. olitorius* is grown on the rainy slopes of India, Bangladesh and Nepal. This upland j. is brown and cannot be bleached without substantial degradation.

The fibers are usually sprayed with 5 % processing oil before carding. For cost reasons, they are not scoured and have a characteristic odor of rancid vegetable or fish oil.

J. needs to be sown by broadcasting into fine seed beds. Due to the fact that fiber quality depends on a fine plant stand, it is important to reach dense populations of up to 500 000 plants per hectare. Harvesting should be done at the "early pod" stage about 3–5 months after sowing. At harvest, the stems, which may be standing in deep water, are cut close to the ground with sickles and are left to dry for a few days until the leaves have withered and can be shaken off. They then need to be retted in water for about three weeks. After retting the bark is stripped by hand from the stems, the fibers are washed free of tissues and hung to dry for several days. Typical length for j. fibers is 3 meters (fiber bundles).

J. is weaker than flax or hemp and less resistant to deterioration and rotting in water.

The main market for j. is sackcloth and is heavily attacked by polypropylene ribbons, which do not have the tendency to rot. This may, however, convert into an advantage because of ecological considerations. J. is still used in substantial quantities in carpet backings.

Until recently, j. was cheap because labor in the main growing areas was plentiful and cheap. Today, these countries not only consume much of their home production of j., but their increasing populations demand more land and labor to grow food, which conflicts with increasing j. output.

World production is about 2.7×10^6 mt/a with main production in Bangladesh and India (about 1×10^6 mt each).

Lit: Purseglove* (1974)

K

Kanamycins

G.: Kanamycine; F.: kanamycines

R[1]	R	R[3]	name
NH$_2$	OH	OH	kanamycin A
NH$_2$	NH$_2$	OH	kanamycin B
OH	OH	NH$_2$	kanamycin C

K. belong to the group of oligosaccharide antibiotics that constist of an aminocyclohexanol and two amino sugar moieties. This antibiotic complex, produced by *Streptomyces kanamyceticus*, is comprised of three compounds: kanamycin A, the major component that is usually designated as kanamycin, and two minor congeners (kanamycins B and C). K. are used as antibacterials against gram-positive as well as gram-negative bacteria.

Lit.: Vandamme "Biotechnology of Industrial Antibiotics" Marcel Dekker inc. NY (1986)
Ullmann* **A2,** 487 + 518
Rehm/Reed* 4, 315
[8063-07-8]

Karaya Gum

Syn.: Indian Tragacanth
G.: Karaya Gummi, Sterculia Gummi; F.: karaya

The dried exudate of the tree *Sterculia urens* Roxb., found in India, especially in the Gujerat region and in the central provinces. It is a partially acetylated polysaccharide that contains about 8% acetyl groups and about 37% uronic acid residues. It is a finely ground white powder with a faint odor of acetic acid. It absorbs water rapidly to form viscous mucilages at low concentrations. Viscosity decreases on addition of acid or alkali.
It can be used as an adhesive, as binder in paper industry, as a stabilizer, thickener, texturizer or emulsifier in food and as a substitute for →tragacanth; it shows some cathartic properties.

Lit.: Kirk-Othmer* (3.) 12, 56

Kenaf

Syn.: Deccan Hemp, Mesta
Hibiscus cannabinus L.
Malvaceae
G: Kenaf, Ambari, Gambohanf;
F: kenaf

K. fiber is obtained from the stems of this representative of the Malvaceae family. K. probably originated in East Africa, where it occurs wild, and has been cultivated there and in Asia (especially in India) for centuries as a fiber crop for domestic use. Since World War II, k. spread to many tropical countries, however with only little importance in world trade.
K. plants may grow as tall as 7 m with a strong tap root and several branches. The beautiful flowers form a pale yellow corolla consisting of 5 large petals, about 10 cm in width across the whole flower. K. may be harvested by hand (common) or with machine harvesters (→harvesting), which cut the stems from the ground. Such harvesting also will need to strip the bark from the stem before or after a retting process (→flax, →jute). Coarse fiber is obtained by cleaning unretted ribbons of bark, which can be used only to make coarse bags, sacks, and coarse fabrics. The fiber is similar to that of jute, and although it is coarser and less pliable, it may be applied for many similar purposes, sometimes even without retting.
In the United States, several research programs have been developed to introduce k. to arid and wet warm zones as a new crop for →paper pulp. K. can also be considered as a source of →cellulose for derivatization.

Lit: Cobley* (1976)
Rehm/Espig* (1994)

Kermes

→Dyes, natural

Keto Fatty Acids (Esters)

G.: Ketofettsäuren; F.: céto-acides gras

One of the few natural sources of k. is →oiticica oil (→licanic acid). K. can by made also by rearrangement of →epoxides.

Long-chain branched k. are available by Claisen condensation of →fatty acid methyl esters:

$$2 \text{ R}-\text{CH}_2-\text{COO}-\text{CH}_3 \xrightarrow{\text{NaOCH}_3}$$

$$\text{R}-\text{CH}_2-\text{CO}-\underset{\underset{\text{R}}{|}}{\text{CH}}-\text{COO}-\text{CH}_3 + \text{CH}_3-\text{OH}$$

R: C_{14-16} alkyl

K. triglycerides, made by rearrangement of epoxidized soybean oil, have shown promising properties as transparent lubricants (→plastics additives) in PVC films.

Lit.: H.Baumann et al. Angewandte Chemie (English edition) **27**, 56 (1988)

Ketofuranoses
→Pentoses

Koch Synthesis
→Hydrocarboxylation

Kojic Acid
G.: Kojisäure; F.: acide kojique

m.w.: 142.11; m.p.: 153–154 °C

K. is produced by the aerobic fermentation of glucose, glycerol, arabinose solutions and from a wide range of other carbon sources with *Aspergillus* sp. K. shows some antibiotic and weak insecticidal activity. It has been converted to →maltol and ethyl maltol as flavor-enhancing additives.

Lit.: J. Sci. Ind. Res. **41**, 185–194 (1982)
  The Merck Index* (11.) 5197
  [501-30-4]

L

Labdanum Resinoid/Cistus Oil

G.: Labdanum Resinoid/Cistusöl;
F.: resinoïde de labdanum/essence de ciste

Labdanum resinoid is a resinous exudation from the leaves and twigs of *Cistus ladaniferus* L. (Cistaceae), a wild shrub found in all mediterranean countries. The main production country for labdanum products is Spain. L. is collected by boiling the leaves and twigs in water and followed by steam-distillation to yield cistus oil or by hydrocarbon solvent extraction for labdanum resinoid.

Cistus oil is an amber-colored liquid with an intense and pungent, longlasting, warm-resinous spicy, balsamic note with a sweet, dry-woody background. The odor characteristics of labdanum resinoid, a generally dark-brown or dark-green semi-solid resin, are similar but with a different intensity. Labdanum resinoid has much less initial impact but a longlasting, more balsamic-warm character.

Both products are frequently used in heavier fragrance types, such as →chypre or →oriental notes, to give depth and volume without adding an overly sweet note. They blend extremely well with all dry-woody notes and are therefore also used in men's fragrances.

Various other products are obtained from the plant material and the resin:

– Cistus oil distilled from the leaves and plants;
– Labdanum absolute through solvent-extraction of the same; and
– Labdanum resin absolute through alcohol extraction of the crude resin.

Their odor characteristics and uses are pretty much the same as l.

Lit.: The H&R Book*
     Gildemeister*
     Arctander*

Lac

→Shellac

Lacquers

→Coatings

Lactates

→Lactic Acid

Lactem

→Lactic Acid Esters of Mono/Diglycerides

Lactic Acid

Syn.: 2-Hydroxypropionic Acid
G.: Milchsäure; F.: acide lactique

$$CH_3-CH-COOH$$
$$|$$
$$OH$$

m.w.: 90.08; m.p.: D(–) and L(+) 53 °C; rac. ≈ 17 °C; b.p.: ≈ 122 °C (1.86–2 kPa/14–15 mm)

There are two enantiomorphic forms, the D(–) and the L(+) isomers (→isomerism), and a racemate (DL). All are occurring naturally in fruits and milk products. The L-form is found in many human and animal tissues, especially in muscles after stress. The salts are called lactates.

Large-scale production is based on fermentation processes (homolactic or heterolactic), starting from →glucose-containing or delivering materials, such as →starch, →molasses, →whey, sulfite liquors (→paper) and →sucrose, with various strains of *Lactobacillus* (*L. leichmannii, L. casei, L. delbrückii, L. brevis, L. bulgaricus*).

L. can also be produced via petrochemical processes (e.g., from hydroxypropionitrile) but biosynthetic pathways are preferred, especially in food applications.

L. is widely used as acidulator and preservative (→food additives) in pickled products, marinated fish, cheese, cider, wine, meat and poultry. Technical applications include →leather auxiliaries, rust remover in metal treatment and →coatings.

Derivatives of importance are Na- and Ca- lactates, which are used in the food industry.

L. is easily transferred into its internal ester, lactoyl lactic acid, which esterified with →stearic acid is widely used, as such or in the form of its Ca- or Na-salt (→lactylic esters of fatty acids), as baking emulsifiers and dough conditioners.

In 1990, 40 000 mt were produced worldwide, 65% of which are manufactured by fermentation, mainly in Europe, and 35% synthetically, mainly in USA and Japan.

Lit.: C.H.Holten, "Lactic Acid" Verlag Chemie, Weinheim (1971)
     Kirk Othmer* (4.) **13**, 1042–1054
     [50-21-5]

Lactic Esters of Mono/Diglycerides

Syn.: LMG; Lactem
G.: Mono/Diglyceridlactylat;
F.: esters lactiques de mono/diglycérides

$$CH_2-O-CO-R^1$$
$$CH-OR^2$$
$$CH_2-O-CO-CH-CH_3$$
$$OH$$

R^1: C_{15-17} alkyl
R^2: H or $-CO-R^1$

s.v.: 230–260; m.p.: 46–48 °C

The formula given above is schematic. In reality, l. is a mixture of numerous compounds because the building blocks are multifunctional and can undergo transesterification during reaction and storage.

Depending on the nature of the fatty acid used (saturation and unsaturation, chainlength) l. are liquid or waxy solids with colors ranging from brownish to white. They are sensitive to temperature and →hydrolysis.

They are produced either by →esterification of glycerol, fatty and lactic acid, or by reaction of mono/diglycerides with →lactic acid.

L. are nonionic emulsifiers, capable of entrapping air in food products and are used in baked goods, shortenings, margarine and salad dressings. US consumption (1981) was ≈ 700 mt, while European consumption in 1993 was 2000 mt.

Lit.: G.Schuster, "Emulgatoren in Lebensmittel", Springer Verlag Berlin/Heidelberg/NY/Tokyo (1985)
N.J.Krog "Food Emulsions" K.Larsson, S.E. Friberg (ed.) Marcel Dekker NY/Basel (1990)

Lactose

Syn.: Milk Sugar
G.: Lactose, Milchzucker; F.: lactose

m.w.: 342.30; m.p.: 202 °C (α-monohydrate), 223 °C (α-anhydrous), 252 °C (β-anhydrous)

L. is obtained commercially from the whey of cow's milk (4.5%). A number of distinct forms are produced by various crystallization and drying processes, which can vary in their contents of crystalline and amorphous l., in the amounts of α- and β-lactose, and in their chemical state (hydrous or anhydrous). It is a white to creamy-white crystalline powder, odorless and sweet-tasting. It is soluble in water, ammonia and acetic acid but only slightly soluble in dilute alcohols. L. is insoluble in chloroform, diethyl ether and absolute alcohol. α-Lactose has an r.s. of 15; β-lactose is sweeter than the α-form; milk at body temperature contains l. as an equilibrium mixture of 2 parts of α-l. and 3 parts of β-l. It is used in pharmaceutical formulations as a diluent, bulking agent, filler and excipient for compressed and molded tablets and capsules; it is an ingredient of infant food, animal feed products and powders presented in sachets. L. can also be used to modify the properties of nondisintegrating polymers in filmcoatings. It is added to freeze-dried solutions to increase plug size and to aid caking, and it is used in combination with sucrose for sugar-coating solutions, as a chromatographic adsorbent in analytical chemistry and for lactic acid fermentation in ensilage and food products. It is worthwhile to mention in this context that certain people cannot tolerate l. in their diet.

World production is 340 000 mt/a, 25% of which is used for pharmaceutical purposes, 30% for baby and instant food, 30% for human food and 15% for animal feed.

Lit.: Ullmann* (5.) A15, 107 + A16, 600
The Merck Index* (11.) 5221
[63-42-3] (anhydrous)
[64044-51-5] (monohydrate)

Lactoyl Lactic Acid

→Lactic Acid

Lactylic Esters of Fatty Acids

Syn.: Stearoyl Lactic Acid; SLA
G.: Stearoylmilchsäure;
F.: stéarate de lactoyl-lactique, acide stéaroyl-lactique

$$R-CO-(O-CH-CO)_n-OX$$
$$CH_3$$

R = C_{15-17} alkyl
X = H (SLA), Na (SSL), Ca (CSL);
n = 1–4 (mainly 1–2)

H: s.v.: 380–400; m.p.: 53–56 °C
Na: s.v.: 230–250; m.p.: 39–43 °C
Ca: s.v.: 220–240; m.p.: 45–48 °C

The esters of →lactic acid and saturated fatty acids with a chainlength of C_{16-18} ("stearoyl") and their sodium or calcium salts (sodium or calcium stearoyl 2-lactylate, also called SSL or CSL) are widely used food emulsifiers (→food additives). Because lactic acid forms internal esters easily, the products are sometimes also called lactoyl-lactates.

L. are white or yellowish waxy materials. The free acid is plastic, while the salts are hard solids. All are rather sensitive to heat and →hydrolysis.

L. are produced by →esterification of lactic acid with the fatty acid in the absence, and the salts in the presence of the respective metal ion.

Because there is a strong interaction with →starch and proteins (→gluten), l. are used mostly as additives in baked goods, especially bread, as dough strengthener, which increases volume, pore size and stability as well as texture. SSL is also used in coffee whiteners, creams and icings.

L. represent almost 10% of all food emulsifiers produced (1981) in USA: $\approx 14\,000$ mt, with almost 90% being consumed in baked goods.

Lit.: G.Schuster, "Emulgatoren in Lebensmittel" Springer Verlag Berlin/Heidelberg/NY/Tokyo (1985)
N.J.Krog "Food Emulsions" K.Larsson, S.E. Friberg (ed.) Marcel Dekker NY/Basel (1990)

Laminaria spp.

→Marine Algae
→Sodium Alginate

Lanolin

Syn.: Wool Wax; Wool Grease; Wool Fat; Adeps Lanae
G.: Lanolin, Wollwachs; F.: lanoline

s.v.: 95–120; i.v.: 13–30

To protect their epidermis and their wool against tough environments, sheep secrete a waxy material called wool wax or l.

After shearing, the wool is washed with a nonionic →surfactant or a →soap solution, resulting in a wax emulsion. By centrifuging or adding sulfuric acid or salt solution, the crude l. is precipitated and further purified by centrifuging, filtering, solvent extraction, bleaching, treatment with →activated carbon, etc.

Various types (degree of purity) of l. are available on the market. Crude l. is a dark brownish material, which melts at 34–38 °C and has an unpleasant odor, while neutral l. is yellow and has a m.p. of 38–42 °C. Adeps lanae is the purest type, slightly yellowish and odorless. Commercial l. may also be a mixture of wool wax (65%), water (20%) and paraffin (15%). L. can emulsify a large volume of water.

It consists of 48% wax esters, 33% esters of stearic acid with long-chain alcohols or sterols and some free acids and free sterols. L. contains a considerable amount of branched acids and alcohols and also hydroxy acids. Nearly all acids are saturated. The sterols are →cholesterol and lanosterol.

All commercial types and the hydrogenated version, as well as the waxy alcohols obtained by →hydrolysis, are used in →cosmetics, leather polishes, temporary →corrosion protection lubricants and →pharmaceutical applications. L. serves also as a source of →steroids.

The amount of crude l. available (based on wool production) worldwide is in the range of 400 000 mt. However, only a part of this volume is used in practice.

Lit.: Kirk Othmer* (3.) **24**, 636
[8006-54-0]
[8020-84-6]

Lard

G.: Schweineschmalz; F.: saindoux

s.v.: 193–202; i.v.: 60–70; m.p.: 36–42 °C

L. is a soft, white fat of faint odor and bland taste. The fatty acid composition is the following:

C14:0	C16:0	C16:1	C18:0	**C18:1**	C18:2
1–2%	20–30%	2–3%	12–18%	**36–52%**	10–12%

It derives from fresh, clean and sound fat tissues of pigs and is gained by dry and wet rendering (→fats and oils). Refining is done by neutralization with lye, followed by bleaching and deodorization. Lard oil can be obtained by cold pressing of lard.

L. is used as edible fat (margarine, shortenings), ointment-base and in cosmetics. Lard oil has many technical applications, such as lubricants, cutting fluids (→metal processing), oiling of wool and in →soap manufacturing. It is also used to make →pommades (extraction of fragrance).

The production in 1984 was 5×10^6 mt.

Lit.: Ullmann* (5.) **A10**, 234
[61789-99-9]

Lard Oil
→Lard

Larixinic Acid
→Maltol

Larix decidua
→Maltol

Lauric Acid

Syn.: n-Dodecanoic Acid; C12:0
G.: Laurinsäure; F.: acide laurique

$CH_3-(CH_2)_{10}-COOH$

m.w.: 200.31; m.p.: 44 °C; b.p.: 225 °C (13.5kPa/
100 mm)

L. is insoluble in water, soluble in alcohol and
ether and is contained in the fruits of laurel (origin
of name). Some species of →cuphea contain high
percentages of l. in their seed oil.

L. derives in large-scale manufacturing from medium-chain trigycerides (→fats and oils), such as
→coconut oil, →palm kernel oil, → babassu oil,
and to a smaller extent →whale oil and →sperma-
ceti, by →hydrolysis and subsequent →distillation.
Up to 99% purity is commercially available by
→fractional distillation.

L. can also be synthesized by → ozonolysis of
→petroselinic acid.

L. is used in →soaps, →metallic soaps, in→alkyd
resins and in synthesis of →lauric esters (→esteri-
fication).

→Hydrogenation yields →lauryl alcohol, an important intermediate to produce →detergent and →cos-
metic raw materials. It is also used to make peroxides.

Lit.: Ullmann* (5.) **A10**, 247
      [143-07-7]

Lauric Oils

→Fats and Oils

Lauryl Alcohol

Syn.: 1-Dodecanol; n-Dodecyl Alcohol
G.: Laurylalkohol; F.: alcool laurique

$CH_3-(CH_2)_{10}-CH_2OH$

m.w.: 186.3; m.p.: 23 °C; b.p.: 260 °C

L. forms colorless crystals of pleasant floral odor.
Pure l. is sometimes used as fragrance raw material. It is only used in special applications as a
starting material for derivatives.

L. is also a synonym for the broader cuts in the
range of $C_{10}-C_{14}$, which are used in large volume
for making →surfactants, especially for body care
applications [→fatty alcohol (ether) sulfates].

Lit.: [112-53-8]

Lauryl Amine

Syn.: Dodecyl Amine
G.: Laurylamin; F.: laurylamine

$CH_3-(CH_2)_{11}-NH_2$

mw.: 185.35; m.p.: 28.2 °C; b.p.: 247−249 °C

L. is a white solid, insoluble in water but soluble in
alcohol.

Used for producing →surfactants (→ethoxylation)

Lit.: [124-22-1]

Lavandin Oil

G.: Lavandinöl; F.: essence de lavandin

Obtained by steam distillation of the flowering
plant of *Lavandula hybrida*, a hybrid developed by
crossing the true →lavender plant (*Lavandula officinalis*; Labiatae) with the spike lavender plant
(*Lavandula latifolia*) to obtain better yields and a
wider possible area of cultivation for the plants.

Important cultivation areas are southern France,
Spain and northern Africa.

Main components include linalol, linalyl acetate
(30−35%), →eucalyptol, camphene, pinene,
traces of →camphor and ethyl-n-amyl-ketone.

It is widely used to add refreshing, natural notes to
fragrance formulations for soap, detergent and
household products.

The oil itself has a fresh herbaceous, slightly
minty, medicinal odor and a yellowish color.

Production yield : 1−2.5% (based on fresh plant
material), depending on origin and distillation
method.

Lit: Arctander*
     The H&R Book*

Lavandula hybrida
Lavandula latifolia

→Lavandin Oil

Lavandula officinalis

→Lavender Oil
→Lavandin Oil

Lavender Oil

G.: Lavendelöl; F.: essence de lavande

The lavender shrub (*Lavandula officinalis*, Chaix/
L. angustifolia Mill., Labiatae) is native to the
Mediterranean countries and grows wild or cultivated at medium altitudes (600−1500 m) of mountain regions. Smaller cultivations exist in Australia,
England, Yugoslavia and Russia, but the bulk of
production takes place in southern France. L. is
produced by steam →distillation of the freshly cut
flowering tops and stalks of the shrub, which are
distinctly recognizable by their striking blue-pur-
plish color. Main components are linalol and its esters, mainly linalyl acetate, the contents of which
increase with the altitude at which the plants grow.

Generally, wild-growing, high-altitude lavender plants produce the finest, most expensive qualities of l., mainly because the flowering tops in these plants cannot be harvested by machines. L. has a typical, fresh, yet sweet-balsamic odor with floral and herbaceous undertones and is widely used as a refreshing, yet delicately rich top note in a multitude of fragrance types, ranging from true lavender fragrances to more complex floral, cologne, fougère and fantasy compositions. Main application areas are perfumes, after-shaves and fragrances for cosmetics and toiletries. For a similar effect in fragrances, where lower cost and a "rougher" note are required (e. g., soaps and detergents), →lavandin oil is more likely to be used.

Production yield: 1.4–1.6% (based on fresh plant material), depending on the production method and origin.

Lit.: The H&R Book*
        Arctander*

Lavender Shrub
→Lavender Oil

Lawsonia inermis
→Dyes, natural

LCA
→Life Cycle Analysis

Lead Soap
→Metallic Soaps

LEAR
→Rapeseed

Leather
G.: Leder; F.: cuir

L. is a product made from animal hides and skins. Many different sources of raw material are known. The largest volume of hides is derived from →cattle, →sheep, →pig and goat. Specialty l. is made from ostrich, snake, crocodile, seal and many other animals.

Animal skin is a three-dimensional network of collagen fibers (→proteins). The raw hide, if not properly treated, is subject to rotting, decomposition or brittling and hardening.

To make a stable, dry and useful material, it is necessary to run through a number of processing steps: soaking, liming, deliming, bating, tanning,

splitting, shaving, retanning, greasing, wet and dry finishing, dying and coating.

For details: →leather auxiliaries.

L. is a material that has been used by human beings for thousands of years. Today, main uses are for shoes, upholstery, clothing and handbags.

World production[1] is $\approx 15 \times 10^9$ square feet $(4.6 \times 10^9 \text{ m}^2)$ of finished l. Out of this, 66% is bovine leather, derived from 1.43×10^9 head of cattle to yield 5.2×10^6 mt of raw hides (wet, salted). The remainder (33%) is mainly sheep and goat leather derived from 1.74×10^9 head to generate 570 000 mt of raw, dry hides.

The future of l. is strongly linked to the number of hides available as a by-product of an increasing meat production.

Efforts to substitute l. by synthetic materials on a large scale have not been successful up to now.

Lit.: E.Heidemann "Fundamentals of Leather Manufacturing" E. Roether KG Darmstadt (1993)
        Ullmann* (5.) **A15,** 259

[1] Leather Dec.1993, 22 and Nov. 1994, 11

Leather Auxiliaries
G.: Lederhilfsmittel;
F.: produits auxiliaires pour cuir

Leather production is a finishing process in several steps by certain techniques and needs many different chemicals that are called leather auxiliaries. Some l. are based on RR; the following procedures of leather manufacturing focus on chemials based on these:

Soaking:
Starting materials are either fresh hides or hides preserved by salt and bactericides. However, thin skins are only dried. All need soaking in water to remove dirt and salt and to get the necessary state of hydration and swelling. This leads also to some degreasing. Soaking is done in drums or mixers with wetting agents, such as →fatty alcohol ethoxylates and →alkylglucosides. The latter are, however, too expensive for routine applications.

Anionic →surfactants, such as →fatty alcohol sulfates, →fatty alcohol ether sulfates and →sulfosuccinates, are also applied. Proteolytic →enzymes and sometimes lipases support the process.

Dehairing and Liming:
Dehairing is accomplished with sulfides and lime. Some RR-based nonionic or anionic →surfactants (e. g., →fatty alcohol ethoxylates or →fatty alcohol sulfates) are used to support the penetration

and dispersion of chemicals, e. g., $Ca(OH)_2$. After liming, defleshing and splitting, deliming is accomplished by ammonium sulfate or chloride and, in modern processes, by gaseous CO_2. Weak acids, e. g., →lactic acid or acetic acid, have a buffering effect. Pankreas enzymes are used for bating at a pH of 8–8.2.

Pickling:

P. is an acidizing step prior to tanning. Inorganic acids and salts are used normally, and formic acid, →lactic acid and acetic acid in special applications. For soft leathers (upholstery, clothing), fat liquors, based on fish or vegetable oils are added already at this stage.

Tanning:

The fundamental process of tanning is accomplished by chromium (III) salts ($\approx 80\%$ of all leather produced). Short-chain aliphatic acids →lactic acid among others are sometimes used for masking Cr-tanning agents, but this dominating tanning procedure needs not much RR-based products.

Vegetable tans, however, are based on wood, bark, leaves and roots of mimosa, quebracho, chestnut and oak. →Tannin is the active ingredient, a polyphenol (sugar derivative of gallic acid), which binds to the collagen (→proteins) by hydrogen bonds. Due to a m.w. of 500–5000, the penetration into the fibrils is slower than with chromium tans and results in heavier and fuller leathers.

An old method, still in use on a small scale, is oil tanning. Unsaturated →fats and oils (e. g., →fish oil or unsaturated oils with an i.v. of 120–160) are bound in the fiber network by oxidation and polymerization. Nonbound fats are pressed out or washed out with soda ash solution. This material is called degras or sod oil and is used as emulsifier in fat treatment of leather (stuffing). The soft leather is used for window cleaning (chamois).

→Ligninsulfonate-based products are sometimes used as auxiliary tanning agents, together with vegetable or chromium tanning agents.

Wet finishing:

The first step is a washing procedure with →fatty alcohol ethoxylates or, especially for "wet blue" (semitanned chromium leather), with →fatty alcohol ether phosphates, which are able to dissolve chromium soaps. The second step is neutralization with sodium acetate or →citric acid salts that contribute complexing effects. The third step is retanning, mostly with vegetable →tannins.

The subsequent step is dyeing. RR-based products are only used as wetting agents, dispersers, level-

ing agents and after-treatment. Anionic, cationic and nonionic →surfactants are usually used.

Natural dyes (→dyes, natural) such as yellowwood or longwood extracts, are not used any more but may be revitalized if more "natural" leather would come in fashion.

The application of fat is the next important step, predominantly responsible for the typical "leather feel". Fat makes the leather soft and hydrophobic, and it increases tensile and tear strength. A fat content of 5–15% is usual.

Stuffing is one application method and fat liquoring the other. In stuffing, warm mixtures of fats are applied, such as fish oil, tallow and degras, acting as an emulsifier, which derives as a by-product of oil tanning (see above).

The most widely used method is fat liquoring, whereby fat emulsions in water (1:3) are applied in a warm bath. The leather takes up 5–8% fat when the emulsion breaks in contact with the leather-chromium complex, and the fat is evenly incorporated.

Many formulations of liquors are in use. Most of them are partially sulfonated (SO_2/O_2; sulfuric acid, SO_3, chlorosulfonic acid, sodium bisulfite) oils, such as →cottonseed, →rice bran, →soybean, →fish oil and →lard. The unmodified oil can be emulsified with surfactants, such as →fatty alcohol ethoxylates/propoxylates, →fatty alcohol sulfates, →fatty alcohol ether sulfates/phosphates, sulfosuccinates, →α sulfo fatty acid esters and →mono/diglycerides (→glycerides).

Other RR-based products, normally found in liquor formulations, are →lecithin, wool fat (→lanolin), →fatty acid/protein condensates and →carnauba wax. Fat liquoring is an area where the use of natural products dominates.

Drying and Staking:

The leather is finally dried and streched simultaneously. There are several techniques in use. In one, the leather is glued to a glass surface (pasting) with →starch- and cellulose (e. g., →hydroxypropyl cellulose)-based adhesives and dried in an oven.

Finishing:

The final performance of leather is obtained by finishing. A base coat (polyacrylate), followed often by plating and ironing, is the first step of finishing. The second coat is again a layer of polyacrylates; however, this one is pigmented and forms a harder film. The third coat is a solvent or emulsion system, based on nitrocellulose (→cellulose nitrate) or polyurethane. →Casein-based finishes are used in aniline leather. Modified casein and blood and egg albumins are also found in finishes.

RR-based → surfactants of various types are used to make emulsion-based, ecologically acceptable finishes, that are almost free of solvents.

Comprehensive market figures about leather auxiliaries are not available due to the plurality and variety of products.

Lit.: E.Heidemann "Fundamentals of Leather Manufacture" E. Roether KG Darmstadt (1993)
Ullmann* (5.) **A15**, 259

Lecithin

Syn.: Phosphatidyl Choline
G.: Lecithine; F.: lécithines

Under the term lecithin (originates from greek: lekithos = egg yolk) two substances are understood: the crude soybean l., which is a mixture of phosphatides, soybean oil and sterols, and phosphatidyl choline, a betain, which is one of the main components of the a.m. mixture. The following phosphatides are contained in soybean l.:

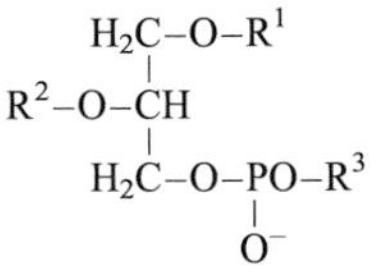

$$H_2C-O-R^1$$
$$R^2-O-CH$$
$$H_2C-O-PO-R^3$$
$$O^-$$

where R^1 and R^2 are fatty acid residues of →soybean oil composition and R^3 may be one of the following structures:

$-O-CH_2-CH_2-N^+(CH_3)_3$ = phosphatidyl choline (PC)
$-O-CH_2-CH_2-N^+H_3$ = phosphatidyl ethanolamine (PE)
$-O-CH_2-CH-N^+H_3$ = phosphatidyl serine (PS)
$\quad\quad\quad COO^-$

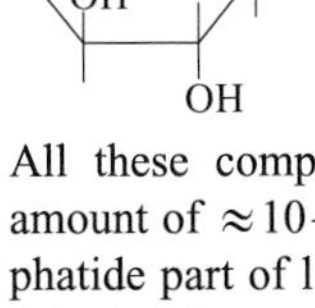

= phosphatidyl inositol (PI)

All these compounds are contained in l. in an amount of ≈ 10–20%, forming together the phosphatide part of l. (≈66%). The rest is soybean oil (30–35%), sterols and a host of minor chemicals, the most important of which is →tocopherol.

L. is produced from fresh, mature soybeans, which are compressed to yield the oil. This contains 2–3% phosphatides. Water (2%) is added, and the mixture is heated to 70 °C for 30–60 minutes under thorough mixing. The hydrate of l. is insoluble in the oil and is separated by centrifuging. The gum can be fluidized or bleached if desired. Thin-film vacuum drying (80–105 °C and 3.3–4 kPa) follows because wet l. is sensitive to microorganisms. For stability reasons the dried product is cooled immediately. A modern process, based on membrane technology, is also used to separate l. from the oil.

There are many commercial grades and rules of classification for l. based on color (unbleached, bleached and double bleached). Fluid and plastic l. is classified by viscosity; they are made by adding diluents, salts and fatty acids. There are also several possibilities to make special l. by solvent treatment: acetone fractionation gives oil-free l. powder, separation of the hexane-insoluble by filtration yields high-grade food l., and the different solubilities of the a.m. phosphatides in ethanol can be used to separate PI from the others. Only unbleached l. types are approved for food applications.

There are several processes to modify l. chemically, e. g., →hydrolysis, →hydrogenation, hydroxylation, →alkoxylation.

Hydroxylation (treatment of l. in lactic or acetic acid with H_2O_2) introduces OH-groups in the fatty acid part and gives l. that disperses faster in cold water. It is the only food-approved modified l.

L. is a W/O emulsifier, and hydroxylated l. is a O/W emulsifier.

Uses of l. are numerous. The most important outlet is →food additives. Due to its interaction with proteins, it is used in baked goods to improve volume, disperse fat easier and to act as an antioxidant. It reduces viscosity in chocolate and prevents crystallization. In instant food, it acts as a wetting, dispersing, emulsifying and stabilizing agent. In margarine, it is used as antispattering and dispersing agent. In animal feed, l. is added as a milk replacer for calves to promote the metabolism of fats and oils in the liver.

Nonfood uses are in →lubricants, fuel oils, leather liquoring (→leather auxiliaries), textile dying, rubber and cosmetics. It functions mainly as disperser and emulsifier in these applications. In pharmaceuticals, it is used as emulsifier and as a source of phosphatidylcholine.

The total commercial l. production potential of USA is estimated to be 140 000 mt/a (calculated from soybean oil production). However, in 1978, the actual use was only ≈33 000 mt there. In food applications, ≈20 000 mt were consumed in USA (1981). European consumptiom in 1993 was ≈75 000 mt.

Lit.: B.F. Szuhaj, G.R. List (ed.) "Lecithins", AOCS (1985)
Ullmann* (5.) **A15**, 293
[8002-43-5]

Legume Starches
→Broad Bean Starch
→Pea Starches

Lemongrass Oil
G.: Lemongrasöl; F.: essence de lemon-grass

L. is used as a raw material in →fragrances and as a starting point for the isolation of →citral, used both in flavors and fragrances. It is steam-distilled from two species of grass, *Cymbopogon flexuosus* W. Wats and *C. citratus* Stapf (Gramineae/Poaceae), yielding l. with slightly different odor characteristics but a generally similar note: lemon-like, slightly grassy and fresh with herbaceous, bitter undertones. L. is a yellow or amber-colored, slightly viscous liquid, which may be slightly turbid due to the presence of water. Due to its strength and low-cost, it is used in lemon, citrus or pine-type fragrances for detergents and household cleaners.

Lit.: Arctander*
      The H&R Book*

Lemon Oil
G.: Zitronenöl; F.: essence de citron

L. is expressed from the ripe fruit peels of the →lemon tree and is, next to →orange oil, the most important citrus oil used in flavors and fragrances. It is a yellow to greenish-yellow liquid with a fresh, yet sweet odor, which is truly reminiscent of the ripe lemon peel. It has a moderate tenacity. The oil is either directly collected from freshly scraped fruit peels or taken from the surface of the pulp/juice mix after expression of the whole fruit.
Main constituents are d-limonene and →citral.
It finds use in all types of fragrances as a juicy-fresh topnote and is essential in the composition of Eau de Cologne and citrus-type perfumes.
Flavor uses are also numerous, with candies and beverages taking the foremost place.

Lit: Arctander*
     Gildemeister*

Lemon Tree
Citrus medica L. spp. *limonum*
Rutaceae
G.: Zitronenbaum; F.: Citronnier

Originating from East India and Burma, this tree was brought to Europe with the returning crusaders in the 12th century. Seeds were brought along with Columbus' ships and were the foundation of California's lemon and orange industry.

Today's main cultivation areas are: California, Florida, Argentina, Brazil, Italy, Cyprus, West Africa.
→Lemon oil is expressed from the l. fruits, and a →petitgrain oil (petitgrain citronnier) is distilled from its twigs.

Lit.: Arcander*

Lesquerella
Syn.: Bladderpod
Lesquerella spp.; *L. fendleri*
Cruciferae
G.: Spaltblume, Blasenschötchen; F.: vésicaire

L. is native mainly to the arid parts of western North America. The genus consists of approximately 70 species of which about 30% are annuals. Species of interest for agronomic evaluation are *L. fendleri* (Gray) Wats., *L. gracilis* (Hook) Wats., and *L. gordoni* (Gray) Wats., of which *L. fendleri* seems to be a prime candidate for domestication. L. was grown worldwide, mostly in experimental plots, at a total acreage of ≈20 ha in 1993.
Limiting for commercial production are: low oil content, small size of the seeds, indeterminate growth habit, and male sterile plants at about 8%.
In Arizona, several planting experiments with *L. fendleri* have been performed with combine-harvested (→harvesting) seed yields of ≈14 dt/ha. (oil content is ≈25%). The dominating fatty acid is →lesquerolic acid (49–65% of the total fatty acids).

Lit.: D.A.Dierig et al.: "*Lesquerella fendleri* and *Vernonia* galamensis as Alternative Sources of Industrial Oils", 2nd Europ. Symp. on Industr. Crops and Products, Pisa, (1993)

Lesquerolic Acid
Syn.: 14-Hydroxy-,*cis*-11-Eicosenoic Acid
G.: Lesquerolsäure; F.: acid lesquerolique

$$CH_3-(CH_2)_5-\underset{\underset{OH}{|}}{CH}-CH_2-CH=CH-(CH_2)_9-COOH$$

m.w.: 326.5

The acid derives from the oil of the →lesquerella plant by →hydrolysis.
The C_{20} acid is similar to ricinoleic acid, which is a C_{18}-hydroxy acid and may be used in similar applications once lesquerella oil becomes available in larger quantities.

Lit.: [4103-20-2]

Leucrose

G.: Leucrose; F.: leucrose

L. is a reducing keto-disaccharide, prepared by action of special bacterial enzymes on →sucrose.

Leucrose
(α (1 → 5) linkage)

(α-D-glucopyranosido)-1→5-β-D-fructofuranose,

m.w.: 342.2 ($C_{12}H_{22}O_{11}$); m.p.: 156–163 °C; DE: 48%; solubility: (30 °C) 64%, (80 °C) 88%; $[\alpha]_D^{20}$: –6.8°; r.s.: 40–50, noncarciogenic, physiological energy content: 17.2 kJ/g.

L. is, together with →dextran, →isomaltose and →isomaltulose, a natural by-product of the action of dextran producing bacteria.

Pilot-plant production (10 mt/a) is performed by action of isolated dextransucrase from cultures of *Leuconostoc mesenteroides* on 65% solutions of 1/3 sucrose and 2/3 fructose at 25 °C. L. is isolated from the reaction mixture by chromatography. The effluent is concentrated and l. is crystallized with 99% purity as anhydrous crystals.

There are attempts to use the noncariogenity and the aroma-enhancing properties in fruit preparations.

Nonfood application relates to the need for hydrophilic building blocks for →surfactants and takes advantage of the high stability of the α-(1→5) linkage and two primary hydroxyl groups. It can form →alkyl glucosides with adjustable hydrophobic/hydrophilic properties, either by the chainlength of the alkyl groups or by defined oxydative degradation of l.

Lit.: Lichtenthaler* 185–195 (1991)

Levans

G.: Laevan; F.: lévane

L. are polymeric components of the extracellular excretions of microorganisms that grow on →sucrose-containing media and naturally occur in the sucrose industry and in dental plaque.

m.w.: 2×10^5–2×10^6; $[\alpha]_D^{24}$: –42.3°/–62.2°; m.p.: 92 °C/>200 °C

The m.w., which directs the solubility, may be adjusted to the application.

They are →fructans, the fructose moieties being bound by β-(2,6) -linkages. Branches are of the

β-(1,2) "inulin"-type. The terminal groups are glucose, (for the formula, see →fructans).

Natural l. have not found any application up to now. Future attention will be directed toward the biotechnological production of fructans of the levan type as easily available sources of →fructose as well as industrial binders and encapsulation agents.

Biosynthesis has been conducted on sucrose substrate with different strains of *Bacillus subtilis, B. polymyxa, Aerobacter levanicum, Streptococcus salivarius* and *Zymomonas mobilis*.

No hydrolysis occurs with →amylases, dextranases and other glucanases. Levanase causes partial hydrolysis of the β-(2,6)-backbone in high-polymer l., but it leaves the branching points unchanged. Total hydrolysis by acids of pH 3.5 leads to almost pure fructose solutions.

Lit.: Lichtenthaler* 169–181 (1991)

Levoglucosan

Syn.: 1,6-anhydro-β-D-glucopyranose
G.: Laevoglucosan; F.: lévoglucosane

L. is a glucose anhydride, formed by pyrolysis of →starch or →cellulose. The formation can be considered as the loss of water from glucose with simultaneous intramolecular 1,6-linking:

Starch

Levoglucosan

m.w.: 162, ($C_6H_{10}O_5$); m.p.: 155–165 °C (97% purity): m.p.: 180–181 °C (purissimum); $[\alpha]_D^{25}$: –67,4° (–66,2°).

Colorless crystals, which are easily soluble in water, slightly soluble in methanol, ethanol and acetone but insoluble in ether.

Production of l. is of synthetic interest because of its three reactive hydroxyl groups, which makes it an important intermediate for regio-selective reactions. The most convenient method of preparation is pyrolysis of predried starch (5% r.h.) at 355–390 °C, distillation and neutralization of the crude distillate. L. is isolated from the aqueous solution (40% d.s.) by chromatographic separation with demineralized water as eluent. The purity of the eluate is sufficiently high for crystallization by evaporation or cooling (yield 35–50%).

Several reactive pathways are starting from l. for utilization as RR, leading to plastics, surfactants, explosives and propellants, resins and highly linear →dextrans. Processing is studied at present on a pilot-plant scale.

Lit.: I.A. Wolff,: Stärke, **20**, 150–158, (1968)
     Lichtenthaler* 82–86 (1991)

Levulinic Acid

Syn.: 4-Keto-valerianic Acid
G.: Laevulinsäure; F.: acide lévulinique

L. is the stable end-product of the acidic and thermal decomposition of hexoses and hexuloses via →5-hydroxymethyl furfural.

$CH_3-CO-CH_2-CH_2-COOH$

m.w.: 116.11; m.p.: 33.5 °C; b.p.: 245 °C

L. is easily soluble in water, ethanol and ether.

L. is formed by hydrolysis of →hydroxymethyl furfural in an equimolecular mixture with formic acid. Starting materials are →fructose, →high-fructose syrups as well as natural or bacterial →fructans. Yields of 70% can be obtained.

Other sources from RR are →glucose or glucose-containing wastes and →furfuryl alcohol.

L. is easily transferred into chemical derivatives, which, due to its high stability, may be utilized as liquid fuel extenders.

Ref.: B.F.M. Kuster: starch/stärke 42, 314–321 (1990)

Levulose

→Fructose

Licania rigida

→Oiticica

Licanic Acid

Syn.: 4-Oxo- 9,11,13- Octadecatrienoic Acid
G.: Licansäure; F.: acide licanique

$CH_3-(CH_2)_3-(CH=CH)_3-(CH_2)_4-\overset{\underset{\|}{O}}{C}-(CH_2)_2-COOH$

m.w.: 292.4; m.p.: 75 °C

L. is a white mass derived from →oiticica oil, which contains 75% of this acid and is one of the few →ketoacids of natural origin.

Lit.: Kirk Othmer* (4.) **5**, 154
     [17699-20-6]

Life Cycle Analysis

Syn.: LCA; Cradle-to-Grave Analysis
G: Ökobilanz, Produktlinienanalyse;
F: bilan écologique

LCA or LC Assessment is a technique with which the stages in the life cycle of an activity, product or process are identified, analyzed and assessed to obtain a LC profile of the subject under study or to compare different products or processes. It is defined as a process to evaluate the environmental burdens associated with a project, process, service or activity by identifying and quantifying energy and material usage and environmental releases to asses their impact on the environment, and to evaluate and implement opportunities for environmental improvements.

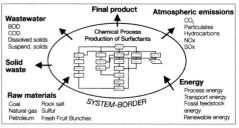

Schematic diagram of the most important input and output factors of the LCA of a chemical process: raw materials, energy raw materials, products, emissions to air and water, and solid waste.

For example, for a chemical product, the stages of its life cycle that need to be considered include:
– acquisition of raw materials;
– processing and production;
– distribution and retail;
– use;
– disposal and final fate in the environment.

The following parameters will have to be inventoried during the different stages of the life cycle of a product:

Raw Materials: agricultural raw materials (such as vegetable oil, straw, cellulose, starch) wood, gypsum, chalk, clay, limestone, metals (such as iron ore, cobalt, manganese, magnesium, tin), sand, water.

Fuels: coal, oil, natural gas, hydropower, nuclear power, bioenergy.

Solid Waste (generated during the different stages of the product life cycle): industrial waste, such as mineral waste, slags and ashes.

Waterborne Emissions: chemical oxygen demand, biological oxygen demand, solids, suspended solids, oil, different metal ions.

Airborne Emissions: particulates, CO, CO_2, NO_x, SO_x, hydrocarbons, methane.

Such inventories are also named LCI because they often deal only with the first steps of a LCA (raw materials acquisition and processing and production of the final product). They are also considered as "cradle-to-factory gate analysis".

After the inventory, it is necessary to do an impact analysis, which measures the scarceness of raw materials (renewable vs. nonrenewable), air and water resources, as well as the effects of the outputs on global warming, the effect of waterborne emissions, the formation of photo-oxidants, environmental toxicity, human toxicity, the quality of waste (hazardous vs. nonhazardous), nuisance and occupational safety.

Presently, such impact assessments have not been performed completely for LCA because the measurement of the different categories of the impact analysis have not been defined, which makes comparisons difficult.

Detailed information about the methodology of LCA is presented in publications of the SETAC (Society of Environmental Toxicology and Chemistry) and ISO.

Lit.: SETAC, Guidelines for Life-Cycle Assessment: A Code of Practice, Workshop Report (1993)

Lignin

G.: Lignin; F.: lignine

m.w.: 5000–10000

The name is derived from lignum, the latin word of wood. L. is contained in →wood (20–40%). It is one of the most abundant polymers in the plant world (estimated volume 3×10^{11} mt) and imparts rigidity and stability to the cell walls of most plants.

L. is the most complex natural polymer. It consists of three hydroxyphenyl propenol building blocks (see formula).

These three monomers are linked together by enzymatic dehydrogenation via phenoxy radicals into a three-dimensional irregular polymer network with C–C and C–O–C linkages. One third of the phe-

p-coumaryl alcohol

coniferyl alcohol

sinapryl alcohol

nylpropane units are linked by carbon-carbon bonds, two thirds by ether bonds. Lignins of different origin differ in their structure.

L. is a by-product of wood pulping (→paper).

Lignin in wood is practically insoluble. Only isolated l. (Kraft or alkali l.) shows solubility in solvents that form strong hydrogen bonds, such as methyl cellosolve and pyridine. It has a T_g of 160 °C.

L. is degraded by sodium hydroxide. Modern pulping processes focus on an accelerated depolymerization of l. while keeping hydrolysis of the polysaccharides at a minimum.

There are two ways to isolate lignin or its sulfonate:

A. Kraft l. (also called sulfate l. or alkali l.) is obtained from wood pulping (→paper) and is isolated from the black liquor (containing ≈40% l.) by precipitation with acids. Only a small part (27000 mt in 1977) of Kraft l. is sold as such. The major volume is sulfonated (→lignosulfonate).

B. In sulfite pulping (→paper), l. is directly obtained as sulfonate (→lignosulfonate). Milled wood l. (MWL) is obtained by extraction of the milled wood with dioxane/water and is used only for scientific investigations.

There have been numerous efforts to make use of the huge volume of l. that result from paper pulping. Only 2% of the l. is separated. Most of the crude material is burned in the power stations of the paper mills. →Lignosulfonate is the only large-volume derivative of l. All other efforts to modify Kraft l. (ozonization, carboxylation, condensation with formaldeyde) and attempts to make cationic l. as emulsifiers for bitumen have led to only small applications. L. has also been used for commercial production of →vanillin.

The production capacity in USA/Canada and Western Europe is estimated at 5×10^6 mt.

It can be expected that modern cellulose pulping processes (e.g., organosolv) will produce new l. types with a different spectrum of applications.

Lit: Ullmann* (5.) **A15**, 305
   Encycl.Polym.Sci.Engng.* (2.) **8**, 795
   K.V.Sarkanen, C.H. Ludwig: "Lignin: Occurrence, Formation, Structure". Wiley-Interscience, NY (1971)
   [8068-00-6]

Ligninsulfonate
→Lignosulfonate

Lignoceric Acid
Syn.: n-Tetracosanoic Acid; C24:0
G.: Lignocerinsäure; F.: acide lignocérique

$CH_3-(CH_2)_{22}-COOH$

m.w.: 368.6; m.p.: 84.2 °C; b.p.: 257 °C (0.532 kPa/4 mm)

L. derives from →peanut oil, lignite, some waxes and beechwood tars.

Lit.: Ullmann* (5.) **A10**, 245
   [557-59-5]

Lignosulfonate
Syn.: Ligninsulfonate
G.: Ligninsulfonat; F.: sulfonate de lignine

L. is the general term for a variaty of structures, purities and concentrations of the sulfonic acid and its salts gained from lignin. The crude product contains mainly Ca- and Mg-salt, which are transferred to the Na- and NH_4- salts by treatment with sulfuric acid, filtration of the precipitated Ca/Mg sulfate and subsequent neutralization with caustic or ammonia.

The $-SO_3H$ (or Na) may be attached to the ring or the side chain of the phenylpropane structure of →lignin but is still part of a polymeric network.

Commercial l., gained from sulfite pulping (→paper), may be either spent liquor or purified l. or their derivatives. They are soluble in water over the total pH range but not in organic solvents.

L. reduces the interfacial tension between liquids only slightly and does not reduce the surface tension of water. It is, however, a strong dispersant by adsorption-desorption and charge formation on the surface of particles. To improve surface activity of

l., long-chain alkylamines are introduced into the lignin structure.

A special reaction of l. is the oxidation by air in alkaline media to vanillin. This process has been abandoned. Reaction of l. with sulfide or sulfur yields dimethyl sulfide and, after further oxidation, dimethyl sulfoxide, which is a useful industrial solvent.

L., gained from the Kraft pulping process (→paper), is sulfonated after isolation from the black liquor with sodium sulfite at 150–200 °C or sulfomethylated in the presence of formaldehyde below 100 °C.

Structure and properties differ somewhat from l. gained from sulfite pulping, but the application areas are mostly the same.

Uses of l. are extremely broad and include the following:
- animal feed binding
- asphalt emulsifier (big outlet)
- binder in →adhesives for board, fibers, wood
- soil conditioning (→agriculture)
- →leather tanning
- oil well drilling (→oil field chemicals)
- →water treatment
- protein precipitation
- dispersants for dyes/pigments in coatings (→coating additives)
- water reducer of cement and concrete (→construction)
- dispersant and foamer in gypsum board
- formulation of pesticides (→insecticides)
- sand binder in →foundry industry
- binder for ore pellets (→mining chemicals)
- collector in fluorspar flotation (→mining chemicals)
- stabilizer for wax and asphalt emulsions
- corrosion inhibitor
- rubber reinforcement
- stabilization of →firefighting foams
- dimethyl sulfoxide manufacturing.

The demand for l. increased from 250000 mt in 1970 to 500000 mt in 1980. The low price of l. ($\approx 10\%$ of other surfactants) frequently compensates for the deficiencies in performance. Availability of this RR-based nontoxic and versatile surfactant is a strong promotor for new applications.

Lit.: Ullmann* (5.) **A15**, 311
   Encycl.Polym.Sci.Engng.* (2.) **8**, 795
   K.V.Sarkanen, C.H.Ludwig: "Lignin: Occurrence, Formation, Structure". Wiley-Interscience, NY (1971)

Lime Oil

G.: Limettöl; F.: essence de limette

L. is commercially available in two distinctly different qualities: expressed l. and distilled l.

The first is expressed from the fruit peels of the green fruit of the →lime tree or collected from the pulp/juice/peel mix of the juice-making process. The second is steam-distilled from the crushed peels or distilled from the juice expressed from the whole fruit.

Expressed l. is a yellowish to dark-green mobile liquid with a fresh, juicy lemon-like and green →topnote and a tenacious sweet-fruity →bodynote. In →fragrances, expressed l. is used as a modifier for citrus topnotes, where its richness and perfume-like bodynote in combination with the natural juicy character produce interesting effects. In flavors, expressed l. is mainly used alongside →lemon oil in soft drinks, ice cream, sherbets and hard candy.

Distilled l. is produced in far bigger quantities and is different in odor and appearance: a pale yellow or almost water-white mobile liquid with a sharpfresh terpene-like and citrusy odor, reminiscent of →pine oil and →orange oil as well as →lemon oil. The oil is volatile and mainly used in flavor work for carbonated beverages, where it is a main constituent of flavors of the Cola type.

In fragrances, traces are used to give lift and pungency to pine, herbal and citrus fragrances.

Lit.: Gildemeister*
      Arctander*

Lime Tree

Citrus aurantifolia Swingle
Rutaceae
G.: Limettebaum; F.: limettier

There are several varieties of l., but only the above-mentioned species, the so-called sour or West Indian l., is used for production of →essential oils. Sweet lime is rarely used and neither is the fruit of *Citrus limetta*, the Italian variety of lime. The l. originated in the Far East and Pacific Islands and is believed to have been brought to the West Indies after the discovery of America. Today, the l. grows wild, semi-wild and cultivated in many tropical and semi-tropical areas.

Expressed →lime oil is produced from the green (unripe) lemon-like fruits, whereas distilled →lime oil is obtained from variable ratios of semi-ripe or ripe (yellow) fruits. Main producing countries today are West Indies, Mexico, Brazil and Italy.

Lit.: Arctander*

Limnanthes alba

→Meadowfoam

Linoleic Acid

Syn.: Linolic Acid; *cis,cis*-9,12-Octadecadienoic Acid;
C18:2 (Δ 9,12)
G.: Linolsäure ; F.: acide linoléique

$CH_3-(CH_2)_4-(CH=CH-CH_2)_2-(CH_2)_6-COOH$

m.w.: 280.44; m.p.: −5.8 °C; b.p.: 228 °C (1.86kPa/14 mm)

L. is a light colored liquid, soluble in most organic solvents but not in water. It is contained in many fats and oils, particularly in drying oils, such as →linseed oil, → sunflower, →soybean and →cottonseed oil, from which it can be obtained by →hydrolysis. Commercial products contain up to 67 % l. The rest consists of saturated and other unsaturated acids, mainly oleic acid.

L. is an essential ingredient of →drying oils and is used in →coatings. It can be hydrogenated partially or totally (→hydrogenation). L. is important for human diet (essential fatty acids, →linolenic acid).[1]

Lit.: Ullmann* (5.) **A10**, 245
      [60-33-3]

[1] Adv.Lipid Res. 18, 203 (1981)

α-Linolenic Acid

Syn.: *cis,cis,cis*-9,12,15-Octadecatrienoic Acid,
C18:3 (Δ 9,12,15)
G.: Linolensäure; F.: acide linolénique

$CH_3-CH_2-(CH=CH-CH_2)_3-(CH_2)_6-COOH$

m.w.: 278.2; m.p.: −11.2 °C;
b.p.: 198 °C (0.532 kPa/4 mm)

L. is a colorless liquid, soluble in organic solvents but insoluble in water.

It derives from →linseed oil, which contains 50–60 % of this acid.

→Hydrogenation yields partially and completely saturated acids.

L. is the main ingredient of →drying oils and is used in →coatings and →inks.

Trigycerides that contain l., →linoleic acid and →arachidonic acid are important for human nutrition (essential fatty acids, →fats and oils) and are sometimes called vitamin F without having really the function of a vitamin.

Lit.: Ullmann* (5.),**A10**, 245
      [463-40-1]

γ-Linolenic Acid

Syn.: *cis,cis,cis*-6,9,12-Octadecatrienoic Acid,
C18:3 (Δ 6,9,12)
G.: γ-Linolensäure; F.: γ-acide linolénique

$CH_3-(CH_2)_4-(CH=CH-CH_2)_3-(CH_2)_3-COOH$

m.w.: 278.2

This isomer of α-l. is found in →borage, →black
currants, →evening primrose, and in oils from different fungal sources. L. is used in the treatment of
atopic eczema. It serves as a substrate for the biosynthesis of →prostagladins, hydroxy fatty acids
and leukotrienes. It plays an important role in the
prevention of heart diseases, e. g., arteriosclerosis.

Lit.: D.J. Jenkins, Med.Sci.Res. **16**, 525 (1988)
    Gunstone* (1994)
    [506-26-3]

Linolenyl Alcohol

G.: Linolenylalkohol; F.: alcool linolénique

$CH_3-CH_2-(CH=CH-CH_2)_3-(CH_2)_6-CH_2OH$

m.w.: 300; i.v.: ≈190

L. is a fatty alcohol derived from α-linolenic acid
and is commercially available by →hydrogenation
of soybean or linseed fatty acid methyl esters with
special catalysts that do not effect the double
bonds. The finished product consists of l. (50%),
oleyl and linoleyl alcohol.
Highly unsaturated alcohols are used for the manufacture of →surfactants ,→emulsifiers, →textile
auxiliaries and reactive plasticizers.

Lit.: Kirk Othmer* (4.) **1**, 885

Linoleum

G.: Linoleum; F.: linoléum

L. is a product that is mainly used as a →flooring
material and is based on →linseed oil (origin of
name from latin "linum oleum"). The oil is blown
with air in the presence of dryers (→drying oils) and
→rosin or →resins (→resins, natural) like copal.
The result is a rubberlike material, which is mixed
with pigments, cork- or wood flour. This material is
coated on →jute and cured at 80 °C for several days.
The surface is coated with wax or a varnish.
The finished floor cover is a good insulator of heat
and noise, is easy to care for and almost entirely
based on RR.
The importance has greatly diminished during the
last decades because PVC-based floor covers can
be produced much cheaper. However, due to environmental considerations, a revival can be observed. Worldwide production of l.:

1965: 60 000 mt
1969: 30 000 mt
1975: 10 000 mt
1995: 40 000 mt

There are four producers worldwide. Developments are concentrated on improved production
techniques.

Lit.: Encycl.Polym. Sci.Engng.* (2.) **7**, 244
    Kirk Othmer* (1.) **8**, 392

Linoleyl Alcohol

G.: Linoleylalkohol; F.: alcool linoléique

$CH_3-(CH_2)_4-(CH=CH-CH_2)_2-(CH_2)_6-CH_2OH$

m.w.: 302; i.v: ≈137

L. is a fatty alcohol derived from linoleic acid. It is
normally contained in a mixture with other unsaturated alcohols (broad cut). Uses are the same as
→linolenyl alcohol.

Lit.: Kirk Othmer* (4.) **1**, 885

Linolic Acid

→Linoleic Acid

Linseed

Syn.: →Flax
Linum usitatissimum L.
Linaceae
G.: Öllein, Faserlein, Flachs;
F.: lin

L. is an ancient annual crop with a long-lasting
cultural importance for fiber and oil production.
Whereas l. describes the application as a vegetable
oil crop, the word →flax is used to describe the application of l. as a fiber crop.

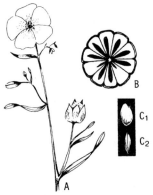

Linseed; A: Inflorescence. B: Cross section of capsule.
C1/C2: Seeds from the top and the side (from Franke*,
1981, with permission).

L. originated from Southwest Asia and North Africa. In ancient Egypt, l. was primarily used as an oil crop; around 2300 B.C. was its application as a fiber crop reported for the first time. In Europe, around 1890, production of the fiber crop started to loose ground against imported →cotton. During the world wars, l. gained substantially in importance. After 1950, its competitiveness with other crops lost again. Today, main producing countries are Canada, USA, Argentina, China, India, Romania, France, Poland, Czechoslowakia, and Belgium. In the last few years, especially in Germany, there has been special interest in growing l. due to its high potential as a RR. Seed yields of l. are between 15–25 dt/ha. Yields may be as high as 40 dt/ha. For flax, 10–15 dt/ha of fiber and 6–12 dt/ha of seed are reported. The extraction of 30–48% of oil results in a valuable meal (quality similar to soybean meal) with about 40% crude protein.

A third application of linseed is the typical "granola" and cereal food business. For this, yellow seed coats, high →seed weights, a typically higher amount of →oleic and →linoleic acid as well as a higher percentage of mucilage in the seed coat is preferred.

Lit.: C.L.Lay, C.D. Dybing, "Linseed", in Röbbelen* 416-430 (1989)

Linseed Oil

Syn.: Flaxseed Oil
G.: Leinöl; F.: huile de lin

s.v.: 188–195; i.v.: 180–185

L. is a yellow to brown oil with a peculiar odor. It thickens and hardens fast upon exposure to air. This is due to the high content of linolenic acid:

C16:0	C18:0	C18:1	C18:2	**C18:3**
5–6%	3–5%	18–26%	14–20%	**51–56%**

The oil derives from →linseed. Various purification and refining techniques are used. Heat-treated l. (260–280 °C) increases viscosity and gives "boiled" l. Blowing air into the oil yields "blown" l. (→drying oils). →Hydrolysis is used to make the fatty acids.

L. is the prototype of a drying oil and is used extensively in →coatings and paints. In →alkyd resins the oil or its fatty acids are important components. Application in food is limited and mainly restricted to dietary uses, mostly of the whole seeds. Another outlet is the manufacture of →linoleum and glazier's →putty. Other uses are in printing

→inks, →caulks and as core-binding resins (→foundry industry).

There are efforts to use the potential of l. for chemical modification such as konjugation (→isomerism), cyclization with maleic anhydride and epoxidation, followed by nucleophilic ring-opening leading to products with functional groups.

World production was in the range of over 1×10^6 mt/a and has declined to 670 000 mt in 1984. Today, production is 700 000mt.

Lit.: Ullmann* (5.) **A10**, 227
[8001-26-1]

Linseed Oil Fatty Acids
→Linseed Oil

Linum usitatissimum
→Flax
→Linseed

Linters
→Cellulose

Lintner Starch
→Thin-boiling Starch

Liquidamber orientalis
Liquidamber styrociflua
→Styrax Resinoid/Oil

Liquid Rosin
→Tall Oil

Lithium Soap
→Metallic Soaps

LMG
→Lactic Esters of Mono/Diglycerides

LM-pectin
→Pectins

Locust Bean Gum

Syn.: Carob Seed Gum
G.: Johannisbrotkernmehl; F.: gomme de caroube

L. is produced by milling the seed of *Ceratonia siliqua* or →carob tree [locust (bean) tree]. L. is contained in the endosperm and milled to a fine powder.

It mainly contains a →polysaccharide of the galactomannan type with a β-(1,4) glycosidyl-mannose backbone and has the following structure:

-man-man-man-man-man-man-man-man-man-
$$\underset{\substack{|\\ \text{gal}}}{} \qquad \underset{\substack{|\\ \text{gal}}}{}$$
man=mannose   gal=galactose

The →mannose units are 1,4-linked while the →galactose is bound 1,6 to the mannose chain. There are regions along the mannose chain that contain no galactose units as shown by the formula.
Commercial l. contains 80–85% galactomannan, 10–13% moisture and small amounts of proteins, lipids, fibers and ash.
The gum has a high water-binding capacity but is not easily soluble in cold water. However, it forms clear solutions when heated to 85 °C, which are highly viscous after cooling. Solutions show pseudoelasticity and are shear-sensitive. They degrade readily with acids or enzymes but are rather stable in alkaline media. The general properties are similar to those of → guar gum, with the main difference that l. is compatible with other plant gums with a synergistic interaction as to viscosity increase.
L. is mainly used in the food industry as an ice cream stabilizer, and in processed cheese, diet food, extruded meat and as a water retarder in doughs. L. is also an emulsifier and stabilizer in dressings (→food additives). Technical applications include textile printing, oilfield chemicals and paper milling but are of minor importance due to the relatively high price of l.
Production in 1990 was ≈20.000mt worldwide.

Lit.: J.K.Seaman "Handbook of Water Soluble Gums and Resins" R.L.Davidson (ed.), McGraw Hill NY (1980)
P.Harris "Food Gels" Elsevier Appl. Sci. Publ. London (1990)
J.E.Fox "Seed Gums" in A.Imeson (ed.) "Thickening and Gelling Agents for Food" 153–170. Blackie Academic and Professional London (1992) [9000-40-2]

Locust Tree
→Carob

Lubricants

G.: Schmierstoffe; F.: lubrifiants

Friction is the force that resists the relative motion of two contacting bodies. L. reduce friction. They are, therefore, a highly important economic factor by saving energy, reducing wear, lowering maintenance costs and extending inspection intervals. The science of friction and lubrication is called tribology. L. are predominantly based on mineral oil. For spe-

cial technical requirements synthetic l. (A) and/or additives (B) are necessary. Some of them are derived from RR. Ecological considerations (C) assist the return of natural resources into modern lubrication techniques.

A. Synthetic lubricants:
By far the most important reason to use l. other than mineral oil is when it -even with the aid of additives- is not able to solve a problem. There are numerous alternatives but the following is dealing only with RR-related products.
A highly sophisticated application is the use of ester oils in jet engine lubricants. Simple ester oils are esters in different combination of the following reaction components:
– branched-chain petrochemical-based alcohols (C_{8-10});
– straight-chain dicarboxylic acids (→sebacic or →azelaic acid);
– monocarboxylic acids, mainly →pelargonic acid;
– neopentylglycol, trimethylolpropane and pentaerythritol;
– several other diols or diacids that are petrochemically based.
Such esters have to show extreme properties. Operating temperatures of 400 °C and more are usual. Compatibility with rubber, plastics and different metals, low viscosity at low temperature, high shear stability, stability to oxidation at high temperature and a low pour point are tough requirements for top applications, such as in supersonic speed engines.
With respect to flash point, pour point and low temperature viscosity, complex esters are superior to the esters described above due to their higher molecular mass. They are polyesters with a construction principle shown in the following example:

M–X– [(O–G–O)–(X–A–X)–(O–G–O)]$_x$ –X–M

M–X = medium chain fatty acid, e. g., pelargonic acid,
O–G–O = polyalkyleneglycol unit,
X–A–X = dicarboxylic acid, e. g., sebacic acid.

The properties of ester oils and complex esters are improved by additives, such as viscosity-index improvers of the methacrylate type. On the other hand, these ester oils are also used as base stock and additive for lithium greases (→lubricating greases). Perfluorinated alcohols (→perfluoro alkyl compounds) are sometimes built in ester oils to increase thermal stability.
Polyalkylene glycols and their ethers and esters are used as special lubricants, as →hydraulic fluids

and brake fluids and in cutting oil (→metal-working fluid). Sometimes, polymers of →tetra-hydrofuran are used instead of polyalkylene glycols because they are lower in toxicity.

B. Additives:

Additives improve the physical or chemical properties of mineral oil-based or synthetic l. in the following areas:

Oxidation inhibitors:

The only RR-based antioxidants are reaction products of sulfur or phosphorus pentasulfide with →terpenes (α-pinene), resin oils and unsaturated esters. Reaction of alcohols (C_8) with phoshorus pentasulfide, followed by a reaction with zinc oxide yields salts of dialkyl dithiophosphoric acid (ZDDP), which are widely used, not only as antioxidants but also as corrosion inhibitors and extreme-pressure additives. Longer-chain alcohols enhance oil solubility.

→Citric acid, →gluconic acid and →lecithin are added as effective chelating agents for metal ions, thus preventing their participation in oxidation reactions.

Viscosity index (VI) improvers:

The decrease of viscosity with increasing temperature, which affects lubrication efficiency, is reduced by these additives. Aside from hydrocarbon polymers, polymethacrylates of fatty alcohols (chainlength C_{12-18}) are used as VI improvers. Copolymers of different alkyl methacrylates or polar comonomers show multipurpose properties (VI improvment, pour point depression and dispersing activities in combination).

Pour point depressants (PPD):

During cooling, precipitation of n-paraffins, which are important for other properties in lubricating oils, may occur. Addition of less than 1 % of a pour point depressant interrupts cristal growth and agglomeration of needle-crystal networks. Again, polymethacrylates of fatty alcohols (chainlength C_{12-22}) are applied.

Detergents and dispersants:

Heavy-duty (HD) additives keep solid combustion and oxydation products in suspension, thus avoiding deposits on metal surfaces, sludge formation and corrosive wear by neutralizing acidic decomposition products. Detergents, some of them RR-based, have sulfonate, hydroxy and/or carboxyl groups and usually contain metal ions or amine functions. More modern HD additives are based on methacrylates of fatty alcohols (C_{12-18}), copolymerized with diethylaminoethyl methacrylate (9 : 1), vinylpyrollidone, N-vinylpyridine and hy-

droxyethyl methacrylate. These ashless dispersants may act also as VI improvers.

Extreme-pressure (EP)additives:

EP additives are added to engine oils, gear oils, →hydraulic fluids and oils used in →metalworking to increase the load-carrying capacity and ability to transmit strong forces. Among other sulfur compounds, sulfurized fatty oils are used.

Chlorinated fatty acids and their derivatives, alkylsuccinic anhydride and zinc dialkyl dithiophosphates (ZDDP) function as EP additives. The latter impart also oxidation inhibiting properties.

Friction modifiers:

→Fatty alcohols, amides and salts of fatty acids (C_{12-18}) and their esters act as friction modifiers, which reduce frictional forces (mild EP agents). Amine salts of →fatty acids or →fatty acid amides have an important function in →metalworking.

Corrosion inhibitors:

Many RR-based products are used for this purpose. Amides of saturated and unsaturated fatty acids with alkanolamine, alkylamine, sarcosine and imidazoline are used in amounts of 0.1 % in lubricants and greases. The salts of →fatty amines (C_{8-18}) with dialkyl phosphates are strong anticorrosion agents. Wool fat (→lanolin) is known as a corrosion inhibitor in combination with oil-soluble sulfonates. →Dicarboxylic acids, unsaturated →fatty acids, hydroxy fatty acids and their esters have also good inhibiting properties. Specialties are →sorbitan monooleate and O-stearoyl alkanolamines.

C. Modern ecological considerations

It is an old technology to mix mineral oils with →fats and oils (compounded oils) or to use them as such for special applications. Today, the tendency to use biodegradable natural oils or simple derivatives as substitutes for mineral oil is increasing, due to environmental considerations. The most frequently cited example of such an application is the use of →rapeseed oil or its methyl ester as a lubricant for saw chains. Other uses are l. for out-board engines, for air compressors in construction sites and for hydraulic fluids as polyol esters. This is an area where utilization of RR may be growing rapidly in volume.

The annual production of l. in USA is $\approx 9-11 \times 10^6$ mt. There are no figures about the share of RR-based products.

Lit.: R.M.Mortier et al (ed.) "Chemistry and Technology of Lubricants" Blackie, Glasgow and London VCH Publishers Inc. NY (1992)
Ullmann* (5.) **A15,** 423

Lubricating Greases

G.: Schmierfette; F.: graisses lubrifiantes

L. are semisolid pastes that are used as lubricants. They consist of:

a.) A base oil, mostly of petroleum origin. Sometimes, ester oils (→lubricants) are used if necessity for high performance or if better ecology tolerates the higher price.

b.) A thickener, which in most cases are →metallic soaps. Lithium amd calcium soaps are by far the most important, while Al, Ba and Na soaps are only used in special applications. Fatty acids employed are →oleic acid, →stearic acid and, especially in lithium l., →hydroxy stearic acid. Sometimes, short-chain acids (acetic) are present in the form of complex soaps with improved performance.

c.) Additives, which are used to improve oxidation stability, rust protection, and extreme pressure performance. In general, the same chemicals are used as in liquid →lubricants. In many l., →glycerol is added to stabilize the soap structure.

The general formulation for l. is: 4–20% soap; 75–95% oil; 0–5% additives. There are many possibilties to adapt the formulation to the desired price/performance relationship.

Aside of lubricating properties, corrosion prevention and stability, an important parameter is the dropping point, at which the l. reaches a certain flow characteristic. The dropping of Ca-soap l. is in the range of 100 °C, while high-performance lithium soap l. reaches 190 °C if →hydroxy stearic acid is the acyl group. Complex soaps have also high dropping points.

Uses of l. are broad, especially if low price and moderate performance are important. The main use is in ball and roller bearings.

The position of RR-based products in this market is rather stable.

Lit.: Kirk Othmer* (3.) **14**, 501
     Ullmann* (5.) **A15**, 489

Lumbang

→Candlenut

Lunaria annua L.

Syn.: Honesty; Moonseed; Silver Dollar
Cruciferae
G.: Silbertaler, Judassilberling, Mondviole;
F.: lunaire

The flat coin-shaped appearance of the silique of this representative of the Cruciferae family is a distinct characteristic of l. Also, the violet color of the flower petals seems to be atypical for a cruciferous crop. This is one of the reasons why l. is widely utilized as an ornamental in the temperate climate. Like many other cruciferous crops, it is native to the Mediterranean countries. Its potential as a new industrial crop lies in the fatty acid pattern of its seeds:

C16:0	C18:0	C18:1	C18:2	C18:3
1–3%	1–2%	16–20%	8–10%	2–4%

C22:1	**C24:1**
38–48%	**22–25%**

The oil content is around 35%. Agronomic pilot experiments have indicated that mechanical harvesting and cleaning of the seeds is a problem. At the moment, commercial production of l. is limited to seed multiplication for ornamentals.

Lit.: L.H.Princen "New Crops Development for Industrial Oils". J.Amer.Oil Chem.Soc. **56**, 845 (1979)

Lupinus spp.

Leguminosae
G.: Lupine; F.: lupin

Among about 300 species, the agronomically most important species in Europe are the white l. (*L. albus* L.), the yellow l. (*L. luteus* L.), the blue l (*L. angustifolius* L.). In South America, it also includes the pearl l. (*L. mutabilis* Sweet) and the garden l. (*L. polyphyllus* Lindl.).

All l. have comparatively large seeds (→seed weight 80–900 g/1000), which may be used as such or after cooking. *L. albus* certainly has become the most prominent, although no l. have really reached the status of major protein and oil crops. They came from North Africa and gained some importance in Germany during the wars, when there was a need for protein and oil crops that would not have to be imported. At that time, also the first varieties of "sweet" l. (low →alkaloid content in the meal) were selected or induced by mutagenesis (→plant breeding). In the following years these developments soon were abandoned when cattle production was started up again and when imported protein sources were again available. Although l. has the potential as a valuable protein and oil crop, its lack of agronomical acceptance and its competitive situation with the →soybean have yet prevented a revitalization. The oil is rich in →oleic acid and →linoleic acid.

L. mutabilis has been evaluated to help fill the protein gap in the nutrition of several South American countries.

Today, most of the ca. 2×10^6 ha of l. grown worldwide is used as fodder crop.

Lit.: Purseglove* (1974)

Lutes

→Putties

M

Mace Oil
→Nutmeg Oil

Macrocystis pyrifera
→Marine Algae
→Sodium Alginate

Madder
→Dyes, natural

Magnesium Soap
→Metallic Soaps

Maize
→Corn

Maize Fibers
→Corn Gluten Feed

Maize Germ Meal
→Corn Germs

Maize Germs
→Corn Germs

Maize Gluten
→Corn Gluten Meal
→Cereal Gluten
→Zein

Maize Gluten Feed
→Corn Gluten Feed

Maize Oil
→Corn Oil

Maize Starch Production
→Corn Starch Production

Malic Acid
Syn.: Hydroxysuccinic Acid
G.: Äpfelsäure; F.: acide malique

$$\underset{\displaystyle \text{HOOC--CH}_2\text{--CH--COOH}}{\overset{\displaystyle \text{OH}}{|}}$$

m.w.: 134.09; m.p.: 100 °C (L-m.); 131 – 132 °C
(DL-m.)

M. is a white crystalline material, which exists in
two stereo isomeric forms (→isomerism): the L-(–)
enantiomer [97-67-6] occurs in many fruits (especially in apples) and is a common metabolite in
plants and animals, and the D-(+) [636-61-3] enantiomer is of scientific interest only. The racemate
(D,L) [617-48-1] is widely used as an organic food
acidulant.
The large-volume production of the racemate starts
from fumaric acid, follows petrochemical pathways
and is therefore not of interest in context with RR.
It can also be produced by fumarase (from *Schizo-
phyllum commune*). The L-(–) m. is produced commercially by biosynthesis, starts also from
→fumaric acid and is connected with immobilized
cells of Brevibacterium sp., *Corynebacterium* sp.,
Microbacterium sp. or *Proteus.*
M. and →fumaric acid are increasingly used as
substitutes for the more expensive →citric acid in
all applications.
Production in USA in 1980 was 7000 mt.
Lit.: Kirk Othmer* (3.) **13,** 103

Maltitol
G.: Maltit; F.: maltitol

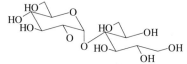

m.w.: 344.37

M. is obtained by treatment of starch (potato or
maize) in succession with α-amylases, β-amylase
and isoamylase, followed by catalytic →hydrogen-
ation. It is also the product of reduction or hydrogenation of →maltose. It is used as a sugar substitute (r.s.: 90) for diabetic patients in food, pharmaceutical formulations and beverages because of the
slow cleavage of m. in the intestine.
Lit.: Fabry in "Developments in Sweetness" Grenby
       (ed.) Elsevier Appl. Sci. London (1987)
       [585-88-6]

Maltodextrins
G.: Maltodextrin; F.: malto-dextrines

M. are concentrated aqueous solutions of nutritive
saccharides, or their dehydrated products, obtained

by controlled →hydrolysis of starch and having a DE-value of <20%. $[C_6H_{10}O_5]_n$. They are mixtures of →malto-oligosaccharides as well as of →starch polysaccharides with reduced m.w. The AGU are linked together by α-1,4- and α-1,6- glucosidic bonds.

Production by hydrolysis is performed primarily by the action of thermally stable α-amylases from bacteria on →corn, →waxy corn, →potato or manioc starches, followed by inactivation of the enzyme, purification, concentration and spray-drying.

Properties of commercial products depend mainly on composition, distribution of m.w., ratio of linear to branched molecules, the starch source and the degree of hydrolysis as characterized by their reduction power (DE values):

Commer-cial type	DE	Glucose	Di-saccharides (all in %)	Tri-	higher	Solubility in cold water
M 440	4–7	0.1	0.9	1.3	97.7	turbid
M 100	9–12	0.7	2.2	3.2	93.9	clear 34–40
M 150	13–17	0.9	3.4	4.5	91.2	clear 60
M 580	16.5–19.5	1.0	6.0	8.0	85.0	clear 70

M. from waxy corn are instantly soluble in boiling water; they form clear, viscous, stable solutions. In m. from amylose-containing starches, turbidity is diminished with increasing DE.

Another type of m., derived from potato starch, DE 5–7%, sets to shiny white, thermally reversible gels at concentrations >12% (w/v).

M. fill the functional gap between the maltooligosaccharides (soluble) and the high-molecular starch polysaccharides (insoluble). For dependence of functionalities on degree of hydrolysis and other molecular properties: →glucose syrups.

Preferred areas of application are:

Food industry:

    stabilizing, water-binding, viscosity-building,
    fat-reducing, bulking, drying adjuvant,
    easy-digestible energy supplier.

Pharmacy:

    baby food formulations, liquid nutrifyers,
    probe nutrition, microencapsulation,
    drug carriers, fillers and binders.

Nonfood:

    sizing for textiles, paper processing,
    core binders, oil-well drilling fluids,
    adhesive for glass-fiber wallpaper.

World production of m. amounts to 200 000 mt/a (USA 88 000 mt).

Lit : Schenk/Hebeda* 233–276 (1992)
    Ruttloff* 597–610 (1994)

Maltol

Syn.: Larixinic Acid

G.: Maltol; F.: maltol

m.w.: 126.11; m.p.: 161–162 °C

Crystals of caramel-like odor; found in the bark of young larch trees (*Larix decidua*), in pine needles (*Abies alba*), in chicory and in roasted malt. Although many synthesic routes are known, m. is still mainly isolated from beechwood tar.

It is used as a component in flavor and taste enhancer in fruit-containing food products and imparts freshly baked odor and flavor to bread and cakes.

Lit.: Ullmann* (5.) **A11,** 205
    The Merck Index* (11.), 5594
    [118-71-8]

Malto-oligosaccharides

G: Maltooligosaccharide;

F: malto-oligosaccharides

Maltosaccharides, containing up to about 10 AGU, $[C_6H_{10}O_5] \approx 10$ are called m. and are components of →hydrolysis products of →starch polysaccharides by means of acids or enzymes (→glucose syrups, →maltose syrups).

In linear m., all AGU are joined together by α-1,4-glucosidic bonds. Branched m. contain additionally single α-1,6-glucosidic bonds, resulting from the branch points of the →amylopectin. Pure grades or products enriched in single oligosaccharides, (maltotriose, maltotetraose, maltopentaose, maltohexaose) have achieved nutritional and dietetical significance as m.-syrups .

Maltotriose-rich syrup is produced from high-maltose syrup by cation-exchange resin chromatography, and →maltose syrup of about 98% purity is obtained as a by-product. The oligosaccharide syrups result from the action of specialized microbial enzymes (exo-maltooligohydrolases) on liquefied starch or debranched starch solutions and various fractionation techniques for purification and enrichment. The pure-grade oligosaccharides are isolated by carbon column chromatographic and gel-chromatographic methods.

Commercially available enriched syrups are those from maltotriose, maltotetraose and maltohexaose. They are used in the production of diagnostic materials for testing →amylase enzyme patterns as well as for their antiputritive activity in the human intestine as components in nutritive and dietetical formulations. Future industrial significance in nonfood applications may result from their moisture-equilibrating and →retrogradation-retarding properties.

Lit: Schenk/Hebeda* 335–366 (1992)
      Ruttloff* 583–585, 600–601 (1994)

Malto-oligosaccharide Syrups

G.: Maltooligosaccharidsirupe;
F.: sirops de malto-oligosaccharides

M. are aqueous solutions of nutritive saccharides obtained by hydrolysis with specialized →enzymes, refining and concentration.

The different types of syrups are enriched with respect to one of the linear oligosaccharides maltotriose, maltotetraose, maltopentaose, maltohexaose $[C_6H_{10}O_5]_n$ (n = 3, 4, 5, 6).

The AGU are joined together by α-1,4-glucosidic linkages. Maltotriose syrup is derived from high →maltose syrup by cation-exchange resin chromatography. A by-product of this separation process is maltose solution of 98% purity. It is also available from hydrolysis of the polysaccharide →pullulan by the debranching enzyme pullulanase.

→Branched oligosaccharide syrups that contain additional α-1,6-glucosidically linked oligosaccharides are available by combined action of β-amylase and transglucosidase.

Most used techniques for linear m. production are based on the action of exo-malto-oligohydrolases:

Substrate	Microorganism	Yield
Starch	*Streptomyces griseus*	55%
Amylopectin		56%
Amylose P17	*Bacillus subtilis*	69%
Amylose P17	*Pseudomonas stutzeri*	75%
Amylose P17	*Bacillus licheniformis*	40%
Soluble starch		
Amylose, Amylopectin		
Amylose P17	*Klebsiella pneumoniae*	40%

When using branched →starch polysaccharides, additional application of the debranching enzymes pullulanase or isoamylase leads to increased yields. Entirely soluble substrates can be reacted economically by continuous flow over immobilized enzyme systems.

From these new enzymes, the exo-maltotrio-hydrolase as well as exo-maltotetrao-hydrolase are commercially available in Japan for production of high maltotriose- and high-maltotetraose-syrups, from which the pure malto-oligosaccharides can be made by chromatographic enrichment and crystallization.

M. show the following functional properties:
entirely digestible;
physiological energy content 17.2 KJ/g;
lower freezing point depression;
less sweetness;
less color formation compared with maltose syrups, dextrose, glucose syrups or sucrose;
prevention of crystallization of sucrose in foods;
uniform moisture level over a wide range of humidity;
prevention or retardation of →retrogradation in starch systems;
favoring glassy states.

These properties are all useful in food industries as well as in pharmaceutical applications.

The maltooligosaccharides in their syrups and solutions exhibit remarkable antimicrobial activity against putrefacient bacteria, such as *Erwinia, Escherichia, Shigella, Enterobacter, Salmonella* and *Bacillus* spp.

The relative antimicrobial effectiveness of the oligosaccharides is in the order:
Maltotriose 105% > Maltotetraose 100% > Maltopentaose 75% > Maltohexaose 40% > maltose 8% (dextrose 0).

Maltotetraose syrups are reported to have a inhibitory effect on putrefactive bacteria, such as *Clostridium perfringens* and *Enterobacter* spp. and to lead to an improvement of human colonic conditions.

Lit.: Schenk/Hebeda* 335–366 (1992)
       Ruttloff* p. 601, 610–612 (1994)

Maltose

G.: Maltose; F.: maltose

m.w.: 342.3 $(C_{12}H_{22}O_{11})$; m.p.: 102–103 °C (hydrate); 160–162 °C (anhydrous); $[\alpha]_D^{20}$: α-m. +168°, β-m. +118°, α-/β-m. +136°

Solubility (20 °C): anhydrous α-m. 175 g and β-m. monohydrate 39 g/100 g water; r.s.: 43; physiological energy content 17.2KJ/g.

Pure m. (1,4-α-D-glucosido-glucose) exists in two anomeric configurations, which form, when dissolved in water, an equilibrium of α-m.: β-m. 42:58.

It crystallizes slowly from water, forming β-m.-

monohydrate ; anhydrous maltose consists of both anomeric forms.

Pure grade commercial products are α-,β-maltose complexes, (solubility 62%, w/w). M. is a reducing sugar (DE 58%). It is entirely digestible by humans and animals and fermentable by yeast and other microorganisms.

Pure-grade m. is crystallized from maltose syrups, produced by enzymatic hydrolysis of liquefied →starches or of soluble →amyloses. M.-forming enzymes are β→amylases from higher plants, such as barley malt, extracts from soybean or sweet potatoes or from microorganisms *Pseudomonas* sp. leading to hydrolyzates of mainly β-m.

The extracts from *Bacillus* sp.*, Streptomyces* sp. *and Clostridium* sp. lead to the α-m. anomer.

Acting together with either debranching →enzymes (pullulanase, isoamylase) in fixed bed or with soluble amylose of low m.w. as substrate it is possible to produce maltose syrups of 70–90%. For crystallization the syrups are enriched in m. content to 95% by various fractionation techniques, such as carbon column or cation-exchange chromatography, solvent precipitation, membrane separation or combinations of the latter with ultra-filtration.

Preferred types of m. in food and nonfood uses are high-m. syrups (→maltose syrups) with 85–90% m. content.

Good color stability, reduced sweetness, low hygroscopicity, improving shelf life, inhibition of microbial growth, prevention of sugar crystallizing and formation of coarse ice crystals are useful properties in breweries, baking and in manufacturing of soft drinks, beverages and confectionary.

Applications in pharmaceutical industries include: intravenously administered nutrients with lower osmolality than dextrose (99% pure m.), carbon source for biotechnological production of special antibiotics (93–95% pure maltose).

Hydrogenation of m. and of high maltose syrups leads to →maltitol and maltitol syrups.

M. and its special anomers may be of future interest as RR for chemical synthesis of biodegradable →surfactants, plasticizers and polymerization compounds .

High purity solid m. is produced only in Japan and the USA. Production (1991/1992) is estimated at 5000 mt.

Lit: Rymon-Lipinski/Schiweck* 147–181, 379–390 (1991)
     Schenk/Hebeda* 335–366 (1992)
     [69-79-4]

Maltose Syrups

G.: Maltosesirupe; F.: sirop de maltose

M. are purified concentrated aqueous solutions of nutritive saccharides obtained by enzymatic →hydrolysis of starch and contain →maltose as the dominating sugar.

M. are produced at varied levels of →maltose with hydrolyzing systems of different maltogenic activity. Typical products contain 45–60% maltose and 0.5–3% →dextrose (high-m., DE: 35–50%) or 70–85% maltose and 1.5–2% dextrose (extra-high m., DE: 45–60%).

They are produced by action of commercial β-amylase preparations from higher plants or microorganism on liquefied starch. A high degree of conversion is reached by an immobilized enzyme combination of β-amylase and a debranching →enzyme (pullulanase, isoamylase). Another type of m. is derived from a →glucose syrup, which is further reacted by fungal α-amylase with maltogenic activity to highly saccharified syrups (62–63% DE) that contain 40–45% →maltose, 30–35% →dextrose, 8–10% →maltotriose and 20–22% higher saccharides.

M. are entirely digestible with a physiological energy content of 17.2 KJ/g. They offer functional properties, that differ from those of glucose syrups of simular DE-value:

Lower viscosity at higher concentration, better prevention of crystal formation on application with sucrose, lower hygroscopicity and browning tendency, mild sweetness and better taste.

Main applications are in food industries: confectionary, fruit-processing industry, ice cream and dairy products, nonalcoholic beverages.

Pharmaceutical uses include: medicated syrups, liquid nutrition products, baby food.

The annual world production of m. is increasing.

Of special importance as RR is the hydrogenation of m. to →maltitol.

Increasing significance is expected as building blocks in nonfood applications for chemical synthesis of biodegradable products.

Lit : Schenk/Hebeda* 335–366 (1992)
     Ruttloff*, 594, 600, 610–615 (1994)

Mandarin Oil

G.: Mandarinenöl; F.: essence de mandarine

M. is produced by expressing the peels of the fruit of the →mandarin tree, yielding an orange-brown, dark yellowish-brown or olive brown, mobile liquid of citrus-fresh, yet sweet and slightly orange-flower-like floral odor. It is used regularly in citrus flavors for candies and soft drinks in combination with orange, grapefruit and lime oils. In perfumery, m. is used as a fresh-sweet nuance in Eau de Cologne and citrus types as well as in floral and herbal notes.

Lit: Arctander*

Mandarin Tree

Citrus reticulata Blanco
Rutaceae
G.: Mandarinenbaum; F.: mandarinier

The m. originates from southern China and was introduced in Europe and the US in the first half of the 19th century. Today, it is cultivated in Italy, Spain and other mediterranean countries, in the US, Brazil and Argentina for fruit production. →Mandarin oil is produced from these fruits.

Lit.: The H&R Book*

Manganese Soap

→Metallic Soaps

Manihot esculenta

→Cassava

Manila Copal

→Resins, natural

Manila Hemp;

Syn. Abacá
Musa textilis Née
Musaceae
G: Manilahanf, Abaka; F: chanvre de Manille

M. is a member of the banana family. The fibers are obtained from its sheathing leaf bases. It is the strongest of all structural fibers. Because it does not deteriorate or rot in fresh- or saltwater and is elastic, light and durable, it is used mostly for the manufacture of cables and ropes, strong sacking, coarse fabrics and strong paper. M. grows best in the low-land, wet tropics with more than 2000 mm annual rainfall. It propagates vegetatively with suckers.

The first harvest of pseudostems for fiber extraction can be made when the crop is 2 or 3 years old, and →harvesting may continue for 5–15 years or even longer, depending upon cultivar and growing conditions. Eventually, after fiber yields decrease, replanting may become necessary. Each plant is harvested at intervals of 4–6 months. The pseudostems are cut off close to the ground, and the laminas are removed and discarded with the stalk of the

→inflorescence because they contain no commercially useful fiber. The leaf-sheaths are separated into 3 or 4 groups, which produce different qualities, or grades, of fiber. The best quality is from the 3 outer, oldest sheaths, and the poorest, soft weak fiber is obtained from the innermost 7–8 sheaths. Only the edges of each sheath (about 15% by weight) contain useful fiber. The center consists of soft tissue (pith) without fiber. The fibrous edges are removed by hand in strips 5–8 cm wide, and from these the fiber bundles are removed by drawing them over a knife blade, mechanically or by hand. After this, the fibers need to be washed with a large amount of water, then dried quickly in the sun, and pressed into bales. A plantation will yield about 4 mt of fibers per ha/a. They contain about 60% →cellulose, 20% hemicellulose and 5% →lignin. The bundles are 1–3 m long, white or reddish yellow, light, stiff and lustrous. Due to their hygroscopic behavior, they absorb up to half their weight of water in a saturated atmosphere.

M. was probably first cultivated in the Philippines, which have always been the leading producer (90% of the world production) with an output of about 90000 mt/a. The name m. was given to the fiber although it is not a true →hemp. It gradually replaced hemp in the cordage industry of Europe during the early 19th century. M. looks similar to the banana tree. Unlike banana, however, it regularly produces viable seeds.

Lit: Cobley* (1976)
  Rehm/Espig* (1991)

Manioc

→Cassava

Mannite

→Mannitol

D-Mannitol

Syn.: Mannite
G.: D-Mannit, Mannazucker;
F.: D-mannitol, mannite

The six-valency polyalcohol (hexite) m. occurs naturally in some brown algae.

$$\underset{\underset{H}{|}}{\overset{\overset{HO}{|}}{H-C}}-\underset{\underset{OH}{|}}{\overset{\overset{H}{|}}{C}}-\underset{\underset{OH}{|}}{\overset{\overset{H}{|}}{C}}-\underset{\underset{H}{|}}{\overset{\overset{OH}{|}}{C}}-\underset{\underset{H}{|}}{\overset{\overset{OH}{|}}{C}}-\underset{\underset{H}{|}}{\overset{\overset{OH}{|}}{C}}-H$$

m.w.: 182.2; m.p.: 165–169 °C; α_D^{20}: 137–146°. Solubility in 100 g water (50 °C) 45 g.

Industrial production is based on 1:1 D-glucose/D-fructose syrup. Hydrogenation is performed under

high pressure with Raney-Ni as catalyst. The reaction product consists of 75% →D-sorbitol and 25% m. The latter is isolated in the pure state by fractional crystallization. Other starting materials are pure →D-fructose, →invert sugar or →isomerized sugars. In most cases, →D-sorbitol is a co-product.
M. is used in the food industry as a low-caloric sweetener (10 kJ/g). It is slowly and incompletely resorbed in the intestine without remarkable influence on the blood glucose level.

Main applications are in the pharmaceutical industry as directly compressible binder for heat-sensitive drugs and as intravenously applied infusion.

Significance as RR results from derivatization to →polyetherpolyols by →alkoxylation used in →polyurethane chemistry.

The annual world production is reported to be 10 000 mt.

Lit.: Rymon-Lipinski/Schiweck* 313–324 (1991)
      [69-65-8]

Maple Sugar
→Sugar Maple Processing

Margosa
→Neem

Marijuana
→Hemp

Marine Algae
Syn.: Seaweed
G.: Meeresalgen; F.: algues marines

M. are heterogeneous group of primitive plant divisions of low order, which have undergone a more or less parallel evolution, specialized for living in water or highly water-saturated environments, with or without pigmentation by chlorophyll and being mostly autotrophic. Marine vegetation consists for the most part of m. They can be divided, according to their biological mode of living, into two large groups: bentic m., the sedentary form; and planktonic m. (phytoplankton), unicellular organisms, that float in the seas. The 17 000 species of m. are systematically classified into seven divisions, from which species of the classes Rhodophyta (red seaweeds) and Phaeophyta (brown seaweeds) are of interest as sources of RR. They can be harvested and further processed for → seaweed extracts, such as → agar, → alginate, → carrageenan and → furcellaran.

A third class, Chlorophyta, the green seaweeds, play an important role at all Pacific coasts for regional consumption as food and feed. Moreover, the species *Chlorella* and *Scenedesmus* have been thoroughly studied for cultivation in aquaculture as sources for fats and proteins. The economically most important organisms and their geographical distribution are:

SOURCES OF ALGINATES
(Brown Seaweeds)

Biological Classification	Geographical Distribution	Location, Habitus, Contents
Class Phaeophyta, Order Laminariales, 30 species	Northern Atlantic coasts, boreal zones of Norway, Ireland, Scotland, Pacific Ocean, Arctic Ocean, White Sea	Thalli of several meters at depths of 2–10 m
Genera *Ecklonia* and *Sargassum* 250 species	Japan, Pacific Ocean, New Zealand, Australia, North Atlantik Sarggasso-Sea populations	Thalli 5m, stipe and blades 35–38% alginic acid; 0,25% I'. "Floating meadows", $4–11 \times 10^6$ mt 7–9% alginic acid; sedentary sp. 13–17% alginic acid
Genus *Macrocystis* *M. pyrifera*	Pacific coasts, California New Zealand l "giant kelp"	Widely branched thalli; 25–50m long, 7–25m in depth; annual harvest 60×10^6 mt; annual growth potential 60–100mt/ha, 14–21% alginic acid; K', Cl', I'.
Order Fucales, 15 species	Northern Atlantic, Baltic Sea	Littoral zones, thalli strap-shaped, often >1m; 18–28% algin. acid.

(all contents are based on d.m.)

SOURCES OF CARRAGEENAN AND FURCELLARAN
(Red Seaweeds)

Biological Classification	Geographical Distribution	Location, Habitus, Contents
Class Rhodophyta, Gigartinales, *Chondrus crispus*	North Atlantic coasts of Europe, North America	Boreal zones in shallow water, 6 m in depth; thalli flat, fanlike, highly branched, 5–15 cm high.
Order Eucheuma, 20 species	Warmer seas of subtropical and tropical Atlantic, South-East Asian coasts	Thalli 5–30 cm high in >2 m, which natives harvest by hand, wash and sun-dry. Applications: paper, textile, food and feed.
Furcellaria fastigiata	Kattegat, North Atlantic, Baltic Sea	Branched thalli, 10–25 cm high; in depths of 4–8m, harvested by nets from trawlers, Furcellaran = "Danish Agar"
Order Gigartina, 90 species	From Europ. Atlantic coasts to South Africa, Malaysia, New Zealand, California, Northern Pacific	Thalli branched or lobed, <90 cm high (low in k-carrageenan)
Order Hypnea, 25 species	Widespread along warmer subtropical and tropical coasts	Thalli 5–30 cm high, ramified; phycocolloid content 24%; "Hypnean", high in k-carrageenan

(all contents are based on d.m.)

SOURCES OF AGAR

Biological Classification	Geographical Distribution	Location, Habitus, Contents
Order Gelidiales Genus *Gelidium*, 40 species	Warmer regions of Eastern and South-Eastern Asian Sea	Littoral habitats and in deeper waters <30m, thalli irregularly branched, <30 cm high. Harvested by raking the thalli from deeper waters. Utilized for "Standard Agar" (25% phycocolloid) and for food
Genus *Gracillaria*, 100 species	South-west coast of Norway, North Sea, Mediterranean coasts, South Africa, Asia, Australia, North- and Central Amer. coasts	Thalli 3–4m long, in depth of 3–14 m, "Agar" 12–19% or "Agaroid" = "Gracillaria gum" 17–30%, used widespread as "Asian Food".

(All contents are based on d.m.)

Most m. used for industrial processing are growing in the different littoral zones of the coasts; they adhere to rocky grounds in closed canopies where they may form plant bodies of several meters. During harvesting, they are cut by machines specially designed for their mode of growing or by drag nets from trawlers. Continuous growth can be achieved by enabling more light to reach the plants by harvesting every four month. Free-floating m. are sampled by trawling in regions of high plant concentration. Harvesting by divers, formerly also for agar production, has only limited regional importance for food. Besides phycocolloids for →seaweed extracts, m. contain constituents of general importance that are widely divergent, depending on botanical origin and environment of growth: antibiotics; amino acids (28 have been identified), peptides and proteins; chlorophylls, carotinoids, phycobilins and other pigments; enzymes; fats and lipids (sterols), up to 6% in d.s.; inorganic components (ash content 4–10% in living matter); starch (Floridean starch); vitamins. The water content of freshly harvested m. may vary between 60% and 88%.

M. are the most important organisms in maintaining ecological equilibrium in the world by their enormous synthetical potential: 90% of carbon dioxide binding by photosynthesis is carried out by m. under the surface of the sea .

Lit.: Levring*, 126, 288, 369

> V.J.Chapman, D.J.Chapman "Seaweeds and Their Uses" (3.), Chapman and Hall, London, (1980).

Marine Oil
→Fish Oil

Mastic
→Resins, natural

MC
→Methylcellulose

MCT
→Fats and Oils

MDG
→Glycerides

Meadowfoam
Limnanthes alba Benth.
Limmanthaceae
G.: Weißer Sumpfschnabel; F.: limnanthe

M. is a small herbaceous winter annual crop, which was first grown on a semi-commercial scale in Oregon, USA in the late 1970's. The name is derived from the typical white blooming canopy of plants, which depicts the image of a white foam covering the soil.

Its origin is in northern California and southern Oregon, and it is endemic to North America.

In 1959, m. was first described as having potential as an oil crop in large-scale chemical screening experiments by the United States Department of Agriculture. Of several species of m., *Limnanthes alba* showed the best agronomic performance in Alaska, California, Maryland and Oregon because of seed retention and erect growth habit. In Europe breeding and production work is reported from the UK and the Netherlands.

Since 1975, cultivars of m. have been described in the US with adaptation to agronomic production.

In 1986, about 160 ha of m. were grown in the Willamette Valley in Oregon. Seed yields range from 1800 to 2500 kg/ha. The oil content of m. seed is as high as 33%. Its fatty acid pattern is:

C20:1   C22:1   C22:2.
52–77%   8–29%   7–20%

The unusual main fatty acid (C20:1) is →cis-5-eicosanoic acid. The C22:2 fatty acid is unique because the double bonds are far removed from each other (*cis,cis*-5,13-docosadienoic acid), and make the acid more stable against oxydation.

The crop may be grown as a winter annual crop with seed production by mid-July.

Products similar to sperm →whale oil can be derived. The oil is also reported to be useful for cosmetics.

Lit: Nikolova-Damyanova, et al.,"The Structure of the Triacyl glycerols of Meadowfoam Oil" J.Amer.Oil Chem.Soc. **67**, 503–507 (1990)

> F.Hirsinger, "Cuphea and Meadowfoam: A Case Study of New Crops",in: "New Crops for Temperate Regions", K.Anthony (ed.) Chapman and Hall London, 197–205 (1993)

Medium-Chain Trigycerides
→Fats and Oils

Melissic Acid
Syn.: Triacontanoic Acid; C30:0
G.: Melissinsäure; F.: acide mélissique

$CH_3-(CH_2)_{28}-COOH$

m.w.: 452.81; m.p.: 93.6 °C; b.p.: 299 °C (0.53 kPa/4 mm)

M. is a long-chain fatty acid, which is found in →waxes, mainly →beeswax, →candelilla wax and →carnauba wax.

Lit.: Kirk Othmer* (4.) **5**, 147
[506-50-3]

Menhaden Oil
→Fish Oil

Mentha arvensis
Mentha piperita
→Peppermint

p-Mentan-3-ol
→Menthol

Mentha spicata
→Spearmint

Menthol

Syn.: p-Menthan-3-ol; 5-Methyl-2-isopropyl-cyclohexan-1-ol

G.: Menthol; F.: menthol

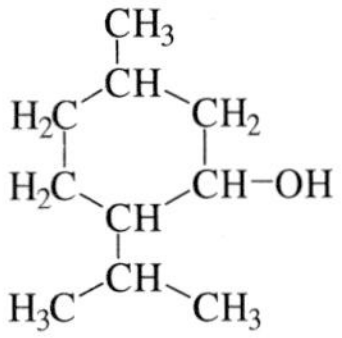

m.w.: 156.27; m.p.: 43 °C; b.p.: 216 °C

M. has 8 possible optical →isomers; however, (–)-menthol is the only one that occurs widely in nature and has the characteristic properties.

M. forms colorless crystals with a fresh, minty and sweet-pungent odor and a cooling effect on mucous membranes. M. is produced by total or part (e.g., from →eucalyptol) synthesis as well as by isolation from →peppermint oils, mainly *from Mentha arvensis* var. *piperescens*. This isolation is carried out by a freezing process and subsequent centrifugation and recrystallization. Another 40–50% are obtained from the resulting mother liquor by →hydrogenation of the menthone present and by →saponification of contained menthyl esters. Main production countries for natural m. are Brazil and China.

M. is used by itself as a cooling agent in cosmetic products (after-shaves and toothpaste) or as a flavor ingredient (cigarettes and chewing gums) with a cooling effect and little or no typical peppermint notes. It is also used in conjunction with →peppermint oil as a fortifier for the above.

Lit.: Ullmann* (5.) **A11,** 167

[2216-51-5]

Metallic Soaps

G.: Metallseifen; F.: savons métalliques

M. are the salts of long-chain saturated or unsaturated fatty acids (C_8–C_{22}). Also, soaps of branched and cycloaliphatic acids as well as of →rosin acids are included in this term today. Theoretically, all metals can form such salts. Only the soaps of the following metals are industrially important: Li, Mg, Ca, Ba, Mn, Fe, Co, Ni, Cu, Zn, Cd, Al and Pb. The soaps of Na and K are described under →soap.

General remarks:

The soaps of saturated fatty acids are solid, unless they are transformed into a paste by addition of water. Unsaturated acids form plastic salts and are normally produced and delivered in aqueous solution.

The molar ratio between metal and acid determines the nature of the soap. There are neutral soaps (1:1) as well as acid soaps or basic soaps, where either acid or metal is in excess. Complex soaps, e.g., of lead, may consist of PbO, $PbSO_4$ and the m. of Pb.

There are three types of reactions in use for manufacturing of m.:

1. Double decomposition: the fatty acid is first transferred into the sodium salt, and the metal soap is precipitated by addition of the aqueous solution of the respective metal salt (method 1).

2. Direct reaction of the fatty acid with the oxide or the hydroxide of the metal: This can either be done (method 2a) in a melt (if the resulting soap is melting below 140 °C and forms a low-viscosity melt); or (method 2b) in an aqueous phase.

3. Rarely is soap formation accomplished in organic solvents (method 3).

There are many variations of these procedures, e.g., continuous versions and special forms of delivery, e.g., dust-free granules that offer convenient and safe handling. The manufacture of m. must be controlled carefully because some of the soaps are toxic and the processes can cause difficulties as to water and air pollution.

The following describes the most important m. (excluding naphtenate and other non-RR-based acids), their manufacture and use:

- Lithium soaps of →stearic [4485-12-5] and →12-hydroxystearic acid [7620-77-1] are made after method 2b and spray-dried because they are difficult to filter. Uses for the stearate are as thickeners for oils, to raise the m.p. of →waxes and in light-metal injection moulding. The hydroxystearate, is as well as the stearate, widely used in →lubricating greases.

- Magnesium stearate [557-04-0] and behenate [43168-33-8] are produced after method 1, from →stearic or →behenic acid. They find use as lubricants and mold releases in the plastic industry, as lubricants for tablet pressing of pharmaceuticals, in →cosmetics, powders and ointments, in →wax formulations and as anticaking and waterproofing agents for hygroscopic substances.

- Calcium soaps of →stearic acid [1592-23-0], →myristic acid and →lauric acid [4696-56-4] are of commercial importance. Oleates [142-17-6] and →rosin soap are also used. Production methods 1 and 2b are applied. Calcium stearate pastes with a solid content of 40–50% are used in large volume in the →paper industry as coating additives. Another large outlet is as lubricant and secondary stabilizer in PVC (→plastics additives). Further uses include: anticaking agents for hygroscopic substances, lubricants and mold releases in tablet pressing, surface treatments of fillers and as waterproofing agents.

- Barium soaps, especially of →12-hydroxystearic acid, are made after method 2b. They are used as lubricants and costabilizers in PVC (→plastic additives), as wetting and dispersion aids in paints and in wire drawing.
- Manganese soaps are produced in neutral (method 1, 2a, 3) and alkaline (method 1) form. Mn-stearate [3353-05-7] and oleate [23250-73-9] are the most used, mostly as →dryers in →coatings and →inks based on →drying oils or →alkyd resins. They act also as catalysts for paraffin oxidation and reduction of fatty acids to →fatty alcohols.
- Iron soaps (trisalt) of →stearic [555-36-2] and →oleic acid [23335-74-2] are produced commercialy after method 1. The are components of →dryer formulations but can only be used in applications where their dark color is of no concern. Another use is as →hydrogenation catalyst.
- Cobalt soaps are in use in neutral or weakly alkaline form and are produced according to 1 or 2a. The →stearic acid soap [13586-84-0] and the →oleic acid salt are the most important. The blue soaps are used in dryers for →coatings and →inks.
- Nickel soaps are produced after methods 1, 2a and 3 and are neutral to weakly alkaline. The stearates [2223-95-2] and oleates [23250-73-9] act as →dryers in →coatings and as oil-soluble →hydrogenation catalysts (e.g., in fat hardening) that liberate finely dispersed Ni upon heating and decomposition. They serve also as stabilizers against resinification in →lubricants.
- Copper soaps are produced with methods 1, 2a and 3. Again, stearic acid [7617-31-4] and oleic acid [10402-16-1] are the main reactants. Fungicides for ship and wood coatings, for textiles and →paper are the outlets.
- Zinc soaps are either neutral (methods 1, 2a, 2b, 3) or alkaline (method 1). Commercial salts are based on →stearic [557-05-1], →lauric [2452-01-9] or →oleic acid [557-07-3]. Uses are as lubricants and costabilizers in PVC (→plastics additives), as release agents and vulcaniza-

tion activator in →rubber, as matting agents and →dryers in →coatings, anticaking agent, waterproofing agent in →textile auxiliaries and in cosmetic powders. The salt of →arachidonic acid is used as grease in metal drawing.
- Cadmium soaps, neutral or alkaline are produced after methods 1, 2a and 3. Main acid components are stearic [2223-93-0], 12-hydroxystearic [35674-68-1], lauric [2605-44-9] and oleic acids [10468-30-1]. Cadmium soaps are PVC stabilizers (→plastic additives) of decreasing importance due to their high toxicity.
- Aluminum soaps are made according to method 1. Because Al is trivalent, there are three different neutralization stages. The most important acid is →stearic acid: Al-tristearate [637-12-7], Al-distearate [300-92-5] and Al-monostearate [7074-84-9] are white powders that are used as gelling agents for mineral oil and in oily preparations in cosmetic and pharmaceutical applications. The gel consistency increases from the trisalt to the disalt and monosalt. Other uses are in →coatings as antisettling, matting and wetting agents, as lubricants in thermoplastics and as waterproofing agents in building protection.
- Lead soaps are made according to methods 1, 2a and 2b if neutral. Method 2 is used for the manufacture of alkaline soaps, and also for making complex soaps. A mixture of PbO and $PbSO_4$ is reacted with the fatty acid. →Stearic acid and →oleic acid [1120-46-3] are the main reactants. Three types of stearates are important: neutral [1072-35-1], monobasic [90459-52-2] and dibasic [56189-09-4]. Main use is as heat stabilizer and lubricant for PVC (→plastic additives). Other uses are as lubricants in pencil manufacture (→inks) and →paper additives.

Lit.: Ullmann* (5.) **A16**, 361
      [→specific soaps]

Metal-working Fluids

G.: Kühlschmiermittel; F.: huiles de coupe

M., used in operations such as cutting, grinding, drawing and rolling of metals, are coolants and →lubricants simultanously.

They remove the metal swarf and reduce friction, wear and heat evolution and act as corrosion inhibitors. These are important functions because temperatures during such operations can raise up to 1000 °C, which causes local welding.

Mineral oils are mainly used as base oils. Ester oils (→lubricants) and fatty oils are also used as base or additive. Anionic and nonionic →surfactants act as emulsifiers. Corrosion inhibitors are amine salts, sulfonates and benzotriazol. →Metallic soaps and →fatty alcohols impart antifoaming. For special purposes, many of the additives for →lubricants are used. There are three types of **cutting** fluids:

- Straight cutting oils (no water) with additive combinations (fatty acid, esters and metallic soaps) are widely used in many applications;
- Cutting emulsions contain 2–5% of oil (O/W). Emulsifiers, corrosion inhibitors (alkanol amides) and antifoaming agents (→fatty alcohols ethoxylates) are added;
- Water-soluble cutting fluids contain no oil but polyalkylene glycols, nonionic surfactants, fatty acids and corrosion inhibitors.

Grinding oils are either straight oil-based or emulsions of oils. Polymethacrylates are added to minimize the formation of oil mists. For **honing** (precision grinding), fatty acids or triglycerides are added for adjustment to higher temperatures.

Compounds for **wire drawing** and **tube drawing** are mineral oils that sometimes contain additives, such as triglycerides, fatty acids, metallic soaps, and corrosion inhibitors.

Deep drawing uses additives such as vegetable oils, polymethacrylates and EP additives (→lubricants).

In **cold extrusion**, powdered metal salts of →stearic acid and →arachidic acid or →lubricating greases, or mixtures of both, are used.

→Gluconic acid is a component in etching formulations.

In hot **rolling** of aluminum (down to a thickness of 1 cm), emulsions are in use that contain oil, anionic or nonionic emulsifiers and active rolling ingredients (fatty acid derivatives). Subsequent cold rolling (to a thickness of 0.5 mm) is accomplished with the aid of oils that contain 2–7% of →lauryl alcohol or 1–4% butyl ester of →palmitic acid. Cold rolling of other metals requires oils that contain 10–15% fatty acids and EP-additives (→lubricants).

Lit.: J.A.Schey, "Tribology in Metalworking" American Soc.for Metals, Metals Park, Ohio (1983)
Ullmann* (5.) **A15**, 479

Metathesis

G.: Metathese; F.: métathèse

Olefin m. is a bond rearrangement reaction, catalyzed by transition metal compounds (molybdenum, tungsten or rhenium):

$$CH_3-CH=CH_2 \atop + \atop CH_3-CH=CH_2 \quad \longleftrightarrow \quad {CH_3-CH \atop CH_3-CH} + {CH_2 \atop CH_2}$$

This reaction is widely used in large scale petrochemical processes (Phillips Triolefin Process and Shell Higher Olefin Process "SHOP") and in some specialty synthesis.

The reaction can also be applied to methyl oleate (→ fatty acid methyl ester), which leads via self-m. to an olefin and a di-carboxylic acid.

$$CH_3-(CH_2)_7-CH=CH-(CH_2)_7-COOCH_3 \atop + \atop CH_3-(CH_2)_7-CH=CH-(CH_2)_7-COOCH_3$$

$$\approx 70\,°C \quad \Big\uparrow \quad WCl_6/Sn(CH_3)_4$$

$$CH_3-(CH_2)_7-CH \atop CH_3-(CH_2)_7-CH \quad + \quad {CH-(CH_2)_7-COOCH_3 \atop CH-(CH_2)_7-COOCH_3}$$

The oleic acid methyl ester derived from high oleic →sunflower oil could be an ideal starting material for this reaction.

Another interesting reaction is the m. of ethylene with →oleic or →erucic acid methyl ester to yield an olefin and a medium-chain fatty acid (after →hydrolysis of the methyl ester).

Both reactions could be interesting new tools in →oleochemistry. Large-scale application is, however, hampered by the high consumption of the expensive catalyst (150 mole of olefin need 1 mole of catalyst). Catalyst research is, therefore, focusing on new systems.

Lit.: R.L.Banks, G.C.Bailey, Ind.Eng.Chem.Prod.Res.-Dev. **3**, 170 (1964)
C.Boelhouwer, J.C.Mol, J.Amer.Oil Chem.Soc. **61**, 425 (1984)

Methanol

Syn.: Methyl Alcohol
G.: Methanol, Methylalkohol; F.: méthanol

CH_3-OH

m.w.: 32.04; m.p.: –98 °C; b.p.: 64.7 °C

Until 1923, the only source for m. was the dry distillation of wood, a technology that is completely abandoned today. Nowadays, m. is made from synthesis gas on a large scale (world capacity 21×10^6 mt). The chemical industry uses 85% of production as solvent and intermediate. The remainder is used in fuels (→fuel alternatives).

Lit.: Kirk Othmer* (3.) **15**, 398
[67-56-1]

Methanolysis

→Transesterification

2-Methoxy-4-allylphenol
→Eugenol

Methylcellulose
Syn.: MC
G.: Methylcellulose; F.: méthylcellulose

M. and its mixed ethers, mainly HEMC, HPMC, EMC, CMMC (→cellulose ethers), are water-soluble if DS is between 1.4–2 (range of commercial products). Lower-substituted types need some alkali to obtain solubility in water, while higher-substituted m. are soluble in organic solvents. Because of nonuniform substitution, pure m. often contains insoluble or only swellable particles. The best method to overcome this problem is mixed etherification. Most important mixed ethers are: hydroxypropyl methylcellulose (HPMC), with a DS for methyl of 1.3–2.2 and a MS for hydroxypropyl of 0.1–0.8, and hydroxyethyl methylcellulose (HEMC), with a methyl DS of 1.5–2 and a MS for hydroxyethyl of 0.02–0.3. Another mixed ether that is commonly used in Great Britain is ethyl methylcellulose (EMC) with a DS ratio of methyl:ethyl of 0.9:0.4. Carboxymethylation with a low DS (0.05) yields mixed ethers with a weak anionic character. M. and most of the mixed ethers are insoluble in hot water. Gelation temperature drops with increasing methyl DS or hydrophobic ether groups (e. g., HP). Hydrophilic groups (e. g. CM or HE) in mixed ethers increase the gelation temperature. The addition of electrolyte to aqueous m. solutions lowers the gelation temperature. Solutions of m. show pseudoplasticity: viscosity decreases with increasing shear forces. Viscosity also decreases with increasing temperature until gelation temperature. Just before the gelation temperature is reached, there is a sharp increase, followed by precipitation. A method to get lump-free solutions is to disperse the m. in water above the gelation temperature and add cold water or cool the slurry.

Due to the presence of both hydrophilic (OH) and hydrophobic (OCH₃) groups, m. can reduce surface tension and support emulsification in two-phase systems. It increases the water retention of aqueous systems, such as plasters, giving the cement or gypsum time to react with the water. Mixed ethers are excellent film formers, the water sensitivity of which can be adjusted by formaldehyde. Higher-substituted ethers are thermoplastic and can be extruded at 120–190 °C to water-soluble sheets (bags).

Production (→etherification) is carried out with a large excess of alkali (3–4 moles/AGU). An excess of methyl chloride is necessary because methanol and dimethyl ether are by-products, which consume reagent. The resulting product can be purified from NaCl easily by washing with water above the gelation temperature. Mixed ethers are made by adding the additional reagent before or after methylation. Retarded dissolution of m. can (as with other starch and cellulose derivatives) be incorporated by slight crosslinking with glyoxal in the water-wet finished product just before drying. This is important to obtain lump-free dissolution in water. Once the product is dispersed, slow hydrolysis of the hemiacetal gives clear solutions.

M. is used in the building industry in plasters, mortar and putties to enhance water retention and adhesion. Another outlet for m. is →wallpaper paste and pasting adhesives for →leather.

Another application is in →paints as thickener and suspension aid. Higher-substituted m. are used in organic-based →paint removers. An addition to →ceramic extrusion improves green strength, e. g., in manufacturing of carriers in car afterburners. M. is a thickener in →textile printing and acts as a soil suspender in →detergent formulations. Seed coating in agriculture is another use of m. and mixed ethers. In vinyl chloride polymerization (→polymerization additives), both act as protective colloid and suspension stabilizers. They are also used as binders for →tabacco dust to form sheets for cigars. In cosmetic, pharmaceutical and food formulations, m. and mixed ethers are widely used as thickening agents and stabilizers. M. is also used in →pencil manufacture (→inks).

Lit.: Ullmann* (5) **A5**, 468
    Encycl.Poly.Sci.Engng.* (2.) **3**, 250
    [9004-67-5]

Methylene Butanedioic Acid
→Itaconic Acid

2-Methylfuran
G.: Methylfuran; F.: méthylfuranne
m.w.: 82.1; m.p.: −88.7 °C; b.p.: 63–65 °C

M. is a colorless liquid with similar properties as →furan.

It is produced by vapor-phase →hydrogenation of →furfural over a copper catalyst.

It can be further hydrogenated to methyl tetrahydrofuran. This and m. are selective solvents and intermediates in organic synthesis.

Lit.: Ullmann* (5.) **A12**, 127
    [534-22-5] + [96-47-9]

5-Methyl-2-isopropyl-cyclohexane-1-ol
→Menthol

Methyl Tetrahydrofuran
→Methylfuran

Microbial Gums
Syn.: Bacterial Polysaccharides
G.: Mikrobielle Polysaccharide;
F.: glucides polymérisés bactériels

M. are a group of different hydrophilic, neutral or acidic, homo- or hetero →polysaccharides with gelling or thickening properties that are produced by the action of bacteria or fungi or their isolated enzymes from sugar-containing culture media.
The following m. have achieved technical and commercial significance:
→Xanthan, →curdlan, →dextran, →scleroglucan, →pullulan and →levan
Erwinia gum has similar properties as xanthan; application is in food industries as suspension stabilizer for pigment colors.
All microbial gums have been intensively examined for food and nonfood applications. This know-how represents a high biotechnological production potential, which will become important if other resources such as land plant gums, exudate gums or seaweed gums (→marine algae) will be-

come short in supply. High versatility in the conditions of production and influence on product formulation allows a rising tendency in this branch of biotechnology.
Lit.: Blanchard/Mittchell*, 251–262, 263–300 (1979)
      Ruttloff* 485–515 (1991)

Microcrystalline Cellulose
G.: Mikrokristalline Cellulose;
F.: cellulose microcristalline

Cellulose contains crystalline zones (→cellulose). A controlled hydrolysis of the amorphous segments by dilute acids, such as HCl or H_2SO_4 at $80-100\,°C$ yields a slurry, which is washed, wet milled, (spray-)dried and dry milled.
M. is highly purified, hydrophilic, crystalline cellulose powder. The DP is in the range of 150–350 and depends on the cellulose source and the mode of treatment. Linters is the best raw material for manufacturing. Yield is $\approx 70\%$, which corresponds to the crystalline part of the cellulose.
M. is transferred by mechanical energy at rather low concentrations (0.5%) into viscous, thixotropic pastes or smooth, stable gels.
M. in various physical forms has opened many new applications due to its nonfibrous gel character.
In the pharmaceutical industry, it is used as tablet binder, diluent, suspending and thickening agent and as base material for powders and →suppositories.
Technical uses are in →ceramics, where it aids green strength. M. acts as a flow and viscosity control in paints, and as an emulsifying and suspending agent in drilling muds. It is also used in chromatography.
For food applications (→food additives), it is approved in many countries and it functions as a noncaloric filler, especially in dietary food, a dispersant in sugar gels, a fat substitute and an agent to control the formation of crystals in ice creams.
Because m. is a relatively new material, further applications can be anticipated.
Lit.: Encycl.Polym.Sci.Engng* (2.) **3**, 86
      Kirk Othmer* (4.) **5**, 482
      [9004-34-6]

Gum	Microorganism	C-Source	Constituents
Bacterial alginate	*Acotobacter vinelandii*	Sucrose, Glucose, Mannose	D-Mannuronic, L-Guluronic acid
Curdlan	*Alcaligenes* spp. *Agrobacterium* spp.	Glucose	D-Glucose
Dextran	*Leuconostoc* spp.	Sucrose	D-Glucose
Erwinia gum (ZANFLO)	*Erwinia tahitica*	Sucrose, Maltose, Lactose, Glucose-syrup	D-Glucose, D-Galactose, D-Guluronic acid L-Fucose, Acetate
Gellan	*Pseudomonas elodea*	Glucose	D-Glucose, L-Rhamnose, D-Glucuronic acid, Acetate
Pullulan	*Pullularia pullulans*	Glucose, Sucrose, Maltose	D-Glucose
Scleroglucane	*Sclerotium* spp.	Glucose	D-Glucose
Xanthan	*Xanthomonas* spp.	Glucose, Sucrose, Glucose-syrup, Whey	D-Glucose, D-Mannose, D-Glucuronic acid, Acetyl Pyruvate

Milk Sugar
→Lactose

Mineral Industry
→Mining Chemicals

Mining Chemicals

G.: Bergbauchemikalien;

F.: produits pour l'industrie minière

Because high-grade ores, which are relatively easy to process, are becoming more rare, new, more selective processes, which employ larger quantities of sophisticated mining chemicals, are being developed. These processes and chemicals are also useful in the recycling of metals and other valuable raw materials. The following products and processes utilize RR.

Flotation

In principle, the functions of →flotation chemicals are classified into three groups:

1. Collectors are chemicals that make the surface of the mineral particle more hydrophobic.

2. Depressants hinder flotation by causing a hydrophilic surface on the particle.

3. Frothers enhance and stabilize the formation of foam.

The selection of such chemicals is based mainly on empirical experience.

Collectors:

Collectors can be further divided into two categories. Cationic collectors for such minerals as iron ore, silicates and potash are based on →fatty amines, fatty diamines, fatty ether amines and diamines.

Anionic collectors, used mainly for the flotation of nonmetallic minerals (so-called industrial minerals), consist basically of →fatty acids and their salts. Special separation problems are being solved by the careful selection of the chainlength of the alkyl groups and the degree of saturation. Recovery is decreased with an improvement in selectivity according to the following molecular changes: →stearic acid, →palmitic acid, →palmitoleic acid, →oleic acid, →linoleic acid, →linolenic acid.

Other anionic collectors are →fatty alcohol sulfates and ether sulfates, →sulfosuccinates, sulfosuccinamates (→sulfosuccinates) and →fatty alcohol phosphates. Minerals, such as baryte, fluorspar and apatite, are upgraded from as low as 10% to a marketable concentration of approx. 80–95% with these reagents.

Depressants:

In addition to inorganic depressants, which also act as dispersants, organic compounds, such as water-soluble polymers, are used extensively in the flotation of metallic and nonmetallic minerals.

→Starch, →dextrine, natural gums (e. g., →guar gum) and their derivatives are used as depressants.

→Tannins and their derivatives belong also to this group.

The interaction and adsorption mechanism of the polymers on the surface of the mineral particle is complex. Their successful use is based on long-term experience rather than theoretical background.

Frothers:

The composition of frothers varies widely. Most frequently used are turpinols (→pine oils), polyglycols, lower →fatty alcohols and →fatty alcohol ethoxylates/propoxylates.

Extraction

Another important technology for the concentration and winning of minerals and metals is solvent extraction. Up to the time of writing there is only one area where RR-based chemicals are being used commercially. Tertiary amines (→fatty amines) with short- to medium-length alkyl chains and their →quatenary derivatives form ion pair complexes with certain metals that are dissolved in an aqueous medium . These reagents are widely used for the recovery of uranium, vanadium and, to a lesser extent, molybdenum.

Other uses

Gold is leached with sodium cyanide solution, whereby a soluble gold complex is formed. Highly →activated carbon, made from →coconut shells, is able to absorb the gold from the solution (carbon in pulp process: CIP). The process is reversible, and the gold can be recovered by electrowinning or zinc precipitation. The carbon can be reactivated. This process is the preferred method for gold recovery today.

Flocculants and coagulants are used in the mining industry to enhance the separation of solids from water. Synthetic polymers dominate this application. Natural polymers are →starch-based, →animal glue or →guar gum, used to solve special separation problems.

To obtain the reverse effect, i. e., the prevention of agglomeration of solids (scale control, filter aids, flotation circuits), dispersing aids, such as →fatty alcohol derivatives (among others), can be used.

To reduce the residual moisture in filter cakes, strong wetting agents, including →sulfosuccinates and →fatty alcohol ethoxylates, are useful.

Dust control in mining operations is necessary to provide a safe and healthy working environment for the miners. The most commonly used dust control agent is water. Its wetting properties can be greatly improved by the addition of wetting agents, such as →sulfosuccinates, →ligninsulfonate, →fatty alcohol ethoxylates and →fatty amines. →Lignosulfonate is also used as a ore pellet binder.

Lit.: Kirk Othmer* (3.) **10,** 523
Ullmann* (5.) **B2,** 23 – 1

Miscanthus sinensis
→China Grass

Modal Fiber
→Viscose

Modified Starches
G.: Modifizierte Stärken;
F.: amidons (fécules) transformés

M. are native starches, treated in such a way as to modifiy one or more of their physical or chemical properties.

The need for modification results from the native →starch properties, which are frequently not sufficient for special applications.

Adaption to final use is performed in different ways, which are distinguished by the kind of action as well as by the resulting change of properties:

Physical modifications comprise methods that change the starch granule in its interior or exterior structure without causing chemical reactions at the monomeric units or depolymerization.

Typical products are:

→pregelatinized starches, in which the granular structure is destroyed and the products exhibit cold-water swelling properties; retarded starches (annealed starches) are starch granules, in which the processes of gelatinization and solution are shifted to higher temperatures, and they are comparable with chemically →crosslinked starches; molding starches, which are overdried in the granular state with small amounts of oil or fat added.

Chemical modifications comprise reaction of oxidants with special sites of the →AGU to give →**oxidized starches;** action of heat in the dry or wet state, leading to **dehydrated starch products** as well as →**dextrins**.; introduction of substituent groups to form →**starch derivatives,** such as →starch ethers, →starch esters, →crosslinked starches, starch graft polymers, →acroylated starches; slight destruction of the polymeric structure by splitting some of the glucosidic linkages while keeping the starch's typical properties, to form →**thinboiling starches** (Lintner and Zulk-

owski starches), →white dextrins and acid-modified starches.

These types of reactions are not clearly separated from each other because of overlapping actions of heat, acid and moisture. So, the →dextrins, based on their production technology, are treated as a uniform group.

The starch components →amylose and →amylopectin are not to be considered as m. because they represent pure, unmodified →starch polysaccharides. They can be converted, however, into modified products, preferentially by the above-cited chemical treatments.

Processes and products of m. differ systematically from →starch hydrolysis products that are obtained by splitting the glucosidic linkages with acids and/or enzymes. Those products comprise the largest group in starch utilization.

M. are preferentially used in nonfood applications (→starch industrial applications). Their use in food and pharmaceutical production is strictly limited by governmental and international regulations with respect to kind and amount of substituents as well as reagent residues.

Lit.: Tegge* 165 – 214 (1988)
Blanchard* 301 – 321 (1929
Ullmann* **A25,** 12 – 21

Molasses
→Sucrose Molasses
→Sucrose Industry By-Products

Molding Starches
→Hydrophobic Starches

Mole Plant
→Euphorbia

Monocotyledonous Plants
Syn.: Monocots, Monocotyledons
G.: Einkeimblättrige Pflanzen;
F.: monocotylédonés

M. represent a class of plants that comprises about 28% of all flowering plants (Angiospermae). Major representatives are grasses and →cereals. Phylogenetically, m. are younger than →dicotyledonous plants. They form only one cotyledon during germination of seeds. Their seeds normally store much →starch, such as in →cereals and →corn, with only little (1 – 4%) fat (exception is rice with 8 – 13%). The less storage capacity exists in seeds, the less is their crop character. For example, orchids produce very small dust-like seeds.

Typical vegetative characteristics are lanceolate leaves with parallel veins. Their flowers will part in threes, often of simple morphology with little coloring; exceptions are popular colorful ornamentals, such as lilies, daffodils, tulips, and orchids.

They form no cambium, which limits the number of trees. The only tree-forming family is the →Palmae (*Arecaceae*).

→Dicotyledonous Plants

Lit.: P.Raven et al."Biology of Plants" (5.), Worth Publish., NY (1992)

Monoecious

G: Monözisch, einhäusig;

F: monoïque, monœcique

The male and the female reproductive organs are together in one plant, but in separate flowers. Normally, plants produce hermaphrodite flowers with male (stamina) and female (pistil) organs in the same flower. Examples for m. plants are pine or hazelnut. The opposite is →dioecious.

Lit.: R.Frankel, E.Galun "Pollination Mechanism, Reproduction and Plant Breeding", Springer, NY (1977)

Mono/Diglycerides

→Glycerides

Monoglyceride Citrates

→Citric Acid Esters of Mono/Diglycerides

Monoglycerides

→Glycerides

Mono Sodium Glutamate

→Glutamic Acid

Montanic Acid

Syn.: Octacosanoic Acid; C28:0

G.: Montansäure; F.: acide montanique

$CH_3-(CH_2)_{26}-COOH$

m.w.: 424.75; m.p.: 90.9 °C; b.p.: 285 °C (0.532 kPa/4 mm)

M. is contained and obtained from →waxes (lignite, peat, candellila wax). The glycerides are used as →lubricants.

Lit.: Kirk Othmer* (4.) **5**, 147

[506-48-9]

Moonseed

→*Lunaria*

Morphine

Syn.: Morphium

G.: Morphin, Morphium;

F.: morphine

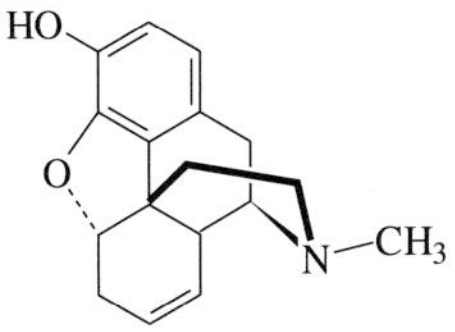

m.w.: 285.33; m.p.: 254–256 °C

M. is an alkaloid and was first isolated in 1803 by F.W. Sertürner by extraction from →poppy. M. is slightly soluble in hot water and chloroform but highly soluble in ethanol. It is the most important of the morphine alkaloids (isoquinoline alkaloids). A good grade of →opium contains between 9–14% of anhydrous morphine.

For industrial production, the crude opium is treated with hot water and calcium oxide, followed by extraction with amyl alcohol. After acidification, removal of the solvents and extraction at pH 9, m. is extracted in purified form with diluted hydrochloric acid. M. is stable as an aqueous solution only for a limited period of time and is used as a narcotic analgesic.

Lit.: Ullmann* (5.) **A1**, 377

[57-27-2]

Morphium

→Morphine

Murex brandaris

→Dyes, natural

Musa textilis

→Manila Hemp

Mustard

Brassica spp., *Sinapis alba* L.

Cruciferae

G.: Senf; F.: moutarde

Sinapis alba (white m.) is the most prominent representative of m. members: Others include *Brassica hirta* (yellow m.), *B. nigra* (black m.), *B. juncea* (brown m.) and *B. carinata* (Ethiopian m.).

M. is planted and harvested similar to →rapeseed. Its taste arises from sulfur-based essential oils and glucosinolates contained in the meal. The oil content of the small brown-, yellow- or white-colored seeds varies between 25 and 40%. Like in most other Brassicaceae, →erucic acid is contained in significant amounts.

16:0	18:0	18:1	18:2	18:3
1–4%	1–4%	20–35%	10–20%	1–5%

20:1	**22:1**
1–5%	**42–52%**

While *B. juncea* in India and Pakistan is preferably grown for its oil, most other m. species are grown for mustard and spice production. In a moderate climate, they often are grown as fodder crops or as green manure crops.

Lit.: Fehr/Hadley* p. 495 (1980)

Mutton Tallow
→Tallow, Mutton

Myristic Acid
Syn: n-Tetradecanoic Acid; C14:0
G.: Myristinsäure; F.: acide myristique

$CH_3–(CH_2)_{12}–COOH$

m.w.: 228.36; m.p.: 54 °C; b.p.: 197 °C (2 kPa/15 mm)

M. is insoluble in water, soluble in most organic solvents.

It is contained in →nutmeg oil. Some →*Cuphea* species have also high content of m. It is derived from medium-chain triglycerides (→fats and oils), e.g., →coconut oil, →palm kernel oil and →babassu oil, by →saponification and →fractionation. Up to 99% purity is available commercially.

M. is used for manufacture of →soaps, isopropyl myristate (→ fatty acids esters) and →myristyl alcohol.

Lit: Kirk Othmer* (4.) **5**, 149
[544-63-8]

Myristica fragrans
→Nutmeg Oil

Myristyl Alcohol
Syn.: 1-Tetradecanol; Tetradecyl Alcohol
G.: Myristylalkohol; F.: alcool myristylique

$CH_3–(CH_2)_{12}–CH_2OH$

m.w.: 214.4; m.p.: 38 °C; b.p.: 172 °C (2.67 kPa/20 mm)

M. forms white crystals. Pure m. (99%) is normally not used in practice. Broader cuts with a high content of m. are used in ointments as a waxy component and as starting material for derivatives (→surfactants).

Lit.: [112-72-1]

N

α-Narcotine
→Noscapine

Neatsfoot Oil
→Tallow, Beef-

Neem
Syn.: Margosa; Nim
Azadirachta indica Juss.
Meliaceae
G.: Nimbaum, Indischer Flieder;
F.: margousier azadirachta

The n. tree is a relative of the mahagony tree. Neem trees are attractive broad-leaved evergreens that can grow up to 30 m tall and 2.5 m in girth of trunk. Their spreading branches form rounded crowns as much as 10 m across. They remain in leaf except during extreme drought, when the leaves may fall off. The short, usually straight trunk has moderately thick, strongly furrowed bark.

The small, white, bisexual flowers are borne in axillary clusters. They have a honey-like scent and attract many bees.

The fruit is a smooth ellipsoidal drupe, up to almost 2 cm long. When ripe, it is yellow or greenish yellow and contains a sweet pulp that encloses a seed. The seed is composed of a shell and a kernel (sometimes two or three kernels), which is about half of the seed's weight. The kernel is used most in pest control. (The leaves also contain pesticidal ingredients, but as a rule, they are much less effective than those of the seed.)

A n. tree normally begins bearing fruits after 3–5 years, becomes fully productive in 10 years and, from then on, can produce up to 50 kg of fruits annually. It may live for more than two centuries.

Its extracts have been applied in India for centuries as a useful repellent and →insecticide. In the 20th century the n. tree has been distributed also to Arabian and African countries as well as to the Caribbean and Pacific region. Since the late 1950's, the n. tree found much interest in research and formulation of insecticides.

Lit.: "Neem, a Tree for Solving Global Problems" National Research Council. Nat. Academic Press. Washington D.C.(1992)

Neo Sugars
→Fructosyl Sucrose

Neral
→Citral

Neroli Oil/Orange Flower Absolute
G.: Neroliöl/Orangenblüten Absolut;
F.: essence de néroli/absolue de fleurs d'oranger

Both of these products are obtained from the flowers of the →orange tree (bitter), by using different production methods.

Neroli oil is a result of steam-distillation of the freshly picked, just opened flowers. A pale yellow, mobile liquid when fresh, it darkens and becomes more viscous upon aging. Its odor combines freshness, a bitter citrus-like note with an intense floral-sweet note of medium tenacity. This combination makes it an extremely useful and important material for all Eau de Cologne types, where it gives floralcy and volume to the citrus-based topnotes.

Orange flower absolute is a hydrocarbon solvent-extracted product from the same plant material. It is a dark brown- or dark orange- colored, somewhat viscous liquid with an intense floral, rich and sweet odor with a typical bitter-fresh nuance, closely reminiscent of orange blossoms. Its long-lasting floral richness, together with the impression of freshness, makes it an universally interesting material for all kinds of floral bouquets, ranging from fresh to heavier oriental (→oder description) types. In Eau de Cologne types, it is often used in combination with neroli oil. Both products are also used as trace ingredients in many types of fruit flavors to round off sharp notes that come from synthetic flavor materials.

A third, less important product of the same family is orange flower absolute from water, which is produced by solvent-extraction of the distillation waters from neroli production. Its odor characteristics are less floral, more herbal and stem-like than both neroli oil and orange flower absolute.

Lit.: The H&R Book*
     Arctander*

Nickel Soap
→Metallic Soaps

Nicotiana tabacum
Nicotiana rustica
→Tobacco

Nicotine
→Insecticides

Nigerseed
Guizotia abyssinica Cass.
Compositae
G.: Nigersaat, Gingellikraut; F.: gouzotie, niger

N. is an important oilcrop in the highlands of
Ethiopia and in central India. It is grown in Africa
mostly in Sudan, Eritrea, Uganda, Zaire, Tanzania
and as far as South Africa. The plant is about 1 m
in height, with numerous yellow flowers that produce tiny, dark seeds with a length of 4 mm and a
width of 1–1.5 mm.

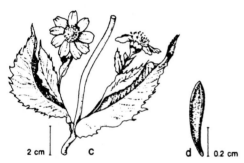

2 cm   c   d   0.2 cm

Niger c: Flowering branch; d: seeds (from Franke* 1981,
with permission)

It is tolerant to poor growing conditions and will
produce seed in a large variety of soils and at altitudes far above 2000 m. Seed yields range from
400–1500 kg per ha. N. is useful in the rotation as
a broad-leaved crop.
Seed oil contents range from 35%–45% with a
maximum of up to 60%. The dominating fatty
acid is linoleic acid at 75%, thus being similar to
sunflower and safflower oils. The seedcake may be
used as animal feed.

Lit.: K.W.Riley, H.Belayneh "Niger", in Röbbelen*
     (1989)

Nim
→Neem

Nitogenin
→Diosgenin

Nitration
G.: Nitrierung; F.: nitration

N. is the introduction of a nitro group as an –C–
NO_2 bond in an organic compound. This type of
reaction is not practiced much in context with RR.
The →esterification of alcohols with nitrous acid
(–C–O–N– bond) is sometimes incorrectly called
nitration. This exothermic reaction is carried out in
the presence of sulfuric acid. Nitrates of interest
are →cellulose nitrate (nitrocellulose) and →glyc-
eryl trinitrate (nitroglycerine).

Lit: Kirk Othmer* (3.) **15**, 841

Nitrocellulose
→Cellulose Nitrate

Nitroglycerine
→Glyceryl Trinitrate

Nonanedioic Acid
→Azelaic Acid

Nonanoic Acid
→Pelargonic Acid

Nonwetting Starches
→Hydrophobic Starches

Noscapine
Syn.: α-Narcotine
D.: Noscapin, Narcotin; F.: noscapine, narcotine

m.w.: 413.43; m.p.: 176 °C

N. is an →opium alkaloid, isolated from poppy and
present in amounts up to 8%, depending on season
and locality. It is extracted from the water-insoluble residue that remains after processing of opium
for the manufacture of →morphine. It is used as an
antitussive.

Lit.: Ullmann* (5.) **A1**, 375
     [128-62-1]

Nutmeg Oil/Mace Oil

G.: Muskatnußöl, Macisöl;

F.: essence de muscade, essence de macis

If produced from the freshly comminuted, dried seeds of *Myristica fragrans* (Myristaceae) this oil is called nutmeg oil; if the dried arillode surrounding the seeds inside the shell of the fruit is used, the oil is called "mace oil".

The plant is cultivated in Indonesia, the West Indies and Sri Lanka for the production of these spices, a part of which is steam-distilled to obtain both oils, which are close in appearance and odor.

N. is a pale yellow or almost water-white, mobile liquid with a typical, spicy, warm, aromatic note. →Topnotes are light and fresh, while the body notes are rich, warm-spicy with woody undertones. N. finds extensive use in flavorings for meat sauces, tomato ketchups, canned soups and spice blends. In perfumery, the oil is used in fantasy floral bouquets or oriental (→odor description) perfumes, where a touch of a fresh-spicy note is required, and as a refreshing aromatic-spicy topnote in fragrances for men's perfumes, after-shaves and toiletries.

Lit.: Arctander*
      The H&R Book*

O

Oakmoss Absolute/Resinoid

G.: Eichenmoos Absolut/Resinoid;
F.: absolue/resinoïde mousse de chêne

Only growing on the trunks and branches of oak trees, the moss *Evernia prunastri* (Usneaceae) is mainly collected in France, Morocco and Yugoslavia, then extracted by hydrocarbon solvents to produce oakmoss resinoid, which is then washed with alcohol to obtain oakmoss absolute. Oakmoss absolute is a semi-solid, dark-green to dark-brown product with a longlasting earthy-sweet, woody-leathery character, giving great richness without sweet notes to fragrances. A whole family of fragrances, the so-called →chypre types is based on the contrast of the dry richness of oakmoss and the crisp freshness of citrus oils, mainly →bergamot oil, together with various floral notes. Oakmoss products are also widely used in masculine fragrances to give long-lasting richness and volume.

Lit.: Arctander*
The H&R Book*

Oats

Avena sativa L.
Gramineae, Poaceae
G.: Hafer; F.: avoine

O. were probably first distributed as a weed mixture with → barley. It was believed to have been first cultivated by ancient slavonic peoples during the Iron and Bronze Ages. O. are best adapted to cooler, more temperate regions with annual precipitation of more than 700 mm. The →inflorescence is a panicle, as in other →cereals (→rice). O. have only limited winterhardiness in special selected varieties that are grown in moderate winter climates in the Mediterranean countries, such as Greece or Turkey. Often, o. are fed whole to horses and sheep; this used to be the most important application of o. before mechanical power moved into agriculture.

O. are mostly used in the food industry as a minor part of grain mixtures or as rolled oats. O. are also used as a raw material for the production of →furfural and, due to the small, uniform starch granules, also for industrial applications.

The most important o.-growing country is FSU. World production in 1995 was 32×10^6 mt.

Lit.: C.M.Brown, "Oat" in: Fehr/Hadley* 427–441 (1980)

Ocimum basilicum
→Basil Oil

Octacosanoic Acid
→Montanic Acid

cis,cis-9,12-Octadecadienoic Acid
→Linoleic Acid

Octadecanoic Acid
→Stearic Acid

1-Octadecanol
→Stearyl Alcohol

cis,cis,cis-6,9,12-Octadecatrienoic Acid
→Linolenic Acid (γ-)

cis,cis,cis-9,12,15-Octadecatrienoic Acid
→Linolenic Acid (α-)

cis,trans,trans-9,11,13-Octadecatrienoic Acid
→α-Eleostearic Acid

trans,trans,cis-8,10,12-Octadecatrienoic Acid
→Calendulic Acid

trans,trans,trans-9,11,13-Octadecatrienoic Acid
→β-Eleostearic Acid

cis-6-Octadecenoic Acid
→Petroselinic Acid

cis-9-Octadecenoic Acid
→Oleic Acid

trans-9-Octadecenoic Acid
→Elaidic Acid

cis-9-Octadecen-1-ol
→Oleyl Alcohol

cis-9-Octadecenyl-1-amine
→Oleyl Amine

Octadecyl Alcohol
→Stearyl Alcohol

1-Octadecyl Amine
→Stearyl Amine

Octanedioic Acid
→Suberic Acid

Octanoic Acid
→Caprylic Acid

1-Octanol
→Caprylic Alcohol

Odor Description
G.: Geruchsbeschreibung; F.: description d'odeur

The development of the smell of a →fragrance or a perfume raw material is generally described in three phases:
Topnote refers to the initial impression, displayed during the first half hour, whereas **bodynote** denotes the main character of the odor in question, lasting up to 12 hours. As evaporation continues, the last phase of residual odor is described as **drydown.**
The odor description vocabulary includes words that refer to the strength or evaporation character of the odor (i. e., pungent, intense, diffusive, tenacious); to terms known from familiar natural smells, such as floral, fruity, leafy or woody; to tactile impressions, such as soft, velvety and rich; and to the gustative vocabulary, such as sweet, mild and sour.
Some terms refer to **fantasy** perfumes, whose structures have become known as classic reference points, such as **chypre**, based on the contrast between →bergamot and →oakmoss, or **fougère**, a bouquet built around →lavender, →vanilla and coumarine, or Eau de Cologne, a mixture of citrus

oils, such as →lemon oil, →bergamot oil, →mandarin oil with →neroli oil and orangeflower absolute. **"Oriental"** describes fantasy fragrances built around sweet notes, such as →vanilla or →benzoin, combined with woody notes, such as →patchouli or →sandalwood, and contrasted with the citrus freshness or →bergamot of other citrus oils.

Oenanthic Acid
→Enanthic Acid

Oenothera biennis
→Evening Primrose

Oil Crops
G.: Ölpflanzen;
F.: plantes oléagineuses, les oléagineux

Many plants synthesize, in their seeds and fruits, oils and fats that supply energy from the cotyledons for the developing seedling, during the first phase of growth.
Normally, these cotyledons contain about 10–70% oil, combined with protein as well as carbohydrates, minerals, hormones, →vitamins, etc., that are necessary for seed development.
Today, about 1400 plant species are known to be useful for obtaining →fats and oils. Out of these, only about 20 are being commercially grown and used as sources for vegetable oils. It is not only a prerequisite of an oil crop to have a high percentage of oil in its seeds, but also a high yield of oil (normally 1–4 dt/ha) and an optimal composition or quality of oil. This quality is determined by the fatty acid composition or fatty acid pattern of the triglycerides that are synthesized by the plant. The distribution of the fatty acids is mainly a matter of chainlength of the →fatty acids and their degree of saturation (number of double bonds per molecule).
In the food sector, the physical properties of a vegetable oil, such as its cloud point, m.p., color, rheology, smell, taste and organoleptic behavior, as well as its nutrional value, decide its application, e. g., margarine, cooking oil, shortening, salad oil, etc.
In →oleochemistry, however, chemical parameters such as fatty acid composition, chainlength of the fatty acids, fatty acid composition, and the number and position of functional groups, such as double bond(s), hydroxy- and epoxy-groups, are the dominant criteria.

Most o. grown commercially have been bred, planted and harvested by focusing for generations on the nutrional aspects. During the last few decades, however, a change towards industrial o. and their development has taken place, thereby giving the farmer broader and safer production options and the industry interesting and new raw materials.

Lit.: Röbbelen* (1989)

Oilfield Chemicals

G.: Erdölchemikalien;

F.: produits auxiliaires pour l'industrie pétrochimique

A large volume and variety of chemicals are used in oil- and gas-field activities for drilling, completion and production. These chemicals are inorganic (bentonite, other clays, $BaSO_4$, salts), petrochemicals (surfactants, emulsifiers, synthetic polymers, mineral oils) and products based on RR. The following discussions are limited to the latter.

Drilling

The main function of drilling fluids or drilling muds is to cool and lubricate the drilling bit and to suspend the cuttings on their way up to the surface. Furthermore, they stabilize the drilling hole. For this purpose. the density of the fluid has to be adjusted to overcompensate for the pressure of the surrounding rock formation. To avoid losses of the sometimes expensive drilling mud (fluid loss), filtrate-reducing chemicals are added to form a dense filter cake on the walls of the hole.

There are three groups of drilling fluids:

A.) Water-based muds

They may contain up to 20% emulsified mineral oil. The following products derived from RR are in use:

1.) Water-soluble polymers:

→Carboxymethyl cellulose (CMC) of different m.w. and DS as well as different purity (salt content) and degree of neutralization;

→Hydroxyethyl cellulose (HEC) of different m.w. and DS;

mixed ethers of cellulose, such as carboxymethylated and hydroxy-ethylated cellulose (CMHEC);

→starch, especially in pregelatinized or degraded and substituted (carboxymethylated) form;

→guar gum (galactomannan); →xanthan gum.

These polymers, as such or in mixtures also with synthetic polymers, are added as viscosity-increasing additives that allow the modification of the rheology of the mud. Another important function

is to avoid fluid loss and to stabilize water-sensitive formations (shale stabilizer).

The decision which polymer will be applied depends on the specific properties mentioned above, the thermal stability, salt compatability and price of the polymer. A volume of 11 000 mt of CMC and 2300 mt of HEC were used in 1978 in USA.

2.) Thinners and dispersers.

Under certain conditions, viscosity-reducing products and dispersing agents are necessary. →Lignin sulfonate and →tannin-containing substances (e. g., quebracho, partly in chemically modified form) have such properties.

To avoid nasty flocculation, deflocculants are frequently applied. The chromium salt or the iron-chromium salt of →lignosulfonate are used. Lignosulfonate has also the ability to reduce fluid loss, like the a.m. polymers. They are therefore frequently used in combinations.

3.) Surfactants.

To emulsify (O/W) mineral oil and other hydrophobic material in drilling fluids, nonionic →surfactants, such as →fatty alcohol ethoxylates, are used. →Lignosulfonate is also effective.

Al-stearate (→metallic soaps), medium-chain →fatty alcohols (C_8–C_{10}), and →fatty acid ethanolamides are used as →defoamers. →Fatty alcohol sulfates and →fatty alcohol ethoxylates function as foamers.

4.) Other additives.

→Fatty acids, their esters and triglycerides have lubricating functions. Under high-pressure conditions, →sulfated fats and oils and sulfonated oils show good performance.

Because a drilling hole is sometimes a corrosive environment →corrosion inhibitors have to be added to drilling muds. →Fatty acid amino amides, →imidazoline derivatives and →fatty amines are effective.

B.) Oil based muds

Less frequently and only in very special situations, drilling fluids based on mineral oil are applied. They may contain up to 50% emulsified water. The following products, derived from RR, are in use:

1.) Emulsifiers (W/O) based on →fatty acid amidoamines, →fatty acids and their Ca/Mg salts, →fatty amines and →quaternary ammonium compounds;

2.) clays that are modified by →fatty amines to make them more organophilic as thickening and suspending agents;

3.) →fatty acid esters, based on palm kernel oil, are used as a readily and completely biodegradable

oil-phase instead of mineral oil. This new development renders oil-based muds less harmful to the environment, especially in off-shore activities. Higher efficiency and better performance compensate for the higher price of the product.
C.) Foam-based muds.
Drilling muds based on foam require foamers and polymers. →Fatty alcohol ether sulfates and the a.m. water-soluble polymers are commonly used for this purpose.

Completion and Stimulation
The next step towards production is cementing the drilling hole. A casing is cemented into the drilling hole to be stabilized. The cement slurry usually contains →water-soluble polymers (to avoid fluid loss) and →lignosulfonate (to delay solidification). This procedure is sometimes followed by the application of techniques that open the oil-containing rock formation to create an easier flow of oil or gas. Treatment with acids (acidizing) remove lime and clay deposits. To avoid corrosion, →fatty amines are added.
Another interesting technique is fracturing. Highly viscous, mostly thixotropic slurries are pumped into the hole under extreme high pressure. The formation fractures and becomes more oil-(gas) permeable, thus increasing the productivity of a well. Guar derivatives (→guar gum), such as hydroxypropyl guar (HPG), and xanthan gum are used. The addition of polymer-degrading →enzymes reduces viscosity of the slurry, once fracturing is accomplished.

Production
Corrosion inhibitors are necessary to protect the drilling hole and pipelines during production. Formulations contain →fatty acids, →fatty amines and their derivatives, →fatty acid aminoamides, →imidazoline derivatives and →fatty alcohol ethoxylates.
An interesting application of products derived from RR, is the use of copolymers of behenyl acrylate (→behenyl alcohol). They prevent solidification and precipitation in high paraffin-containing crudes during production, storage and in pipelines (pour point depressants: PPD)

Enhanced oil recovery
Tertiary recovery focuses on the exploitation of wells after primary (natural pressure of the well) and secondary recovery (use of water and gas under pressure) have been accomplished. →Water-soluble polymers are used in polymer flooding. The biopolymer →xanthan gum has proven to have high shear-stability and is insensitive to high electrolyte concentrations at high or low pH.

→Surfactants are used in miscellar flooding to form a microemulsion of the residual oil. They must have high stabilty against hydrolysis. →Sulfonates and (especially in high salt containing formations) →fatty alcohol ether sulfates, →fatty alcohol ether sulfonates and ether carboxylates (→carboxymethylated fatty alcohol ethoxylates) have shown good performance.
Tertiary methods are expensive and are therefore used only in times of high crude oil prices.

Further processing of oil
The oil coming out of the well contains normally a large volume of water or brine. Clarification, flocculation and deemulsification are common measures to purify the oil. Nonionic and anionic surfactants, such as →fatty acid ethoxylates and their derivatives, are used.

Oil spills
Off-shore accidents, e.g., tanker collisions, may cause oil spills. Surfactants of various structure and absorbants that are made hydrophobic by treatment with, e. g., fatty amines are used.

Lit.: Kirk Othmer* (3.) **17**, 143+168
    G.R.Gray, H.C.H.Darley "Composition and Properties of Drilling and Completion Fluids" (5.) Gulf Publishing Comp. Houston Tex.

Oil Palm
Elaeis guineensis Jacq.
Palmae
G.: Ölpalme; F.: palmier à huile

O. is a tree of the palm family and produces palm oil (PO) from its fruit pulp (mesocarp) and palm oil (PO) and palm kernel oil from its seeds (endosperm) (PKO). Besides the →coconut, o. is the other economic source of →lauric oils.
Unlike the → coconut, o. will flower after only 3–5 years and produces its fruits in bunches, containing about 1000 to 4000 egg-shaped fruits, each 3–5 cm in length. A fruit bunch weights about 15–25 kg.
The bunches will have to be cut with sickles and transported within 24 h (otherwise, lipase →enzymes will produce considerable amounts of free fatty acids) to the oil mill where, after a sterilization process, the pulp will be separated from the palm kernels, and the →palm oil will be pressed from the pulp. Kernels will be cracked and shipped to an oil mill, where →palm kernel oil and palm kernel meal will be gained after mechanical pressing. In a palm nursery, hybridization between varieties of *E. guineensis* and the American o., *E. olei-*

fera H.B.K.Cortés, is a common technique for gaining new high-yielding varieties. As for →coconut, yields are highest with dwarf types. The highest yields are reported from the most important producing and exporting country, Malaysia, with 10–12 mt of oil per ha and an average of about 4.5 mt PO and 0.45 mt PKO per ha.

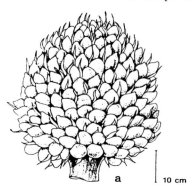

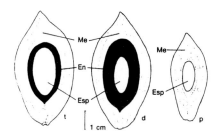

a | 10 cm

b

Oil palm fruit; a: Fruit bunch. b: Cross section of fruits of different oil palm varieties, Tenera (t), Dura (d) and Pisifera (p). Me: Oil containing mesocarp, En: Hard endocarp, Esp: Endosperm (from Rehm and Espig*, 1995, with permission)

Production 1994 ($\times 10^6$ mt)	Palm kernels	Palm kernel oil	Palm oil
Malaysia	2.40	1.05	8.00
Indonesia	1.13	0.50	4.00
Nigeria	0.28	0.14	0.57
Columbia	0.07	0.03	0.04
Ivory Coast	0.07	0.02	0.31
Papua New Guinea	0.07	0.02	0.14
Benin	0.08	0.04	0.04
Venezuela	0.12	0.06	0.02
World	4.59	2.01	14.68

O. is thus the highest-yielding oil crop. Planting density should be 8–9 m, with legume intercropping in the first three years. Pollination is enhanced by the Cameroon weevil. Typical applica-

tions of PO are for cooking oil and margarine, wheras PKO is used in the same applications as →coconut oil. Both oils find wide applications in →oleochemistry.

Lit.: F.D.Gunstone "Palm Oil", Wiley & Sons (1987)

Oils
→Fats and Oils

Oiticica
Licania rigida Benth.
Chrysobalanaceae
G.: Oiticica; F.: oïticica

O. is a large, spreading evergreen tree, growing wild in North-East Brazil. The seeds are contained in a fibrous, hard shell, 3–5 cm in length. They need to be split to gain the long reddish kernel. A tree will produce 150–500 kg of kernels per year, which can be picked from the ground from December to March. Oil content of the seeds is around 60 % (→oiticica oil).

Lit.: Gunstone* (1986)

Oiticica Oil
G.: Oiticica Öl; F.: huile d'oïticica

s.v.: 185–195; i.v.: 230–240

O. is yellowish and lardlike. It is stabilized for export by heating for 30 minutes at 210–229 °C. It derives from the nuts of the Brazilian →oiticica tree and contains 78 % of a highly conjugated unsaturated keto acid (→licanic acid). Other fatty acids include 7 % C16:0 and 5 % C18:0.

Uses in →coatings are the same as for →tung oil. In South America (1981), ≈14000 mt were produced.

Lit.: Ullmann* (5.) **A10**, 232 + **A9**,57
[8016-35-1]

Olea europaea
→Olive

Oleic Acid
Syn.: *cis*-9-Octadecenoic Acid; C18:1 (Δ 9)
G.: Ölsäure; F.: acide oléique

$CH_3-(CH_2)_7-CH=CH-(CH_2)_7-COOH$

m.w.: 282.47; m.p.: 14 °C; b.p.: 223 °C (1.33 kPa/ 10 mm)

O. is a colorless liquid, which oxidizes easily in contact with air and gets darker gradually.

O. is contained in many →fats and oils. It is commercially obtained from tallow by →hydrolysis and →crystallization. Concentrations of up to ≈70% can be obtained with this technology. Such qualities are also called →olein. O. is a key oleochemical for many further reactions.

Oils with high oleic acid content are obtained from →olive, →*Euphorbia lathyris* and high-oleic →sunflower. Simple →hydrolysis yields o. concentrations up to 90%. This new quality o. will bring new aspects to old products and processes (→glyceryl monooleate, →metathesis, →ozonolysis, →epoxidation, →hydroformylation and →hydrocarboxylation).

O. can be transferred by →hydrogenation to →stearic acid. Bromine and other nucleophilic agents can be added to the double bond.

O. is used in mixtures with other fatty acids in the manufacture of →soaps and →metallic soaps.

Nitrous gases transfers o. into →elaidic acid.

Lit.: Kirk Othmer* (4.) **5,** 147
      [112-80-1]

Oleic Acid Amide

→Fatty Acid Amides

Olein

G.: Olein; F.: oléine

O. represents the mono-, di- and triesters of oleic acid with glycerol (→gycerides).

In commercial language, however, o. means the oleic acid qualities (<70%), which contain also →palmitoleic acid, →linoleic acid and some saturated acids. O. derives from tallow by →hydrolysis and subsequent →crystallization.

Oleochemistry

G.: Oleochemie, Fettchemie; F.: oléochimie

The term o. is gaining increasing importance. O. is the industrial chemistry of →fats and oils (→triglycerides). It can be compared with the term "petrochemistry", which is based on crude oil and natural gas.

O. has a long history, is based on RR, uses many modern technologies today (→splitting, →hydrogenation, →sulfonation, →epoxidation, →ozonolysis, →alkoxylation and many others), and comprises a huge number of product groups (→fatty acids, →fatty alcohols, →fatty acid esters, →fatty amines and their derivatives) as well as the wide field of oleochemical-based →surfactants (→fatty alcohol sulfates, →fatty alcohol ethers and their sulfates, →estersulfonates, →alkylpolyglucosides and many others). O. opens many areas of application mainly →detergents, →cosmetics, →food and →plastic additives, →textile and →leather auxiliaries, →mining and →oilfield chemicals, →paper and →coatings (raw materials and additives), →fibers and molded goods.

There is some tendency to intensify o. and to give old empirical experience a modern scientific backing.

Lit.: J. Amer. Oil Chem. Soc. **61,** 271 (1984)

Oleoresins

→Resins, natural
→Essential Oils
→Rosin
→Turpentine Oil

Oleyl Alcohol

Syn.: *cis*-9-Octadecen-1-ol
G.: Oleylalkohol; F.: alcool oleïque

$CH_3-(CH_2)_7-CH=CH-(CH_2)_7-CH_2OH$

m.w.: 268.4 m.p.: −7.5 °C b.p.: 208−210 °C (2 kPa/ 15 mm)

Pure oleyl alcohol is hard to obtain and only of scientific interest. Normally there are mixtures marketed under the term o. that have a broad C-chain distribution and contain no, one, two and more double bonds.

Uses of such o. are in cosmetics, plasticizers for stencil manufacture based on cellulose nitrate, and as a foaming agent.

Lit.: [143-28-2]

Oleyl Amine

Syn.: 1-Octadecenyl-1-amine
G.: Oleylamin; F.: amine oleïque

$CH_3-(CH_2)_7-CH=CH-(CH_2)_8-NH_2$

m.w.: 267.52; m.p.: 21 °C; b.p.: 200−210 °C

O. is a gasoline additive.

Lit.: [112-90-3]

Oligomers

G.: Oligomere; F.: oligomères

Molecules that can form long-chain, high-m.w. polymers are called monomers. They are also able to form o. In o., only a few (2−100) monomer units are linked together. O. are frequent in saccharide chemistry (→oligosaccharides).

Lit.: Encycl.Polm.Sci.Engng.* (2.) **10,** 432

Oligosaccharide Syrups, Branched
→Branched Oligosaccharide Syrups

Oligosylfructoses
Syn.: Coupling Sugars
G.: Oligosylfructosen; F.: oligosylfructoses

O. are mixtures of, preferentially, nonreducing tri-, tetra- and higher saccharides from →fructose, →glucose, →maltose and →sucrose units.
They are prepared by the action of cyclodextrin-glucosyl-transferase from different *Bacillus* sp. (→cyclodextrins) on mixtures of soluble starch and sucrose. During reaction, short-chain →starch hydrolysis products are transfered with their reducing end group to the C4 of the sucrose-AGU.
The crude reaction product is isolated from the unreacted components, glucose, sucrose, maltooligosaccharides, by chromatographic separation. The yield of o. is 53–67%.
Pure o. are nonreducing, r.s. 60–80, thermostable, meltable and there is no tendency for browning reactions. The competitive inhibition of glucan formation from sucrose in the oral region by *Streptococcus mutans* in the presence of o. is the reason for its application as a noncariogenic sweetener in Japan since 1980.

Lit.: Sh.Suzuki: starch/stärke **30**, 217–223,(1978)
       Ruttloff* p. 591, 622–623 (1994)

Olive
Olea europaea L.
Oleaceae
G.: Olivenbaum; F.: olivier

Through the centuries, the o. tree has been a unique attribute of the Mediterranean culture and landscape. Because of its long life and hardiness and its adaptation to areas where it is often only possible to grow tree crops, different cultivation techniques, beliefs and traditions have evolved. Of more than 800 million registered olive trees worldwide, about 97% are located in the Mediterranean region.
The o. tree is a →perennial evergreen with a height of 12–15 m. Its →inflorescences are formed in the axils of leaves on the previous year's growth. The flowers are small, regular and yellowish white in color with a tendency for self-pollination. The fruit is a fleshy, elliptical drupe with a stone, which at maturity is brownish-black in color. Depending upon the cultivar, the fruit contains 15–35% oil at maturity, which is reached in the Mediterranean region in late September or October. This depends

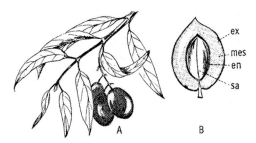

Olive; A: Branch with fruits. B: Longitudinal section of fruit (ex: Exocarp, mes: Mesocarp, en: Endocarp, sa: Seed) (from Franke*, 1981, with permission).

on the use of the olives. Table o. will be harvested earlier in a green state, whereas oil o. are not harvested until February of the following year when a maximum oil content is reached. An o. tree yields on average 60–65 kg of fruit.
The trees are normally propagated by cuttings or by graftings. Normally, it takes 3 to 4 years to grow an o. tree from a grafted seedling. After 8–10 years, the first olives may be harvested, with the highest yields between 60–100 years. For more than 6000 years, o. oil has been gained by crushing the olives, pressing the crushed material and decanting the oil after allowing the material to settle. Oil yields of 15–35% are normally obtained with modern crushing equipment. For a conventional o. yard, 40–60 liters of oil are calculated per 100 kg of fruit.
Typical qualities of o. oil are so-called "virgin oils", which sometimes have high acidity levels due to the mechanical processing of the oil. "Pure olive oil" often is a mixture of virgin oil and refined o. oil. Refining leads to lower acidity levels.

Lit.: G.Brousse, "Olive" in Röbbelen* 462–474, (1989)

Olive Oil
Syn.: Sweet Oil
G.: Olivenöl; F.: huile d'olive

s.v.: 185–196; i.v.: 80–88

O. derives from the →olive tree and is a yellowish oil, which gets cloudy below –5 °C and has a characteristic odor. The oil contains:

C16:0	**C18:1**	C18:2
7–16%	**64–86%**	4–15%

It is gained by cold and hot pressing, followed by extraction with hexane. Thus different levels of quality are produced.
The oil is almost completely used as edible oil.

The high oleic acid content seems to be an attractive source for this acid. The high price, however, has hampered such oleochemical uses, especially because high-oleic →sunflower and →*Euphorbia lathyris* promise to be cheaper sources in the future.

Production in 1994 (× 10⁶ mt)	Olive oil
Italy	0.40
Spain	0.59
Greece	0.29
Tunisia	0.10
Turkey	0.09
Syria	0.05
Morocco	0.05
Portugal	0.05
World	1.67

Lit.: Ullmann* (5.) **A10**, 219

Opium

G.: Opium; F.: opium

O. is the air-dried exudate from incised, unripe capsules of →poppy (content ca. 2 mg of o. per capsule). It constitutes approx. 20 alkaloids (ca. 25% opium) but also meconic acid, some lactic and sulfuric acids, sugar, resinous and wax-like substances. →Morphine is the most important alkaloid and occurs to the extend of 10–14%. Other alkaloids include →noscapine (4–8%), →codeine (0.8–2.5%), papaverine (0.5–2.5%) and →thebaine (0.5–2%). O. is used largely for the manufacture of morphine, codeine and other opium alkaloids. It acts as an hypnotic and narcotic analgesic. Annual world production is ca. 2000 mt (import to Germany is ca. 14 mt).

Lit: Kirk Othmer* (4.) **2**, 731
      Hagers Handbuch der pharmazeutischen Praxis **6 a**, 441 Springer Verlag Berlin (1990)
      The Merck Index* (11.) 6809

Opuntia coccinellifera

→Dyes, natural

Oral Hygiene Products

G.: Zahn- und Mundpflegemittel;
F.: produits d'hygiène dentaire

Important →cosmetics for teeth are toothpastes and mouthwashes.

Toothpastes:

T. are used for cleaning teeth and to remove stain and debris from the mouth. In addition, toothpastes provide cleaner, fresh breath. The main ingredients of toothpastes are water, abrasives, such as dicalcium phosphate dihydrate, anhydrous dicalcium phosphate, insoluble sodium metaphosphate, calcium pyrophosphate, silicon dioxide, aluminium hydroxide, calcium pyrophosphate, calcium carbonate, →surfactants, such as the sodium salt of → fatty alcohol sulfates, humectants (e. g., →glycerol, →sorbitol), gelling agents (e. g., →carboxymethyl cellulose, →carrageenan), sweetener and flavor (e. g., →spearmint oil, →menthol). For anticaries and antiplaque, toothpastes contain fluorides, such as sodium fluoride and antimicrobial agents, such as chlorohexidine.

Mouthwashes:

The function of a mouthwash, like that of a dentifrice, is to freshen the mouth and breath by swilling the product around the mouth, followed by expectoration (spitting out). Furthermore, a mouthwash appears to be the best means to apply a therapeutic ingredient to the mouth, gum or teeth. The ingredients of mouthwashes are mainly water, →alcohol, flavor, humectants (e. g., →glycerol, →sorbitol), and →surfactants (e. g., →fatty alcohol polyglycol ethers). As with →toothpastes minor ingredients can include therapeutic agents such as sodium fluoride, antimicrobial agents, such as chlorohexidine, and colors.

Lit.: Ullmann* (5.) **A18**, 209–214
      Kirk Othmer* (4.) **7**, 1023–1030
      Hilda Butler (ed.) "Cosmetics" Poucher's (9.), **3**, 64–90 Chapman & Hall, London/Glasgow/NY/ Tokyo/Melbourne/Madras (1993)

Orange Flower Absolute

→Neroli Oil/Orange Flower Absolute

Orange Oil (sweet)

G.: Orangenöl; F.: essence d'orange

The orange tree (*Citrus aurantium* L. var. *dulcis*; Rutaceae) originated in southwestern China and was brought to Africa and Europe in the early part of the 16th century. From there it travelled with the Spanish and Portuguese explorers to the Americas and is today one of the most widely produced and used →essential oils in the fragrance and flavor industry. Main production areas are: USA (California, Florida, Texas), Brazil, South Africa, Guinea, Italy, Spain, Israel and Cyprus.

The oil is obtained by →expression of the fruit peels and contains a large amount of monoterpenes

(→terpenes), mainly α-limonene and pinene, next to smaller amounts of fatty aldehydes (C_8–C_{10}).

Its color ranges from pale orange-yellow to dark orange or olive orange, occasionally with a brownish tinge. The odor is distinctly reminiscent of freshly scratched orange peel with fresh, sweet, fruity aldehyde notes.

O. is primarily used in flavors, often as concentrated or terpeneless oils to avoid problems such as deterioration and rancidity, poor solubility or bleaching of the flavor due to formation of peroxides from the monoterpenes. Main application areas are candies, soft drinks, ice creams, and pharmaceutical preparations. In perfumery, o. is used in Eau de Colognes as a main contributor to the note, and in smaller amounts in all kinds of perfumes and after-shaves to give lift, freshness and a pleasant light juicy note. Due to its low cost and high volatility, o. also finds use in air fresheners, cleaning products and dishwash fragrances where an impactful clean, fresh note is required.

The distillation of expressed oranges (by-product of the orange juice industry) also produces an essential oil, which is mostly further distilled under vacuum to obtain pure α-limonene, used as starting material for other perfume and flavor ingredients (e.g., synthetic carvone) and as a cheap, pleasant-smelling masking agent in industrial perfumery.

Lit.: Arctander*
　　　The H&R Book*
　　　Gildemeister*

Orange Tree (bitter)

Citrus aurantium L. subsp. *amara* or *Citrus bigaradia* R. Rutaceae

G.: Bitterorangenbaum; F.: bigaradier

Originated in China, this plant is today cultivated in France, Spain, Egypt and Morocco.

Almost all parts of the tree are used to produce →essential oils and →absolutes: the fruit peels are expressed to produce bitter orange oil, the flowers are either subjected to distillation to produce →neroli oil or extracted with solvents to obtain →orange flower absolute. → Petitgrain oil is produced by distillation of twigs and leaves.

Lit.: Arctander*
　　　The H&R Book*
　　　Gildemeister*

Orbignya sp.

→Babassú

Orotic Acid

Syn.: Whey Acid

G.: Orotsäure; F.: acide orotique

m.w.: 156.10; m.p.: 345–346 °C

O. is an intermediate in the biosynthesis of pyrimidine nucleosides and is produced either by a condensation of urea with oxalacetic acid monomethyl ester or microbially with pyrimidine-deficient mutants of *Brevibacterium ammoniagenes* or *Corynebacterium glutamicum* or a *Micrococcus glutamicus* mutant. O. and its esters with choline or its metal salts (Ca, Cr, Mn, Na, Fe, K, Co, Cu, Ni, Mg, Zn, Sn) are used as uricosurics, in dietetics and in geriatrica.

It has been proposed as a feed supplement in combination with methionine to aid the growth of calves.

Lit.: Ullmann* (5.) **A12**, 156
　　　[65-86-1]

Oriental

→Odor Description

Orris

Iris pallida Lam.

Iridaceae

G.: Iris, Schwertlilie; F.: iris

Wild-growing or cultivated in gardens all over Europe, this plant is mainly grown for fragrance and flavor use in Italy (Florence region); its rhizomes are the starting material for →orris oil/absolute. Smaller amounts are produced in Morocco.

O. is a small plant with a single stem that supports a typically shaped, deep-blue to purple flower with white and yellow markings on the inside of the petals.

Lit.: Arctander*
　　　The H&R Book*

Orris Absolute/Oil

G.: Iris Absolut/Öl; F.: absolue/essence d'iris

The rhizomes (subterranean stems) of the →orris flower are the starting material for the most expensive fragrance and flavor raw material. After harvesting, the rhizomes have to be washed, decorticated and stored for three years; fresh material is practically odorless. The material is then pulverized and steam-distilled with an extremely low yield (<0.1%). The resulting →essential oil solidifies in the receiver to a wax-like, cream-colored mass, which explains why this product is frequently termed orris concrete or orris butter. This phenomenon is due to the high content (85–90%) of →myristic acid, which is removed by alkaline washing of the orris oil in ether solution to yield orris absolute.

This water-white or pale-yellow oily liquid has a delicate, floral violet odor with little initial impact but extreme lasting power and volume. Fruity and woody characteristics are also present in the →dry down. Orris absolute displays a beautiful, rich powdery note.

Due to its high price, it is only used in fragrances for high-class perfumes, where its rich, powdery, yet nonsweet notes lend volume and depth to all kinds of perfume. It is occasionally used in fruit flavors due to its violet-raspberry tonalities.

Lit.: Arctander*
    The H&R Book*

Oryza sativa
→Rice

Ouricury Wax
→Waxes

Oxazolines
→Alkyloxazolines

Oxidized Starches
G.: Oxydierte Stärken; F.: amidons oxydés

O. are →modified starches by oxydation of native starches, leading to formation of some aldehydic and/or carboxylic groups at the sites of primary or secondary hydroxylic groups.

Out of the many known oxidizing agents, only sodium hypochlorite and periodic acid have gained industrial significance.

Oxidation by sodium hypochlorite is performed in an aqueous slurry of refined starch milk and sodium hypochlorite. A pH 9.5–10 is maintained during the reaction by addition of sodium hydroxide. The extent of the reaction is governed by the "available chlorine" in relation to the kind and concentration of starch, length of time and temperature. The reaction is terminated by addition of sodium bisulfite to a pH of 4.0. The slurry is then washed, dewatered and dried. The reaction is nonspecific and may go in the following directions (see formula).

Functional properties are: starch granules that are insoluble in cold water; on heating, dissolution to thin-boiling and thin-setting clear pastes (depending on starch kind and degree of oxidation); superior film-forming properties; anionic in character.

Major applications are in the paper industry (→paper additives) for surface sizing, and as →textile auxiliaries (warp sizing and finishing). Minor uses are in laundry finishing and as →adhesive.

The **periodate oxidation** leads to dialdehyde starches by specific action on C_2 and C_3 of the AGU. Continuous production is run in a series of stirred tanks in a heterogeneous system of a neutral starch slurry and sodium periodate. During reaction, the reagent is reduced to iodic acid, which is

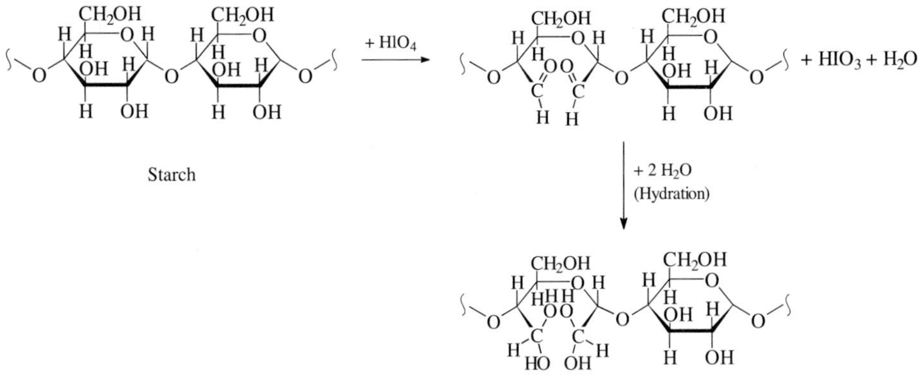

CH$_2$OH CH$_2$OH   CH$_2$OH CH$_2$OH

Starch   + HIO$_4$ →   + HIO$_3$ + H$_2$O

+ 2 H$_2$O
(Hydration)

CH$_2$OH CH$_2$OH

separated, reoxidized in an electrolytic cell and re-cycled to the process. The dialdehyde starch (DAS) is washed and dried in the usual manner.
The dialdehyde groups are highly reactive, leading to dicarboxylic groups by further oxidation, to di-hydroxymethyl starch by reduction, to inter- or intramolecular hemialdalic or hemiacetalic forms. Dried dialdehyde starch is soluble only in boiling water or in solutions of sodium acetate or sodium borate at 60 °C.
Dialdehyde starch by its protein crosslinking activity, is toxic and therefore only usable in nonfood applications: wet-end →paper additives, leading to increased wet and dry tensile strength; seasoning of cotton textiles; tanning agent for skins; hardening for photo-gelatine; and glues.
→Hydrolysis splits the monomeric groups to chemicals, such as glyoxal, erythrose and derived products, for future production of RR. Oxidation to dicarboxylic starches or hydrolysis products will lead to biodegradable cation-binding agents, suitable for heavy-metal binding.

Lit.: Tegge* 168–172 (1988)
      Blanchard* 307–311 (1992)

1,8-Oxido-p-menthol
→Eucalyptol

Oxirane
→Epoxides

4-Oxo-9,11,13-Octadecatrienoic Acid
→Licanic Acid

Oxosynthesis
→Hydroformylation

Ozonization
→Ozonolysis

Ozonolysis
Syn.: Ozonization
G.: Ozonisierung; F.: ozonolyse

O. is the splitting of a double bond by ozone. The reaction was and is used in the laboratory to identify structures of chemical compounds (e. g., structure of natural →rubber by Harries, 1905).
Today, the reaction is used in small-scale pharmaceutical synthesis.
The only large-scale application is the splitting of →oleic acid into →pelargonic and →azelaic acid and of →erucic acid into →pelargonic and →brassylic acid. →Petroselenic acid yields →lauric and adipic acid.

$$CH_3-(CH_2)_x-CH=CH-(CH_2)_y-COOH$$

O$_3$	x= 7; y= 7: oleic acid
	x= 7; y=11: erucic acid
	x=10; y= 4: petroselenic acid

$$CH_3-(CH_2)_x-COOH + HOOC-(CH_2)_y-COOH$$

x= 7: pelargonic acid
x=10: lauric acid
y= 7: azelaic acid
y=11: brassylic acid
y= 4: adipic acid

An ozonide is formed as intermediate during this reaction. The yields are high. There are only few impurities in the finished products.
The economy depends on the costs of electricity for the generation of ozone. The danger of explosions is nowadays under control.
New aspects may arise when high-oleic →sunflower oil is available in sufficient volume and at a competitive price. Higher quality of the final products, improved economy due to only one splitting,

operation and simpler work-up procedures will result.

Lit.: C.G. Goebel et al., US Pat. 2 813 113 (1957) Emery Ind.

R.Criegee, Angew.Chemie **87**, 765 (1975)

P.S.Bailey, "Ozonization in Org. Chemistry" Academic Press NY (1978)

L.H. Princen, J.A. Rothfus, J. Amer. Oil Chem. Soc. **61** (1984) 291

P

PA
→Polyamides

Paint Additives
→Coating Additives

Paint Removers
Syn.: Paint Strippers
G.: Abbeizmittel; F.: décapants

Prior to applying a fresh →coating to a surface, the old paint often has to be removed.
P. are viscous liquids, which are either strongly alkaline, aqueous solutions that contain →methylcellulose as thickener or are solvent-based (e. g., methylene chloride) and neutral. The latter are thickened by →methylcellulose of high DS or by →ethylcellulose.
Lit.: Ullmann* (5.) **A18**, 472

Paints
→Coatings

Palmitic Acid
Syn.: n-Hexadecanoic Acid; Palmitinic Acid; Cetylic Acid; C16:0
G.: Palmitinsäure; F: acide palmitique

$CH_3-(CH_2)_{14}-COOH$

m.w.: 256.4; m.p.: 62.9 °C; b.p.: 210.6 °C (1.33 kPa/10 mm)

P. is a white powder that is insoluble in water and soluble in hot alcohol.
It is derived, together with →stearic and →oleic acid, from most animal and some vegetable fats. P. is also contained in several →waxes, such as →beeswax and →spermaceti.
It is produced by →hydrolysis of →triglycerides and →waxes. Commercial →stearin contains up to 50 % p. →Fractionational distillation gives up to 99 % purity.
Uses are similar to →stearic acid and are focused on →soaps, →lubricants and as intermediate for esters, etc.
Lit.: Ullmann* (5.) **A10**, 245
[57-10-3]

Palmitin
G.: Palmitin; F.: palmitine

P. is a mixture of the glycerol esters (mono, di and tri) of palmitic acid. (→glycerides).
Lit.: [3486-67-7]

Palmitinic Acid
→Palmitic Acid

Palmitoleic Acid
Syn: *cis*-9-Hexadecenoic Acid; C16:1 (Δ 9)
G.: Palmitoleinsäure; F.: acide palmitoléique

$CH_3-(CH_2)_5-CH=CH-(CH_2)_7-COOH$

m.w.: 254.4; m.p.: +1 °C; b.p.: 218 °C (17.6 kPa/14 mm)

P. derives, together with →oleic acid, from →palm oil and soybean oil. Uses are the same as for oleic acid.
Lit.: Ullmann* (5.) **A10**, 245
[373-49-9]

Palmityl Alcohol
→Cetyl Alcohol

1-Palmityl Amine
Syn.: Hexadecyl Amine
G.: Palmitylamin; F.: amine cétylique

$CH_3-(CH_2)_{15}-NH_2$

m.w.: 241.46; m.p.: 46 °C; b.p.: 321.9 °C

P. is normally not used in high purity. It is part of a broad cut. For properties and uses: →fatty amines.
Lit.: [143-27-1]

Palm-Kernel Oil
G.: Palmkernöl; F.: huile de palmiste

s.v.: 242−254; i.v.: 16−19; m.p.: 23−30 °C

The properties and the fatty acid composition are similar to those of →coconut oil:

C8:0	C10:0	**C12:0**	**C14:0**
3−5%	3−7%	**47−52%**	**14−18%**

C16:0	C18:1
6−9%	10−19%

It derives from the kernels of the fruits of the →oilpalm. The oil is recovered by pressing and extraction.

P. is, next to →coconut oil, an important source for medium-chain triglycerides (→fats and oils) and fatty acids.

Top grades of p. are used in the →food industry (couvertures, margarines and confectionary products). In these applications also hydrogenated p. or pure palmkernel stearin or olein (gained by fractional →crystallization of pre-refined p.) are used. The technical uses of the fatty acids are identical to those of →coconut oil.

The estimated world production in 1993 was $1,86 \times 10^6$ mt. Production figures for the important producer countries: →oil palm.

Lit.: Ullmann* (5.) **A10**, 221
[8023-79-8]

Palm-Kernel Oil Fatty Acids
→Palm-Kernel Oil

Palm Oil
G.: Palmöl; F.: huile de palme
s.v.: 195−205; i.v.: 44−58; m.p.: 36−40 °C

Crude p. is dark yellow/red and contains variable amounts of free fatty acid (<5%). The fatty acid composition is:

C14:0	**C16:0**	C18:0	**C18:1**	C18:2
1−3%	**32−45%**	4−6%	**38−53%**	6−12%

P. is gained from the fruits of the →oilpalm tree. Refining processes are manifold, but mainly →distillation (270 °C at 0.5−0.8kPa/4−6 mm) is used. During distillation, neutralization, bleaching (decomposition of colored carotenes) and deodorization are also accomplished. Bleaching earth and oxydative bleaching are used. The oil can be separated in a solid (25%) with a m.p. of 50−52 °C, called palm stearin, and a liquid fraction, called palm olein, by →crystallization techniques.

P. is used predominantly in food. Refined oil and palm stearin are common in the manufacture of margarine and shortenings, while the liquid palm olein is used in frying fats and shortenings.

Technical palm oil fatty acids, obtained by →hydrolysis, contain 50% palmitic and 40% oleic acid.

Technical uses are in →soap production and in metal tempering (→metal processing).

World production was $\approx 14 \times 10^6$ mt in 1993 and is strongly increasing due to well-organized plantations in Malaysia.

For production volumes in the important countries: →oil palm.

Lit.: Ullmann* (5.) **A10**, 218
[8002-75-3]

Palm Oil Fatty Acids
→Palm Oil

Palms
G.: Palmen; F.: palmiers

P. are any species within the order (Arecales or Palmae) of tropical or subtropical →monocotyledonous trees or shrubs that have a woody, usually unbranched trunk and large, evergreen, featherlike or fan-shaped leaves that grow in a bunch at the top. P. are the most important woody →perennials of the →monocots. Flowers are formed in comparatively large →inflorescences, surrounded by spatha. Often the fruits are berries or nuts. Their seeds contain a well-established endosperm and a comparatively small embryo. The palm family is comparatively large, with more than 2000 different species in about 200 genera.

Many palms produce in their fruits high amounts of lauric oils (→fats and oils). Examples are →oil palm, →coconut, →babassú and date. From *Copernicia cerifera* (→carnauba), a wax is obtained. There are also numerous p. that contain sugars (→sugar palms processing).

Lit.: M.J.Balick "Palms, People and Progress" Horizons **3**, 32 (1984)

Panning and Pressing
→Crystallization

Panose
G.: Panose; F.: panose

m.w.: 504 ($C_{18}H_{32}O_{16}$)

P. is formed as one of the reversion products during acid →hydrolysis of →starch, and it is enriched in the mother liquor (→hydrol) of glucose crystallization.

Together with the other reversion saccharides, →isomaltose and →gentiobiose, it lowers the yield on crystallized dextrose. P. is furthermore a transglycosylation product when transglucosidase (α-glucosidase) acts on →maltose and is thus a component of →branched oligosaccharide syrup.

Lit.: Tegge* 68, 262 (1988)

Papain

Syn.: Papayotin
G.: Papain; F.: papaïne

m.w.: 23.400

P. is a proteolytic →enzyme that cleaves proteins nonspecifically and has esterase, amidase and endopeptidase activity; the p. molecule consists of one folded polypeptide chain of 212 residues.

It is a white or greyish-white, slightly hygroscopic powder.

It is produced by drying the juice of the unripe fruits of *Carica papaya* L. (Caricaceae) and is used in food technology to prevent turbidity in beer and fruit juice during chilling; it is used for partial →hydrolysis of →gluten in the food industry (e. g., cheese making or meat processing), as a ripening agent, for the recovery of silver from photographic processes, as a digestive, for tenderizing meat and for bating skins. It is also used for the prevention of adhesions and as an antihelminthic, for the treatment of contact lenses to prolong wearing time in keratoconic patients with papillary conjunctivitis.

Lit.: Whiteaker "Principles of Enzymology for Food Science" 524–530, Marcel Dekker NY (1972)
Ullmann* (5.) **A9**, 396 and **A11**, 578
[9001-73-4]

Papaver somniferum

→Poppy

Papayotin

→Papain

Paper and Pulp

G.: Papier und Zellstoffgewinnung;
F.: papier et pâte chimique

P. are ancient products based almost exclusively on RR. Paper is a sheet material that consists of small bonded fibers, usually and mainly cellulosic in nature. An aqueous suspension of this fibrous ("pulp") material, together with added chemicals (→paper additives), is transferred into a sheet by pouring onto a fine screen and draining the water. Sheet formation is followed by drying and coating

procedures. All these steps are completed today on huge continuous, fast-moving paper machines.

A large volume of water is used in papermaking. However, due to economical and ecological considerations, the neccesary volume has been reduced recently (≈ 1900: 500–1000; 1974: 50 and 1985: 21 mt of fresh water per kg of paper). Many measures are also taken or under way to release water from the pulping and papermaking operations that is ecologically acceptable.

The main raw material for paper is wood (>95%). Other raw materials are or were cotton →linters, →bagasse, →hemp, old rags and, to a large extent, recycled paper. There are some new annual crops in development or use (→kenaf, →hemp,→China grass).

Pulping of wood is done mechanically in a large variety of pulping procedures, which yield papers with a broad spectrum of properties.

Mechanical pulp contains all the by-products (→lignin) and gives a paper that is poor in light stability and strength. Thermomechanical pulping (120 °C), sometimes in combination with bleaching (alkaline H_2O_2), results in a much better paper quality but with some sacrifice in yield (85% instead of 92%).

In **chemical pulping**, the by-products of wood are more or less eliminated. Semichemical processes use neutral sulfite, sodium carbonate and sulfide and sodium hydroxide in various combinations (yields are 70–80%).

Top quality pulp, which is necessary for high-standard paper as well as for chemical derivatization or modification (→cellulose), is produced by two traditional and several modern (pilot) processes:

A: The Kraft (sulfate) process: Wood chips are digested in an aqueous liquor, containing sodium hydroxide and sodium sulfide, at 170 °C for several hours. →Lignin, sodium soaps of →rosin acids and fatty acids (→tall oil) are solubilized and form the "black liquor". The cellulose pulp (yield $\approx 50\%$) can be used as such or is bleached before starting the papermaking process or derivatization.

B: The acid sulfite process is another chemical pulping process. The wood chips are treated at 125–145 °C with SO_2-containing solutions of sodium, calcium or magnesium hydrogensulfite. The by-products of wood are solubilized and separated (→lignin; →lignosulfonate).

Both processes have advantages and disadvantages in raw material selection, in pulp quality, in yields, in possibility to use the by-products and in environmental acceptance. The Kraft process is still the favorite. Almost 100×10^6 mt of pulp were made by this process and only 10×10^6 mt by the sulfite process, which tends to shrink further with time.

C: New processes[1] have been intensively investigated recently to resolve the severe ecological problems inherent in the old technologies. These mostly solvent-based pulping methods are called "organosolv processes" and use ethanol, ethanol/water, methanol, methanol/anthrachinon, acetic acid/ethyl acetate/water or →tetrahydrofurfuryl alcohol for break up the wood structure, yielding good cellulose and partly very pure lignin. The solvents are recycled and pollution of air and water is reduced to a minimum.

Another main source of paper pulp is **recycled paper.**

There are many different types of wastepaper and also many different techniques to make use of it. Pure-grade, unprinted waste paper can be used directly while lower grades must be cleaned by washing or **deinking**, or they can be used for lower-quality paper (paperboard, etc.). The traditional washing process yields a large volume of waste water. The deinking technology is therefore gaining importance.

The waste paper is dispersed in water (12–15%), and the pH is adjusted to 10–11 with sodium hydroxide. Hydrogen peroxide is added to prevent yellowing. The suspension is treated by →flotation. →Soaps and →fatty acids (forming Ca-soaps with the hardness of the water or the fillers) are used as dirt collectors and flotation agents. The resulting pulp is transferred to the paper manufacturing process, while the ink particles are rejected. The use of waste paper is limited because all recycling processes reduce the quality of the fiber, which must be improved by adding fresh, primary fiber.

The production of paper and paperboard in 1993 was 247×10^6 mt worldwide[2]. Roughly a quarter of the fiber material used derives from mechanical or semi-mechanical wood pulp, another quarter from the use of waste paper, while the rest is bleached or unbleached chemical pulp.

Lit.: Encycl.Polym.Sci.Engng.* (2.) **10**, 720
      Ullmann* (5.) **A10**, 600

[1] Tappi J. **74**, 3, 113 and 4, 87 and 5, 191 (1991)
[2] Vital Signs 1994, World Watch Inst. 78, W.W.Norton NY

Paper Additives
G.: Papierhilfsmittel;
F.: auxiliaires de fabrication du papier

Numerous chemicals are added to paper pulp and during paper making. The following is limited to RR-based additives.

P. are divided into two groups: Processing aids (to improve the process technically, economically and ecologically) and functional additives (to enhance the quality of the finished product).

Processing aids:

Retention aids are necessary to coagulate and flocculate the pulp and to avoid losses due to fibers, fillers and additives passing through the mesh of the forming screen. →Cationic starches are the only RR-based polymers used to retain mineral fillers and cellulose-reactive sizes. The market is dominated by polyacrylamide, which is applied as W/O emulsion. The aqueous polymer solution is dispersed in kerosene, the substitution of which by RR-based oils (e.g., rape seed oil) is subject of intensive investigations.

Formation aids improve fiber distribution and sheet formation. →Guar and →locust bean gums can be used.

The same gums are used as **wet-end additives**, which improve the strength of the wet sheet leaving the wire.

→**Defoamers** are based on →fatty acids, →fatty alcohols and →fatty acid esters and are available in paste, solid or liquid form. Fatty acid ethoxylates, fatty alcohol ethoxylates and fatty acid ethanolamides with only a few EO are also used.

Aids for creping paper control the adhesion of the sheet to the creping device. Among others, animal glues (→glue, animal) are used.

Functional additives:

Dry-strength resins are based on polyacrylamide or on natural gums. Again, →guar and →locust bean gums, which are capable of forming hydrogen bonds with the cellulose fiber, are used.

→Starch, cooked in the paper mill or supplied as →pregelatinized starch, is used also. →Modified starches (→oxidized starch, →cationic starch or →starch ethers) are common in that application. Sodium →carboxymethyl cellulose is also effective, but as an anionic polymer, it requires a retention aid such as alum.

The increasing use of waste paper enhances the the application of dry-strength improvers.

Wet-strength resins are mainly based on formaldeyde and epichlorohydrin. The only product derived from natural sources is dialdehyde starch (→oxidized starch). Its use is, however, limited. Another product under consideration is chitosan which is derived from shellfish waste (→chitin and chitosan).

Sizes are additives that decrease the wettability of the paper for certain applications (printing with aqueous inks, manufacturing of milk cartons and paper cups). →Rosin (gum rosin, wood rosin or tall oil rosin) in the form of its sodium or potassium salt or as free acid are used as powder, solution or dispersion. They are excellent sizes, which are used together with alum to develop full sizing power. Reaction products of rosin with maleic anhydride or fumaric acid (fortified sizes) are more effective. They are saponified and used as pastes or dispersions.

Free rosin emulsions are the newest and most effective sizes. For ordinary rosin 15–20 kg are necessary per mt of paper, fortified rosin requires 5–8 kg/mt and free rosin emulsion only 2–4 kg/mt to get the same effect.

Another group of effective sizes include alkyl ketene dimers (→fatty diketenes) and alkenylsuccinic anhydrides. These products react with the OH-groups of the cellulose fiber. While rosin-based sizes can only by used in acid paper making processes, the cellulose-reactive sizes are used in neutral and alkaline paper making. The result is better sheet strength, reduced corrosion on the equipment, less salt in the effluent and better stability of the paper towards aging (archival paper).

Surface treatment

Sometimes it is not possible to add the chemicals to the slurry due to negative effects on the process or final property of the paper. A size press is then the normal equipment for application. It is integrated in the paper machine.

→Starch (in all forms) and natural gums are used to improve strength of printing grades. →Carboxymethyl cellulose is also used, sometimes in combination with →starch derivatives.

For sizing, all products used as functional additives (see above) are applied with chromium complexes of fatty acids.

Paper coatings

A smooth paper surface is necessary for good printabilty. Therefore, coatings are applied in the coater section of the paper machine. Such coatings consist of water, pigments, binders, rheology modifiers and other additives.

The tradional binder is →casein, which is frequently replaced by →starch derivatives, →dex-

trins and soy protein isolates (defatted soy meal is extracted with water, and the protein is precipitated by acids).

→Carboxymethyl cellulose and →sodium alginate of high m.w. are providing the coating rheology. Ca stearate (→metallic soaps) is used as lubricant. →Defoamers are in principle the same as used as processing aids.

Lit.: Kirk Othmer*(3.) **16**, 803
Encycl.Polym.Sci.Engng.* (2.) **10**, 761
Ullmann* (5.) **A10**, 611

Paramorphine

→Thebaine

Parthenium argentatum

→Rubber, natural

Patchouli Oil

G.: Patchouliöl; F.: essence de patchouly

P. is steam-distilled from the dried and fermented leaves of *Pogostemon patchouli*, also known as *Pogostemon cablin* Benth., a small plant from the Labiatae family, originating from the Philippines and Indonesia. Nowadays, p. is produced also in China, Malaysia and Madagascar.

The crude oil ranges from dark orange to brown-red color, due to the presence of traces of iron. Frequently, p. is subjected to further processing (i. e., molecular distillation or washing with acids) to remove these traces and yield a much lighter-colored oil with greater stability. The odor is strong and typically woody, slightly earthy and herbal, accompanied by a sweet, balsamic note of outstanding tenacity.

P. is used in many types of fragrances, both for functional products as well as fine fragrances. Its note blends well with other woody notes, oriental (→odor description) accords, rich florals and all types of men's fragrances.

Due to its interesting odor characteristics and outstanding cost/performance ratio, p. is one of the most widely used →essential oils in fragrances, but its uses in flavors are limited to certain tobacco flavors and specific applications where a "perfumy" flavor is looked for.

Lit: The H&R Book*
Gildemeister*
Arctander*

Peanut

Syn.: Groundnut
Arachis hypogaea L.
Leguminosae
G.: Erdnuβ; F.: arachide

P. is one of the most widespread and probably most important food legumes of the tropics and sub-tropics. It competes as edible oilseed crop with →sesame, →sunflower and →soybean.

It has been identified also as useful for industrial applications. Unique is the maturing of the fruit 2–7 cm under the soil. This is caused by a nee-dlelike tissue (the peg or gynophore), which de-velops and elongates quickly about a week after fertilization by a positive geotropism, thus grow-ing into the soil. The peg protects the maturing ovary.

P. is indigenous in South America and was already utilized in Peru around 1200 B.C. First domestica-tion likely took place in the Chaco region of the valleys of the Paraguay and Parana rivers. The Por-tuguese then spread it to Europe, Asia, Africa and the Pacific islands. Much later, p. was introduced to the USA and the Carribbean Islands.

The whole p. contains 48% oil, 26% protein, 24% carbohydrates and 2.7% minerals. The oil content may be as high as 56%.

Nut yields may be as high as 50 dt/ha. The world average is 11 dt/ha, with the highest average in the USA (28 dt/ha).

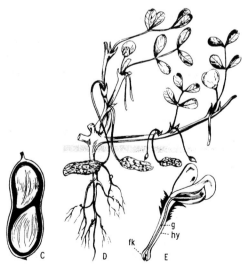

Peanut; C: Longitudinal section of fruit. D: Plant habit with flowers and fruits; E: Longitudinal section of inflorescence (g = Stylus, hy = Hypantium, fk = Ovary) (from Franke*, 1981, with permission).

About 50% of all p. is prepared directly into food products, such as peanut butter and confectionary products. The rest is used for oil production.

Production in 1994 ($\times 10^6$ mt)	Peanuts	Peanut oil
India	8.40	1.94
China	9.68	1.35
USA	1.93	0.15
Indonesia	0.88	0.02
Senegal	0.74	0.10
Burma	0.46	0.10
Sudan	0.39	0.06
Zaire	0.39	0.04
World	26.44	4.12

Lit.: F.A.Coffelt "Peanuts", in Röbbelen* (1989)

Peanut Oil

Syn.: Arachis Oil; Groundnut Oil
G.: Erdnußöl; F.: huile d'arachide

s.v.: 185–196; i.v.: 84–105; m.p.: 2–3 °C

P. is a yellowish, odorless oil with a fatty acid com-position:

C16:0	C18:0	**C18:1**	**C18:2**
6–12%	3–6%	**35–42%**	**39–44%**

C20:0	C22:0
5–8%	3–4%

P. derives from the kernels of the →peanut plant. The oil is expelled by screw press from the kernel. The rest is extracted with hexane. Neutralization, bleaching and deodorization yields the refined oil.

P. is used as salad oil, cooking oil and, in smaller amounts, in margarine. Hardened oil (m.p.: 31–38 °C) is used in shortenings.

Industrial applications are in →soaps and as an ointment base in cosmetics.

World production in 1993 was 3.55×10^6 mt.

Lit.: Ullmann* (5.) **A10**, 230
      [8002-03-7]

Peanut Oil Fatty Acids

→Peanut Oil

Pea Starches

Syn.: Smooth Pea Starch; Wrinkled Pea Starch
G.: Erbsenstärken; F.: amidons de pois

P. are starch granules that are deposited in the coty-ledons of different pea species as reserve polysac-charide, together with oligosaccharides and pro-teins. Industrial or pilot-plant products, obtained by →wet milling and raffination from either *Pisum sativum* convar. *sativum* L. (smooth pea) or *Pisum sativum* convar. *medullae* ALEF (wrinkled pea). The first one contains 47–54% starch, 5–7% oligosaccharides, 21–34% protein and fiber, the second contains 30–37% starch, 25–36% protein, oligosaccharides and fiber.

P., until now still far away from being bulk starch sources, have attained growing interest in research and development in recent years with respect to breeding, cultivation, conversion, characterization and future utilization as RR:

– as a valuable by-product of the isolation of plant protein for novel food products (smooth pea),
– as a starch source with unusual high amylose content (→high-amylose starch) for cultivation under European climatic conditions (wrinkled pea).

Isolation from the ripe seeds is started with steeping, for swelling and softening of the semen, as well as for extraction of solubles; dehulling, wet milling and sifting removes most of the fiber. Further refining by separating the starch from proteins and thoroughly washing is easily accomplished with smooth pea raw materials, but with wrinkled pea, the strong protein-starch complex needs introduction of mechanical energy or slightly alkaline reaction conditions to yield starches with protein contents of <0.5%.

Both kinds of p. belong to the crystalline x-ray C-polymorph type, which is a mixture of the A- and B-unit cells.

They are different with respect to composition and properties:

	Granule diameter, μm	Amylose, %	Crude protein, %	Starch yield %	Extraction coefficient
Smooth p.	20–40	30–40	0.5–0.8	48–50	0.85–0.9
Wrinkled p.	5–30	**60–90**	0.6–1	31	0.89–0.94

The m.w.[1] of the starch components are as follows:

M.w. $\times 10^{-6}$:	Starch	Amylose	Amylopectin
Smooth p.	33.5	5.45	53.8
Wrinkled p.	30.5	2.62	77.9

Smooth p. show distinct onset of swelling between 55° and 70 °C, reaching a consistency maximum at 100 °C at a low level but they have high cooking and mechanical paste stability. At cooling, they set rapidly to rigid white gels. Application was successful in the food industry (pudding mixes, confectionary, instant food after pregelatinization) and in the paper industry. Wrinkled p. begin to gelatinize between 57° and 98 °C without any consistency gain. They remain undissolved by pressure cooking until <140 °C. The promising aspect of

wrinkled p. is its high amylose content, which may make it a substitute for →amylomaize starches in Europe. Containing >80% amylose, it is well suited for the development of thermoplastic →biodegradable materials for packaging, films, fibers and, furthermore, as basic compound for pharmaceutical and agricultural carriers by utilizing its property of forming inclusion complexes with numerous hydrophobic organic agents.

Lit.: F.Schierbaum, starch/stärke **44**, 234–236, (1992)
　　　F.Meuser.et al.: starch/stärke **45**, 56–61, (1995)

[1] Th.Aberle et al.: starch/stärke **44**, 331, (1994)

Pectins

G.: Pektinstoffe; F.: pectines

P. are heteropolysaccharides, consisting predominantly of partially methylated galacturonic acid units. They occur in all higher plants, seaweed and certain freshwater algae as the major structural basis of the primary cell wall and the middle lamellae of young plants. The commercial product, obtained by extraction from the by-products of apple, citrus, beet and sunflower processing, is mainly used as gelling agent for →food additives and to some extent for the pharmaceutical industry.

Preferred raw materials are apple press cake or peels from the extraction of citrus juice (dried apple or citrus pomace). They are extracted with excess acid water (pH 1.5–3, 60–100 °C, 0.5–6 h). The raw extract, containing 0.3–1% p., is desludged by centrifugation and filtration. Insolubles are dried for cattle feed. Prior to precipitation by ethanol (>45% v/v), the extract is concentrated to 3–4%. Following steps are separation and washing of the precipitate, final drying, grinding and sifting.

P. are polymers of D-galacturonic acid and form linear chains of poly-1,4-α-D-galacto-pyranosyluronic acid.

Additional sugars are attached to the galacturonic backbone: L-rhamnose by α-1,2-linkages; β-D-xylose as side chain of O-3; D-galactose and L-rhamnose as long side chains, attached to the rhamnosoyl groups; there are also O-2 and O-3 acetyl groups.

The m.w. exhibits a broad statistical distribution with an average range of 100000.

The carboxyl groups are esterified in part with methanol, and residual carboxyl groups carry cations or protons.

Fragment of a polgalacturonan chain

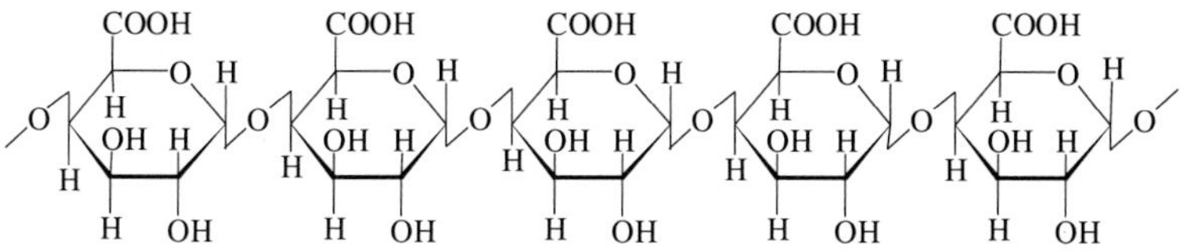

α-D-galactopyranosyl-
uronic acid

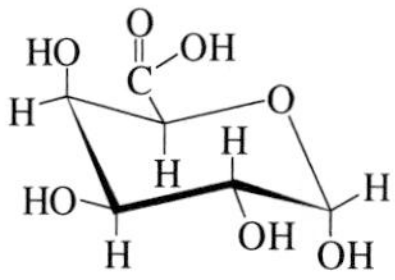

L-rhamnopyranosyl unit

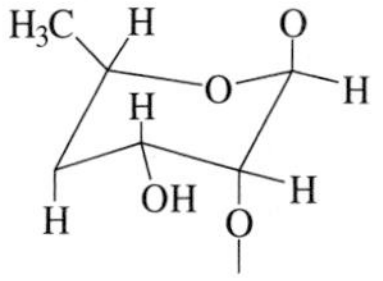

The main types of commercial p. are distinguished by their degree of methoxylation: high-methoxyl p. (HM-p.) has >50%; low-methoxyl p. (LM-p.) has <50%.

The degree of esterification largely influences the functional properties: solubility, gel formation and gel properties.

HM-p. needs high concentrations of sucrose or other polyhydroxy compounds and acidic conditions to form networks of "sugar-acid p. gels" by hydrogen bonds; LM-p. needs Ca-ions to form junction zones between neighboring chains by attraction of the carboxyl groups according to the "egg-box model" (Ca-pectate gel).

P. may be degraded by several →enzymes from higher plants as well as from microbial sources. The latter are responsible for p. degradation in the human large intestines and colon. Therefore, p. is considered to be a dietary fiber with important influence on cholesterol levels, on low-density lipoproteins and on glucose metabolism.

Extended utilization is in the food industry, mainly as gelling agent with high water immobilization activity for fruit preparations, in confectionary and jelly products, in dairy products, and as emulsion and foam stabilizer. Pharmaceutical applications take advantage of the water binding and regulating properties as well as the protecting activity in the gastrointenstinal tract.

Annual world production in 1994 has been estimated at 25000 mt (75% citrus p.). The annual growth rate in food application (1980–1995) on volume basis is +3.6%.

Lit.: Encycl.Polym.Sci.Engng.* (2.) **7**, 602–604,
Ullmann* (5.) **A25**, 24–34,
Intern.Food Ingred. **5**, 55–59. (1992)

Pedilanthus aphyllus
Pedilanthus pavonis
→Candelilla Wax

Pelargonic Acid
Syn: n-Nonanoic Acid; C9:0
G.: Pelargonsäure; F.: acide pélargonique

$CH_3-(CH_2)_7-COOH$

m.w.: 158.2; m.p.: 12 °C; b.p.: 254 °C

P. can be made by oxidation of the corresponding alcohol or aldehyde. Commercial production starts from →oleic or →erucic acid, which are split by →ozonolysis.

Technical grades have up to 94% purity, but 99% is also available.

The acid is used as intermediate in organic synthesis (e.g., for flavor and fragrance raw materials and pharmaceuticals). It is also employed in ore flotation (→mining chemicals) and for the manufacture of hydrotropic salts. An important outlet is the use of p. esters for jet →lubricants.

Lit: Ullmann* (5.) **A10**, 245
[112-05-0]

Pelargonium graveolens
→Geranium Oil

Pencils
→Inks

Penicillins
→6-Amino-penicillanic Acid

Penicine
→6-Amino-penicillanic Acid

Pentosans
→Hemicelluloses

Pentoses
G.: Pentosen; F.: pentoses

P. are reducing sugars, built up of five-carbon systems. The most important members are:

Aldopyranoses (I) and Aldofuranoses (II):

I:

II:

D-Ribose   D-Arabinose   D-Xylose

Ketofuranoses:

D-Ribulose   D-Xylulose

m.w.: 150.13 ($C_5H_{10}O_5$).

P. exhibit the typical chemical properties of other reducing sugars (hexoses): oxidation to the salts of the corresponding acids or their lactones, reduction to polyalcohols (pentites), substitution of the alcoholic groups to esters or ethers, dehydration splitting (water elimination) to furfural, fermentation by microorganisms.
Industrial relevance has only been gained by the xyloses and arabinoses. They are prepared by slight hydrolysis of the corresponding pentosans,

Important physical indices include the following:

	β-L-Arabinose	α-D-Xylose	D-Ribose
d^{20}	1.585	1.525	
m.p. (°C)	160	145	87
Solubility in:			
Water (%)	59.4 (10 °C)	117 (20 °C)	
Ethanol (%)	0.42		
Ether	insoluble	insoluble	insoluble
$[\alpha]_D^{20}$	+190.6° − +104.5°	+92° − +19°	−23.7°
Equilibria of anomeric structure in aqueous solution:			
α-Pyranose	61	35	21.5
β-Pyranose	35	65	28.5
α-Furanose	2	−	6.5
β-Furanose	2	−	13.5

They are not to be found as free sugars but only as monomeric units of polymers (arabans and xylans) in the structure of →hemicelluloses and →pectins, or as constituent parts of enzymes and →saponins (ribose). Furthermore, they occur as intermediates in metabolic cycles (plant photosynthesis).

xylan for →xyloses and araban for →arabinoses, using →hemicelluloses as raw materials. Xylan is found enriched in the wood of angiosperms as well as in the tissues of annual plants (birch wood, →bagasse, corn cobs, →straw). The preferred carrier of arabans is →beet sugar pulp. Crude sugar-containing hydrolysates may be fermented to yield yeasts for the preparation of proteins, fats, amino acids or vitamins.

Enriched solutions are used for →furfural production. Purified arabinose and xylose are hydrogenated to arabitol or xylitol as noncariogenic, low-caloric, insulin-independent sweeteners for confectionary and pharmaceutical uses.

World production, based on data from the industrial suppliers: xylose, 1000 mt/a; xylitol, 3000 mt/a. Future utilization of both sugars is expected to increase as building blocks with industrial application profiles.

Lit.: J.Lehmann, "Chemie der Kohlenhydrate", 279, G.Thieme Verlag, Stuttgart (1976)
Rymon-Lipinski/Schiweck*, 295–311, (1991)
Lichtenthaler*, 208–210, (1991)

Peppermint

Mentha piperita L., *M. arvensis* L.
Labiatae
G.: Pfefferminze; F.: menthe poivrée

Only *M. piperita* produces true →peppermint oil, but it has become common practice to also term *M. arvensis* (cornmint) as p.

M. arvensis originates in China and Japan, which until today are amongst the world's largest producers of mint oil, whereas *M. piperita* is a →hybrid that originated in southern Europe and was first cultivated in Mitcham, England. In the early 19th century, the plant was brought to the USA and cultivated there on a large scale.

The plant is a small leafy, flowering herb with pink/purplish flowers and is normally harvested at the beginning of the blooming season.

Both plants are cultivated on a large scale for the production of →peppermint oil and →menthol.

Lit.: Arctander*
The H&R Book*

Peppermint Oil

G.: Pfefferminzöl; F.: essence de menthe poivrée

P. is steam-distilled from the partially dried leaves and stems of the →peppermint plant and then further fractionated by steam or vacuum distillation to yield a rectified oil that is free from bitter (menthone) or grassy smells and tastes as well as from residues and traces of water.

The resulting oil is water-white or almost colorless and has a cool, fresh, typically minty →topnote, followed by a sweet-balsamic →drydown note.

Main constituents are →menthol, menthyl esters and traces of the bitter-tasting ketone menthone. Today's main producing countries are the USA, Brazil, China, Spain, France, Italy and Japan.

P. is intensively used in flavors, mainly for oral hygiene preparations, chewing gums, candies and cigarette flavoring, where its cooling effect is particularly appreciated. Perfume use is restricted to traces in after-shaves and men's colognes where it adds freshness and lift to citrus, lavender and fougère (→odor description) notes.

Mint oils, i.e., peppermint and cornmint (*Mentha arvensis* L.), which are used as starting material for the production of natural →menthol, are one of the largest and most important essential oils produced today, both in volume and value.

Lit.: Arctander*
Gildemeister*

Perennial

G: Mehrjährig; F: pluriannuel, vivace

P. plants have a life cycle of more than two years. The term is especially used for herbaceous plants that produce flowers and seed from the same root structure year after year. Examples are many ornamentals, such as roses or oleander, p. grasses for meadow and lawn formation, or p. crops, such as →grapes, →palms or hops.

See also: →annual; → biennial.

Lit.: V.B.Youngner, C.M. McKell ed. "Physiological Ecology. The Biology and Utilization of Grasses" Academic Press, NY/London (1972)
P.Raven et al. "Biology of Plants" (5.) Worth Publish. NY (1992)

Perfluoro Alkyl Compounds

G.: Perfluoralkylverbindungen;
F.: composés alkyl-perfluorés

Short- chain fatty acids (C_6–C_{12}) are perfluorinated by electrofluorination (Simons process). All H atoms are substituted by F atoms.

Uses are as special agents for wetting, dispersing, emulsifying and foaming. Main outlet is for anti-soil finishing of textiles.

Perfluoro alcohol is used in ester oils (→lubricants)

Lit.: Ullmann* (5.) **A11,** 371
E.Kissa, "Fluorinated Surfactants" Surfactant Series nr. 50 Marcel Dekker Inc. NY/Basel (1994)

Perfume

G.: Parfüm; F.: parfum

P. is a dilution of a →fragrance in alcohol and water, destined to be applied on the skin.

Depending on the fragrance concentration level in the formula, the product will be called p. or extrait (15–30%), Eau de Toilette (8–15%) or Eau de Cologne (<8%). In men's perfumes, Eau de Toilette and after-shave are produced, after-shave being the equivalent to Cologne in concentration.

Persea americana

→Avocado

Pesticides

Syn.: Crop protecting agents

G.: Pflanzenschutzmittel; F.: pesticides

P. are chemicals that are used to protect plants/crops from rodents (rodenticide), molluscs (molluscicide), insects (→insecticide), fungi (fungicide), bacteria (bactericide), viruses (virucide) or weeds (herbicide).

Lit.: Ullmann* (5.) **A8,** 61

Pests

G.: Pflanzenkrankheiten;

F.: maladies des plantes

Besides wildlife, rodents, slugs and snails, there are several small animals, especially from the classes of the insects and the Arachnida that may cause severe damage to crops.

They include aphids (Homopterae), bugs, beetles and their larvae (Coleopterae), moth and butterfly larvae (Lepidopterae), fly larvae (Dipterae) and wasp larvae (Hymenopterae). Among the Arachnida, the spidermites are most notorious for their damage to plants.

See also: →diseases

Lit.: F.G. Jones, M.G. Jones. "Pests of Field Crops". Edward Arnold. London (1974)

Petitgrain Oil

G.: Petitgrain Öl; F.: essence de petit-grain

Steam distillation of the leaves, twigs and unripe fruits of the →orange tree (bitter) yields p. Most commercially available oils are produced from the bitter-sour variety of this tree, which is extensively cultivated in Paraguay. The oil is therefore often called "Petitgrain Paraguay". The oil obtained from the mediterranean bitter orange tree is called "Petitgrain Bigaradier" (from the french name for this tree).

P. is a pale to dark yellow colored mobile liquid with a fresh, initially slightly harsh bitter-sweet, floral, green odor of low to medium tenacity.

It is extensively used in Eau de Cologne types and all types of fragrances for after-shaves and men's fragrances, where its bitter-fresh, slightly metallic effect successfully enhances the overall freshness. Due to its affordable price, it also finds use in citrus-type fragrances for cosmetics and bodywash products. Traces are used to impart sweetness and naturalness to some fruit flavors and fantasy spice blends.

Lit.: Arctander*
      The H&R Book*

Petroselinic Acid

Syn: *cis*-6-Octadecenoic Acid; C18:1 (Δ 6)

G.: Petroselinsäure; F.: acide pétrosélinique

$$CH_3-(CH_2)_{10}-CH=CH-(CH_2)_4-COOH$$

m.w.: 282.45; m.p.: 30°C; b.p.: 237°C (3.2kPa/24 mm)

P. is contained in the oil of →fennel, →coriander and anis. P. is an isomer of →oleic acid with similar properties. However, derivatives show some interesting differences due to the fact that the location of the double bond is closer to the carboxylic group compared to oleic acid (higher polarity).

→Ozonolysis yields →lauric acid and adipic acid.

Lit.: Ullmann* (5.) **A10,** 245
      J. Amer. Oil Chem. Soc. **68,** 604 (1991)
      [593-39-5]

PGE

→Polyglycerol Esters of Fatty Acids

Phaeophyta

→Marine Algae

→Sodium Alginate

PHB/PHBV

→Poly-3-hydroxybutyric Acid

Phellonic Acid

→Cork

Phloioic Acid

→Cork

Phosphatidyl Choline

→Lecithin

Photosynthesis
Syn.: Assimilation
G.: Photosynthese; F.: photosynthèse

P. describes the production of organic substances, chiefly sugars and starch, from carbon dioxide and water, which occurs in green plant cells that are supplied with enough light to allow chlorophyll to aid in the transformation of radiant energy into the chemical form (see also: →C$_4$-plant):

$$6\ H_2O + 6\ H_2O \xrightarrow{\text{Light energy}} C_6H_{12}O_6 + 6\ O_2$$

Accumulation of fundamental information about how plants grow, improved techniques and instrumentation, recognition of the opportunities in interdisciplinary research and continuing interest in higher yields all enhance the potential for crop improvement through breeding for physiological components of yield. Because of the fundamental relationship between p. and yield, there is considerable interest in enhancing p. capacity through rates and traits, such as leaf area index, leaf angle, leaf orientation and stomatal frequency, which affect light utilization and CO$_2$ entry into the plant.
Cultivar differences in p. rates have been reported in many crops, including →maize, →soybeans, →wheat, →barley, ryegras and − sugar cane.

Lit.: V.B.Younger, C. M. McKell (ed.): "Physiological Ecology. The Biology and Utilization of Grasses." New York/London, Academic Press. (1972).
P.Raven et al. "Biology of Plants" (5.) Worth Publish. NY (1992)

Phytic Acid
Syn.: Myoinositolhexaphosphate
G.: Phytinsäure; F.: acide phytique

m.w.: 660.08

All six OH-groups of inositol

are esterified with phosphoric acid. The viscous, yellowish sirup is soluble in water and contained (as Mg or Ca salt) in many plant parts, especially seeds. It is used as chelating agent for heavy metals, in clarification of red wine and as an anticorrosion agent.

Lit.: Ullmann* (5.) **A11**, 568
[83-86-3]

Phytolacca americana
→Dyes, natural

Phytopathology
Syn.: Plant Pathology
G.: Phytopathologie, Phytomedizin, Pflanzenheilkunde; F: phytothérapie

P. is the biological science that deals with →pests and →diseases of crop plants, their diagnosis, their →taxonomy, as well as how to protect and cure crops.

Lit.: F.G.Jones, M.G.Jones "Pests and Field Crops" Edward Arnold London (1974)

Phytosterols
G.: Phytosterine; F.: phytostérols, phytostérines

Sterols (→steroids) derived from plants are called p. The term is more specificly used for a mixture of sitosterol, camposterol, →stigmasterol and other components, which is obtained, either as such or as their fatty acid esters, from the deodorizer distillates of soybean oil refining.
P. or their ethoxylates are used in skin preparations to reduce irridations and in hair preparations to improve gloss in finishes.

Lit.: R.Wächter, B.A.Salka, A.Magnet, Parfümerie und Kosmetik **75**, 755 (1994)

Pig
Syn.: Hog
G.: Schwein; F.: cochon

Pigs, mainly grown for meat production, also generate RR. →Lard is used as a source for →fats and oils. The skin is used for →leather production and for making high-quality →gelatin. The pancreas is a source of →insulin:

Production in 1994 ($\times 10^6$)	Hog, total slaugther inventory
China	390.0
USA	90.4
FSU-12	59.5
Germany	40.9
Spain	27.1
France	25.2
Danmark	20.4
Poland	19.6
World	864.3

Pimaric Acid
→Rosin

Pine Oil

G.: Pineöl; F.: essence de térébenthine

P. is produced by steam →distillation of the roots, heart- and stump-wood of various pine species, especially *Pinus palustris* and *Pinus ponderosa* (→rosin). The crude oil is then vacuum-distilled and fractionated or steam-distilled under atmospheric pressure to yield p. The lighter fraction is wood →turpentine. P. shows some disinfectant property.

P. is used in perfume compositions for low-cost applications, such as household →cleaners and disinfectants, where its pungent, fresh, camphoraceous and lime-like odor provides excellent masking properties and blends well with pine and citrus notes.

Lit.: Arctander*

Pine Resin
→Rosin

Pinus spp.
→Pine Oil

Pistacia lentiscus
→Resins, natural

Pisum spp.
→Pea Starches

PLA
→Polylactic Acid

Plant Breeding

G.: Pflanzenzüchtung; F.: cultures des plantes
See also: →hybrid, →heritability, →cytology, →polyploidy, →taxonomy.

P. is a method for improving the yield of qualitative and quantitative characteristics of crop plants, such as seed yield, plant height, stability, flower color, content and quality of fatty acids, sugars, starch, proteins or fibers, resistence to pests, diseases, different temperature and water regimes, time of flowering and seed set and terminated growth. P. actually was already developed before Gregor Mendel discovered the 4 principles of hereditary phenomena. The methods were mostly those of mass selection, by selecting the best individuals or eliminating the worst, preferably before flowering, thus reaching well adapted land varieties, e.g., in → cereals. More sophisticated are individual selec-

tion methods, where single plants are separately propagated and newly combined with possible testing of their progenies, thus gaining important information about the →heritability of the tested characteristics of such individuals. It is important to consider the self-(e.g., wheat, barley, oats) or open-pollinated (e.g., corn, rye, rapeseed) mode of reproduction of a crop and the level of polyploidy, as detected by cytological methods (→cytology). This determines if and which emasculation or isolation technique is necessary.

P. methods may also create new variation of characteristics, either by crossing individuals from far distant gene pools or by inducing mutants with nuclear irradiation or with chemomutagens, such as ethylmethanesulfonate or sodium azide. Such methods need to select for comparatively rare appearing positive mutants, e.g., hairless plants, new resistance against mildew or rust →diseases in barley and →wheat, and to find ways of detecting such mutants and combine them into other genetic backgrounds.

Comparatively new are the →hybrid breeding method and →genetic engineering methods. Both may provide more rapid progress than conventional p. methods. While the first provides access to never-before reached yields (e.g., in →corn), the latter enables variation beyond the natural barrier of a →genus or even a →family of plants.

Besides →fertilizers and →pesticides, p. is probably the most important tool for improving the status of agricultural production. Estimates in Germany indicate that, during the last 30 years, annual yield increases of 1–2% were reached in cereals; 45% of these increases are due to the progress by p.

Lit.: F.N.Briggs, P.F.Knowles "Introduction to Plant Breeding" Reinhold Publishing (1967)
P.Raven et al. "Biology of Plants" (5.) Worth Publish. NY (1992)

Plant Diseases
→Phytopathology

Plant Pathology
→Phytopathology

Plastics Additives

G.: Kunststoffhilfsmittel;
F.: additifs pour les matières plastiques

Modern plastics can be adjusted to specific needs by carefully selecting the type of plastic, by modifying their chemical structure, e.g., by use of a comonomer, or by blending. To modify and process plastics, it is necessary to employ additives. This is especially true for PVC, which requires the largest volume and variety of p. Chemicals may be added by the manufacturer of the base polymer or, in many cases, by the plastics processor. Numerous variations and combinations are possible. Many p. are based on petroleum or coal, but also a variety of products are made from RR. The following discussion concentrates on these:

Plasticizers: [1]

Plasticizers are inert, organic compounds (mainly esters) with a low vapor pressure. They increase flexibility, elongation and workability when incorporated into a rigid plastic material. They lower the glass transition temperature (Tg) and the melt viscosity, and they improve the elastic modulus of the polymer. They form a homogeneous mixture with the polymer, which is sometimes accompanied by gelation. Those plasticizers causing gelation are known as primary p.

Used alone, a secondary plasticizer has little plasticizing effect and limited compatibility with the polymer. However, their inclusion enhances the performance of a primary plasticizer.

Special plasticizers need to be selected for plastisols (→coatings). They should, ideally, be a nonsolvent for PVC at room temperature but a good solvent at elevated temperature.

Phthalates, made by →esterification of phthalic acid anhydride with two moles of an alcohol (C_7–C_{13}), are the largest-volume plasticizers for PVC. The alcohols are predominantly of petrochemical origin (e. g., ethylhexanol). However, straight-chain alcohols, such as octanol (→caprylic alcohol) or decanol (→capric alcohol), which can be derived from RR, are also used. Esters of these have better low-temperature performance and a lower viscosity than the branched phthalates. Their lower volatility has led to their application in flexible PVC, especially in the automotive industry, to avoid "fogging", which is the unwanted condensation of plasticizers on the windshield. These properties often compensate for their higher price. Dodecanol is employed in phthalates, where good high-temperature performance is demanded.

Esters of →azelaic acid, →brassylic acid and →sebacic acid are used where good low-temperature performance and low viscosity are required. Again, the alcohols employed are mainly derived from petrochemicals. However, sometimes they are used together with straight-chain alcohols of oleochemical origin.

→Azelaic and →sebacic acids are the basis for some polyester plasticizer, which are made by reaction of these acids with 1,2 propanediol. Their low volatility and extraction resistance compensates for the disadvantages caused by their higher viscosity.

Lower-volume plasticizers based on RR are the esters of →oleic, →stearic, →citric and →ricinoleic acid with mainly butanol.

An interesting class of plasticizers are →epoxides, which are secondary plasticizer and (co)stabilizers simultaneously. They are formed by →epoxidation of unsaturated triglycerides, generally →soybean oil and →linseed oil, or of unsaturated →fatty acid esters (e. g., butyl or ethylhexyl esters of →oleic or →tall oil fatty acids).

They are nonvolatile, resistant to migration and are approved in food packaging (films, bottles). They also stabilize PVC and other chlorinated plastics because they absorb HCl, which is released from the polymer by thermal or light-induced degradation:

$$-CH-CH- + HCl \longrightarrow -CH-CH-$$
$$\underset{O}{\diagdown\diagup} \qquad \qquad \underset{OH \; Cl}{|\quad|}$$

The Western European plasticizer market is estimated to be in a range of 1×10^6 mt (1989). The dominant products are the phthalates (70%), followed by epoxides (5–10%).

There are many specialty plasticizers for polymers other than PVC, but they are of minor importance.

Antioxidants/Stabilizers

Antioxidants/stabilizers are additives that stabilize plastics against aging and degradation during processing and while in practical use. Degradation can be caused by light, heat, mechanical stress, oxidation and hydrolysis.

Numerous chemicals with different structures are in use. In the context of RR, only a few are of interest. There are metal-containing heat stabilizers for PVC and metal-free types. Alkaline or neutral lead stearates (→metallic soaps) are so far the most important. They have a good price/performance relationship and are added at a level of 0.3–5% Mixed metal stabilizers, such as →stearates, →oleates and →laurates of Ba/Cd or Ba/Zn, are also effective. Lower toxicity favors Ca/Zn combinations, but these systems require costabilizers.

An entirely different group are the organotin compounds. These are organotin thioesters or carboxylates, sometimes with →dodecanol or →oleic acid as the ester component.

The costabilizer function of →epoxides has been noted already.

There is a move away from Ba/Cd to organotin compounds or Ca/Zn costabilizer systems in PVC window frames. Pb-stabilizers are still in use, predominantly in PVC pipes and cables. In beverage bottles, approved tin mercaptides or Ca/Zn in combination with epoxidized soybean oil are commonly employed. Liquid stabilizers are dominant in flexible PVC with a trend toward the use of Ba/Zn and, more often, Ca/Zn.

A large volume of stabilizers are sold in so-called "one pack" systems to provide the customer with convenient, nondusting blends, which also contain other p., such as lubricants.

The world market for PVC stabilizers is in the range of 250 000 mt (1990) with lead-based products still representing half.

Lubricants

Many oleochemical compounds are used in this group of p. They function as internal lubricants and reduce the friction between polymer particles during melting and processing by enhancing flow, reducing energy consumption, and increasing output and quality by minimizing the mechanical stress within the melt.

External lubricants lower the friction and adhesion of the melt to metal surfaces of the machinery during processing (extrusion, calendering, molding). They also improve the quality of the surface of the finished product. Rigid PVC contains 1–4% and flexible PVC 0.5–1% of such lubricants.

Mold →release agents, slip agents, antiblocking agents and antislip agents have similar functions and structures.

→Stearic acid is a cheap external lubricant used in extrusion of Pb-stabilized PVC pipes and profiles, while →12-hydroxystearic acid is used in Ba/Cd stabilized profiles.

Oleylamide and erucamide (→ fatty acid amides) are external lubricants and important slip agents in polyolefins.

Ethylenediamine reacts with two moles of stearic acid to yield an excellent external lubricant and antiblocking agent ("amide wax").

→Fatty alcohols are internal lubricants, which are being substituted increasingly by ester lubricants due to their volatility.

→Butyl stearate (→fatty acid esters) is used in semi-rigid or flexible compounds. The esters of adipic and phthalic acid with →stearyl alcohol are

another class of lubricants. More important are the esters of polyols and fatty acids. Partial esters of glycerol (→glycerides) and also of trimethylolpropane and pentaerythrol with →oleic, →12 hydroxystearic, →stearic, →ricinoleic and →keto fatty acids are a large group of internal lubricants, which sometimes have a synergistic effect on the stabilizing system. Wax esters, e.g., cetyl stearate and behenyl behenate have a high stability towards heat and hydrolysis.

A relatively new class are the complex esters based on dicarboxylic acids (e.g., adipic acid), polyols (e.g., pentaerythritol) and fatty acids (e.g., →stearic acid). They are mainly used as external lubricants in PVC calendering. Liquid complex esters are used in the processing of plastisols and flexible PVC.

Metal soaps (mainly Ca, Zn, Pb) are a further group of lubricants. They have multiple uses, ranging from external and internal lubrication to stabilization.

The volume of lubricants used worldwide is ≈ 130 000–140 000 mt/a (1992), with RR-based products representing 60–70%.

→Antistatic agents.

Plastic surfaces can become electrostatically charged, resulting in dirt pick-up, adhesion problems, spark-induced explosions and electric shock to persons walking on plastic floors. Surface-active agents, used as p. (internal antistatic agents) or sprayed on the surface (external antistatic agents), prevent such difficulties. The following products are in use:

Cationic: →quaternary ammonium compounds.

Anionic : alkyl sulfonates and phosphonates.

Nonionic: →Fatty amine ethoxylates or →fatty alcohol ethoxylates, →fatty acid amides, esters of fatty acids with →glycerol, polyethyleneglycol and →sorbitol.

They are applied in many polymers (polyolefins, polystyrene, PVC, and acrylonitrile/butadiene/styrene copolymers ("ABS")).

Worldwide 10 000 mt (1992) are used with ethoxylated fatty amines being the leader.

Lit.: Encycl.Polym.Sci.Engng.* (2.) **1**, 472
Ullmann* (5.) **A20**, 439 + 459

[1] Kirk Othmer* (3.) **18**, 111

Plasticizers

→Plastics Additives

Plastisols
→Coatings

Ploidy
→Cytology

Pogostemon cablin
Pogostemon patchouli
→Patchouli Oil

Polianthus tuberosa
→Tuberose

Polishes
G.: Polituren; F.: cirages, polissure, vernis

Polishes produce or restore a glossy or smooth protective finish on surfaces. There are mainly five areas where p. are in use: Furniture, floors, shoes, automobiles and metals, and there are three princi-

The US production of floor p. alone is ≈ 320 000 mt/a.

Lit.: A.Davidson, B.M.Milwidsky "Polishes" C.R.C. Press, Cleveland Ohio (1968)
M.G.Halpern "Polishes and Waxy Compositions", 82, Noyes Data Corp.(1982)
Kirk Othmer* (3.) **18,** 321
Encycl.Polym.Sci. Engng.* (2.) **7,** 247

Polyalkoxycarboxylates
→Carboxymethylated Fatty Alcohol Ethoxylates

Polyamides
Syn.: PA
G.: Polyamide; F: polyamides

P. are polycondensates that contain an amide group $R^1-CO-NH-R^2$, where R^1 and R^2 are $-CH_2$-blocks, which connect to the next amide groups.
P. are formed either by reaction of a diamine with a dicarboxylic acid: (Type I)

$$HOOC-(CH_2)_x-COOH + H_2N-(CH_2)_y-NH_2 + HOOC-(CH_2)_x-COOH + H_2N-(CH_2)_y-NH_2$$
$$\downarrow$$
$$-CO-(CH_2)_x-CO-NH-(CH_2)_y-NH-CO-(CH_2)_x-CO-NH-(CH_2)-NH-$$

or by reaction of an amino acid (or its lactam) with itself: (Type II)

$$H_2N-(CH_2)_x-COOH + H_2N-(CH_2)_x-COOH + H_2N-(CH_2)_x-COOH$$
$$\downarrow$$
$$-HN-(CH_2)_x-CO-NH-(CH_2)_x-CO-NH-(CH_2)_x-CO-$$

ple forms of products: Solvent-based creams or pastes, emulsions, and liquid p.
The following RR-based raw materials are used for formulations:
→Waxes, such as →carnauba and →candelilla, which impart high gloss, especially to wood floors; →turpentine is frequently used as the solvent of choice, especially because of its clean, typical p.-odor; →fatty acids are reacted with basic nigrosine dyes in dispersions for shoe p. The *in situ* produced salts of →fatty acids with ammonia and morpholine are emulsifiers for p. offered in emulsion form. Other emulsifiers, such as →sorbitan esters of fatty acids are, applied frequently. Small amounts of nonionic or anionic →surfactants are used in some p. as leveling agent.
As a polymeric binder in liquid p., →shellac was used, and →rosin, modified with maleic acid, is (among many synthetic polymers) used today.

More than 90% of all polyamides are based on short-chain $-CH_2$-blocks where x and y are 6 (polyamide 66/nylon 66). Some p.-types are, however, formed by condensation of longer-chain dibasic acids or diamines or amino acids and are of interest in context with RR.
Longer chains make the p. more hydrophobic and lower melting.
The following products should be mentioned:
A. Polyamide 11:
This p. is based on →undecylenic acid, which derives from →castor oil and represents a p. of Type II. Undecylenic acid is reacted with HBr in the presence of peroxides or air in, e.g., toluene at ≈ 30°C. 11-Bromo-undecanoic acid is formed in over 95% yield. Further reaction with 25% aqueous ammonia gives the desired 11-amino-undecanoic acid. Repeated crystallization enhances the purity of the monomer. The polycondensation is

carried out in a column with three zones: melting, main reaction and equilibration of m.w. Phosphorous and phosphoric acids and their salts act as catalysts. The resulting p. (m.p.: $\approx 190\,°C$) is very hydrophobic, has a low moisture content and is therefore used in electrical insulation. Other applications are as filament, powder →coatings and →molded goods. Castor oil consumption for this area was 35 000 mt (1988).

B. Polyamides with longer-chain dibasic acids.
There are a number of p.(type I) based on longer-chain dicarboxylic acids derived from RR.
Hexamethylenediamine with →suberic acid forms polyamide 6,8; →azeleic acid forms polyamide 6,9; and →sebacic acid polyamide 6,10.
The polyamides 6,13 or 13,13 are based on →brassylic acid and hexamethylenediamine or the diamine derived from this acid.
All these p., based on longer chains than C_6, have higher tensile strength and elongation, better toughness, hydrophobicity and electrical properties than normal C_6-polyamides. There are also mixed polyamides, where longer-chain acids are used together with adipic acid.
Commercial importance of these polymers is rather low and limited to special applications.

C. Dimer acid-based polyamides
P. containing →dimer acid are of great importance in speciality applications.
They are formed by a condensation together with a diamine and may contain other dibasic and monobasic acids as modifiers. Reactive types result if polyamines (e. g., diethylenetriamine, triethylenetetramine) are used instead of a linear diamine (e. g., hexa- methylenediamine).
Polycondensation is carried out in a stainless-steel reactor. Heat is applied after mixing to promote the endothermic dehydration and formation of a polymer melt. The m.w. and the desired properties of the p. are controlled by the kind and ratio of the base reactants, the use of coreactants (monobasic acids, amines, different kind of difunctional monomers, etc.) and by the heating rate. The diamines are normally used in excess to avoid undesired branching and are stripped off after the reaction.
The product is cooled for finishing and transferred in the final form for delivery: pellets, extruded ropes, crushed resins, viscous liquids or solutions.
P. based on dimer acids can be easily tailored to specific needs.
Dimer-based p. are used as →hot-melt adhesives in product assembly with excellent adhesion to hy-

drophobic surfaces (plastics, greasy metals) but also to paper and wood. They are used in the shoe and automotive industries, in cable joints and other elctrical applications. The higher price, compared to normal hot melts, limits the application to the solution of special problems.
Another large field of application is in flexographic and gravure →inks on flexible, nonabsorbent surfaces, such as films and foils. They are of low m.w. and applied in (alcoholic) solution, as high solid or as aqueous dispersion.
Reactive p. with amine values of 350–450 (polyamino amides) are used as curing agents in combination with epoxy resins (diglycidyl ether of bisphenol A) in →adhesives, castings and →coatings. Casting technology is mainly used in the electrical industry.
The total market is split into 35% reactive p., used as curing agent for epoxies, 30% in inks and 35% in hot-melt applications.
Production (1987) was ≈ 27000 mt with an annual increase of 5%.

Lit.: Kirk Othmer* (3.) **7**, 776
Encycl.Polym.Sci.Engng.* (2.) **11**, 476

Polydextroses
Syn.: Glucose Polymers
G.: Polydextrosen; F.: polydextroses

These artifical low-polymeric carbohydrates are produced by a polycondensation process of →dextrose in the presence of small amounts of →sorbitol and →citric acid under vacuum at high temperatures. Randomly branched chains of glucose polymers result with >90% α-1,6 bonds (some sorbitol- and citric acid monoester groups) and minor amounts of glucose (<4%), free sorbitol (<2%), citric acid, levoglucosan, →hydroxymethylfurfural.
m.w.: 162–18 000 with an average distribution: 88.7%, 162–5000; 5%, 5000–10000; 1.2%, 10000–16000; residue, >18 000. This corresponds with a DS of 1–110.
P. is a white or creamy colored, hygroscopic powder (4% water). m.p.: 130 °C. It is nonsweet, acidic (pH 2.5–3.5), noncariogenic and suited for diabetics. The physiological energy content is 4–12 kJ/g (different test methods). Digestion in the small intestine occurs randomly.
P. gives texture, consistency and water regulation to filled dessert, instant products, fruit prepara-

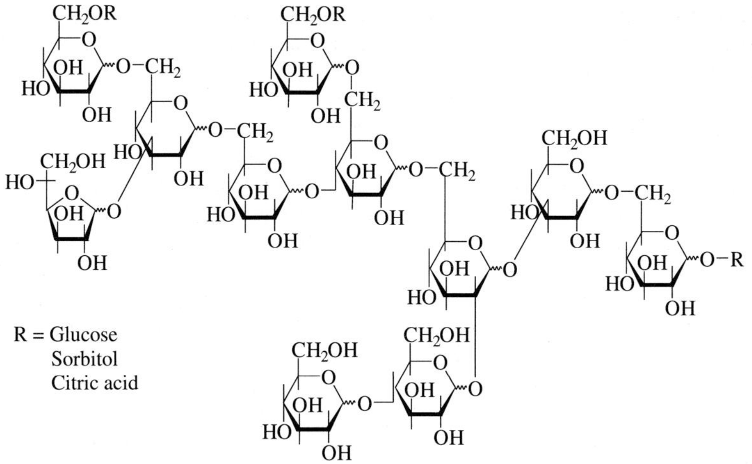

R = Glucose
  Sorbitol
  Citric acid

tions, ice cream, chocolates, and baking goods. It is a low-energy sugar and fat replacement.

Lit.: Rymon-Lipinski/Schiweck* 256–258, (1991)
      S.Radosta et al.: starch/stärke **44**, 150–153 (1992)

Polyesters

G.: Polyester; F.: polyesters

→Alkyd resins are the most important p. based on RR. The esters of →azeleic acid, →brassylic acid and → sebacic acid with polyols are other commercial p. used in hydrophobic molded goods. →Ester oils are used as →plastics additives and in →lubricants.

Polyetherpolyols

G.: Polyetherpolyole; F.: polyétherpolyols

P. carry hydroxylic groups to be reacted with di/tri-isocyanates for producing →polyurethanes, such as thermoplasts, →coatings, elastomers and foams. Traditional starting materials for p. are glycol and the RR-based →glycerol, →sucrose, →dextrose, →sorbitol, →molasses, xylitol, →mannitol, and products of →hydrogenolysis of sucrose.

P. are prepared by →alkoxylation of the starting polyols with ethylene oxide or propylene oxide. The length of the polyether chains is regulated by the amount of alkylene oxide and the reaction time (acidifying). The kind of isocyanate and the structure of the p. command the properties and the uses of the resulting polyurethanes:

diol long chains	thermoplastics, →coatings,
triol long branches	crosslinked elastomers coatings
polyol short	branches rigid foams
triol long branches	flexible foams.

Based on an annual world production of polyurethanes of 4.6×10^6 mt (1991), the possible demand of RR could be 88 000–100 000 mt/a.

Lit.: Eggersdorfer* 383–388. (1992)

Polyfructoses

→Fructans

Polyglycerol

G.: Polyglycerin; F.: polyglycérine

P. is always the by-product of →glycerol manufacturing due to self-condensation whereby several glycerol molecules are linked together under elimination of water (etherlinkages). Color problems and the lack of constant quality are, however, the reason not to use this by-product but to make p. in a separate process (alkali-catalyzed condensation of glycerol). Isolation and purification are accomplished by ketal formation with acetone. The properties and uses are similar to glycerol. →Polyglycerol esters of fatty acids find use as emulsifiers in cosmetics and food. Esters of p. with polyhydroxy stearic acid are modern solubilizer in oily components of skin preparations.

Lit.: Ullmann* (5.) **A12**, 487

Polyglycerol Esters of Fatty Acids

Syn.: PGE

G.: Fettsäurepolyglyceride;

F.: esters polyglycériques d'acides gras

$$R^1-CO-[O-CH_2-CH-CH_2]_n-O-R^2$$
$$O-R^2$$

R^1: C$_{11-17}$ alkyl
R^2: H or $-CO-R^1$
n: 2–10

s.v.: 115–135; m.p.: 48–52 °C

The formula shown above is schematic because the chemical structure is very complex and depends on the degree of condensation of the glycerol and the number, purity and kind of fatty acid. Free fatty acids, polyglycerol and, depending on the kind of manufacture, also mono- and di→glycerides are always present.

P. may be plastic, waxy or rather brittle and hard in consistency, and color may vary from white to brown, reflecting the kind and purity of the starting materials, the ratio of the reaction components and the manufacturing procedure.

P. are made by →esterification of →polyglycerol with →fatty acids or by →transesterification of →fats and oils with polyglycerol.

Uses are as emulsifier in margarine, desserts and cakes. The main application is, however, in coatings and glazes and icings.

Almost 2000 mt were used in USA in 1981.

Lit.: N.J.Krog "Food Emulsions" K.Larsson, S.E.Friberg (ed.) Marcel Dekker NY./Basel (1990)
G.Schuster "Emulgatoren in Lebensmittel", Springer Verlag Berlin/Heidelberg/NY/Tokyo (1985)

Poly-3-hydroxybutyric Acid

Syn.: PHB; PHBV

G.: Poly-3-hydroxybuttersäure;

F.: acide poly-3-hydroxybutyrique

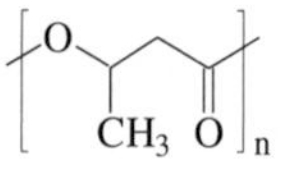

m.w.: $5 \times 10^5 - 2 \times 10^6$; m.p.: 157–188 °C

This →biopolymer is produced by *Alcaligenes eutrophus* mutants from →sucrose or →glucose (yield 75–90%).

The polyester is deposited within the bacterial cell membranes as a storage polymer. The highly crystalline granules (80–90% d.s.) are isolated from the biomass by rupturing the cell walls with steam or extraction, aided by enzymes and/or detergents. The polymer is precipitated and further processed

to granulates, fine powders or emulsions of technical or pharmaceutical purity. The residual biomass is utilized as fertilizer.

P. is stiffer and more brittle than polypropylene and is usually used for similar applications. This drawback is eliminated by using β-hydroxyvalerianic acid as a comonomer (PHBV). The m.p. is reduced to 135 °C, and crystallinity is also less pronounced. Both polyesters exhibit outstanding properties: thermoplastic, biocompatible and biodegradable. The latter (composting) needs some weeks or month, in aerobic waste water six to sixty days.

They can be processed on standard equipment for thermoplastic processing. Handling and properties are similar to polypropylene.

Applications have been successful in food, cosmetics, hygienic and agricultural materials, in pharmaceutical packaging and in medical surgery (nails, sutures), and as a matrix for drugs in long-term therapy.

About 100 mt were produced first in 1990; 5000 mt were planned in 1995. Future large-scale production in EC focuses on a sugar consumption of 200 000 mt/a.

Lit.: M.Korn,"Nachwachsende und Bioabbaubare Materialien im Verpackungsbereich" 153–178, Verlag Roman Kovar, München (1993)
Ullmann* (5.) **A13,** 515
Eggersdorfer* 287–289 (1993)

Polyisoprene

→Rubber, natural

Polylactic Acid

Syn.: PLA

G.: Polymilchsäure; F.: acide polylactique

Thermoplastic and biodegradable poly-α-hydroxy acid, produced by a multistep process: fermentation of sugars to →lactic acid, which is catalytically polymerized via lactide to p. in a yield of 98%.

P. is crystalline, thermoplastic and rigid and exhibits high mechanical stability. It can be processed on standard equipment. The products are physiologically compatible and may be resorbed by human body fluids; they are biologically degradable in composting as well as in aerobic waste water treatment.

Utilization is in food, cosmetic and pharmaceutical packaging. At present, it is mainly used for sutures, implants, controlled- release drugs and in microencapsulation.

Annual world production at present is 20 000 mt. Future large-scale applications are estimated at a level of 100 00–130 000 mt/a.

Lit.: M.Korn, "Nachwachsende und Bioabbaubare Materialien im Verpackungsbereich" 178–191, Verlag Roman Kovar, München (1993)

Polymerization Additives

G.: Polymerisationsadditive;
F.: additifs pour la polymérisation

In the emulsion polymerization of vinyl compounds, such as vinyl chloride, vinyl acetate, styrene and of acrylates, such as the esters of acrylic or methacrylic acid and acrylonitrile, as well as of butadiene and isoprene, additives are necessary. Some of them are RR-based.

The most important group of chemicals are **emulsifiers.** Their function is threefold: solubilization of monomers in micelles where the polymerization starts, emulsification of the nonsolubilized monomer, and protection of the latex particles formed against coagulation during and after polymerization.

→Fatty alcohol sulfates and →fatty alcohol ether sulfates are dominating in PVC emulsion polymerization. Other anionic emulsifiers used in specialties are →oleic acid sulfonates, →sulfosuccinates and →fatty alcohol ether phosphates. To a smaller extent, cationic (e. g., laurylpyridinium chloride) and nonionic (→fatty alcohol ethoxylates) surfactants are used as emulsifiers. →Sorbitan fatty acid esters are mainly used for inverse emulsion polymerization. →Alkyl polyglucosides may also find some use in the future.

Soaps of C_{12-18} saturated or monounsaturated fatty acids are used in synthetic rubber polymerization, especially in SBR (styrene butadiene rubber) manufacture. The resulting latex coagulates easily when acidified, leaving the fatty acid in the polymer, which may remain there as active ingredient of the final compound. Good tackiness of the finished rubber is reached with soaps of hydrogenated or disproportionated →rosin acid as emulsifiers.

Another group of additives used in polymerization, especially in suspension polymerization of PVC, are **protective colloids.**
The polymerization takes place in the monomer droplets, which are stabilized in the aqueous phase by water-soluble polymers derived from RR, e.g., cellulose ethers, such as →methylcellulose, →carboxymethyl cellulose, and →hydroxyethyl cellulose. →Starch and →sodium alginate are also used.

There are some other additives of minor importance that are based on RR:
→Peroxides made from fatty acids (e.g., lauryl peroxide) are used as initiators. The vinyl esters or ethers of long-chain fatty acids or alcohols, such as vinyl stearate or lauryl vinyl ether are used in small amounts as comonomer for special polymers. However, even a few long side chains in a polymer have a negative effect on the physical properties of the polymer, such as hardness and tensile strength.

There are no market figures for emulsifiers used in polymerization. A rough estimate can be made, however, because $\approx 1\%$ of emulsifier is generally used in emulsion PVC, other latices and dispersions.

The markets for PVC and elastomers are rather stagnant, while the use of dispersions in →coatings, →adhesives and binding agents is increasing due to the substitution of solvent-based systems.

Lit.: G.Odian "Principles of Polymerization" John Wiley & Sons (1981)
Encycl. Polym. Sci. Engng.* (2.) **6** 1

Polynosic Fibers

→Viscose

Polyoses

→Hemicelluloses

Polyploidy

G.: Polyploidie; F.: polyploïdie
See also: →plant breeding, →cytology

P. describes organisms with more than 2 complete sets of chromosomes. It occurs in evolution as a result of a "mistake" in mitosis (cell division), in which chromosomes divide but the nucleus does not. It happens naturally in plants at higher elevations due to increase in natural radiation. In this way, a cell with twice the usual chromosome number is produced. If such a cell divides properly, it can give rise to a new individual or species (→hybrid). P. can be induced experimentally, especially for the stabilization of interspecific hybrids, with colchicine. Examples of polyploid crops are →wheat, →cotton, →tobacco, →sugarcane, →potato, →safflower, banana. Several degrees of p. exist in wheat:

Diploid wheat has 2 normal sets of chromosomes (2n=14), tetraploid wheat has 4 sets of chromosomes (2n=28), and hexaploid wheat has 6 sets of chromosomes (2n=42).

Lit.: N.W.Simmonds, "Evolution of Crop Plants", Longman London/NY (1976)

P.Raven et al. "Biology of Plants" (5.) Worth Publish. NY (1992)

Polyrotaxanes
→Cyclodextrin Polymers

Polysaccharides
G.: Polysaccharide; F.: polysaccharides

P. are polymers of simple monomeric units, hexoses or pentoses, linked by glucosidic bonds. Monomeric groups may be arranged in long, linear chains (→amylose, →cellulose) or may be branched (→amylopectin, →pectins).

Molecules with a long main chain (backbone) and a few short side chains are called linearly branched p. The majority of all p. are formed by biosynthesis in green plants or in fungi and bacteria. Some important p. are synthesized by animals, such as →chitin, hyaluronic acids or glycogen. P. can be hydrolized (→hydrolysis). P. that yield only one monomer are called homopolysaccharides; if there are two or more different monomers, they are heteropolysaccharides.

P. may be classified into three broad groups according to their most important biological functions :

– Skeletal p., which serve as mechanically rigid building structures, such as →cellulose, →hemicelluloses, cell wall, membrane p., moisture-regulating p., such as →pectins, →agar, →chitin.

– Nutrient p., which act as metabolic reserves for the reproduction processes, such as →starches, glycogen, and →guar gum.

– Protecting p., as exudates of higher plants, such as →gum arabic, →tragacanth, or as exopolysaccharides of bacterial growth, such as →dextran and →xanthan.

Most native p., though hydrophilic in character, do not easily dissolve in cold water. They may take up water and swell to form viscoelastic gels. Dissolution in water needs introduction of thermal or mechanical energy and produce colloidal, highly viscous solutions or dispersion with plastic or pseudo-plastic flow properties. Some of these dispersions, on cooling, set back to gels of different structural types. Solubility is generally increased when the average m.w. is decreased by hydrolysis.

The limit between →oligosaccharides and p. is primarily determined by the molecular properties. Solubility in an 80%(v/v) ethanol-water mixture is frequently used as an empirical characteristic of oligosaccharides.

Versatility of p. in industrial use is highly increased by introduction of neutral, acidic, alkaline substituents, by crosslinking, by oxidation, by copolymerization or by physical modification.

Lit.: W.Burchard "Polysaccharide, Eigenschaften und Nutzung", Springer Verlag, Berlin/Heidelberg/NY/Tokyo (1985)

Ullmann* (5.) **A25**, 21–23.

Polyurethanes
→Polyetherpolyols

Pomace Oil
→Grape Seed

Pomade
G.: Pomade; F.: crème capillaire

P. is a perfume material prepared by a traditional fat extraction method called enfleurage. The extremely labor-intensive and costly process is only applied to certain flowers (e.g., jasmine and tuberose), where it yields sufficiently different results compared to hydrocarbon solvent extraction.

During the enfleurage process, the freshly picked flowers will be laid out on a fatty base (→lard or beef →tallow), spread on a plate, which will absorb the odorous principle present in and exhaled from the flower. After 24 hours, the flowers are removed, and a new batch is sprinkled onto the fat. This process is repeated several times (up to 35) until the fat is saturated. This odor-saturated fat is known as p. and is then further processed to →absolute in the same way as →concrete is treated.

Lit.: Arctander*

Pontianak Copal
→Resins, natural

Poppy
Syn.: Opium Poppy
Papaver somniferum L. ssp. *somniferum*
Papaveraceae
G.: Mohn, Schlafmohn;
F.: pavot somnifère, œillette

P. is an annual crop of various climates grown for →opium, oil, confectionary ingredient and as an ornamental. The most important cultivated variety is

Poppy; Inflorescence (D) with capsule in cross section (E) (pw: swollen placenta) (from Franke*, 1981, with permission).

opium p., grown for both oil and opium. Because opium is only formed in warm temperate climate, this is also where most of the p. seed oil is produced. Opium does not occur in the seeds but is obtained from the unripe capsules after manual scarring.

→Morphine, the principle pharmaceutical, is gained more cheaply from straw after harvesting, especially in the USSR. P. is harvested by combine (→harvesting), and the tiny seeds are expelled and solvent-extracted. The oil content varies between 36 and 50%. Main producing countries are the USSR, China, India, Turkey, Czechoslowakia and several South East Asian countries. For baking and as bird seed, p. is also grown in some European countries and in North America.

Composition of p. fatty acids:

16:0	18:0	18:1	**18:2**	18:3
10%	2%	11%	**72%**	5%

P. seed oil is a typical semi-drying oil (→drying oils), which is also applied in artists' paints, in white paint after boiling of the oil, in soap after →hydrogenation and in oleochemical processing as a source of →linoleic acid.

Lit.: J.Krzymanski, R.Jönsson "Poppy" in Röbbelen* (1989)

Potato

Solanum tuberosum L.
Solanaceae
G.: Kartoffel; F.: pomme de terre

Although p. is native to the Andean mountains in South America, it is not considered a tropical crop. It was introduced to England and Ireland around 1590 by Sir Walter Raleigh, and it took another sixty years until it was grown by Irish farmers as a fruit crop.

From there it spread all over England, and by 1750 also to Prussia. In 1845, it caused the great famine in Ireland, which was effected by the fungal wilting →disease *Phytophthora infestans*. This forced many immigrants to move to the new world. P. is a herbaceous, branched, →annual crop, about 0.5–1 m tall, with fine, fibrous, adventitious roots and swollen stem tubers. Flowers are white, red or purple. The fruit is a small inedible (poisonous) berry. P. is propagated vegetatively by means of tubers (seed potatoes). Chemically, the p. tubers contain ≈80% water, 2% proteins and 17% carbohydrates.

P. is one of the most important vegetable and starch crops in the world today; the tubers are cooked in various ways. It is also used as an industrial feedstock for the manufacture of →starch, spirits and industrial alcohol. Today, in volume and value, the p. exceeds all other crops worldwide, including →wheat.

Potato production, 1993	($\times 10^6$ mt); (FAO data)
FSU	38.0
Poland	36.3
China	35.1
USA	19.0
India	15.7
Germany	12.1
Netherlands	7.7
UK	7.1
World	288.2

Lit.: Purseglove* 560 (1974)

Potato Protein

G.: Karoffelprotein;
F.: protéine de pommes de terre

The soluble and insoluble protein complex of the fresh potato tuber is derived from the concentrated fruit water of → potato starch production.

Fresh potatoes may contain 0.5–1.2% (d.b.) of pure protein in the dissolved state. It makes up about 50% of the entire N-substance. The amino acid content is − in contrast to most plant proteins − highly balanced as to its physiological value for human and animal nutrition (such as egg or fish protein). To utilize this valuable product, as well as for environmental reasons, p. is isolated as a by-product of →potato starch production.

The continuous two stage-process (disintegration of the potatoes and decantation of the rasped potatoes) yields 90% of the fruit water solids at a concentration of 5% d.b. The warmed fruit water is adjusted to pH 5 (isoelectric point) by hydrochloric acid, and the protein is coagulated by steam injection (115 °C, 1.2MPa/12 bar). The coagulate is cooled and dewatered by decanter centrifugation. It may be shipped in that semi-dry state (45% d.s.) or dried to 92% in pneumatic ring dryers or ultra-rotor dryers.

The resulting bright fine or coarse granule powders have the following composition :

Protein	80–85%,
Crude lipids	5–10%,
Carbohydrates	2–4%,
Ash	1–3%,

(Alkaloids 300–1000 ppm).

A yield of 1.3% (13kg/1 mt of fresh potatoes) is realistic.

The decanter effluent, still containing 2% of the potato dry substance, is used as →fertilizer or may be clarified by anaerobic fermentation.

The use as a high-protein feed component takes advantage of the physiologicallly balanced amino acid composition, which makes p. well suited for the production of piglets and calves.

Attemps for upgrading human food (maize) by enrichment with p. were successful (coextrusion).

Lit.: Tegge* 146–148 (1988)
Hutterer, starch/stärke 30, 56 (1978)

Potato Pulp Feed

G.: Kartoffelpülpe; F.: pulpe de pommes de terre

P. is the commercial term for the mainly fibrous and starchy by-product of →potato starch production.

It is the residue of the wet-screening of starch from the rasped potatoes and consists mainly of the fragments of the cell walls, peels and residual starch. After the last washing step of the pulp and admixture of the fine fibers from the refining, a highly hydrated mass with 95% water results, which may be worked up for dry feed. Mechanical dewatering by decanter centrifuges or screw- or roller presses leads to a water reduction to 75%. Further dewatering may be accomplished by admixture of either kieselguhr (70% water) or calcium chloride (60% water) or by treatment with pectolytic →enzymes (65% water). Final drying to maximum 13% water is done in flash or rotary tubular dryers. The composition of dried p. (d. b.):

Starch (cell-included, bound starch)	33–39%,
Crude fibers (cellulosic material)	14%
Cell wall substance (pectic material)	34%
Crude protein	4%
Crude lipids	0.5%
Minerals	2%

About 11–13% of the potato input (d.b.) becomes p., a low-protein, medium-energy feed. Attempts to use the highly fibrous material as bulk fillers for plastics are promising.

Lit.: Tegge* 141–146 (1988) (1988)

Potato Starch Production

G.: Kartoffelstärkegewinnung;
F.: production de fécule de pommes de terre

P. results from processes that lead to the production of starch for food and industrial uses as well as to by-products →potato protein and →potato pulp feed.

Generally, continuous →wet-milling processes with low input of fresh water are applied to farina potatoes as starting material. P. runs in campaigns (4–5 months) from the beginning of harvesting in August, first with a continuous supply of fresh potatoes and later from silos or store-houses.

Process outline: The potatoes, after being cleaned from contaminants, are disintegrated first by rotary rasp mills to open the potato cells and release the starch granules.

Potato cell juice, containing most solubles, is removed by centrifugal decanters and may be further worked up for →potato protein.

Starch granules and pulp (components of the peel and the cells) are separated from each other by washing on jet extractors with recycled water streams.

The starch-free wet pulp may be shipped as such for fodder or worked up for →potato pulp feed.

Crude starch milk is refined by removing residual fine fibers (jet-screening), washing and concentration to pure starch milk ($\approx$420g/liter), which is ready for further processing (to →modified or hydrolyzed starch (→starch hydrolysis products).

In the last washing and concentration step (hydrocyclones or nozzle-disc centrifuges), fresh water (<1 mt/1 mt fresh potatoes) is introduced. The washings of all other steps are recycled and provide the processing water for starch and pulp separation.

For gaining dry starch, the concentrated starch milk is dewatered (40% water) by rotary vacuum filters or channel centrifuges and finally dried to a water content of 20% by flash driers. The yield of starch depends on the initial starch content of the potatoes (17–21% fresh weight).

Recent innovations:

Disintegration by means of novel rotating vacuum mills avoids air input and reduces foaming and discoloration of the smashed potato slurry without use of sulfur dioxide.

Application of the high-pressure disintegration technique shows the same advantages and leads to larger amounts of opened cells (less bound starch in the pulp).

Fresh water input is reduced to 0.3–0.5 mt/1 mt of fresh potatoes by employment of two-stage, high-powered centrifugal decanters in the separation of >90% of the fruit water in the initial step.

Bound starch may be reduced from 31 to 23%, the starch yield is enhanced from 96% to 98%, energy input is reduced from 5.7 kWh to 4.9 kWh/mt of potatoes and preparation of →potato protein from the concentrated fruit water becomes economic.

The annual world production of potato starch is $>2 \times 10^6$ mt.

The seasonal character and the geographical location in Europe restrict the prospects of growth. Nevertheless, the unique →starch properties and composition, as well as better environmental performance of manufacture, are responsible for novel growth of more than 1×10^6 mt in 1992 in the E.C. P. is the typical technology for manufacture of starch from other root and tuber crops, such as tapioca (→cassava), which takes the second place in world production (3.5×10^6 mt), arrowroot, yam and sweet potato (the latter is produced at 100 000 mt in Japan).

Lit.: F.Meuser et al., starch/stärke **36, 73**–77 (1984)
Tegge* 140–152 (1988)

Pregelatinized Starches

Syn.: Starch Pregelatinized; Cold Swelling Starches
G.: Quellstärken;
F.: amidon gonflant, amidon pré-gélatinisé

P. are physically →modified starches, obtained by drying a →starch paste, with or without the admixture of chemical reagents, additives or fillers and having the property of marked swelling on contact with cold water or giving a colloidal dispersion.

Generally, p. exhibit lower water-binding power (8–20 w/w) and peak viscosity in cold water, as compared with their parent starches, after thermal gelatinization and cooling (20–100 w/w). P. are prepared for consumers' convenience to avoid thermal gelatinization before application.

Cold water-swelling may be attained by several techniques, but only thermo-mechanical treatment by steam-heated roller-dryers as well as →extrusion cooking have gained industrial significance. Dry milling in ball mills or vibratory ball mills are techniques for pilot plant or laboratory use.

The most applied technique is to spread starch slurries, (e.g., 40 parts starch, 60 parts water) on single or double drum roller-dryers (surface temperature 120–160 °C, speed of rotation 4–10 per min). The starch is quickly cooked, forms a starch paste and, on further rotation, is dried to a film of variable thickness, which is finally cut from the surface and exposed to grinding and sifting.

The functional properties of the dried flake products depend largely on the feed, which can be varied from a slurry of nonpregelatinized starch to a thoroughly cooked starch paste, on the parameters of the rolls and drums (speed, temperature, gap between the drums) and on the particle size, which is controlled by grinding and sifting. No severe chemical changes occur when pure starch-water systems are treated.

Additives (alkali, mineral salts, surfactants, chemically modifying or hydrolyzing agents) may be introduced before and during the run to influence the gelatinization behavior during processing as well as the functional properties in applications. Frequently, nonreactive additives are introduced, such as colors, sweeteners, nutrients and aroma-giving compounds for production of convenience food intermediates, as well as clay and other minerals, pigment colors, saw-dust and carbon for nonfood applications.

Industrial nonfood applications are as foundry core binders, as stabilizers for oil-well drilling muds, as components in glue mixtures, in the processing of paper and cardboard, as carrier for water in soluble colors and printing inks.

Applications in food industries are for ready-to-prepare, not-to-be-cooked or convenience foods, such as sauces, puddings, custards, filling creams, flour blends for cake and cookies.

Spray-drying of pasted starch slurries is another possibility for preparation of p., but it is applied to a much lesser extent because of higher energy consumption.

Lit.: T.Galliard (ed.) "Starch: Properties and Potential" 99–100 John Wiley & Sons, Chichester/NY/Brisbane/Toronto/Singapore, (1987)
Tegge* 166–167 (1988)

Pregnenolone
→Steroids

Pressing
→Expelling

Primrose

→Evening Primrose

Printing Inks

→Inks

Progesterone

→Steroids

Propanetriol (1,2,3-)

→Glycerol

Propellants

→Explosives

Propoxylation

→Alkoxylation

Propylene Glycol Esters of Fatty Acids

Syn.: Fatty Acid Propyleneglycol Esters

G.: Propylenglycolfettsäureester;

F.: mono/diesters de propylèneglycol et d'acides gras

$$CH_2-O-CO-R^1$$
$$CH-O-R^2$$
$$CH_3$$

R^1: C_{11-17} alkyl

R^2: H or $-CO-R^1$

s.v.: 170–185; m.p.: 36–38 °C

P. may be a mixture of 55% mono- and 45% di-ester or, if molecularly distilled, 90% monoester. Properties, such as consistency, color and odor, depend on the kind and quality of fatty acids used. P. are rather stable against temperature and →hydrolysis.

P. are made by →esterification of propylene glycol with fatty acids (200 °C, vacuum) or by their propoxylation (→alkoxylation)

They are used in cake shortenings, fats for whippable emulsions, toppings, glazes and icings. US consumption in 1981 was 6500 mt.

Lit.: N.J.Krog "Food Emulsions" K.Larsson, S.E.Friberg (ed.) Marcel Dekker NY/Basel (1990)
G.Schuster, "Emulgatoren in Lebensmittel", Springer Verlag Berlin/Heidelberg/NY/Tokyo (1985)

Prostacyclin

Syn.: Prostaglandin I_2; Prostaglandin X

G.: Prostacyclin; F.: prostacycline

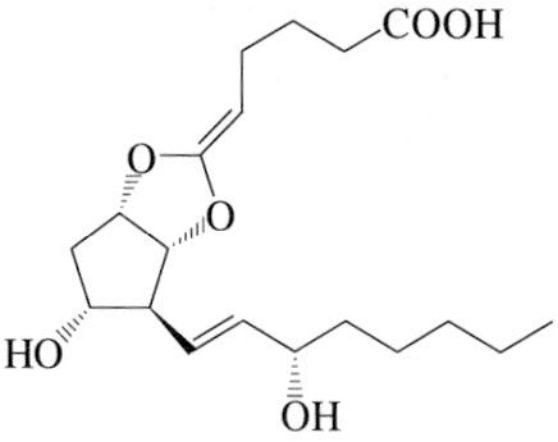

m.w.: 352.48

P. is →prostaglandin that is produced by enzymatic transformation of prostaglandin endoperoxides and which dilates blood vessels. P. is chemically unstable in aqueous solutions, but the mono sodium-salt is stable as a solid and in solution. In addition, p. shows antimetastatic effects and has been used for patients with acute myocardial infarction. It has also been postulated that p. acts to stimulate platelet adenylate cyclase and to prevent the action of thrombi in phospholipid breakdown as well as platelet aggregation. In clinical practice, it is used for preventing the loss of platelets in cases where the blood comes into extracorporal contact with artificial surfaces, i.e., in hemodialysis, cardiopulmonary bypass, and charcoal column perfusion for treatment of liver failure.

Lit.: J. R. Vane, S. Bergstrom "Prostacyclin", Raven Press, New York, 1979
Ullmann* (5.) **A5**, 301
[35121-78-9]
[61849-14-7] mono sodium salt

Prostaglandins

G.: Prostaglandine; F.: prostaglandines

The p., leucotrienes and thromboxanes are endopensus compounds with hormone-like activities. They are a family of biologically potent lipid acids that were first discovered in seminal fluid and extracts of accessory genital glands of man and sheep. The single nonmammalian source of p. intermediates is the gorgonian sea whip or sea fan (*Plexaura homomalla*). P. are divided into types A, B, C, D, E, and F, based on the functions in the cyclopentane moiety.

P. have the prostanoic acid skeleton in common:

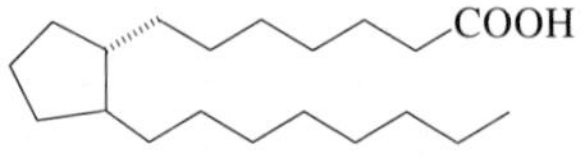

They share →arachidonic acid as their common precursor. Biological activities include stimulation of smooth muscle, dilation of small arteries, bronchial dilation, lowering of blood pressure, induction of labor, abortion and menstruation. They are implicated also in inflammatory reactions, kidney function and in automatic neurotransmission. The economically most important applications are the treatment of cardiovascular and gastrointestinal diseases and of patients with chronic arthritis.

Lit.: S. M. Karim "Prostaglandins" (3 vols), Park Press, Baltimore, 1975, 1976, 1976
Ullmann* (5.) **A13**, 102 + **A22,** 261

Protein/Fatty Acid Condensates
→Fatty Acid/Protein Condensates

Proteins
G.: Proteine, Eiweiße; F.: protéines

P. are polymers that are fundamental to life (reproduction, growth, movement, metabolism and sensory perception). Various →amino acids (≈ 20) are the building blocks and are linked together by amide bonds, which join the amino group of one amino acid with the carboxylic group of the next. The kind, the number and the sequence of the various amino acids lead to a myriad of theoretical and practical combinations. There are estimates that 10^{14} different p. exist in our world.

Chains with <100 amino acids are called (poly)-peptides; larger molecules are p., which form also higher spatial structures (e. g., helix).

There are globular p., such as albumin, globulin, histones and protamines, as well as fibrillar p. (sclero-p.), such as collagens, elastins and keratins. Some p. are combined with lipids (lipo-p.), with sugars (glyco-p.) or with phosphorous compounds (phospho-p.)

P. are important in all physiological functions and for nutrition. Plants are able to produce p. to satisfy their own demands. Animals need supplementary amino acids. Sources of p. for food and feed as well as technical applications are meat, →fish, →soybean, gluten (→gluten cereal), →potato p., pea (→pea starches).

With respect to RR, the following items are important:

P. are widely used in clothing (→wool, →silk →leather). Collagen, the p. of animal skins and bones, is the raw material for making →gelatin and animal glue (→glue, animal). →Casein is used in →adhesives. P. have important functions as →enzymes for biotechnological processes, in →detergents, in food processing and as pharma-

ceuticals. P. hydrolysates are used in hair and skin cosmetics. P. hydrolysates are condensed with fatty acids (→fatty acid/protein condensates) to yield →surfactants.

Lit.: Kirk Othmer* (3.) **19**, 314
Encycl.Polym.Sci.Engng.* (2.) **13**, 531

PS
→Sorbitan Esters of Fatty Acids

Pseudocereals
→Cereals

Pullulan
G.: Pullulan; F.: pullulane
m.w.: $5 \times 10^4 – 2.5 \times 10^6$

P. is a →microbial gum that consists of 1,6-linked maltotriose units. P. is produced by submerged fermentation with the fungus *Pullularia pullulans* and →starch, →sucrose or monosaccharides as medium. After →fermentation the mixture is diluted, the microorganisms are filtered off, the filtrate is decolorized and concentrated, and p. is precipitated by addition of organic solvents (e. g., methanol, ethanol).

P. is a white, tasteless and odorless powder that forms viscous solutions in water. It is a polymer, which can be formed to molded goods, fibers and films and can be combined with polypropylene and cellophane to form physically very stable fibers. In food and drug technology, it is used as an edible airtight wrapping or coating. Because it is not degraded in the gastrointestinal tract, it is used in low-calory food. Technically, p. and its derivatives are used as adhesives and flocculants.

P. has been produced at a level of 12 mt/a. Its high price has made p. a specialty product.

Lit.: Encycl.Polym.Sci.Engng.* (2.) **13**, 650
[9057-02-7]

Pulping
→Paper

Purple Fleabean
→Vernonia

Purple, Tyrian
→Dyes, natural

Putties

Syn.: Lutes; Sealants
G.: Kitte; F.: mastics

P. are products with an →adhesive function but are used predominantly to fill and seal gaps. The traditional product is glazier's putty, which is based on →drying oils and is a mixture of 450 parts whiting and 40–80 parts of (boiled) →linseed oil.

Modern reactive sealants are mainly made from synthetic products (silicone, polyurethane, polyacrylics and polysulfides). There are, however, a series of specialty products based on →resins (natural), wax, rubber, glue and starch.

Lit.: Kirk Othmer* (3.) **20**, 554

Pyrethroids

→Pyrethrum

Pyrethrum

Syn.: Pyrethroids (incl. synthetic p.)
G.: Pyrethrum; F.: pyrèthre

P. is gained mainly from the →pyrethrum plant (*Chrysanthemum cinerariifolium* and *C. coccineum*) and is a powerful →insecticide based on RR.

The following six chemicals and structures are the active ingredients:

	R^1	R^2		
	CH$_3$–	–CH=CH$_2$	Pyrethrin I	[121–21-1]
	CH$_3$-COO–	–CH=CH$_2$	Pyrethrin II	[121–29-9]
	CH$_3$–	–CH$_3$	Cinerin I	[25402–06-6]
	CH$_3$-COO–	–CH$_3$	Cinerin II	[121–20-0]
	CH$_3$–	–C$_2$H$_5$	Jasmolin I	[4466–14-2]
	CH$_3$–	–C$_2$H$_5$	Jasmolin II	[1172–63-0]

P. is one of the oldest insecticides. Because it is highly active and quite safe for humans and animals, extensive effort was made during the period 1910–1945 to determine its chemical structure. These efforts culminated in the discovery and production of some of the most active synthetic insect control agents. The first product, allethrin, which was only partially synthetic, was introduced commercially during the early 1950s. Entirely synthetic products followed later and substituted the natural product to a large extent.

P. is a valuable →insecticide, which causes, frequently in combination with piperonyl butoxide as a synergist, a rapid paralysis ("knock-down") of flying insects. P. is deactivated when exposed to light, thereby loosing its toxic properties. This is highly desirable from a consumer safety point of view and has lead to its broad use as household insecticide (pressurized aerosols containing 0.04–0.25% of active material), in sprays to protect livestock and stored grain and to control insect pests on fruits and vegetables prior to harvest.

In 1979, 23 000 mt of dried flowers were produced. Kenya, Japan and Dalmatia are the main suppliers.

Lit.: J.E.Cassida (ed.) "Pyrethrum the Natural Insecticide", Academic Press N.Y. (1973)
Ullmann* (5.) **A14**, 273
Kirk Othmer* (3.) **13**, 424

Pyrethrum Plant

Syn.: Dalmatian Pyrethrum
Chrysanthemum (or *Pyrethrum*) *coccineum* Willd., *C. cinerariifolium* (Trev.) Vis., *C. roseum* (Adam) M.Bieb.
Compositae
G.: Pyrethrum, Persische bzw. Dalmatinische Insektenblume; F.: pyrèthre rose

P. is a perennial shrub with composite-type yellow flower heads, which reaches about 70 cm in height. *C. cinerariifolium* originated in the Mediterranean region. It is produced in these countries as well as in Kenya and Ecuador. *C. roseum* comes from oriental regions, where it has been known for more than 400 years because of its →insecticidal effects. Sun-dried flower heads contain about 1.3–2.4% pyrethrin. Harvesting is done by picking the flowers, and drying and grinding them.

In former times, the powder was used as insecticide or it was suspended in water and sprayed over the plants to be treated against insects. Today, →pyrethrum insecticides are gained by solvent extraction of the flowers.

Lit.: J.E.Cassida (ed.) "Pyrethrum the Natural Insecticide", Academic Press N.Y. (1973)

Pyrodextrins

→Dextrins

Pyroxilin

→Celluloid

Q

Quaternary Ammonium Compounds

Syn: Quaternary Ammonium Salts
G.: Quartäre Ammoniumverbindungen;
F.: composés d'ammonium quaternaires

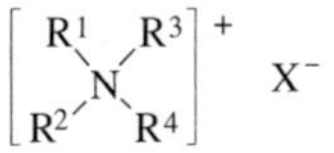

$$\left[\begin{matrix} R^1 & R^3 \\ & N^+ \\ R^2 & R^4 \end{matrix}\right] \quad X^-$$

R^1: >C_{10} alkyl
R^2: >C_{10} alkyl or $-CH_3$, $-CH_2-CH_2-OH$, benzyl
R^3 and R^4: $-CH_3$, $-CH_2-CH_2-OH$
X: Cl^-, OH^-, $CH_3-O-SO_3^-$

Q. are cationic →surfactants and disinfectants. They have a strong affinity to negatively charged surfaces, such as textile fabrics.
Q. are marketed in aqueous or aqueous/alcoholic solutions.
They are produced by alkylation (quaternarization) of tertiary amines with methylchloride, dimethylsulfate and ethylene oxide at 50–100 °C in a polar solvent, such as ethanol, within several hours or much faster under pressure with an excess of the alkylating agent.
The benzalkonium compounds (R^2 = benzyl) are good disinfectants.
The main product used to be ditallow dimethyl ammonium chloride as a fabric →softener in household products and in industrial applications. Because it has shown substantial toxicity to fish, it has been replaced by products with ester groups ("ester quats"), which are readily hydrolyzed in water. They are obtained by alkylation of difatty acid esters of triethanolamine with dimethylsulfate:

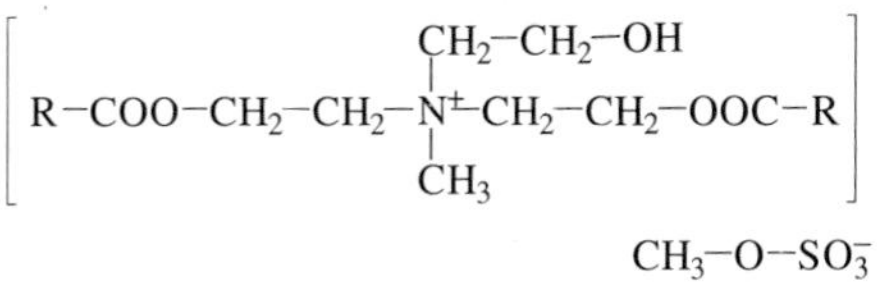

$$\left[\begin{matrix} & CH_2-CH_2-OH \\ & | \\ R-COO-CH_2-CH_2-N^+CH_2-CH_2-OOC-R \\ & | \\ & CH_3 \end{matrix}\right]$$

$$CH_3-O-SO_3^-$$

An estimated 300000 mt of q. were produced worldwide in 1990.

Lit.: Ullmann* (5.) **A25**, 747

Quebracho

→Tannins

Quercus coccifera
Quercus ilex

→Dyes, natural

Quercus suber

→Cork

Quinine

G.: Chinin; F.: quinine

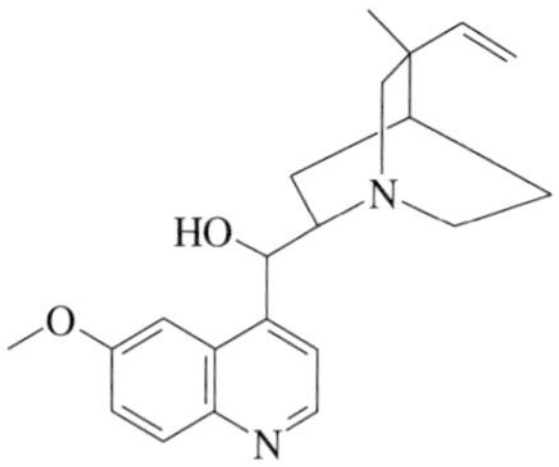

m.w.: 324.41; m.p.: 177 °C

Q. is the most important →alkaloid of the bark of *Cinchona officinalis* L. (Rubiaceae), which contains about 8% of q. Cinchona trees grow wild in South America and are cultivated in Java. The finely powdered cinchona bark is treated with lime solution and 5% NaOH solution, followed by extraction with aromatic hydrocarbons at elevated temperatures. The raw alkaloid is removed from the organic solvent by shaking with dilute sulfuric acid, followed by neutralization and crystallization of the alkaloid.
Q. forms triboluminescent, orthorhombic needles.
Since q. exhibits activity against plasmodium species, it is used in human therapy as an antimalarial (quite often substituted by chloroquine) and as an antipyretic. Main market today is in tonic water manufacturing.

Lit.: Ullmann* (5.) **A1**, 395
      [130-95-0]

R

Ramie

Boehmeria nivea (L.) Gand.
Urticaceae
G.: Ramie, Chinagras; F.: ramie

R. is native to Eastern Asia where it has been cultivated for thousands of years. It is a perennial herb with storage roots and a number of slender stems arising from the rootstock. It is normally propagated vegetatively from rhizomes or less commonly from stem cuttings. Up to four crops may be harvested from each plant per year while a succession of stems is produced from the rootstock. Harvesting may continue for several years until yield and quality of fibers are reduced. Fibers are coated with a pectic, gummy substance, which must be scraped from the unretted bark or fibers by hand or with machines. This substance may not be removed by retting.

R. fiber is one of the best vegetable fibers because of its strength, luster and durability. The strength of the fiber is up to three times that of cotton. It is a strong, fast-drying fiber of high wet-strength and good water absorbance and it can be spun or woven. However, it is costly to extract and clean the fibers, which is the main reason why r. has not achieved the importance among fiber crops that its quality might suggest.

It is used in high-quality paper and fabrics (apparel, seat covers).

Lit.: Martin/Leonhard* (1967)

Rapeseed

Syn.: Canola; HEAR; LEAR
Brassica napus L. ssp. *oleifera* (Metzg.) Sinsk.
Cruciferae
G: Raps; F: colza

R. is native to the southwest European climate. Several centuries ago it appeared for the first time by spontaneous hybridization of turnip rape (*Brassica campestris*) with wild cabbage (*Brassica oleracea*). This means that r. is of amphidiploid nature (→cytology). In North America, r. is also called Canola, which is a quality brand for zero-erucic and/or low-erucic r. varieties, combined with the trait for low-glucosinolate meal. Reducing the glucosinolate content was necessary for a better acceptance of r. meal as animal feed. Such varieties are also named double zero or 00-r. In North America, the variety description for r. also includes →turnip rape.

Among all the world oil crops, r. is in 4th position with regard to oil production. In Middle Europe, especially in Germany and France, r. and other crops from the Cruciferae family [→turnip rape (*Brassica campestris*), →mustard (*Brassica nigra*), oil radish (*Raphanus sativus*)] as well as sugar beets and →potatoes, are considered the most important "leaf crops" for intercropping with the comparatively high percentage of cereal crops ("straw crops") in the rotation. R. is being grown mostly as a winter crop (planted in the fall and harvested in the summer). However, there is also a spring variety for the northern hemisphere where the vegetation period is comparatively short. Of all the world r. production, 62% comes from China, Canada and India.

The primary product of r. is the seed, which contains around 45% oil. R. oil is being produced by mechanical →milling and solvent →extraction of the seeds. The world average seed yield is only 12.8 dt/ha. In European countries, the seed yield is around 30–40 dt/ha. With breeding efforts since about 1975, the comparatively high content of →erucic acid, at 40–60% of the fatty acids (→rapeseed oil), has been reduced to less than 3% (so-called zero-erucic varieties). Erucic acid is said to be harmful in higher concentrations in human nutrition, causing certain liver malfunctions and heart lesions. Today, the application of r. oil for chemical and energy purposes is being practiced again. For this purpose, the breeding goals are especially directed at rising the content of erucic acid and of oleic acid as well as the total oil yield. High erucic acid r. varieties are also abbreviated "HEAR". In normal r. (LEAR, canola or 00-r.), the erucic acid (C22:1) content is <3%, whereas HEAR varieties may reach as much as 58%. In earlier varieties, the C22:1 content used to be only as high as 45%. After quality programs for canola were established in Europe and Canada for low glucosinolate and low C22:1 contents in the 1970's, HEAR was established as an alternative source for technical applications. About 35000 ha/a of HEAR are grown in Europe. It may be complicated to raise the C22:1 content in rapeseed above 66.6% because C22:1 seems to be linked only to the two outer positions (1+3) of the →trigylyceride in cruciferous crops. The only exemption is *Tropaeolum majus* L. (Indian cress) where C22:1 contents of >70% have been observed. With →genetic engineering techniques, the gene for forming such trierucates are sought to be transferred into rapeseed.

In 1995 the first successful crop of 1100 ha of another →genetically engineered r. variety was produced, which synthesizes about 40% of →lauric acid in its seedoil was reported. The gene was derived from the California Bay Laurel.

Production in 1994 ($\times 10^6$ mt)	Rapeseed	Rapeseed oil
China	7.49	2.26
India	5.35	1.61
Canada	7.23	0.98
Germany	2.86	1.11
France	1.83	0.47
UK	1.11	0.46
Poland	0.68	0.24
World	29.63	9.83

Lit.: R.K.Downey, G. Röbbelen "Brassica species" in: Röbbelen* (1989)

Rapeseed Oil

G.: Rapsöl; F.: huile de colza
s.v.: 167–174; i.v.: 97–100

Depending on the intensity of refining, the oil is a brown to yellowish liquid. Fatty acid compositions of the high-erucic acid type HEAR and the low-erucic acid type LEAR are:

	HEAR	LEAR
C16:0	1–5%	1–5%
C18:0	1–4%	1–2%
C18:1	**13–38%**	**50–65%**
C18:2	10–22%	15–30%
C18:3	1–10%	5–13%
C20:1	5–8%	1–3%
C22:1	**40–64%**	**0–2%**

R. derives from various →rapeseed species. Fatty acids are obtained by →hydrolysis. These fatty acids are used as such or are fractionated. R. is mainly used for nutrition.

Technical uses of r. range from →metal-working fluids to →lubricants (biodegradable chain saw oil) and →factice. The oil and fatty acids are used in →alkyd resins. The acid is a starting material for →dimerization.

The oil and the methyl ester are considered as →fuel alternative and diesel substitute. Some applications of the methyl ester as solvent for insecticides, household products and lubricants in outboard engines and construction machinery are considered.

Worldwide, 8.8×10^6 mt were produced in 1992. There has been a strong and constant increase over the past few years.

Lit.: Ullmann* (5.) **A10**, 231
[8002-13-9]

Rapeseed Oil Fatty Acids

→Rapeseed Oil

Rayon

→Viscose

Release Agents

G.: Trennmittel; F.: agents de démoulage

R. are used to reduce the adhesion between surfaces in contact. Adhesion problems of that kind arise in the plastics and rubber processing industry, in glass manufacturing, in metal casting, in woodworking and in food processing.

The following RR-based product groups are of interest:

→Waxes, e. g., ester waxes and ethoxylated wax alcohols, are used in water-based systems. →Metallic soaps are widely used in plastics processing (→plastics additives) and in injection molding of polyurethane foams as internal lubricants and r. →Fats and oils and some of their ester derivatives can be used also if the processing temperature is below 150 °C. They are mainly used in woodworking and in food applications (→food additves).

Lit.: M.McDonald "Release Agents" Noyes Data Corp. New Jersey (1972)
Ullmann* (5.) **A23,** 67

Rendering

→Fats and Oils

Renewable Resources

Syn.: RR
G.: Nachwachsende Rohstoffe;
F.: matières premières renouveables

For details and definition: →preface

Reppe Reaction

→Hydrocarboxylation

Resinoid

G.: Resinoid; F.: résinoïde

A r. is a perfumery and flavor material that is prepared from natural resinous substances by extraction with hydrocarbon solvents, e.g., petroleum ether, benzene or butane. Resinous materials are balsams, gum-resins, natural resins (→resins, natural), oleo-gum resins, etc. R. contain all hydrocarbon-soluble matters from these substances, which are the odorous matters, plus rosin acids, rosin acid anhydrides, sesquiterpenes, plant colors and waxes. They are generally viscous liquids, semi-solid or solid, homogeneous, noncrystalline masses. Like their equivalents, obtained from "live" (cellular) plant material, the →concretes, r. can be further purified to →absolutes.

Lit.: Arctander*

Resins, natural

G.: Naturharze; F.: résines naturelles

Natural resins are various solid or semisolid, amorphous organic substances with a m.w. of ≈ 10 000. They are yellowish to brown, transparent and soluble in organic solvents. Above the glass transition temperature (T_g), they are oily liquids or rubberlike masses; below the T_g, they behave like brittle glass.

Most of them are exudates of terpenoid- (→terpenes) or flavonoid- containing trees. Amber, which is used as a gemstone, is the oldest type. R. have been used for generations in varnishes and are mostly harvested through collection by peasants.

The largest group of r. are →rosin and its derivatives. Rosin is harvested and processed under industrial conditions, especially in the South East of the USA.

Another important r. is →shellac, which is a secretion of an insect.

Lower-volume specialty resins are the following:

Dammar is tapped from trees of the Dipterocarpaceae family in Indonesia (Batavia dammar) or Malaysia (Singapore dammar). There is a freshly tapped white dammar and a recent-fossil brown dammar. Dammar resins have a low a.v. of 17–35, s.v. of 25–60 and m.p. between 80–90 °C for white dammar and 160–180 °C for brown. They are soluble in hydrocarbons and terpenes. Uses are in →coatings because of their high gloss, superior color, high flexibilty and excellent resistance and durability. Sometimes, a small fraction of the resin has to be removed to get spirit solubility. This is done by precipitation of this fraction by adding

ethanol to a hydrocarbon solution of dammar. This "dewaxed dammar" is used as a thin protective, easily renewable topcoat on valuable oil paintings. Production in 1965 was 3000 mt and 600 mt in 1974.

Copal is a collective term for various r. of different geographical and botanical origins:

Manila Copal is exuded from *Dammara orientalis* (Indonesia, Philippines), while **Kauri copal** derives from *Agathis australis*. They are soluble in alcohols and ketones. Copal softens at 90–115 °C and has an a.v. of 110–141. Uses are in →paper coatings, finishes for various substrates, →adhesives and →inks. Dispersions in water, formed by salt formation, are used in floor →polishes. In 1974, 500 mt were produced.

Pontianak is a resin of *Agathis borneensis*. It is semi-fossil and has a higher softening temperature. It is therefore used in specialty coatings.

East India r. is not obtained by tapping but collected by finding. It originates from dammar trees, has a high softening range of 110–130 °C and a low a.v. of 20–30. The r. is soluble in hydrocarbons and used in →coatings, →adhesives and cements.

(Gum) elemi is a sticky, soft exudate of *Canarium luzonicum*, Burseaceae, growing in the Philippines, with a softening point of 80 °C. It contains 20–25% essential oils, which can be isolated and used in pharmaceutical, cosmetic and fragrance formulations. This film-forming, plasticizing r. is used in coatings with excellent adhesion, flexibility, water resistance and gloss.

Sandarac originates from Morocco, North America and Australia and is exuded from *Tetraclinis articulata* (Vahl) Mast (*Callitris quadrivalois* Vent.). It softens at 100–130 °C and has an a.v. of 117–155. Solutions in alcohol have been used for generations as preserving finishes for fine paintings. The r. is a fumigant.

Japanese Lacquer is tapped from the lacquer tree (*Rhus verniciflua* Stokes), which grows in Japan, China and India, and is a viscous, gray to brown milk used to make the well-known Japanese lacquerwork. The latex dries by enzymatic oxidation. Today, the term japanese lacquer is more a quality specification.

Mastic is harvested from *Pistacia lentiscus* in the Greek island of Chios. It is a soft (55 °C) resin, which has been used for hundreds of years in the mediterranean area as a chewing gum to give a fresh breath. Solutions in alcohol or aromatic hydrocarbons are used in wood →coatings, printing →inks and →adhesives (theater make-up).

The modern tendency to use RR-based products has drawn new attention to some of these resins.

Lit.: Kirk Othmer* (3.) **20**, 197
 Encycl.Polym.Sci.Engng.* (2.) **14**, 438
 Ullmann* (5.) **A23**, 73

Respiration

G: Veratmung; F: respiration

R. represents the processes by which a living organism or cell takes up oxygen from the air or water, distributes and utilizes it in oxidation, and releases products of oxidation, especially carbon dioxide.

It generalizes the process of biodegradation of all organic material under aerobic or anaerobic conditions.

$$C_6H_{12}O_6 + 6 O_2 \longrightarrow 6 CO_2 + 6 H_2O$$

The opposite process of r. is assimilation or →photosynthesis.

Lit.: Baldwin, "Dynamic Aspects of Biochemistry" Cambridge University Press (1959)

Retrogradation

Syn.: Set-back
G.: Retrogradation; F.: rétrogradation

R. is the gradual and irreversible insolubilization of an aqueous starch dispersion, obtained by gelatinization of starch or its derivatives, with the formation of a precipitate or a gel, depending on concentration.

R. is a molecular association process of fundamental importance for starch functional behavior. Linear α-1,4-glucosides are the active components of the →starch polysaccharides when undergoing r. The process is understood as an alignment of linear chains of DP 10–12 by formation of parallel, left-handed double stranded helices, which partly form the crystalline unit cells (→starch granules), stabilized by hydrogen bonds. In →amylose or →starch pastes the B-polymorph is mostly found. When r. takes place at temperatures of >50 °C, the A-polymorph will be formed; V-crystalline forms result in the presence of complexing agents. R. depends on the chainlength and advances with maximum velocity and stability if DP amounts to 80–200. Such retrograded aqueous systems are resistant to enzymic (amylolytic) actions; they need temperatures of >150 °C to become resolubilized. Towards lower DP, the gels formed become thermally reversible at temperatures of <100 °C (→maltodextrins, →dextrins). The short chains of →amylopectin retrograde slowly at high concentrations.

At low temperatures, they are easily dissolved by heating. High-molecular amyloses of DP >1200, don't set back because of molecular entanglements and thermal motion of the long chains, which cause difficulties in them lining up for molecular association. R. is strongly restricted to neutral and slightly acidic pH of the system. It is inhibited in the alkaline milieu of pH >10; retrograded systems may be redissolved by alkali or other hydrogen bond-splitting agents. Introduction of substituents (→starch derivatives, →modified starches), oxidation (→oxydized starches), thermal treatment (→dextrins), hydrolysis (→maltodextrins) gradually or entirely inhibit the processes of r. R. is a highly starch-specific property with positive as well as negative consequences for application.

Prominent examples for the practical action of r. are:

skin forming on starch pastes upon cooling, formation of insoluble, enzymatically undegradable components, viscosity enhancement of pastes, set back to gels or puddings, binding and surface coating, weeping (syneresis) of starchy systems as well as freeze-thaw instability, formation of aggregated and stabilizing systems, staling of wheat bakery goods.

Lit.: Van Beynum/Roels* 39–42 (1985)
 T.Galliard (ed.) "Starch: Properties and Potential" 120 John Wiley & Sons, Chichester/NY/Brisbane/Toronto/Singapore, (1987)

Rhodophyta
→Marine Algae

Rhus succedanea
→Waxes

Rhus verniciflua
→ Resins, natural

Ribes nigrum
→Black Currant

Riboflavine
→Vitamin B_2

D-Ribose
→Pentoses

Rice

Oryza sativa L.
Gramineae, Poaceae
G.: Reis; F.: riz

R. is an annual →cereal that has been cultivated for several thousand years as the principal food crop of Asia. R. is thought to have been domesticated in India more than 4000 years ago from the wild species *O. perennis*. African rice (*O. glaberrima* Steud.) is the only other cultivated species of the genus.

Paddy or swamp r. accounts by far for the major part of world r. production. The traditional method of growing paddy r. is to transplant seedlings, which are raised for 5–7 weeks in a nursery, after reaching a height of 20–30 cm. In industrialized countries, this method has been changed to presoaking the seeds until radiculas have been developed after a few days and then sowing the seeds by commercial methods, such as sowing machines, helicopters or airplanes onto the flooded (1–5 cm) fields.

Soft or "glutinous" r. contains mostly →amylopectin, giving it a gelatinous and sticky appearance after overcooking. Hard r. consists of →amylose and →amylopectin (1:3), giving it a glass-like structure, and which does not become sticky after overcooking. Hard r. is grown more extensively and is much more important in world trade than soft r.

The flower morphology is that of a panicle (→inflorescence), like in →oats. Besides the many applications of r. in the kitchen, it also is used for noodle and malt production. R. seed contains 77% →starch, 8% protein, 0.5% fiber, 1.5% oil and 13% water. Polishing of r. results in r. bran, which contains several useful products: →r. bran oil, r. wax and protein meal.

It can be used as raw material for making →furfural. World production of r. in 1994 was 359×10^6 mt.

Lit.: Rehm/Espig*

Rice Bran Oil

G.: Reisschalenöl; F.: huile de riz

s.v.: 183–194; i.v.: 125–144

R. is a by-product of rice processing. The bran is extracted with hexane to yield 8–16% of an oil with a high nutritional value and the following composition:

C16:0	**C18:1**	C18:2
13–18%	**44%**	30–40%

In rice-growing countries, r. is also used for technical purposes.

Lit.: Ullmann* (5.) **A10**, 226
[68553-81-1]

Rice Bran Wax

→Waxes

Ricinenic Acid

Syn.: Dehydrated Castor Acid; 9,11,(12)-Octadecadienoic Acid; C18:2 (Δ 9,11/12)
G.: Ricinensäure,
F.: acide gras du ricin

m.w.: 280.44; m.p.: 53–55 °C

R. is predominantly a mixture of the conjugated 9,11-linoleic acid (30–40%) $CH_3-(CH_2)_5-(CH=CH)_2-(CH_2)_7-COOH$ and nonconjugated →9,12-linoleic acid (50%).

R. is produced from →ricinoleic acid by →dehydration. The reaction is carried out at 200–320 °C without or with a catalyst (e. g., Al_2O_3). This goes together with an increase of the *trans* isomer content (→isomerization). The degree of →conjugation can be increased.

R. is used in →alkyd resins.

Lit.: Ullmann* (5.) **A10**, 270
[61789-45-5]

Ricinoleic Acid

Syn.: 12-Hydroxy-, *cis*-9-Octadecenoic Acid
G.: Rizinolsäure; F.: acide ricinoléique

$$CH_3-(CH_2)_5-CH-CH_2-CH=CH-(CH_2)_7-COOH$$
$$|$$
$$OH$$

m.w.: 298.45; m.p.: 4 °C b.p.: 215 °C (2 kPa/ 15 mm)

R. is a yellow, viscous liquid, derived from →castor oil, which contains 80% of this acid, by →hydrolysis.

R. is a versatile starting material for making →soaps, →turkey red oil, →textile auxiliaries, →sebacic acid, →ricinenic acid and →undecylenic acid. Esters are used in →plastics additives.→Hydrogenation yields →12-hydroxystearic acid. High pressure →hydrogenation of the methyl ester is a possibility to produce ricinoleyl alcohol [540-11-4].

Lit.: Kirk Othmer* (4.) **5**, 155
[141-22-0]

Ricinoleyl Alcohol
→Ricinoleic Acid

Ricinus communis
→Castor

Rosa centifolia
Rosa damascena
→Rose

Rose
Rosa centifolia L., *R. damascena* Mill.
Rosaceae
G.: Rose; F.: rose

Two species of r. are cultivated for the production of flavor and fragrance materials:
a) *Rosa centifolia* (so-called "Rose de Mai"), cultivated in France, Algeria, Morocco and Egypt. This pink, typically rosebud-shaped flower is used for the production of →rose de mai concrete/absolute.
b) *Rosa damascena,* cultivated in Bulgaria, Turkey, Morocco, India and China, produces both →rose oil and absolute.
The flower itself is rather large, multi-petalled when open and pink in color.

Lit.: The H&R Book*
Arctander*

Rose de Mai Concrete/Absolute
G.: Rose de Mai Concrete/Absolut;
F.: rose de mai concrète/absolue

Produced by volatile solvent extraction from rose petals (*Rosa centifolia*), yielding first r. →concrete, then, after extraction with alcohol, r. →absolute (Yield from fresh plant material is 0.25% for concrete, 60–70% of which is obtained as absolute).
R. absolute is an orange-yellow to orange-brown viscous liquid with a long-lasting, deep-rich rosy odor and honey-like, spicy tonalities.
Due to its high price, it is only used in fragrances for high-class perfumes, where it is used to increase the floral character and is frequently combined with →jasmine absolute and →rose oil.
In flavors, it is used in traces to round off and enhance fruit flavors and tobacco flavorings.

Lit.: Arctander*
The H&R Book*
Gildemeister*

Rose Oil/Rose Absolute
G.: Rosenöl/Rose Absolut;
F.: essence/absolue de roses

Rose oil is steam-distilled from the petals of *Rosa damascena* and is one of the most expensive raw materials used in flavors and fragrances due to its low yield (0.02–0.05% of the plant material). Sometimes called "Otto of Rose/Rose Otto", r. is a pale yellow or slightly olive-yellow liquid, which separates white or colorless, odorless blades of crystals at temperatures below 21 °C.
Its odor is warm, very floral, reminiscent of red roses, slightly tea-like and fruity, with a hint of green notes and a honey background. Rose absolute is an orange-yellow or orange-reddish viscous liquid with a longer-lasting, sweeter, spicier and less green-tea-like odor profile than rose oil.
Both products are used in all kinds of floral high-class perfumes, where the oil makes its effect felt even in traces in the →top notes. R. blends well with →jasmine and fruity notes. In flavors, main uses are fruit flavors and tobacco flavorings. Main constituents: phenylethyl alcohol, citronellol, geraniol and esters.

Lit.: Arctander*
The H&R Book*

Rosin
Syn.: Gum Rosin; Pine Resin; Colophony; Turpentine;
G.: Kolophonium, Balsamharz; F.: colophane

Rosin and its derivatives are the most important natural resins. They are isolated from pine trees (*Pinus palustris*, *P. ellioti*, *P. maritima* and others). In former times, rosin was used in caulking the bottom of wooden sailing vessels. This is the reason of calling the r.-processing industry and their products "naval stores". There are three kinds of rosin.
Gum rosin [8050-09-7] is the collected exudate of trees, stimulated by cutting the bark. Distillation yields →turpentine oil (30–15%) and gum rosin (70–85%).
Wood rosin is gained from wood stumps pulled from the ground and transformed into chips, which are extracted by solvents (hydro-carbons). Distillation recovers the solvent and yields a volatile fraction (→turpentine oil and →pine oil) and a dark rosin, which is further purified from its solution by fractionated precipitation with another solvent and by absorption on activated clay.

Tall oil rosin [8052-10-0] is a by-product of the manufacturing of →paper pulp by the Kraft process. The so-called "black liquor" contains the sodium salts of →ligninosulfonate and rosin acids as well as fatty acids. For details of further processing: →tall oil. The tall oil rosin is separated from the tall oil fatty acids by →distillation.

The three types of rosin gained by these processes are somewhat different, resulting in different performances in end use.

In general, r. is brittle and has a T_g of 30 °C. The m.w. is in the range of 300, and the a.v. is 162. Above the softening point (≈ 70 °C), the resin shows a sharp drop in viscosity, which is an important characteristic used in many rosin applications. R. is soluble in hydrocarbons, esters and ethers and also compatible with many other polymers.

R. is a complex mixture of alkylated hydrophenanthrenes, mainly (90%) based on two structures:

abietic acid   pimaric acid

R. is frequently modified by →hydrogenation, disproportionation and dimerization. Hydrogenation gives better color stability. Disproportionated acids and soaps are made by heating (270 °C) without or with a catalyst (sulfur, strong acids, noble metals on carbon). Dehydro- and dihydro-acids are formed. At higher temperatures and acidic conditions, dimerization occurs to yield a higher-melting rosin of good stability. Rosin can be transferred into the Na and K salts (→soaps) or the Zn and Ca salts (→metallic soaps) and esters. Methyl, glyceryl and pentaerythryl esters are made by →esterification. Rosin acid may be ethoxylated (→alkoxylation) and has →surfactant properties. Alcohols are produced by →hydrogenation, and amines via the nitriles. An important derivate is modified r., made by the Diels Alder reaction with maleic acid anhydride.

The largest use of r. is in paper sizing (→paper chemicals), mostly as maleic anhydride-modified and partly neutralized r. Modified r. salts and esters are used in printing →inks. Another application of r. and derivatives is in pressure-sensitive and hot-melt →adhesives as well as in core binders in the →foundry industry. The glycerol ester of hydrogenated r. is used in chewing gums. The soaps of disproportionated r. function as emulsifiers in styrene-butadiene rubber (SBR) synthesis. They remain in the rubber as tackifier (→polymerization additives). Use in →coatings is reduced nowadays and substitued by →alkyd resins and →drying oils. R. and its derivatives are used as fixatives in certain types of perfume.

Worldwide production of r. is 90000 mt/a with the United States and the People's Republic of China each manufacturing one third. The rest originates from Scandinavia, Indonesia, Mexico, Brazil and Portugal.

Lit.: Encycl.Polym.Sci.Engng.* (2.) **14**, 438

Rosmarinic Acid

G.: Rosmarinsäure; F.: acide rosmarinique

R. is commonspread in rosemary, oregano, salvia, thyme and →peppermint. R. is extracted from plant cell cultures of *Coleus blumei* or produced in its racemic form by chemical synthesis.

R. shows activity for the treatment of rheumatic diseases and infections and is an inhibitor of the →prostacyclin and →leucotriene biosynthesis; it is used in the therapy of shock-syndromes and shows activity against influenza, herpes and polio viruses; additionally, it is applied in liniments.

Lit.: C.J.Kelley et al. J.Org.Chem. **40**, 1804 (1975) [20283-92-5]

Rotenone

→Insecticides

RR

→Renewable Resources

Rubber, natural

Syn.: Polyisoprene
G.: Gummi, Kautschuk; F.: gomme, caoutchouc

$$H-(CH_2-C=CH-CH_2)_x-H$$
$$\quad\quad\quad\; | $$
$$\quad\quad\quad CH_3$$

Natural r. is derived from the →rubber tree (*Hevea brasiliensis*) and is polyisoprene of high m.w. Another possible source is guayule. The structure was defined by Harries through →ozonolysis. All other rubbers are petrochemically based.

The name "rubber" is derived from the fact that the first bouncing balls, which were brought from the New World to Europe could be used to "rub out" marks made on paper. Technically, r. soon became increasingly interesting after the discovery of vulcanization in 1838 by the American chemist Charles Goodyear, which was the first step towards improving the properties of crude r. Contrary to the situation of many alternative sources for other RR products, the r. tree is about the only commercially utilized source of natural r. More than 99 % of all natural r. is derived from it.

Several attempts (Mexico in ≈1910, California during the Second World War and again in the 1970's and 1980's) have been made to commercialize guayule, derived from a shrub (*Parthenium argentatum* A.Cray, Compositae). There is no structural difference between guayule and hevea rubber. Only a resinous by-product of guayule, which retards vulcanization, has to be removed.

R. is traded as latex (emulsion), pale crepe and smoked sheets.

R. is compounded (→rubber chemicals) and vulcanized at elevated temperatures in molds or during extrusion to yield highly elastic molded goods (e. g., tires, hoses, etc.).

Reaction with chlorine gives chlorinated rubber (contains 65 % Cl), which is a base material for →coatings that are highly resistant to chemicals and environmental damage, and for →adhesives.

R. may be cyclized by reaction with sulfuric acid and its derivatives. It is used in printing →inks and →coatings.

Reclaimed r. is replasticized vulcanized r., which can be added to rubber compounds (recycling).

Despite the invention and large-volume production of synthetic elastomers, r. has maintained its technical importance in certain applications and is gradually increasing in production volume worldwide: ($\times 10^6$ mt)

1960	1970	1980	1991	1992	1993
2.01	3.10	3.81	5.07	5.69	5.58

Lit.: Ullmann* (5.) **A23**, 225
     Encycl.Polym.Sci.Engng.* (2.) **14**, 687

Rubber Chemicals

G.: Gummichemikalien;
F.: composés entrants dans la fabrication des caoutchoucs

There are several RR-based products used in rubber compounding and processing. The →rubber may be natural or of synthetic origin.

Active ingredients:

Vulcanization accelerator systems, based on rubber-sulfur-organic accelerator, need activators, e. g., ZnO and →stearic acid or Zn-stearate (→metallic soaps). →Fatty acids also improve processability. They may be already contained in synthetic rubbers, where their salts are used as →polymerization additive similar to the use of the salts of disproportionated rosin acid (→rosin), which impart improved tackiness to the rubber compound.

Another example of product combination is the coating of fillers with →stearic acid to improve distribution.

In rare cases, naturally occurring products, such as →cellulose, →lignin and especially →cork (in rubber based sealants), are used as organic fillers.

Plasticizers used in rubber compounding are, as far as RR-based products are concerned, the same as in plastics (→plastics additives).

Less important are products, such as →lanolin, →palm oil, →linseed oil, →gelatin, pine tar and →rosin, which are added to the rubber compound more because of local availabilty than for technological reasons.

Processing aids:

These are products that are added in small amounts to improve processability without substantial changes in the physical properties of the final product.

→Fatty acids, →metallic soaps, →fatty acid esters and →fatty alcohols act as dispersing agents, which improve the dispersion of fillers and the uniform distribution of other ingredients.

Lubricants and flow improvers lower the friction within the compound as well as between compound and the metal surface of the compounding machine. Again →fatty acids and their esters as well as →fatty acid amides are used.

→Rosin gives tackiness to the compound during processing.

→Factice is a product that improves extrudability and stability during processing, and gives a surface finish to the final product. For mold release: →release agents.

Latex chemicals:

Aside of the usual r. used in solid rubber, latex needs some special chemicals:

→Casein is used as stabilizer and dispersing agent.
→Soaps and nonionic surfactants (e. g., → fatty alcohol ethoxylates are emulsifiers. →Fatty alcohols are added to prevent foaming.

Latex thickeners are →polysaccharides, albumens, →casein, →tragacanth and →agar agar.

Lit.: W.Hofmann "Rubber Technology Handbook" Hanser Verlag, München (1989)
J.v.Alphen "Rubber Chemicals" D.Reidel Publ. Dordrecht (1973)

Rubber Tree

Hevea brasiliensis (H.B.K.) Muell. Arg.
Euphorbiaceae
G.: Gummi-, Kautschuk-, Parakautschuk-Baum;
F.: hévéa brésilien, caoutchouc de Para

H. brasiliensis is native to the Amazon region. It was utilized from wild r. populations in a monopoly up to 1876 in Brazil until, during a dramatic adventure, seeds were "stolen" and multiplied in Kew Botanical Gardens in Richmond near London and transferred via Ceylon to Malaysia. Out of 2700 seedlings, only 22 survived the shipment to Malaysia.

The r. is up to 30 m tall with a 4.5 m-long root. The milky sap used for producing rubber, is derived from the outer cell layers of the cambium (bark) of the trunk. This sap production is typical for all representatives of the →Euphorbiaceae family (→*E. lathyris* and →*E. lagascae*). The milky sap contains about 30% rubber (maximum 40%), which is a dispersion of small rubber spheres (1–2 micrometer) in water. R. seedlings are planted at densities of 140–400 trees per ha. They are often cultivated in → intercropping with coffee, cocoa or food crops. Yields of rubber range between 500 and 3000 kg/ha. Tapping of the r. may start at a trunk diameter of 45 to 50 cm at a height of 1 m above the ground. There are special tapping techniques to make the best available use of the genetic potential of the r. and to do the least damage to the tree while hurting its bark.

The most important rubber-producing countries are:

Rubber production in 1993	($\times 10^6$ mt)
Thailand	1.50
Indonesia	1.37
Malaysia	1.21
India	0.44
China	0.32
Philippines	0.17
Nigeria	0.13
World	5.58

Lit.: E.P.Imle, "Crop Resources" D.S.Seidler (ed.) 119–136, Academic Press (1977)

Rubia cordifolia
→Dyes, natural

Ryania speciosa
→Insecticides

Ryanodine
→Insecticides

S

Sabadilla
→Insecticides

Saccharose
→Sucrose

Saccharum sp.
→Sugar Cane

Safflower
Syn.: Dyer's Saffron; False Saffron
Carthamus tinctorius L.
Compositae
G.: Saflor, Färberdistel;
F.: safran bâtard, carthame des teinturiers

S. belongs to a group of species of the Compositae, commonly called "thistles". Centers of origin of s. are in Afghanistan and in the Mesopotamian area. Since ancient times it has been cultivated in Egypt and the Middle East, mainly for its red dye, which is obtained from the dried flowers.
It is a branched herbaceous annual, 0.5–1.5 m tall, with a strong, somewhat thickened tap root and many thin lateral roots, which grow more or less horizontally in the upper layers of the soil. Growth after germination is comparatively slow while a ro-

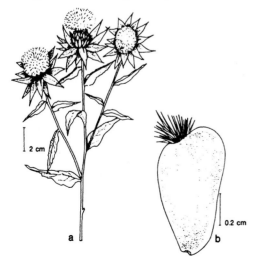

Safflower; a: Flowering branch b: Fruit (from Rehm and Espig*, 1995, with permission).

sette of leaves is produced close to the ground, followed by several direct branches.
The →inflorescence of s. is a head or capitulum, about 4 cm in diameter with up to 90 single flowers each. The flowers have yellow to deep orange petals. The fruits are achenes, similar to those of →sunflower, 8 mm in length and light grey. They consist of 40–60% pericarp, 26–37% oil and 20–55% protein. The oil is easily gained by mechanical pressing, followed by solvent extraction, leaving up to 50% protein in the meal. In recent times, s. has become an important oilseed, which is well adapted to semi-arid climates due to its thistle character. Main producing countries are the USA, Mexico, Australia, China, North Africa and India.
Until the late 1980s 1.4×10^6 ha of S. yielded 760 000 mt/a of seeds worldwide.

Lit.: P.F.Knowles, "Safflower", in: Röbbelen* (1989).

Safflower Oil
G.: Safloröl, Färberdistelöl; F.: huile de safre

s.v.: 180–194; i.v.: 136–152

S. is a drying oil similar to →linseed oil but less yellowing.
It derives from the →safflower plant and is produced by pressing, expelling and extraction of the expeller cake with hexane. Fatty acid composition:

C16:0	C18:0	C18:1	**C18:2**
5%	2–3%	12–20%	**70–80%**

There are also special varieties with 80% C18:1 and 15% C18:2.
The oil is used in dietary nutrition as salad oil and industrially in →coatings and in →linoleum production. The fatty acids gained by →hydrolysis are used in nonyellowing →alkyd resins.
World production amounts to about 260.000 mt/a (1984).

Lit.: Ullmann* (5.) **A10**, 229
      [8001-23-8]

Safflower Oil Fatty Acids
→Safflower Oil

Saffron
→Dyes, natural

Saint John's Bread
→Carob

Salvia sclarea
→Clary Sage

Sandalwood Oil
G.: Sandelholzöl; F.: essence de santal

S. is steam-distilled from the coarsely powdered wood and roots of *Santalum album*, a comparatively small tree of the Santalaceae family. The tree grows in India (Mysore), Sri Lanka and Indonesia. It is counted amongst the oldest perfume raw materials and was originally used in incense sticks.

The oil is a pale to yellow colored, viscous liquid with an extremely soft, sweet-powdery woody note that possesses almost no →topnote but extremely long-lasting →body notes.

S. is used in all types of fragrances as a soft powdery background note, especially in woody-floral and →oriental notes.

Lit: Gildemeister*
    Arctander*

Sandarac Resin
→Resins, natural

Santalum album
→Sandalwood Oil

Sapium sebiferum
→Chinese Tallow

Saponification
G.: Verseifung; F.: saponification

S. is one of the oldest reactions practiced by human beings. It is the →splitting of →fats and oils into →soap and →glycerol.

$$R^1\text{--CO--O--CH}_2$$
$$R^2\text{--CO--O--CH} \quad + \ 3\,NaOH$$
$$R^3\text{--CO--O--CH}_2$$

$$\longrightarrow \quad \begin{matrix} R^1\text{--COONa} \\ R^2\text{--COONa} \\ R^3\text{--COONa} \end{matrix} \ + \ \begin{matrix} HO\text{--CH}_2 \\ HO\text{--CH} \\ HO\text{--CH}_2 \end{matrix}$$

R^{1-3}: C_{7-21} alkyl

In the old process, the fat is heated in open pans for several hours with caustic soda. Other alkali hydroxides can be used also. This technology has been modernized in various continuous processes.

In modern soap production, the fat is split by →hydrolysis first and subsequently is neutralized by the caustic. This can be done in a two step process in one vessel or by using isolated →fatty acids. Many different processes are practiced. The neutralization of fatty acid is sometimes also called s. Other (than alkali) metal salts of fatty acids (→metallic soaps) can be produced by s. or reaction of →fatty acids with the respective metal(hydr)oxide. For further details about fatty acid salts: →soaps and →metallic soaps.

Lit.: Ullmann* (5.) **A10**, 254
     Kirk Othmer* (4.) **5**, 168

Saponins
G.: Saponine; F.: saponines

S. are widespread, occurring in higher plants, starfishes and sea cucumbers, and consist of an agluconic part (sapogenin) and a gluconic moiety joined by glucosidic linkages.

Sapogenins may consist of triterpenoids, →steroids or glyco-alkaloid components. Glucons are hexoses (β-D-glucose, β-D-galactose, α-L-rhamnose) or pentoses (α-L-arabinose, β-D-xylose). Mostly, the sapogenins are linked either with one or two glucosidic units, monodesmoides, which are concentrated in the outer tissues of seeds, roots or bark as plant-specific barriers against microorganisms, or bidesmoides, which are found in leaves and stems.

Occurrence: Solanaceae, Hypocastanaceae, Primulaceae, Rosaceae, and Caryophyllaceae are rich in genera, which contain mainly triterpene and glycoalkaloid s. *Digitalis* sp. and genera of monocotyledones are producers of steroid s.

Animal sources of triterpene s. are sea cucumbers (Holothuridea) and starfish (Asteroidea) produce steroid s.

An example for the structure of a typical steroid s. from *Digitalis purpurea* and *D. lanata*, consisting of digitonin (glucosidic part) and digitogenin (aglucon):

$R, R^2, R^4, R^5, R^6, R^7 = H$; $R^1 = \beta\text{-OH}$; $R^3 = CH_3$; $R^8 = OH$;

Isolation is performed after defatting by solvent extraction (methanol, ethanol, their aqueous mixtures or slightly acidified water), distribution of the crude extract between water and an immiscible solvent phase, and precipitation of s. from the latter. Further concentration and purification may be achieved by chromatographic techniques.

Most important properties are:

- surface activity, caused by the amphiphilic structure, the intensity of which depends on the length and the branching of the glucon moiety;
- hemolytic activity, the ability to lyse erythrocytes is stronger in the case of monodesmoides and may be decreased by substitution of the aglucon hydroxyl groups;
- steroid-complexing ability, pronounced in the steroid s. and glycoalkaloid s., leads to the formation of insoluble complexes with cholesterol and other sterols;
- biocidal activity (fungitoxic, fungistatic, mild antimicrobial);
- specific action is known from steroid s. and glycoalkaloid s., whereas triterpene s. exhibit broad spectra of effectiveness;
- toxic action against certain animal families is known : molluscs and other invertebrates, termites, leaf-cutting ants, potato beetle, and aquatic vertebrates, such as fish and tadpoles.

Toxicity upon parenteral or intravenous application shows a broad spectrum of LD_{50} $0.7-50$ mg/kg. Hemolytic actions prevent their intravenous administration; membrane penetration may cause lesions with pore diameters of 8 nm; complexing has been shown with erythrocytes and liposomal membranes.

Uses are based on specific therapeutic activities in pharmacy:

- enhanced permeability and increased secretory effects;
- anti-inflammatory activity;
- anti-exudative and anti-granulomatous action by activation of cortisone secretion;
- anti-allergic actions;
- lowering the uptake of cholesterol by complexing;
- anti-acid and peptic ulcer treatment;
- increase of bile salt excretion ;
- psychotropic, analgesic, sedative cardiovascular activation;
- anti-tumor activity;
- anti-viral action against herpes and polio;
- increase of immune response to vaccines;
- improved uptake of β-lactam antibiotics after oral application.

Other special uses are : steroid saponins as starting material for the industrial synthesis of hormones (pregnosolone, progresterone, steroid hormones); →steroid synthesis from glyco-alkaloid s.; surfactants and O/W emulsifiers (cosmetics); flavoring agent in tobacco and food; expectorant uses for triterpene s.

Animal s. have not found any use.

Lit.: Ullmann* (5.) **A23**, 485–497

Sarcosinates

Syn.: N-Acylsarcosinates; Sarcosides
G.: Sarkoside; F.: sarcosinates

$$R-CO-N-CH_2-COO^-\ Na^+ \qquad R:\ C_{11-17}\ \text{alkyl}$$
$$\qquad\qquad |$$
$$\qquad\quad CH_3$$

S. are less sensitive to water hardness and →hydrolysis than soaps. They are good wetting agents and dispersants.

They are made by reaction of fatty acid chlorides and sarcosine.

S. are used as soap-like →surfactants and in shampoos. A special application is in dentifrices, where it is claimed to inactivate certain enzymes in the mouth.

Lit.: Kirk Othmer* (3.) **22**, 348

Sargassum spp.

→Marine Algae

Schoenocaulon officinale

→Insecticides

Scleroglucans

G.: Scleroglukane; F.: scléroglucanes

S. is the general designation for a number of similar neutral homo-polysaccharides, produced by the action of some fungi of the genus *Sclerotium* on glucose-containing culture media.

They consist of linear main chains of 1,3-β-glucosidically linked →AGU, in which every third of them carries another AGU, which is 1,6-β-glucosidically bound. Occasionally, longer side chains, always 1,6-β-connected with the main chain, have been found. The DP ranges between 110 and 1600. Quantity and length of the side chains depend on the fungal origin.

Production is accomplished by submerged, aerobic fermentation of glucose by *Sclerotium rolfsii*; after removing the mycelium, s. is isolated, either by spray-drying or solvent precipitation, and is further purified.

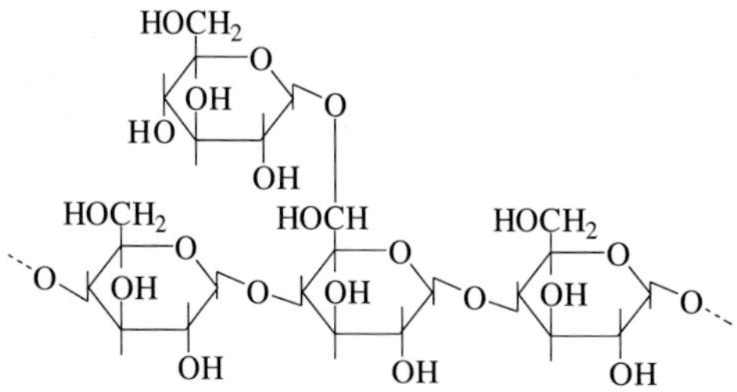

S. are easily soluble in water to yield pseudoplastic solutions of high stability against heating and changes of pH. They may be derivatized to film-forming products.

Besides uses in the food industry (gelling, stabilization of suspensions, covers and coatings), application in cosmetics and in pharmaceuticals (hair sprays, lotions, tablets), in ceramics production and in oil-well drilling, have been described.

Lit.: Ruttloff* 508–509, (1991)

Sealants
→Putties

SCP
→Single-Cell Proteins

Sealing Wax
G.: Siegellack; F.: cire à cacheter

S. is an ancient product, orginally based on →shellac dissolved in →turpentine oil, which is dyed red by addition of cinnabar. Cheaper versions use other natural or synthetic resins.

S. was and sometimes still is used to seal, close and stamp official documents.

Lit.: Ullmann* (3.) **9**, 575

Seaweed
→Marine Algae

Seaweed Extracts
G.: Meeresalgenpolysaccharide;
F.: polysaccharides presents dans les algues marines

S. is a group of →polysaccharides that occur as structural components of the cell walls of the phylum *Phycophyta* (→marine algae). Up to 40% of the dry substance is s. which is preferentially extracted by slightly alkaline aqueous treatment, purification, followed by precipitation or direct drying of the extracts.

Important food and nonfood products are derived from Phaeophyceae, brown seaweeds, and Rhodophyceae, red seaweeds. The third class, Chlorophyceae, green seaweeds, is up to now of minor industrial importance. The latter have been considered as potential source of nutritional fat and proteins by means of large-scale aqueous cultivation.

Most important s. are →agar, alginates (→sodium alginate), →carrageenans and →furcellaran.

Industrial utilization growth, as revealed for the period 1988–1995, is: carrageenan, + 2.8%; alginates, + 1.3%; agar, − 2.8%. Additional amounts of alginates may be produced as →microbial gums in future.

Lit.: Levring* 288–289 (1969),
Intern.Food Ingredients **5**, 55–59 (1992),
Ullmann* (5.) **A25**, 34–35

Sebacic Acid
Syn.: Decanedioic Acid; Sebacinic Acid
G.: Sebacinsäure; F.: acide sébacique

$HOOC–(CH_2)_8–COOH$

m.w.: 202.24; m.p.: 134 °C b.p.: 295 °C (13.3 kPa/100 mm)

S. is insoluble in water, soluble in alcohol and ether.

Commercial production[1] starts from →ricinoleic acid, which is split into s. and octanol-(2) at 300–320 °C in the presence of $NaOH/KOH/H_2O/CdSO_4$:

$CH_3–(CH_2)_5–CHOH–CH_2–CH=CH–(CH_2)_7–COOH$
$$\downarrow$$
$CH_3–(CH_2)_5–CHOH–CH_3 + HOOC–(CH_2)_8–COOH$

Yields are in the range of 75–85%.

S. is used in →alkyd resins, →polyesters, →polyamides. The esters of s. are used as →lubricants, →hydraulic fluids, plasticizers (→plastics additives).

The production is estimated to be 30000 mt/a. There is one producer in the USA and several in China. Brazil and India are said to enter the market too.

Lit: Ullmann* (5.) **A8**, 531
[111-20-6]
[1] SFÖW **89**, 771 (1963)

Sebacinic Acid
→Sebacic Acid

Seed Weight

G: Tausend-Korn-Gewicht (TKG);
F: poids de mille grains

Measure of the size of seeds, measured as the
weight of 1000 seeds of that plant in g.
Examples:

wheat	40–50 g
oats	3–50 g
corn	150–400 g
soybean	100–240 g
rapeseed	3–7 g
poppy	0.25–0.6 g
sunflower	70–120 g
linseed	8–15 g
sugarbeet	13–20 g

Lit.: Martin/Leonard* (1967)

Sesame

Syn.: Benniseed
Sesamum indicum L., *S. orientale* L.
Pedaliaceae
G: Sesam; F: sésame

S. is probably the most ancient oil crop, described
already in Iran more than 4000 years ago. It was
domesticated in Africa, spreading early to India
and other Asian countries. Today, the main producing countries are India, China and Burma, with
more than 50% of the world crop of 2×10^6 mt of
seed.

S. is considered a specialty crop in the semi-arid
tropics, with good drought resistance and high sen-
sivity to excessive rainfall. Often it needs to be ir-
rigated and is grown preferably in rotation with
rice and vegetables. S. is a traditional peasant agriculture crop.

S. is erect, more or less branched, annual, about
1–3 m tall, with leaves, stems and flowers covered
with glandular hairs.

Leaves are formed opposite at the base of the plant
and alternate in its upper parts. Normally, flowers
are formed singly but rarely in groups of 2 or 3 in
the axils of leaves in the upper stem region and the
branches. The flower is of dorsiventral symmetry
with a two-lipped tubular, bell-shaped corolla, 3–
4 cm in length of white or pink color with red or
purple spots inside the tube, and with hairs outside.
Normally, s. is self-pollinated due to premature
pollination of the anthers in the closed flower.

The fruit is a rectangular, deeply grooved capsule
of about 3 cm in length with a short beak. It de-
hisces by two apical pores through which the maturing seeds may be lost before or during harvest
("Open sesame" of the famous Ali Baba Fairy
Tales is derived from this). Although indehiscent
mutants were selected, the traditional method of
overcoming this obstacle in s. is to select for erect
growing types as well as to prevent shaking of the
plants during harvesting and, on the other hand, to
make use of the indehiscence by harvesting the
seeds while turning them upside down into collecting devices. Seeds are about 3 mm long, flat and

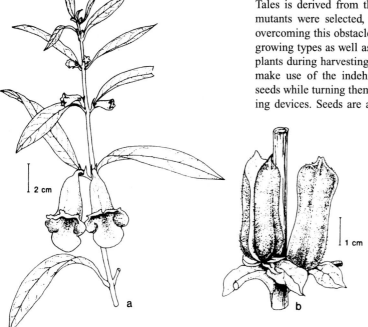

2 cm

1 cm

a

b

Sesame; a: Shoot tip with flowers. b: Maturing capsules (from Rehm and Espig*, 1995, with permission).

ovate with a white to brown testa. Best oil qualities are gained from white seeds.

The oil content is between 45−58%. The oil is gained by pressing, followed by solvent extraction by which cold-pressed oil and solvent-extracted oil are segregated. It is comparatively stable and does not turn rancid. S. oil is mainly used for food but also for producing soaps and lamp oil, especially in India and China. The seeds are used as confectionary or even as ornamental and flavoring dressing for cakes, pastry and rolls. The defatted meal, remaining after extraction, is a good cattle feed (35% protein, rich in methionin).

Lit: A.Ashri, "Sesame" in: Röbbelen*, 375 (1989)

Sesame Oil

Syn.: Benni Oil; Gingelly Oil; Teal Oil
G.: Sesamöl; F.: huile de sésame

S. is obtained from the seeds of one or more cultivated varieties of *Sesamum indicum* by expression or extraction. It is subjected to a refinement process to obtain refined sesame oil for pharmaceutical use. A typical analysis of refined s. shows the following fatty acid pattern of the triglyceride:

C16:0	C18:0	**C18:1**	**C18:2**	C20:0
9%	4−5%	**45−48%**	**40−42%**	0−2%

Sesamine, a complex cyclic ether and sesamiline, a glycoside, are antioxidants present in small amounts. S. is slightly miscible in alcohol and soluble in carbon disulfide, chloroform, ether and petroleum ether; it is insoluble in water and is used in pharmaceutical formulations as a solvent and oleaginous vehicle. It is widely used as an edible oil and for the manufacture of margarine.

Lit.: Ullmann* (5.) **A10,** 227
[8008-74-0]

Sesamum indicum
→Sesame

Shea Butter Tree

Vitellaria paradoxa Gaertn.
Butyrospermum parkii (G. Don) Kotschy;
Sapotaceae
G.: Sheabutter; F.: beurre de Karité

S. is gained from the fruits of a medium-sized deciduous tree, which is a common natural constituent of the open savanna woodlands of West Africa where annual rainfall is around 1000−1300 mm.

It also occurs eastwards across the savanna zone of Africa to northern Uganda and the southern Sudan, where a distinct type, var. *nilotica*, is recognized.

The tree grows up to 10 m tall with a thick trunk and a spreading crown of branches. The oblong leaves are clustered at the ends of stout, young branches. The flowers are sweet-scented and white; they occur in dense groups in the axils of the leaves. The fruit is egg-shaped and about 5 cm long, surmounted by the remains of the style. The single, large seed is embedded in a sweet pulp within the fruit. It contains about 45−60% oil, which is solid below 32 °C. It has the following fatty acid pattern:

C16:0	C18:0	C18:1	C18:2
5−7%	3−5%	16−26%	14−24%

C18:3	C20:0
50−65%	6%

S. is used locally in foods, cosmetics and for illumination. Most of it, however, is exported from West Africa to Europe, where s. is used as a substitute for cocoa butter in confectionery and cosmetic applications as well as for soap and candle making. Special mention should be made here of the UV-protecting abilities that are natural to s. The press cake contains little protein, but may be used as a constituent of compound feed cakes for cattle.

Lit.: Cobley* (1976)

Sheep

G.: Schaf; F.: mouton

S. is a rather important RR.

The meat is used for nutrition and the →tallow (mutton), obtained as a by-product, is used as an oleochemical feedstock. S. are not only bred for meat but also for →wool, leather and fur production. By-product is wool wax or grease (→lanolin).

Production in 1995	Sheep, total slaugther inventory ($\times 10^6$)
China	125.0
India	61.8
FSU	29.2
Australia	29.6
New Zealand	30.8
Turkey	23.5
Spain	19.5
UK	18.8
World	427.0

The world s. population is $\approx 1.74 \times 10^9$ with Australia and New Zealand being the main producing countries.

Lit.: Kirk Othmer* (3.) **24,** 612

Shellac

Syn.: Lac, Gum Lac
G.: Schellack; F.: gomme-laque

m.w.: ≈ 1000; s.v.: 190–230; a.v.: 67–90;
m.p.: 65–90 °C

S. is a hard, tough, bright orange to brown, nontoxic, amorphous resin, which forms water-resistant films of high gloss. The best solvents for s. are alcohol and acetone containing a small amount of water. It is not soluble in chlorinated solvents, ethers, esters and hydrocarbons.

S. is the hardened, resinous secretion of the shell louse (*Kerria lacca* or *Laccifer lacca*), domestic in India, Burma and Thailand, and functions as a protective cover for the larvae.

The small insects attach themselves to the twigs of the host tree (Palas, Ber, Ghont, Kusum), and their larvae suck juices from the tender bark. Farmers cultivate these colonies and harvest them four times a year, depending on tree type and insect strain. The larvae secrete the lac in such an amount that it forms a continuous coating on the twig. Twigs are cut off and/or twisted so that the resin falls off. It is said that 3 million insects are necessary to harvest 1 kg of this crude resin, called "stick lac". It is washed and kneaded with water to remove trash, carbohydrates and a water-soluble red dye. Woody material and the insect bodies are skimmed off. The resulting product is "seed lac".

Bleaching with hypochlorite gives "bleached lac", while "orange lac" is made by pressing the molten resin through a filter cloth. There are several other processes to make different grades.

The complex composition of s. is only partly known. A main component is aleuritic acid, which is a 9,10,16-trihydroxy palmitic ester (43%) that forms, together with other polyhydroxy acids, lactones and (poly)esters. Cyclic sesquiterpenes (→terpenes) are also contained, e.g., shelloic acid and an aldehyde-acid, jalaric acid.

S. has a long history but is still a readily used fast-drying protective →coating for floors, furniture and panels.

Modifications with urea resins and oils (french polish) are well known. S. was used in the past to a large extent to make phonograph records. Vinyl copolymers and CD-recordings have taken away that market. Paper coated with s. is used for electrical insulation (motor windings). The pharmaceutical industry makes use of the property that tablets coated with s. are insoluble in the acidic stomach but release the drug in the alkaline intestines. Smaller uses are as stiffener for hat felt and playing cards and as a coating for citrus fruits. Shellac soaps of borax, ammonia, morpholine and ammonium hydrogen carbonate are used as protective colloids in drawing →inks. The esters of methanol and ethanol are used as plasticizer for cellulose lacquer.

More than 50% of the world production comes from India, which still controls the world trade to a large extent. The rest comes from other East Asian countries, mainly from Thailand, which started in the 1950's to compete with India. Production fluctuates, depend on growing conditions, and levels out by storage and local differences in harvest. In 1956, 50000 mt were exported from India. US imports over the years range between 5000 and 6000 mt/a.

Lit.: Kirk Othmer* (3.) **20,** 737
Encycl.Polym.Sci.Engng.* (2.) **14,** 450
[9000-59-3]

Shoot Inhibitors
→Tobacco

Siccatives
→Drying Oils

Silk
G.: Seide; F.: soie

The domestication of the silk worm in China and the processing of silk (sericulture) derived from it goes back for about 5000 years. The glossy protein fiber is triangular in cross-section, partly transparent, and has an average diameter of 10 μm. Degummed silk is a strong fiber, with wet strength being 80–90% of dry strength. Elongation is 20–30%. S. has a high modulus and is hygroscopic (10% water content at 20 °C and 60% RH).

S. is a thread spun by the silk moth or silk worm (*Bombyx mori*). It is composed of two fibroin fibers surrounded by a thin layer of sericin. Fibroin is a →protein that consists mainly ($\approx 90\%$) of the →amino acids glycine, alanine, serine and tyrosine, while sericin consists of 37% serine, and 14.7% each of glycine and aspartic acid. Both contain small portions of other amino acids. S. is highly crystalline (60%).

S. is derived from the mulberry silk worm, the life cycle of which consists of four stages: the moth lays up to 500 eggs (1–2 mm in diameter). A larva (3 mm) hatches from these eggs and grows fast on the diet of fresh leaves of the mulberry tree (*Morus alba*) up to a size of 8–9 cm. Within 4–6 weeks, it consumes 30 g of leaves. It starts to spin the protective cocoon from its glands. The aqueous protein solution is pressed out and hardens fast in the air and forms the two fibroin filaments surrounded by the sericin cementing layer. A pupa is formed inside the cocoon in the third stage, from which a moth emerges after a week. It has a life span of only a few days, time enough for mating and laying the eggs.

In industrial production, the silkworm is grown on trays, and the larvae are fed with freshly harvested mulberry tree leaves. Thirty grams of eggs (40 000–60 000) consume one ton of leaves in the 35 days of their life time. The pupae contained in the cocoons are killed by steam. In the filature (reeling plant), the cocoons are treated with hot water, whereby they are opened to show reelable thread ends. The continous filaments (up to 3000 m long) are wound onto a hank, which is finally dried. Wastes are used either as spun yarns (schappe) or bourette silk, which is made from the wastes of schappe spinning. Dead pupae are used as fish feed.

For further processing, the s. is partially or completely degummed to yield s. of different properties for specific uses. This is done with solutions of soap or salts (phosphates) or with →enzymes. During degumming, the sericin (up to 30 %) is dissolved. The weight loss of degumming can be compensated by weighting (incorporation of metal salts, plant gums or →dextrin and by graft polymerization with methacrylamide.). This is done only for certain articles. Bleaching with H_2O_2 is the next step. S. can easily be dyed or printed with conventional wool dyes.

Uses of s. are numerous: Shirts, ties, high-quality clothing (kimonos), underwear, lining materials, but also upholstery, tapestry, covers and carpets. Technical application are typewriter ribbons and surgical articles. Silk powder is added to lipsticks, ointments and soap. Fibroin and sericin are added to shampoos and lotions.

There is another silk-producing moth of commercial importance (production volume is 10 % of normal s.), called tussah (several *Antharaea* species). It is not cultivated like the mulberry silk worm but is harvested in oak forests.

Production of s. was 77 000 mt worldwide in 1991 Main producers are China (48 000 mt), India (12 000 mt) and Japan (6000 mt).

Silk is highly appreciated by the consumer, has never suffered under the meteoric development of synthetic fibers, and the future outlook for this natural fiber, whose production is constantly improved as to economy and ecology, is bright.

Lit.: Ullmann* (5.) **A24**, 95
Encycl.Polym.Sci.Engng.* (2.) **15**, 309

Silver Dollar
→*Lunaria annua*

Silver Grass
→China Grass

Simondsia chinensis
→Jojoba

Sinapis alba
→Mustard

Single-Cell Proteins
Syn.: SCP
G.: Einzellerproteine;
F.: albumen unicellulaire

The total →biomass of protein-producing algae, bacteria, yeasts and fungi, containing 40–70 % (d.b.) raw →proteins, are called s. They are seen as a potential protein source for animal feed and human nutrition, which eventually could fill the global protein gap.

All kinds of C-sources can by used as media, e. g., CO_2, methane, paraffins, methanol and carbohydrates such as →sucrose molasses and black liquor from →paper pulping. Protein production is fast, and yields are rather high: 250 kg of yeast (67.5 kg d.s.) produces within 15 hours 2 mt of protein, while the same amount of soybean plants yields only 18 kg/d.

Lit.: Science **219**, 740 (1983)
Rose (ed.) "Microbial Biomass" Academic Press, London (1979)

Sisal
Syn.: (Green) Agave
Agave sisalana Perrine
Agavaceae
G.: Sisal, Agave; F.: sisal

S. is a fiber crop, which like →manila hemp, produces stuctural fibers in bundles of sclerenchyma that occur in the leaves. The name s. is derived from the small Mexican port of Sisal, from which the fiber was first exported. It originated in Central

America and Mexico, where it is present as a wild as well as a cultivated plant. S. grows best in savanna zones, although it is a xerophyte. It is normally grown in plantations. The stem, without the inflorescence may be 1 to 1.5 m tall and 20 cm in diameter during the time of bloom. The stem is swollen and serves as a storage organ for nutrients for the rapid-growing inflorescences. The lanceolate leaves are about 120 cm long. They are dark green and covered with a waxy bloom. In a plantation, the growth may last 7–10 years before the generative phase is initiated. Once a s. plant has flowered, it dies. The →inflorescence is a large panicle, which is borne by a pole, 5–6 m in height, which ends in a panicle with about 250 branches that carry the flowers in clusters. The first harvest of leaves may take place 2–4 years after planting. Normally, 100 out of about 120 leaves are harvested per plant per year.

The sap of s. contains 0.5–1% hecogenin, which is a valuable starting material for making corticoids (→steroids).

The cut leaves are taken to a factory, where the fibers are extracted by machines (decorticator). This machine works mostly with large quantities of water, which are used to wash away the waste. After extraction, the fiber bundles are washed and dried in the sun or with drying machines, graded and further processed to threads, ropes and twine. In former times, s. fibers were mostly used for so-called "binders" for tying harvested temperate cereals into sheaves before stacking. This ended suddenly when combine →harvesters replaced binders. S. fibers are coarser than bast fibers but more resistant to rotting. They find application in cordage production.

Increasing competition arose in the 1960's when synthetic fibers caused a large drop of world market prices for s. After the price shocks of the mid-1970's, a dramatic recovery took place.

Lit.: Martin/Leonard* (1967)

Sitosterol

→Phytosterol
→Stigmasterol

Skin Preparations

G.: Hautbehandlungsmittel;
F.: produits pour les soins de la peau

→Soaps, bath and shower products, skin creams and lotions, sunscreen products, deodorants, anti-transpirants, lipsticks, make-up products and per-

fumes are the most commonly used →cosmetics for cleaning, care and improvement of appearance of skin. Furthermore, face masks, genital hygiene products and tanning products are worth mentioning.

Bath and shower products

Bath and, especially shower products are applied more and more with changing habits of life with increasing sport activities. →Surfactants, their main raw materials, and other ingredients are similar to those in shampoos (→hair preparations). Shower preparations often are recommended for skin cleaning as well as for shampooing. Further bath preparations are bath oils, which contain similar oily substances as skin creams and lotions.

Skin creams and lotions

Skin creams and lotions, which are either O/W or W/O →emulsions, are used for skin care after cleaning. The natural skin grease washed out must be substituted; the Romans did so already. Especially dry skin is intended to remain soft and smooth. This may temporarily be secured by moisturizing, because lack of water is the reason of rough skin. So, a lot of different moisturizers are recommended for such products, e. g., mineral oil; polyols, such as →sorbitol; →lanolin; →fatty alcohols, such as →cetyl alcohol and →stearyl alcohol; and →triglycerides, such as →sesam oil and →cocoa butter. Other important raw materials are emulsifiers, e. g., →soaps, →glycerides, →sorbitan esters, and many other anionic, cationic and amphoteric →surfactants, and recently rediscovered fruit acids, e. g., →citric acid, →lactic acid and →tartaric acid. Furthermore, special active agents are added, such as →antioxidants, e. g., tert.butylmethoxyphenol (BHA) and butylhydroxytoluene (BHT); humectants, such as →glycerol; and preservatives, such as →sorbic acid or a mixture of esters of p-hydroxybenzoic acid. Meanwhile, it is possible to produce stable W/O skin creams and lotions without preservatives. Skin care is supported by skin cleaning through mild facial cleaners, make-up removers and baby oils. These products contain either no or small amounts of →surfactants and oily substances, just as used in skin creams and lotions.

Sunscreen products

Sunscreen products are intended to protect the skin against sunburn by UV radiation. The active agents are cosmetic UVB absorbers, such as 4-amino-benzoic acid, octyl methoxycinnamate or octyl salicylate, and UVA absorbers, such as 2-hydroxy-4-methoxy-benzophenone, and physical sunscreens, such as titanium dioxide or zinc oxide. Formulations of such products can be →emulsions, gels, sticks, oils, mousses or aerosols. The basic raw materials of these formulations are similar to those in skin creams and lotions, i. e., water, →fats and oils, solvents, thickening agents and emulsifiers. The sun protection factor expresses the effectiveness of a sunscreen by comparing the time taken for the skin of an individual to redden without the use of a sunscreen compared to the time when a sunscreen is worn. For people, who desire a tanned appearance, there exist ways of achieving a tan without radiation. The best known method is to use dihydroxyacetone. After-sun preparations are similar as the a.m. skin creams an lotions. They contain additionally soothing and healing ingredients, e. g., allantoin.

Deodorants

Deodorants mask and reduce odor through the use of antimicrobial agents, e. g., trichlosan (2,4,4'-trichloro-2' hydroxydiphenyl ether), chlorhexidine [1,6-di-(4 chlorophenyl-biguanido) hexan], farnesol (3,7,11-trimethyl-2,6,10-dodecatrien-1-ol) or mixtures, such as triethyl citrate (→citric acid) and butylhydroxytoluene (BHT). They also contain →oils, such as isopropyl myristate (→fatty acid esters), and fragrances. D. are sold as roll-ons, aerosols and sticks, based on sodium stearate as the gelling agent for either →ethanol, →sorbitol or 1,2-propanediol.

Antitranspirants

Antiperspirants actively reduce the amount of underarm perspiration. Aluminium salts, such as aluminium dichlorohydrate, are especially suitable. Similar to the a.m. deodorants, they are sold as sticks and aerosols.

Lipsticks

Purpose of lipstick is to color, to camouflage, to protect from wind and water, the effects of cold, dryness and light, or a combination of these. Formulations are dispersions of pigments in an oil-wax gel. The ideal lipstick should be longlasting and easy to apply, and it should have a neutral taste. The main ingredients of lipsticks are liquid →oils, e. g., mineral oil and →castor oil, semisolid constituents, e. g., →lanolin, and solid constituents, e. g., →bees wax and high-melting →waxes, such as →carnauba wax and ozokerite. Colors are either inorganic or organic pigments or oil-soluble colors as permitted by local regulations.

Make-up products

Make-up products (rouges, blushes and highlights) are colored preparations that are generally applied to the cheeks, either as liquids, →emulsions or powders. The liquid types contain →thickening agents, e. g., →methylcellulose; polyols, e. g.,

→glycerol; colors and preservatives, e. g., p-hydroxybenzoic acid methyl ester, all dissolved in water. Creams are compounded similar to skin creams and lotions but with a high percentage of →glycerol. Make-up powders mainly contain talc and as binders, mixtures of →oils, e. g. mineral oil, →castor oil and →stearyl alcohol. Colorants are white pigments, e. g., zinc oxide, kaolin, titanium dioxide, and colored inorganic pigments or pigments, which produce a pearl-gloss effect, e. g., titanium dioxide and coated mica. Similar colors are used for eye cosmetics and nail cosmetics.

Lit.: Ullmann* (5.), **A24**, 219–245
Kirk-Othmer (4.), **7**, 572–619
Hilda Butler (ed.) "Cosmetics" Poucher's (9.), **3**, 212–242, 335–392, 405–439 Chapman & Hall, London/Glasgow/NY/Tokyo/Melbourne/Madras (1993)

SLA
→Lactylic Esters of Fatty Acids

Smooth Pea Starch
→Pea Starches

SMS
→Sorbitan Esters of Fatty Acids

Soaps
G.: Seifen; F.: savons

S. are the alkali metal (Na, K) and ammonium salts of →fatty acids (in specialties also of →rosin or →tall oil fatty acids).

They have a chainlength between C_8–C_{20} (mainly C_{12}–C_{18}) and may be saturated or unsaturated. Some of the hydrogens of the ammonium cation may by substituted by organic groups in ammonium soaps.

All other salts of fatty acids are called →metallic soaps.

Soap, known for at least 5000 years, is still the most important anionic →surfactant.

S. are readily soluble in soft water. A soap molecule has the characteristic of bearing a hydrophilic group attached to a hydrophobic long chain. On the surface of a soap solution, the molecules form a monolayer with the hydrophilic groups turned toward the solution, while the hydrophobic parts are turned toward the air. Within the solution, the molecules form micelles with the hydrophilic ends turned outside and the hydrophobic chains form the interior. They are able to take greasy soil into the core of the micelle, thus dispersing and emulsifying it.

Na-salts are hard s., while K-salts are pastes. Long chains (C_{18}) make the soap less soluble in water, medium chains (C_{12}) may cause skin irritation if present in high concentration. Too much unsaturation makes the product sensitive to oxidation.

S. are readily soluble in hot alcohol. In aqueous solution, s. react weakly alkaline, and fatty acid precipitates in acidic media. Washing power is therefore limited to alkaline or neutral environments. In hard water, s. form insoluble Ca and Mg salts. This sensitivity to water hardness has lead to a strong reduction of soap use in →detergents and industrial applications (→textile auxiliaries), after alternative →surfactants appeared on the market around 1930. In personal care, e. g., body cleansing, s. are still the favorite product.

Starting materials for manufacturing are →fats and oils mainly →tallow and →coconut oil. →Lard and →palm oil are substitutes for tallow, while →palm kernel oil is sometimes used instead of coconut oil. "Foots", i. e., the crude soaps obtained from refining edible oils, can be used also to a limited extent (→fatty acids, →fats and oils). Normally, mixtures of 80 % tallow and 20 % coconut oil (imparts foaming) are used, which are refined, bleached or hydrogenated prior to →saponification.

Before 1938, s. were produced exclusively by the old open steel kettle batch process. →Saponification was carried out by treating the fat with 50 % sodium hydroxide solution and live steam for heating and agitation. After the saponification was complete, the s. were salted out by addition of NaCl. Most of the glycerol and salt was removed with the aqueous lye. Washing and graining operations followed. The quality of the soap depended -aside from the raw material composition- on how much glycerol, water and NaCl remained. This old batch method is almost completely abandoned today.

However, continuous processes based, on this saponification technology, are numerous. Most of them use counter-current techniques for washing the crude soap and adjusting the glycerol content. A modern technology (Mazzoni process) starts from fatty acids gained by →hydrolysis. Again, several variations exist. The process has the advantage that various fatty acid combinations are possible to give greater flexibility to cost and performance considerations. Impurities can be eliminated by →distillation, and soda ash is (partially) satisfactory for neutralization.

Only 8 % of s. production starts from fatty acid due to stability problems during transportation and storage of the fatty acids.

One s. producer uses fatty acid methyl esters as raw material.

The final steps of all these processes are: crutching (incorporation of additives), framing (bar formation), drying (neat soap contains 30 % of water and must be dried to 10–15 % to make bars), mixing and milling, flaking and spray-drying.

S. are marketed mainly as bars but also in flakes and needles for technical applications and as pastes or liquids (K-salts).

There are many possibilities to formulate final products. In addition to **toilet and pasty s.**, which contain pigments, dyes, →fragrances and antioxidants, there are many specialty s.: **Deodorant s.** (containing antibacterial agents), **superfatted s.** (containing still unsaponified fats and oils or other fatty material), **floating s.** (with air bubbles in the bar), **transparent s.** (containing alcohol and more glycerol), **scouring s.** (with a mild abrasive component) and **shaving cream** (mixed Na-and K-s. of mainly stearic acid and not fully neutralized to have a pearly luster).

Especially in the United States, there are also soap bars on the market, called "syndets", which are not salts of fatty acids. In contrast with normal s., they are not sensitive to pH and water hardness. The following (mostly RR-based) →surfactants are used: →fatty alcohol sulfates, → fatty alcohol ethersulfates, →betains, →fatty alcohol ether sulfosuccinates and →fatty acid/protein condensates. Commercial importance of syndets is limited.

Aside from personal care and detergent applications, there are many technical uses for s. As →textile auxiliaries, s. are used in kier-boiling →cotton, in scouring →wool and degumming →silk. Food emulsifiers (→food additives) can be made self-emulsifiable by adding a small amount of s.

S. (especially rosin-based) play an important role in synthetic rubber (SBR) manufacture and in vinyl- and acrylic →polymerization. Na- and Li-s. are used in →lubricants (greases). Leather may be cleaned with saddle s., which contains →beeswax or →carnauba wax. S. also function as wetting and dispersing agents for →pesticides. Ammonium soaps are used in cosmetic preparations and function as emulsifiers in →polishes and emulsion paints (→coating).

The s. of mono-, di- and triethanolamine find use in →detergents and bar soaps due to their mild alkalinity and excellent detergency.

In 1990, 8.9×10^6 mt were produced worldwide (56 % of the total →surfactant market). Per capita use worldwide is 0.5 kg, in industrialized countries it is 2.1 kg.

The market position is and will remain stable.

Lit.: Kirk Othmer* (3.) **21**, 162
Ullmann* (5.) **A24**, 247

Sodium Alginate

Syn.: Algin
G.: Natrium Alginat; F.: alginate de sodium

β-D-mannuronic acid   α-L-guluronic acid
ManUA   GulUA

m.w.: 30 000 – 200 000

S. is the purified carbohydrate product extracted with dilute alkali from the following brown →seaweeds (Phaeophyceae):

Macrocystis pyrifera (gigant kelp), *Laminaria gitata* (horsetail kelp), and *Laminaria sachcharina* (sugar kelp).

It is a linear glycuronoglycan that consists mainly of β-1,4-linked D-mannuronic and L-guluronic acid units in the pyranose ring form, the relative proportions of which vary with the botanical source and state of maturation of the plant.

Strains of *Azotobacter vinelandii* or *Pseudomonas aeruginosa* are used for fermentative production.

S. is a white to yellowish-white, fibrous powder and is practically odorless and tasteless. It is slowly soluble in water to form a viscous, colloidal solution; it is insoluble in alcohol and in aqueous-alcoholic mixtures with an alcohol content of >30% (w/w).

It is used as a hemostatic agent in surgical dressings, as a suspending and viscosity-increasing as well as a disintegrating agent. Other uses are as dental impression material and as a tablet binder.

In food, it is used in the manufacture of ice cream, it produces a creamy texture and prevents the growth of coarse ice crystals. It is also used as stabilizer and suspending agent in soft drinks.

Technical uses are for sizing paper (→paper additives), as a →flocculating agent in waste water treatment, as a thickener in textile printing and dying, and as an ingredient of drilling muds in oil fields. Production countries are USA, UK, Japan and the FSU. ≈ 20 000 mt/a are produced.

Lit.: Ullmann* (5.) **A4**, 226 + **A25**, 34–40
Kirk-Othmer* (3.) **12**, 48–51
Levring* (1969)
[9005-32-7]; [9005-40-7] salt

Sodium Stearoyl Lactylate
→Lactylic Esters of Fatty Acids

Solanum tuberosum
→Potato

Solid Glucose

G.: Stärkezucker; F.: sucre d'amidon

S. is a mixture of D-glucose, small amounts of maltose and other maltooligosaccharides of low m.w.

It is obtained from almost complete acid hydrolysis of starch (→hydrolysis, glucosidic linkages), followed by desludging, decoloring, concentration and solidification. The latter is performed either by bulk crystallization and shredding or by spray crystallization.

In the US, such products with DE< 80% are known as "70 or 80 DE ship sugar", in Germany as "70er Stärkezucker". Enzyme produced and solidified starch hydrolysates with about 95% DE are shipped as "total sugar" or as "solid glucose".

S. is used in the food industry as sweetener in the same manner as →glucose syrup or →dextrose .

Lit.: Schenk/Hebeda* 121–176 (1992)

Soluble Starches
→Thin-boiling Starches

Solvent Extraction
→Extraction

Sorbic Acid

Syn.: *trans,trans*-2,4-Hexadienoic Acid
G.: Sorbinsäure; F.: acide sorbique

$CH_3-CH=CH-CH=CH-COOH$

m.w.: 112.13; m.p.: 132–135 °C b.p.: 228 °C (d.)

Naturally occurring s. has been extracted as the lactone (parasorbic acid) from the berries of mountain ash (*Sorbus aucuparia* L.). Today, s. is produced synthetically. It is a white crystalline powder, tasteless and has a faint characteristic odor. Sorbic acid and its calcium and potassium salts are the most used preservatives in food (→food additives) and beverages. It is also used in cosmetics and in pharmaceutical formulations.

Lit.: Ullmann* (5.) **A24**, 507
[110-44-1]

Sorbitan Esters of Fatty Acids

Syn.: SMS; STS; PS (polysorbate)
G.: Sorbitanfettsäureester
F.: esters de sorbitane d'acides gras

$$CH_2 \quad HC-CH-CH_2-O-CO-R^1$$

with OR^2 above, and below:

$$R^2O-CH--CH-OR^2$$

SMS: R^1: C_{15-17} alkyl; R^2: H
STS: R^1 and two $R^2 = C_{15-17}$ alkyl; one R^2: H
PS: R^1: C_{15-17} alkyl; R^2: $-(CH_2-CH_2-O)_x-H$
     x: 20

m.p.: $55-60\,°C$

→Sorbitan is derived from →sorbitol by →dehydration and is subsequently esterified with fatty acids or their methyl esters at $200-250\,°C$ with 0.1% NaOH. Depending on the amount of fatty acid sorbitan monoester (SMS), diesters or triesters (STS) are formed.

SMS is dispersable in water and soluble in fats and oils, while STS is only soluble in fats and oils. They are nonionic →surfactants with low →HLB values. Products with better water solubility and higher →HLB value may be obtained by ethoxylation (→alkoxylation). Such ethoxylates are also called polysorbates (PS) and are plastic or liquid products, soluble in water.

SMS is used as an emulsifier in desserts, often in combination with ethoxylated s. STS is used as a crystal modifier in fats, such as cocoa butter substitute or margarins. Roughly 1000 mt were used in the USA (1981) in food applications.

Other uses are in fiber lubricants, textile processing and in cosmetics.

PS find use as emulsifiers, softeners, fiber lubricants, solubilizers and in many food applications, such as bread, desserts, whipped toppings and milk products. US consumption in the food area (1981) was almost 4000 mt.

Lit.: Ullmann* (5.) **A25**, 747
     N.J.Krog "Food Emulsions" K.Larsson, S.E.Friberg (ed.), Marcel Dekker Inc. NY/Basel (1990)
     G.Schuster "Emulgatoren für Lebensmittel" Springer Verlag Berlin/Heidelberg/NY/Tokyo (1985)

D-Sorbitol

Syn.: D-Glucitol; Glucit
G.: Sorbit; F.: D-sorbitol

S. is a six-valency polyalcohol (hexite), naturally occurring in numerous higher plants, their fruits and in algae.

The industrial product is obtained by →hydrogenation of pure D-glucose solutions or high-level →glucose syrups.

$$H-C-C-C-C-C-C-H$$

with $OH\ OH\ H\ \ OH\ OH\ OH$ above and $H\ \ H\ \ OH\ H\ \ H\ \ H$ below

m.w.: 182.2 $(C_6H_{14}O_6)$; m.p.: $92-96\,°C$;
$[\alpha]_D^{20}$: $4-7°$.

It has a physiological energy content of 17,5 kJ/g, an r.s. of $50-60$ and has no influence on the blood sugar level of diabetic persons.

Industrial production is based on 50% solutions of D-glucose (dextrose). The yield of s. depends strictly on the purity of the sugar. Hydrogenation is performed continuously under high pressure on fixed-bed Cu-oxide or Ni-oxide catalysts. After refining, the reaction fluid is evaporated to 70% d.s. (main commercial product). S. powder may be prepared by (spray)crystallization of highest-purity s. syrups. Noncrystallizing syrups are prepared by hydrogenation of high-conversion →glucose syrups.

Fields of application are in the food industry as sweetening, softening and humidity-regulating agents (confectionary, chewing gums, baked goods, jams, desserts, creams, dressings, ice creams, diabetic's food); in pharmacy as laxative, as stimulant for the gall bladder, to enhance the detoxifying function of the liver for osmotherapy and in numerous medical powders, tablets, capsules and fluids as preferred carrier, stabilizer and bulking agent, in cosmetics as nonirritating emollient that does not dehydrate the skin, as humidifying and moisture-protecting agent, and as emulsion stabilizer.

Traditional nonfood applications derive from the moisture conditioning, softening and plastifying properties, such as needed in adhesives, paper, textiles, leather, cellulose-based films and foils, as emulsion stabilizer in environmentally sensitive roles, as complexing agent for sequestering metal ions, as binder and stabilizing agent in the foundry and plastic industries.

Future importance as RR results from chemical intermediates that can be derived from s. by esterification, etherification as well as from polycondensation reactions. They lead to biodegradable →surfactants (→sorbitol esters of fatty acids), →alkyd resins (coatings), →polyetherpolyols (soft polyurethane foams), melamine formaldehyde resins and phenol resins.

In 1990, the world production (70% solution) was 650 000 mt. One third of this was used for the synthesis of →L-ascorbic acid and pharmaceuticals, one third for food and cosmetics and one third in the chemical industry.

Lit.: A.Rappaille; starch/stärke 40 356−359 (1988)
Rymon-Lipinski/Schiweck* 265−280 (1991)
Ullmann* (5.) **A25**, 413−437

L-Sorbose

G.: L-Sorbose; F.: L-sorbose

S. is a monosaccharide of the L-configuration, naturally occurring in the juice of plant species in the genus *Sorbus*. It is a keto-hexose:

$$\begin{array}{c} \text{OH OH H} \quad \text{OH O} \quad \text{OH} \\ \text{H−C−C−C−C−C−C−H} \\ \text{H H OH H} \quad \text{H} \end{array}$$

m.w.: 180.16 ($C_6H_{12}O_6$); m.p.: 159−161 °C; $[\alpha]_D^{20}$: 43.4°; r.s.: 60−75.

The colorless crystals are soluble in water, 44.5% at 20 °C and 61.3% at 80 °C (w/w).

Industrial large-scale production is performed by biological oxidation of D-sorbitol in solution by the action of *Acetobacter xylinum*, and is the first step in the technical sequence of →ascorbic acid production.

Chemical properties are nearly identical to those of →D-fructose. As an L-sugar, it cannot be energetically metabolized in the human body nor by the microorganisms of the human gut.

Its usefulness as a noncaloric sweetener as well as nondiabetic food is not sufficiently investigated. Its suitability as a component of food and beverages has been proven.

Possible applications as RR may be based on the need for highly purified building blocks of L-configuration.

World production is 25 000 mt/a.

Lit.: Eggersdorfer*0.151−168 (1993)
Rymon-Lipinski/Schiweck* 243−245 (1991)

Sorghum Sugar

→Sugary Sorghum Processing

Soybean

Glycine max (L.) Merr
Leguminosae
G: Sojabohne; F: soya

S. is the typical representative of the legume family. It is the most important world protein crop and, as a co-product, also the most important oil crop.

The s. was already cultivated in China around 2800 B.C. Only in the 17th century was it first introduced to Indonesia and then to Polynesia. Later, it was spread to India, Sri-Lanka and North-Africa. In 1829, it was first brought to the USA, and around 1840, the first agricultural experiments were reported from France, Austria, Hungary, Italy and also from Germany. In the USA s. was first grown as grazing and hay crop, then as an oilseed crop and finally as a protein crop (with oil being the by-product). S. gained world importance around the 1920's in North-America, Argentina and Brazil, with some limited importance in southern European countries, especially Italy, Romania, France and Hungary.

The seeds of s. contain 40% protein, 21% fat, 34% carbohydrate and 5% ash. S. protein is one of the few vegetable proteins that come close to egg proteins. Because of this, the major application of soybean meal is as a feed.

As a by-product of soybean oil refining, so-called deodorizer distillate (0.2−0.5% of the oil) is obtained, which contains 25−55% ffa, 20−40% trigycerides, 7−11% →tocopherol, 10−15% sterols (→stigmasterol) and 25−30% unsaponifiable matter. →Lecithin can also be gained from the oil.

The extraction of s. seed goes through the normal procedure of so-called hard seed solvent extraction. The s. meal may undergo the normal proce-

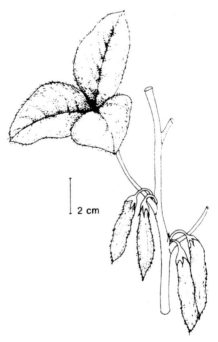

2 cm

Soybean; part of a shoot with pods (from Rehm and Espig*, with permission).

dures of feed production but may also be processed with special extraction methods for food applications, such as soymilk, soy protein concentrates and isolates, and several other soyfood products, or for cosmetic applications, such as protein hydrolysates (→hydrolyzed vegetable proteins).

Soybean production in 1994 (10^6 mt)	seed	oil
USA	69.63	6.97
Brazil	25.50	3.70
Argentina	12.70	1.59
China	16.00	1.21
India	3.30	0.54
Paraguay	2.20	0.13
Canada	2.25	0.18
Indonesia	1.60	0.05
Italy	0.62	0.31
World	138.14	19.54

Lit.: W.R Fehr, "Soybean", in: Röbbelen* 283–301, (1989)

Soybean Oil

G.: Sojaöl; F.: huile de soya
s.v.: 188–195; i.v.: 120–135

S. is a light yellow oil with bland flavor.
It derives from the →soybean and is gained by crushing and extracting with hexane. The oil is degummed and →deodorized. Its fatty acid composition is:

C16:0	C18:0	**C18:1**	**C18:2**	C18:3
2–10%	2–6%	**23–32%**	**48–52%**	2–12%

S. is a versatile edible oil, used from salad oil to margarine. →Hydrogenation yields a more stable oil because the →linolenic acid content is reduced. →Hydrogenation under isomerization (→isomerism) conditions gives an oil that melts at 36–43 °C, which is used in shortenings.
S. is a semi-drying oil, used as such or as fatty acid in →alkyd resins, →coatings and →putty.
Technical s. fatty acids consist of ≈50% →linoleic, ≈25% →oleic and some →linolenic and saturated acids. The acids are used for soap manufacture and for many other technical applications.
In 1994 19.54×10^6 mt were produced.

Lit.: Ullmann* (5.) **A10,** 230
      [8001-22-7]

Soybean Oil Fatty Acids

→Soybean Oil

Spearmint

Mentha spicata
Labiatae
G.: Krauseminze; F.: menthe crépue, menthe verte

S. is a plant that is native to and grows wild in many European countries. It has been used as a kitchen herb for a long time and is now commercially cultivated for the production of →spearmint oil, mainly in the midwestern US and in China, Brazil and Japan.
Its pink flowering tops and small leaves are partially dried before distillation.

Lit.: E.Guenther, "The Essential Oils" Van Norstrand/
      Reinhold, NY (1952)
      Arctander*

Spearmint Oil

G.: Krauseminzöl; F.: essence de menthe crépue

S. is steam-distilled from the flowering tops of →spearmint, yielding a pale olive or pale yellow, mobile liquid with a powerful, warm minty-herbaceous odor, which is truly reminiscent of the crushed herb (main odorous principle: d-carvone).
The main use of s. is in →flavors, especially for chewing gums and toothpastes, where it either constitutes the main note or is used as a modifier in →peppermint flavors. In perfumery, s. is used in traces in men's cologne fragrances, lavender types and certain floral notes as a →topnote booster.

Lit: Ullmann* (5.) **A11,** 233

Species

→Taxonomy

Spermaceti

Syn.: Cetaceum
G.: Walrat; F.: blanc de baleine, spermacéti
s.v.: 120–136; i.v.: 3–4.4; m.p.: 42–50 °C

S. is a white, waxy material, which is derived from the oil gained from the sperm whale (especially head and bones) by cooling and pressing. S. is a wax ester (mainly cetyl palmitate).
It was used in former times for ointments in cosmetics and medicine, in candles and soaps and for the manufacture of →fatty alcohols (→cetyl alcohol).
Due to the restrictions of whale catching, it is almost entirely substituted by esters of fatty acids and alcohols from other sources.

Lit.: [8002-23-1]

Sperm Oil

Syn.: Sperm Whale Oil

G.: Spermöl, Walratöl; F.: huile de spertmacéti

s.v.: 123–147; i.v.: 79.5–84

S. is a yellowish oil of low viscosity which consists of 75% wax esters and 25% glycerides. It is derived from the fat (sometimes also called sperm oil) of the sperm whale (*Physeter macrocephalus*) after the wax esters are separated by cooling and pressing (→spermaceti).

The oil was used as a high-quality →lubricant and in the manufacture of fat liquors (→leather auxiliaries). The economic importance has diminished greatly during the last decades due to whale protection regulations.

S. is substituted by synthetic esters and by →jojoba oil.

Lit.: Ullmann* (5.) A**10**, 236

[8002-24-2]

Spirit of Turpentine

→Turpentine

Splitting (Fat-)

G: (Fett)Spaltung; F: hydrolyse des corps gras

S. of →fats and oils is the first important step into →oleochemistry. Commercially, the most important process is →hydrolysis, which leads to →fatty acids. Another significant process is →transesterification with methanol to yield →fatty acid methyl esters. The two other splitting technologies →saponification, yielding →soaps, and →aminolysis, yielding (substituted) →fatty amides, are less important and nowadays limited to special situations. In all four reactions, →glycerol is the by-product.

Lit.: Ullmann* (5.) A**10**, 254

Stabilizer

G.: Stabilisator; F.: stabilisateur

S. are substances that stabilize a chemical or physical state. S. for plastics: →plastics additives. S. for ice cream: →Food additives.

Starch

G.: Stärke;

F.: amidon (cereal s.), fécule (non-cereal s.)

S. is a carbohydrate polymer that occurs in granular form (→starch granule) in the organs of higher plants and is composed almost exclusively of anhydro-α-D-glucose units (AGU).

It is also an industrial product that consists essentially of the →starch polysaccharides, minor constituents and moisture and is obtained by industrial →wet-milling, refining and drying.

m.w.: $17 \times 10^6 - 80 \times 10^6$ ($[C_6H_{10}O_5]_n$; $n \approx 10^4 - 10^5$ AGU)

The m.w. varies with the plant origin of the s. and approaches that of →amylopectin more than that of →amylose, with the exception of →high-amylose starches.

S. is the most abundant storage polysaccharide in the plant kingdom. It is digestible in the human and animal intestines, which makes it the most important nutritional component in food and feed. The physiological energy content is 4.1 Kcal/g.

Typical s. carriers: (average starch contents in %, d.b.)

Maize (Corn)	(71)	Rye (whole)	(72)	Yams root	(80)
Wheat (whole)	(74)	Triticale	(74)	Sago stem	(90)
Sorghum(Milo)	(75)	Potato	(82)	Bananas(fruit)	(35)
Rice (broken)	(89)	Maniok	(77)	Smooth peas	(40)
Barley(whole)	(75)	Sweet potato	(72)	Wrinkled peas	(40)
Oats (whole)	(72)	Arrowroot	(83)	Broad bean	(43)
				Lentil	(40)

The annual world production of s. by →wet milling is $>15 \times 10^6$ mt. Production is concentrated on just a few s. carriers, which are →maize, →wheat, →potato, →maniok, sweet potato, sorghum and →rice.

In the EC, 1995 production of 6.6×10^6 mt was covered for 95% by maize, wheat and potato, with maniok being the fourth crop. Sweet potato and sorghum s. are also of industrial significance. Production and application of s. from barley, oats, rye and triticale have been tried as well as s. of legumes, such as pea, beans or lentils. Some other tropical s. sources are of local interest, mainly to "third world countries". Special varieties with extraordinary high contents of →amylose or →amylopectin are of rising interest in special industrial and food applications.

Despite the leading position of maize, wheat and potato, the amounts of raw materials utilized for s. production is small compared to the entire world production of these three crops:

5% of the maize, 1% of the wheat and 2% of the potatoes harvested. Surplus production of s. crops allows a great increase in starch utilization as RR to substitute for fossil resources if necessary.

This is possible because of the unique properties of native s., which can be characterized as follows:

- Easy handling of the starch granules by their ready dispersibility in cold water;
- swelling, gelatinization and molecular dissolution in aqueous media on heating or depending on pH;

- separation of whole s. into its individual poly-saccharides, resulting in raw materials for new and unusual applications;
- higher reactivity, compared to cellulose, due to its crystallinity and the possibility of tailored crystallinity;
- great versatility by chemical modifications based on the primary and secondary alcohol groups by classical organic chemical reactions;
- great variation of useful products by splitting the glucosidic bonds via hydrolysis by means of acids, enzymes, heat, radiation, mechanical energy and the further reaction of the resulting products;
- possibility of modifying the granular structure of s. and changes in the swelling and solubility behavior;
- the α-glucosic bonds are the only glucosidic bonds in nature that are digestible by human and higher animals as well as by numerous strains of lower plants; this property is the basis for biodegradability, the most important prerequisite for industrial applications as RR.

Therefore, s. and its derivatives and hydrolysis products are utilized in food as well as in nonfood applications and in feed after being processed by physical modification, by chemical modification or by hydrolysis to sugars.

Average production numbers are: $16-17 \times 10^6$ mt of →starch hydrolysis products, 7×10^6 mt of physically and chemically modified s. and $10-12 \times 10^6$ mt of (native) unmodified starches.

In Germany, about 50% of the entire production of s. and derivatives is used for industrial applications.

The present ratio between food and nonfood uses worldwide is 54% : 46%, with a strong upward trend in traditional industrial consumption and in novel areas as a renewable, biodegradable raw material.

Out of the total s. consumption of 6.0×10^6 mt (1995) in EC and EFTA 3.1×10^6 mt were used in food and 2.8 mt in nonfood applications. The latter has grown by 4.6% and food uses by 3.6% since 1981. In the EC, the future need has been estimated at about 3.6×10^6 mt, only as RR for novel or extended traditional nonfood applications. Similar statistics are known from other highly industrialized countries.

Lit : Schenk/Hebeda* 23–45, 417–439 (1992)
      Van Beynum/Roels* 15–46 (1895)
      Fachverband Stärkeindustrie, Statistical Materials, Bonn (1996)

Starch Acetate
→Starch Esters

Starch, Chemical Properties
G.: Stärke, Chemische Eigenschaften;
F.: amidon/fécule propriétés chimiques

The chemical reactivity of starches is based on the primary and secondary OH groups and/or the glucosidic α-1,4- and α-1,6- linkages.

Reactions of classical organic chemistry on hydroxyl groups:
Substitution:

→starch ethers	AGU–O–R
→starch esters	AGU–O–CO–R
→cross-linked	R–O–AGU–O–R
starch	R–OC–O–AGU–O–CO–R
	(or combinations of both).

Oxydation:
Oxidation of primary OH (→oxidized starch, yielding –COOH and –CHO groups), of secondary OH (periodic acid-oxidized starch) to –CHO and –CO groups, accompanied by C_2–C_3 ring cleavage.

Thermolysis:
1,6-elimination of one water molecule and formation of →levoglucosan.

Splitting and formation of glucosidic linkages:
Hydrolysis:
Hydrolysis of glucosidic linkages by enzymes or/and acids: →thin-boiling starches, acid →dextrins, →malto-dextrins, different →maltooligosaccharides, →maltose syrups, →glucose syrups, →dextrose.

Reversion: leading to nontypical glucosidic linkages and structures by thermolytic reactions during the preparation of →dextrin or in the final phase of starch hydrolysis.

Reaction principles usually applied:
Wet processing, bi-phase:
Carrier for the reactions is water, organic solvents or mixtures of both. Starch and the final products are insoluble. Working up of reaction products is carried out on typical equipment, such as wet-sieving, refining and drying; it is the most frequently used technique in the starch industry, leading to low DS-products.

Wet processing, single-phase:
Starch is dissolved in water, organic solvents or mixtures of both. High DS-values (>1) are possible; separation of final products needs precipitation or crystallization.

Reactions in a homogeneous phase are rather unusual in starch derivatization, but nearly almost all technical hydrolysis are carried out in a homogeneous aqueous phase.

Dry processing:

Granular starch is contacted with solutions or suspensions of reactants, inhibitors or catalysts by soaking or spraying. After predrying, the reaction is carried out at elevated temperatures to yield →dextrins, →starch phosphates; →cationic starches. Successful attempts have been made to react semi-dry systems by →extrusion cooking. Sometimes, excess reactants or by-products must be removed by subsequent extraction.

Lit.: VanBeynum/Roels* 73–99, 101–142. (1985)

Starch, Composition

G.: Stärke, Zusammensetzung;

F.: amidon/fécule, composition

Commercial native, high-purity starches of different plant origin contain minor amounts of organic and inorganic components, in addition to the main constituents of →starch polysaccharides (→amylose, →intermediate fraction and →amylopectin), which account for up to 99% of the d.s.

These minor compounds may be divided into natural (accompanying) plant materials and into foreign matter, taken up during processing, both in technologically inevitable amounts. The water of hydration, which adapts to the water activity of the environment, is an essential structural component. Maximum allowable values of minor constituents and water of →starch granules are frequently regulated by governmental regulations, international standards (ISO) as well as by commercial agreements. Nevertheless, all starch indices published must be considered as average numbers, with differences resulting from the natural materials and the analytical techniques applied.

The most important starch indices for minor constituents and equilibrium humidity are listed below:

Lit.: VanBeynum/Roels* 20–30,(1985)
      Tegge* 301–308,(1985)
      Ullmann* (5.) **A25**, 6–8.

Starch origin	Humidity (%)	Protein	Lipids	Minerals	Phosphorus
			Minor constituents in % (d.s.)		
Corn	11–14	0.3–0.5	0.6–0.7	0.1–0.2	0.015–0.02
Waxy corn	11–13	0.2–0.3	0.1–0.2	0.07–0.1	0.007–0.02
Amylomaize	11–13	–	0.4	0.2	0.07–0.30
Wheat[2]	11–14	0.4–0.5	0.5–0.8	0.15–0.2	0.05–0.07
Rye[2]	11–14	0.14–0.3	0.4–1.5	0.15–0.25	0.24–0.3
Triticale[1]	11–14	0.2–0.3	0.2–1.8	0.04–0.13	0.03–0.05
Sorghum	11–14	0.3–0.5	0.1–0.7	0.08–0.3	n.d.
Rice	12–14	0.4–0.5	0.5–0.8	0.5	0.1–1.0
Barley	11–14	0.5	0.1	0.2	n.d.
Oat	11–14	0.5	0.1	0.2	n.d.
Potato*	17–21	0.06–0.2	0.05	0.4–0.5	0.05–0.1
Manioc	12–18	0.1–0.3	0.01	0.2–0.5	0.01
Sweet potato	13–18	0.2	traces	0.1–0.5	n.d.
Sago	14–16	0.1	0.1	0.2	0.01
Broad bean	14–17	0.35–0.6	<0.01	0.05–0,25	n.d. fibers
Smooth pea	14–16	0.55–0.8	0.08	0.15	0.45
Wrinkled pea	14–16	0.55–1.0	0.08	0.45	2.0

n.d.= not detected

* Potato starch is the only one in which inorganic phosphate exists as a natural →starch ester. Other phosphate is found as organic, solvent-extractable phospholipid, in part complexed by the amylose moiety of the starch granule, especially in starches of the triticeae group.

[1] M.Palasinski et al., Nahrung/Food **29**,257–266 (1985),

[2] F.Schierbaum et al.,starch/stärke **43**,331–339 (1991),

Starch-containing Plastics

G.: Stärkehaltige Kunststoffe;

F.: plastiques contenant des amidons

S. are biodegradable, modified polymers that contain various amounts of starch in combination with traditional synthetic polymers and →plastics additives.

According to the amount of starch and its physical or chemical state, two types of s. need to be distinguished

– plastics that are mechanically blended (filled) with granular starch (6–24%);

– composites of plastic materials and disintegrated starch, the latter forming the continuous phase, up to 65% of the entire mass, compounded by physical and chemical forces.

Starch-filled polymers:

Polyolefines and s. are blended in a ratio of at maximum 1 : 1 to form stable intermediates, called masterbatches. Both components are thoroughly distributed in the batch without chemical reaction or physical changes in the starch granules. →Corn starch with <2% moisture is the preferred starch source. Further processing of the masterbatch with granulated polyolefines and additives (autoxidants, unsaturated fatty acids) is performed on a production line for thermoplastics by coextrusion, injection moulding, film blowing and foaming. Resulting products are polyolefin bottles with <15% starch, films for bags (<6% starch) or agricultural mulching foils. The starch-filled plastics are already used on a worldwide level at more than 100 000 mt/a. In landfills the starch granules are attacked by microbial enzymes in the presence of humidity. The films become porous, and the plastic component becomes susceptable to oxidative degradation. Degradation is reported to occur within 6–12 months under favorable conditions.

Starch-plastic composites:

Biodegradability is enhanced by increasing the starch moiety and reaching its homogeneous distribution in the gelatinized state. This has been successfully achieved in different ways:

– An aqueous dispersion of starch (50%) and a copolymer of ethylene and acrylic acid are processed by casting, blowing, milling and rolling to biodegradable films of high clarity, elasticity and water resistance for packaging materials and agricultural mulch (annual consumption in USA 50 000 mt).

– Dry-processing variants are produced with starches of 5–30% moisture, plastic compo-

nents and additives by coextrusion under conditions that favor the thermoplastic behavior of the blend.

– Composites were produced, containing 60% starch, 16% polyvinyl-alcohol, 20% →glycerol as plasticizer and additional polyvinyl-chloride as water administering coating for use as packaging materials and as agricultural mulch.

– Thermoplastic properties can be varied by mixing low- or high-density polyolefines, corn starch and plasticizers. They all can be processed by typical plastics technologies, such as molding or film-blowing. Such resins contain up to 60–75% natural materials and up to 40% synthetic polymers.

In such products, starch forms the continuous phase in the thermoplast under conditions of elevated pressure and temperatures above its T_g and m.p. The components are bound together by strong physical forces and/or covalent linkages; the m.w. of the plastic moiety is between 5000 and 50 000.

Commercial products cover a broad application range of packaging or packaging aids : bags for shopping and trash, containers for dry powders and oils, single-use table settings, drinking straws, typewriter cartridges, pen barrels, cosmetic boxes, disposable razors, packaging foams, medical disposals, lawn and leaf compost bags, bottles and other containers. Several products have been available since 1990. Investigation of biodegradability, preferentially by the modified Warburg test, by bacterial or fungal attack, composting, sea and lake water ageing, has confirmed the wanted rapid degradability of the starch phase, leaving the minor, slower degrading synthetic part.

Lit.: H.Röper et al., starch/stärke **42**,123–130 (1990),
W.M.Doane, Cereal Foods World **39**,556–563 (1994)

Starch, Cross-linked

→Cross-linked Starches

Starch Derivatives

G.: Stärkederivate; F.: dérivés de l'amidon

S. are chemically →modified starches, where the primary/secondary OH-groups are substituted in part or entirely by ether and/or ester groups.

In contrast to →cellulose derivatives, which are modified to a greater extent to make them water-soluble, low →DS are sufficient in most of the s. to have the wanted effect on their functional properties. Derivatization reactions are mostly performed in heterogeneous aqueous phase at alkaline pH, keeping the s. in the granular state. After completing the reactions, the starch granules are washed free from reagent residues and other solubles, dewatered and dried.

S. comprises the following products:

→ starch esters, neutral or acidic with organic or inorganic substituents, which cause →anionic or neutral behavior; → starch ethers, with neutral, →anionic or →cationic starch behavior; →cross-linked starches, in which the →starch polysaccharide chains are linked together or reacted with another macromolecule by ether- or ester-bonds of bi- or trifunctional reagents.

Mixed derivatives of the three groups are prepared for combined functional effects.

For practical uses see: →anionic starches, →carboxymethyl starch, →cationic starches, →cross-linked starches. →hydrophobic starches, →oxidized starches, →starch industrial applications.

Lit.: van Beynum/Roels* 73–84, 89–99 (1985)
Ullmann* (5.) **A25**, 12–21

Starch, EC Market

G.: Stärke, EG-Markt;
F.: marché européen de l'amidon

Starch production and processing in 1994 (in $\times 10^6$ mt) surpassed 6.0; 3.2 for food utilization (annual growth rate 3.6%) and 2.8 for nonfood utilization (annual growth rate 4.6%).

Industrial nonfood application was largely favored by the EC-market regulations that became effective in the financial year 1989/90. The fundamental step was the drift of annual production refunding from a starch-producing platform to a chemico-technical application platform. Production refunding serves to smoothen the discrepancies between the fluctuations of the world market prices and the interior European price regulations as well as to protect against dumping by third countries. The EC nonfood starch consumption potential in 2000, with the exception of starch as energy source, is estimated to increase to 3.6×10^6 mt.

The EC starch market (1991) was supplied by (in $\times 10^6$ mt):

3.8 corn starch	from	6.1 corn ,
1.2 wheat starch	from	2.4 wheat,
1.2 potato starch	from	6.7 potatoes, and

small amounts of maniok, rice, →waxy corn and legume starches.

Practically, all starch sources are produced in the EC. Former imports of corn from the USA have been diminished to 5% by the corn supply from Spain and Portugal.

The shares of EC and EFTA countries on all starch supplies (1992) are as follows:

Country	$\times 10^6$ mt	% of all	Crops processed		
			Corn	Wheat	Potato
Germany	1.3	22	2	8	9
United K.	1.00	17	3	5	–
France	0.94	16	3	4	3
Italy	0.53	9	4	2	–
Netherlands	0.52	9	–	1	5
Spain	0.47	8	2	2	2
other EC	0.53	9			
EFTA	0.59	10	2	1	6
Σ	5.90				

These amounts are distributed by processing and consumption as food and nonfood products of native, →modified and hydrolyzed starches, in $\times 10^6$ mt (1992) as follows:

EC + EFTA		Native	Modified	Hydrolyzed
All	5.9	29%	16%	55%
Food	3.1	14%	7%	79%
Nonfood	2.8	27%	27%	46%

The most important sectors of consumption were (1992):

Non-Food	%	Food	%
Paper	20	Confectionary	14
Chemicals and fermentation	13	Beverages	12
Corrugated paper and cardboard	7	Food preparations	7
		Fruit products	6
Other	7	Other	14

High starch-producing potential are found, besides EC and EfTA, in Poland, the Baltic Republics and the GUS states. Predominant starch sources are potato and corn, with less production from wheat, waxy corn and legumes.

Lit.: Fachverband der Stärkeindustrie, Statistical Materials, Bonn,(1993)

Starches, Anionic
→Anionic Starch

Starches, Cationic
→Cationic Starches

Starches, Extrusion-cooked
→Extrusion-cooked Starches

Starches, Industrial Application

G.: Stärken, Industrieanwendung;
F.: amidon/fécule, application industrielle

Starch uses are traditionally split into food and nonfood, or into native, modified and hydrolyzed starches, starch sugars and fermentation products (→starch, EC market, →starch industry, world). Strongest growth is in →starch hydrolysis products for bulk sweeteners as well as for sugars as C-source in fermentation processes to produce solvents, starting and auxiliary materials for chemical reactions and for fuel ethanol.

Use of physically and chemically modified starches will gradually overcome native starch utilization. Methods of physical modification are favored for environmental reasons.

The outlook for starch processing and main fields of utilization until the year 2000 is presented below:
Development of novel plant raw materials includes:
– fine granule starches from oats or rice;
– high-amylose starches from wrinkled pea and amylocorn;
– amylopectin starches from waxy corn, waxy rice, waxy sorghum, potatoes (breeding);
– conversion technology for new and old starch sources;
– utilization of the entire starch plant biomass.

→Starch hydrolysis products, primarily glucose, are equifunctional with →sucrose in fermentation, biotechnology and as reaction adjuvants. Preferences are determined by regional availability and market price. Thus, glucose is the preferred economical material as well as energy-economical RR in the USA and parts of Asia, whereas sucrose dominates in the EC and some regions in Asia and South America.

Starch cannot be substituted because of its highly specific properties and versatility as natural polymer.

The outlook of for the need of industry starches in the EC for the year 2000 is 3.6×10^6mt.

Lit.: "Stärke im Nichtnahrungsbereich" Reihe A/ Heft 388, 5–40, Schriftreihe des Bundesministers für Ernährung, Landwirtschft, Forsten, Landwirtschaftsverlag GmbH, Münster-Hiltrup (1990) Nachwachsende Rohstoffe, Konzept zur Forschungsförderung 1990–1995, 33–39, 65–68, 82–84; Bundesminister für Forschung und Technologie, Bonn (1990),

Starches, Processing and Industrial Utilization
(Corn, Wheat, Potato, Maniok, Sorghum)

Mode of process.	Physical	Chemical	Hydrolysis	Fermentation (of hydrolysis products)
	hydrothermal, thermal, thermolytical, extrusion, blending	substitution,(esters, ethers,crosslinks), oxidation, copolymerization	acids,enzymes, and combinations thereof	biotechnological production; low-molecular, polymeric compounds.
Important products and properties	pastes, dextrins, extruded products, levoglucosans, HMF, inhibited starches	hydrophilic, hydrophobic, freeze-thaw stable, heat/pressure resistant starches, super slurper, high compatible starches, hydrogenated products	soluble,thin-boiling starches, maltodextrins, maltose and syrups, glucose syrups, dextrose, fructose and syrup	nutritional yeasts, amino acids, organic acids, solvents, ethanol
		Material Economics + Energy		
General fields of use	coatings, drilling muds, adhesives, binders, starting materials and intermediate products for syntheses, filling and auxiliary products, protective coatings, foils, films, covers, biodegradable plastics, starch plastic compounds, packaging material			fuel ethanol, biogas processing, residues for combustion
Utilizing industries	papermaking, textile, glues, binders, paint thickeners, geological prospecting, metal foundry, plastic materials, cosmetics, pharmaceuticals, chemicals, hydrogenated products			motor fuel, fuel additives, local energy production

Starches, Oxidation
→Oxidized Starches

Starch Esters
G.: Stärkeester; F.: esters de l'amidon (fécule)

S. are →starch derivatives in which some or all of the available hydroxyl groups are esterified. Inorganic or organic esters may be prepared:
- starch sulfuric acid esters, starch nitric acid esters, starch orthophosphoric acid esters (monostarch phosphates, distarch phosphates), starch xanthogenic acid esters;
- starch acetates, starch adipates, starch citrates, starch laurates, starch maleates, starch succinates.

From the great variety of s. described, only a few are marketed to a high extent, due to the special properties arising from the nature of the substituent group.

The most important s. are:

Starch monophosphates are prepared by reacting of granular starch with monosodium phospate, disodium phosphate or sodium tripolyphosphate at pH 5–8.5 in the dry state at 120–170 °C:

$$AGU-OH + Na_2HPO_4 \longrightarrow AGU--O-\overset{\displaystyle ONa}{\underset{\displaystyle ONa}{P}}{=}O + H_2O$$

The reaction can be performed in a belt dryer, by spray-drying or by extrusion. The esters formed with a low DS (<0.1) belong to the →anionic starches. They exhibit enhanced paste viscosity, clarity, freeze-thaw reversibility and reduced retrogradation. The pastes are drastically thinned at acidic pH, in the presence of salts and high cooking temperatures. Stabilization is achieved by →crosslinking.

Fields of application are in the paper and textile industry, in adhesives, as scale inhibitors, flocculants and paste stabilizers.

Distarch-phosphates belong to the →cross-linked starches with extremely high paste stability against acids, pressure cooking and shearing.

Starch nitrates were widely applied as explosives during the building of the Panama channel; at present they have no importance.

Starch acetates and some other **carboxylate esters**, symbolized by the introduction of R–CO–O– groups, are the most important organic esters. They are prepared by heterogeneous phase reaction of granular starch in slightly alkaline media (pH

8–8.5) at 15–25 °C with the respective acid anhydrides

$$AGU-OH + (H_3C-CO)_2 \xrightarrow[+ NaOH]{}$$

$$AGU-O-\underset{\displaystyle O}{\overset{\displaystyle ||}{C}}-CH_3 + CH_3COONa$$

These granular, weakly substituted products (DS 0.02–0.1) exhibit reduced swelling temperatures, enhanced swelling power (water-binding capacity) and freeze-thaw reversibility and provide thickening, bodying and texture-enhancing properties. Instability of the paste consistency against cooking, shearing and acid may be overcome by crosslinking. Starches that contain 0.5–2.5% acetyl groups are permitted (FDA) without declaration in canned, baked or instant goods, baby food, frozen food, ice cream and fillings.

In nonfood applications, such low-substituted esters serve as warp sizing in textiles, as glass-fiber sizing, surface sizing in paper making and as adhesive component.

Starch acetates lose their water-swelling and paste-forming properties when DS increases. With up to 15% acetyl groups, they are still swellable in water at 50–100 °C; at 40% acetyl content, they are only soluble in organic solvents, such as acetone, trichloromethane and aromatic hydrocarbons. Such peracetylation is best reached in homogeneous-phase reactions after dissolution of the starch in N-containing solvents. The solutions may be cast, spun or blown like →cellulose acetates, but they are of inferior quality. Such s. may reach novel significance as RR because of their compatibility with synthetic polymers in thermoplastic molding and film plasticising. →High-amylose starches are the preferred starting materials for esters as biodegradable plastics materials.

Starch maleate of low DS exhibits lowered onset temperatures of gelatinization and high water-binding capacity. Besides food application, it is used as foundry core binder.

Starch succinates and, to a higher degree, alkenyl succinates have, due to their hydrophobic groups, surface-active properties. They are used for preparation of emulsions and for encapsulation of lipophilic substances. Additionally, starches with poor water-binding capacity and low paste viscosity in the native state, like triticeae starches[1], are upgraded by succinylation to highly stable pastes.

Lit.: Van Beynum/Roels* 91–93, (1985)
        Ullmann* (5.) **A25**, 17–19

[1] F.Schierbaum et al. starch/stärke **46**,331–339 (1994).

Starch Ethers

G.: Stärkeether; F.: éthers de l'amidon

S. are →starch derivates in which some or all of the hydroxyl groups of the polysaccharide are etherified.

Etherifying of starch is essentially possible with any reagent capable of forming an ether bridge with alcoholic OH-groups. Out of a large variety of possible s., only three groups have gained technical significance:

– hydroxyalkyl starches (hydroxyethyl- and hydroxypropyl ethers),
– →carboxymethyl starches (starch glycolic acid ethers),
– →cationic starches (amino-alkyl, alkyl ammonium ethers).

Preparation of hydroxyalkyl ethers is mostly run in an aqueous, alkaline heterogenous-phase process, keeping the starch in the granule state if DS <0.1. Higher DS needs organic solvent/water systems.

$$AGU-OH + H_2C-CHR$$
$$\underset{O}{\diagdown\diagup}$$

$$\xrightarrow[+ \text{NaOH}]{} AGU-O-CH_2-\overset{\overset{H}{|}}{\underset{\underset{R}{|}}{C}}-OH$$

R = Alkyl

Reagents applied are alkylene oxides (and alkylhalogenides). Alkylene oxide gas is introduced into the alkaline starch slurry of 35–45% solids content and reacted until the wanted DS is reached. The ether is filtered, washed and dried. Gelatinization of the reacted starch is frequently inhibited by neutral salts (sodium sulfate), which are added up to 30% to the reaction mixture. Ether groups, dependent on DS, enhance remarkably the hydrophilic character of the starch granules and lead to decreased swelling temperatures down to cold-water swellability, increased water-binding capacity, peak viscosity, stability against heating, acidic pH and mechanical forces, clearness, freeze-thaw reversibility of the pastes. The tendency for aggregation (→retrogradation) and setting to gels is greatly inhibited. The high solubility causes good film forming properties. Further enhancement of mechanical and pressure stability may be achieved by cross-linking. This is done preferentially for applications in processed food for thickening, stabilizing and mouth-feel enhancement.

Nonfood hydroxyethyl starch applications are in the paper industry for surface sizing and coating, as quality enhancer for pigmented papers, as wet-end additives; in textiles for warp sizing; as an adhesive component for bag pastes, case sealing and enveloping materials. Hydroxypropyl starches with higher DS are used routinely in pharmacy as a blood plasma extender.

Ethers of →high-amylose starches achieve increasing relevance for the production of biodegradable →starch-plastic films and filaments.

Lit.: VanBeynum/Roels* 94–96, (1985)
        Ullmann* (5.) **A25**, 9–21.

[1] F. Schierbaum et al. starch/stärke **46**, 331–339 (1994).

Starch-Filled Polymers

→Starch Containing Plastics

Starch Glycolic Acid

→Carboxymethyl Starch

Starch Granule

G: Stärkekorn; F: grain d'amidon (de fécule)

S. is the state in which starch occurs in the plant cell and as commercial starch after isolation from its plant raw material or after modification without changing the granular form.

Starch granules are typical for each species and variety of the plant from which they are derived. They differ in

shape: round or ellipsoid, edged or compounded,
size: 1–120 μm,
size distribution: random, normal (Gaussian), bimodal.
density: 1.5–1.6;
color: white to creamy;
solubility: insoluble in cold water < 45 °C ; onset of swelling > 45–65 °C, depending on starch species and pretreatment; gelatinization and dissolution with rising temperature; total dissolution requires pressure heating for cereal and legume starches; dissolution in aqueous media without heating takes place in the presence of alkali or salts of monovalent cations and anions or organic N-containing bases.

Commercial s. contain 99% of the d.s. as →starch polysaccharides and not more than 1% minor constituents (→starch composition), such as crude protein, lipids, cations (Na^+, K^+, Mg^{++}, Ca^{++}), anions (Cl^-, SO_3^-, SO_4^{--}, PO_4^{---}, SiO_2) and trace elements; in cereal starches, most phosphorus is part of the lysophosphatide fraction of the lipids, which are bound to the amylose by complexing.

Water is bound depending on the water activity of the environment according to the sorption isotherms with a sigmoid shape and hysteresis between the ad- and desorption branches.

Equilibrium water contents under storage conditions differ between cereal, legume and some tuber starches: 12–14% for cereal and 18–21%. for potato starches.

S. are partially crystalline when investigated in a water- saturated state. This level is reached as follows:

Maize 39%, rice 38%, wheat 36% ; tapioca 38%, sweet potato 37%, potato 25%.

Drying to zero humidity eliminates the crystalline pattern in x-ray diffraction studies.

The crystalline regions of the α-1,4-D-glucans exhibit polymorphism, which is the reason for different behavior of the starch species:

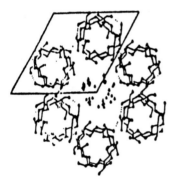

B-polymorph

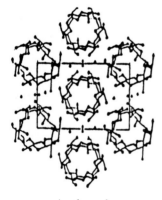

A-polymorph

Unit cells and helix packing of crystalline parts of starch granules (by Blanchard, 1987)

- Starches from cereals – the A-polymorph – show left-handed, parallel-stranded double helices that are crystalized in the monoclinic space group B 2 .
- Starches from roots and tubers – the B-polymorph – show double helices that are crystalized in the hexagonal space group P 6.
- Starches from legumes – the C-polymorph – show A- and B-structures existing together in various portions, depending on the plant species.

Very pure A- or B-polymorph types are found in maize or potato starches, whereas small amounts of the B-type (<8%) have been detected in the starches of the triticeae-group (rye, →wheat, triticale).

Comparing the three-dimensional order of the double helices in the unit cells it becomes obvious that the central channel in the hexagons is filled either with another double helix (A-type) or with water molecules (B-type).

This leads to the principal differences between the densely packed cereal type and the higher hydrated root and tuber types of the starch granules. The structural differences become evident in functional properties, such as gelatinization, solubility, viscosity, solution properties and gelling (set back). Crystallization of some parts of the gelatinized starches takes place on cooling, leading to the B-type, independent of the polymorph type of the original starch. The B-pattern, together with a sub-crystalline diffraction spectrum, the V-spectrum, is the typical structural basis of the processes known as starch →retrogradation.

The native structure of the s. may be changed with severe consequences to the properties: dextrinization (solubility, hydrolysis, dehydration and formation of new linkages), radiolysis (solubility, destruction of linkages), annealing, heat-moisture treatment (inhibition of swelling, reduced solubility, changes in viscosity formation, higher stability of solution properties). The latter principles of physical treatment can be used instead of chemical methods of crosslinking without any discharge of waste water.

Numerous treatment practices of chemical modification are performed in the granular state, with the advantage of working in biphasic systems that allow easy separation of the reaction products and the effluents.

One typical starch granule property is its resistance against enzymatic attack. Up to now, all current processes of enzymatic modification of starch in industry need – at minimum – gelatinization. Otherwise, the enzymic processes will run with unacceptable velocity and completeness. Starch hydrolysis in the granular state is performed completely only in nature.

Lit : Galliard* 16–54 (1987)
      Ch.Gernat et al. starch/stärke **42**, 175–178 (1990)
      A.Imberty et al. starch/stärke **43**, 375–384 (1991)

Starch Hydrolysis Products

G: Stärkehydrolysate;
F: produits d'hydrolyse de l'amidon

S. is the generic name for such products as →maltodextrins, →glucose syrups, →maltose syrups, →maltooligosaccharides, →dextrose, →maltose.
These starch sweeteners are produced by acid →hydrolysis or enzymatic →amylolysis. Additional modification of the product spectrum is achieved by isomerases (glucose isomerase) or by transferases (cyclodextrin glucosyltransferase).
The worldwide production of $16-17 \times 10^6$ mt of starch sweeteners (entire starch production $>37 \times 10^6$ mt) is highly concentrated in the USA, E.C., Japan and South-East Asia.
Rising demand is reported for nonfood applications in chemical synthesis, biotechnological production of organic acids, solvents and →biopolymers and for fuel ethanol.
Slightly hydrolyzed products with behavior of high-molecular starches are not termed s. but →thin-boiling starches (Lintner and Zulkowsky starches) or →white dextrins.

Lit : Schenk/Hebeda* 79–366 (1992)
     Ruttloff* 577–645 (1994)

Starch Industry, World

G.: Stärkeindustrie,Welt;
F.: amidonnerie (féculerie), monde

The starch industry has the following objectives worldwide:

– extract natural starch as pure and as unchanged as possible from their sources,
– isolate and process all other non starch components (entire biomass) as completely as possible,
– modify native starches or →starch polysaccharides by physical or/and chemical treatment to upgrade or adapt their functional properties to special fields of application,
– hydrolize native starches to products with different DE, ranging from slightly diminished m.w.

to total hydrolysis to →dextrose, as well as process the latter further to →fructose products.
The **Entire world production** of starch and starch products is difficult to determine. It is estimated to be about $>37 \times 10^6$ mt/a. →Corn starch accounts for about 74 % (26×10^6 mt) of the production, followed by 15 % (5.5×10^6 mt) of →wheat and →potato starches, the first of which has the higher growth rate. Maniok (tapioca, cassava) starch is produced in an amount of $\approx 3.5 \times 10^6$ mt, and →waxy corn starch and →amylocorn account for 0.2×10^6 mt. All other starch crops from cereals, such as sorghum, →rice, →barley and →oat, from roots or tubers, such as arrowroot, yam and sweet potato, from legumes, such as →broad bean or →pea, and from stems of palmae, such as sago, contribute more than one million mt (1995).
Main production countries for commercial starches and the main uses of them are:

Corn starch	USA, Japan, EC/EFTA, FSU	Sweeteners, food industry, paper, corrugated board, ethanol and other fermentations.
Wheat starch	Australia, EC/EFTA, Japan	sweeteners, bakery products, food industry, paper, adhesives.
Potato starch	EC/EFTA, FSU, Poland	food industry, paper, adhesives.
Maniok starch	Brazil, Thailand	food industry, adhesives, ethanol.
Waxy corn st.	USA	food industry, adhesives.

Of the entire corn starch production 70 % is converted into caloric sweeteners, such as HFCS (→isomerized sugars), →glucose syrups, →dextrose, →maltose and other →starch hydrolysis products.
More detailed production and consumption figures are given in →starch, EC market for Europe.
In the USA (1994/95), 17.272×10^6 mt corn were processed by →wet-milling (7.4% of the harvest) to produce 12.75×10^6 mt of corn sweeteners. For ethanol production 15.44×10^6 mt corn were used.
Of HFCS world production 77% is located in the USA, followed by Japan; here, the prices for corn sweeteners are lower than for →sucrose-based products. Continuous growth is now taking place in starch and sweetener production in China, South-East Asia, South America and FSU. Development of sweeteners in the EC is blocked by the

EC quota for HFCS production to protect the beet sugar industry.

Lit.: J.J.M.Swinkels, starch/stärke **37**, 1–5, (1985)
F.O.Lichts Internationaler Alkohol- und Melasse-bericht **31**, 52, 280, 281, (1994)
Fachverband der Stärkeindustrie, Statistical Materials Bonn, (1993)
Ruttloff* 597–603 (1994)

Starch, Intermediate Fraction

G.: Stärke, Zwischenfraktion;

F.: produit intermédiare de l'amidon (de la fécule)

S. is the →starch polysaccharide fraction of the native starch granule that consists of mostly α-1,4-glucosidically joined AGU and of small amounts (3–20) of α-1,6-glucosidic branch points, thus leading to a slightly branched →amylose or an →amylopectin with high average chainlength.

It is generally isolated and estimated together with amylose. Its detection results from the observation of incomplete splitting of certain pure amyloses because of incomplete degradation with highly purified β-amylases.

Lit.: Tegge* 31–35 (1988)

Starch Modification

→Modified Starches

Starch Pastes

G.: Stärkepaste, Stärkekleister; F.: colle d'amidon

S. are viscous masses, resulting from swelling, →gelatinization and colloidal dispersion of →starch granules in aqueous media, in the presence or absence of chemical reagents at normal or elevated temperatures.

The physically defined state of "starch paste" is the most employed application form of starches. It covers a broad range of different flow properties. Viscosity depends upon kind and concentration of starch, temperature of heating and holding, shearing force by agitation, stirring, flow through pipes, and on pressure.

Mostly, s. are biphasic: a colloidal dispersion of →starch polysaccharides, mainly →amylose, undissolved swollen starch granules and fragments thereof. Viscous flow and viscoelastic behavior are further determined by the degree of →retrogradation and gel-setting of the dissolved amylose, including dissolved and undissolved →amylopectin, in the network.

Pastes of native starches differ with respect to their origin:

– root, tuber and waxy starches are opalescent or clear, high but instable in viscosity with respect to cooking, mechanical shear and acidic pH;
– cereal and legume starches give pastes that are turbid, low in viscosity and set back easily to shiny white gels of high viscoelasticity; they undergo syneresis.

All pastes of native starches are rendered insoluble by freezing; when thawed, they exhibit hardly soluble, sponge-like structures. →Retrogradation-based behavior can be overcome by maintaining the paste at alkaline pH or at temperatures above 60 °C. Introduction of substituent groups or oxidation diminishes or removes the tendency to retrograde.

Pastes can be produced without heating by dispersing of the starch in alkali of suitable concentration, an operation that may lead to monophasic dispersions.

Alkaline pastes are stable, but they undergo slight oxidation on prolonged storage. When neutralized, they undergo rapid retrogradation.

Lit.: Ullmann* 85.) **A25,** 4–6
J.M.V. Blanchard, J.R.Mitchell, "Polysaccharides in Food" 139–152, Butterworths, London/Boston/Sydney/Wellington/Durban/Toronto (1979)

Starch Phosphates

→Starch Esters
→Cross-linked Starches

Starch, Physical Properties

G.: Physikalische Stärkeeigenschaften;

F.: amidon (fécule), propriétés physiques

The useful (functional) properties of native starches depend on both →starch granule structure and →starch polysaccharide composition. Additional influence is created by the minor constituents (→starch composition).

The →starch granule structure, organized by crystalline and amorphous regions and stabilized by hydrogen bonds, leads to insolubility in cold water. Absorption and desorption of moisture follows sigmoid isotherms with hysteresis. The B-polymorph type hydrates the most, the A-polymorph type the least. The "mixed" C-type has an intermediate position. Water exchange of starch granules is accompanied by a reversible increase or decrease of the granule diameter.

The reversibility of granule swelling is lost if, in the presence of excess water, the gelatinization temperature is reached. This onset temperature of swelling is different for various starches. Granule swelling without heating may be achieved by increasing the pH to >10. At the swelling point, the granules begin to take up water, loose their crystallinity and their original shape, and increase in volume in quite different degrees. Beginning from this "gelatinization temperature range", diffusion of soluble polysaccharides, especially amylose, into the surrounding liquid begins. Increasing the temperature to 100 °C leads to more granule swelling and dissolution of both linear and branched polysaccharides. The most-swollen granules burst, fragment and dissolve.

From the onset of swelling most starch slurries exhibit increased viscosity and reach the consistency of utmost granule swelling, but the consistency decreases if heating and agitation are continued because of fragmentation and dissolution.

A boiled starch slurry, →starch paste, is characterized as a colloidal solution of amylose and amylopectin, in which undissolved starch particles or small starch granules are suspended. Complete solution is achieved by further increasing the temperature, violent agitation or homogenization or pressure cooking until 150 °C. The intensity needed for complete solution as well as for maximum paste clarity largely depends on the kind of starch. Rapid thinning of most pastes occurs in an acidic milieu of pH<5. On cooling most starch pastes become turbid and finally set back to chalky-white or opaque, thermally irreversible gels (puddings).

Only →potato starch and →waxy starches yield clear or opalescent nonsetting pastes. Further cooling below 0 °C and formation of ice crystals causes irreversible aggregation of all polysaccharide material, which results -after thawing- in the separation of phases into porous, sponge-like bodies and pure solvent phase (freeze-thaw irreversibility).

The gel-network formation is thought to be the consequence of initiating inter-chain interactions of amylose molecules (retrogradation, aggregation) to form an amylose network, which stabilizes the amylopectin phase: a partially crystalline B-structure is rapidly formed by the amylose; a second, slow-structuring phase of the amylopectin forms reversible crystalline structures.

This generally adopted model of an amylose gel, filled with amylopectin, includes possible interactions between amylose and sufficiently long amylopectin chains.

The native starches of different plant origin obey these general principles to a different extent as shown by their **pasting indices**:

Starch origin	Gelatinization temperature, visual, °C	Pasting temperature Viscograph °C	Peak consistency, Viscograph Units, BU	Water binding capacity, 95 °C (w/w)
Corn	65–76	75–80	700	24
Waxy corn	63–76	65–70	1100	64
Amylocorn	67–92	90>95	≈0	6
Wheat	51–59	85–90	400	21
Rye	48–54	85–90	450	12
Triticale	50–59	85–90	200	15
Sorghum	68–78	75–80	700	22
Rice	68–78	70–75	500	19
Potato	58–68	60–65	>1000	>100
Maniok	59–69	65–70	>1000	>70
Sweet potato	58–72	65–70	>1000	46
Sago	60–72	65–70	>1000	100
Broad bean	56–81	70–95	300	8–10
Smooth pea	55–70	90–96	200	8–10
Wrinkled pea	57>100	96>100	≈0	3–6

According to the listed pasting indices, the **functional properties** of the pastes, gels and films differ widely, thus permitting a choice of the wanted functional properties for a specific application.

Starch origin	Peak heighth	Consistency stability	Paste texture	Film clarity	Rate of retrogradation
Corn	medium	medium	short	opaque	high
Waxy corn	high	low	long	high	zero
Amylocorn	zero	–	–	turbid	high
Wheat, Rye, Triticale	low	high	short	turbid	high
Sorghum	medium	medium	short	opaque	high
Rice	medium	medium	short	opaque	high
Potato	very high	very low	long	high	very low
Maniok	high	low	long	opaque	low
Sweet potato	high	low	long	opaque	low
Sago	high	low	long	opaque	low
Broad bean	low	high	short	turbid	high
Smooth pea	low	high	short	turbid	high
Wrinkled pea	zero	–	–	turbid	high

Starch pastes may be dried to films, especially for nonfood applications, such as adhesives and coatings for paper sheets or textile fibers. The desired technical properties are plasticity, strength, water solubility, response to humidity, transparency and gloss. They result from the type of paste texture, clarity and gelling, which are consequences of amylose-amylopectin interactions. Only starches from waxy corn, potato or other root and tuber crops exhibit the desired properties, whereas native cereal starches are not so well suited in this respect because of their short texture and rapid setting to gels, which favors insolubility, turbidity and brittleness.

Drawbacks of most native starches are:
- necessity of heating for gelatinization and dissolution (except for alkaline dispersion);
- instability of consistency against prolonged heating, violent agitation and acids;
- turbidity on cooling and set-back to gels;
- thermal irreversibility of the gels;
- lack of freeze-thaw reversibility;
- formation of brittle and rather instable films.

They may be overcome by physical, chemical and/or biochemical processing methods (→modified starches, →starch derivatives, →starch hydrolysis products).

Lit.: Van Beynum/Roels* 0.30–46 (1985)
S.Radosta et al. starch/stärke **44**, 8–14, 1992),
F.Schierbaum et al. starch/stärke **46**, 2–8, (1994)

Starch Plastics

G.: Thermoplastische Stärken;
F.: amidons thermoplastiques

Biologically degradable s. consist of merely →starch, →high amylose starch, esters or ethers of starchy materials and added plasticizers.

The most important variables in production are high pressure, temperature and product moisture. The latter, in combination with plasticizing chemicals (→glycerol, glycol, →sorbitol), regulates T_g as the most important index for plasticity. Low starch moisture (<20%) and high extrusion temperatures (140–170 °C) lead to entire disintegration of the starch and formation of a continuous polymer phase of desired elasticity, which is influenced by the nature and amount of the plasticizer. Standard equipment for thermoplastics processing can be used to produce biodegradable films and other disposables from starchy materials:
- Injection molding to prepare starch disposables, such as drinking cups, plates and other single-use dishes;
- medical capsules;
- foaming extrusion of potato starch and plasticizers to produce rigid plastic foams, which may replace the nondegradable polystyrene foam pellets and foamed upholstery materials;
- extrusion processing of →amylomaize starch to granules as raw material (master batch) for the production of foils and films of high flexibility, firmness and clearness;
- utilization of traditional wafer baking equipment to produce starch-based foam items for one-time use and subsequent disposal; the product spectrum includes a multitude of fast-food service items, blisters and trays for confectionary.

Plant fibers and plasticizers are added to the native starch to enhance mechanical stability. Temporary water resistance is achieved by coatings.

All cited materials are entirely biodegradable in landfills, composting, water treatment, biogas devices and in animal feed. Large-scale production will increase in the near future, together with effective marketing, because of environmental demands and decreasing prices.

Lit.: H.Röper et al.: starch/stärke **42**,123–130 (1990),
W.M.Doane: Cereal Foods World **39**, 556–563 (1994).

Starch Polysaccharides

G.: Stärkepolysaccharide;
F.: polysaccharides de l'amidon

S. consist of α-1,4- and α-1,6-glucosidically linked AGU polymers. Purified native starch contains ≈99% of three polysaccharides: →amylopectin, →amylose and →starch intermediate fraction.

The relationship between amylose and amylopectin depends on plant origin. It may vary from ≈80% amylose in →amylomaize to zero in →waxy corn. Most kinds of starch contain 15–35% amylose and 65–85% amylopectin (intermediate fraction is estimated with one of the other fractions).

Amylose content (%) of starches of different plant origin:

% amylose = 100 − % amylopectin.

Corn	28	Waxy corn	0
Barley	29	Amylomaize	50–80
Wheat	28	Wrinkled pea	50–78
Potato	21	Barley, high-amylose	42
Banana	16	Smooth pea	35
Maniok	17	Arrowroot	20
Sorghum	28	Sago	27
Rice	17	Rye	28
Oats	27	Parsnip	11

Lit : Van Beynum/Roels* 27–29 (1985)
Y.Takeda et al.: Carbohydr.Res. **165**, 139–145 (1987)

Starch, Pregelatinized

→Pregelatinized Starch

Steam Distillation

→Distillation

Stearate

→Soaps

Stearic Acid

Syn.: n-Octadecanoic Acid; C18:0
G.: Stearinsäure; F.: acide stéarique

$CH_3–(CH_2)_{16}–COOH$

m.w.: 284.47; m.p.: 69.3 °C; b.p.: 383 °C

S. is a colorless waxlike material, soluble in alcohol and other organic solvents. S. derives from →tallow (main source) and many other animal and vegetable →fats and oils. S. is a by-product of many oleochemical operations, e.g., refining of vegetable oils for nutritional purposes and of fats and greases for technical applications.
→Hydrolysis is normally the first step in production. →Saponi-fication is nowadays of minor importance. Unsaturated acids may be converted to saturated acids by →hydrogenation before or after →splitting or are separated by cooling, pressing and filtering (→winterization) or other →crystallization processes.
A lot of different qualities are commercially available. Normal technical-grade s. contains 45% →palmitic acid, 50% s. and 5% →oleic acid.
Uses are in manufacture of →candles, →soaps, cosmetics, crayons (→inks), →polishes. A large outlet is its use in rubber compounds (→rubber chemicals).

Lit.: Ullmann* (5.) **A10**, 245
[57-11-4]

Stearin

G.: Stearin; F.: stéarine

S. is the ester of →stearic acid with →glycerol, whereby one, two or three of the hydroxy groups of glycerol may be esterified with stearic acid (→glycerides).
In commercial terminology, s. is also a mixture of mainly →stearic acid and →palmitic acid, which results after the separation of →oleic acid from crude →tallow fatty acids or acids derived from other oils by →hydrolysis and →crystallization. S. consists mainly of stearic and palmitic acid and some other saturated acids depending of origin.
S. is widely used in →candles and for the manufacturing of →metallic soaps and →emulsifiers. It is also used in production of →soaps, →rubber compounds, →leather auxiliaries, →lubricants and →textile auxiliaries.

Lit.: Ullmann* (5.) **A20**, 273

Stearoyl Lactic Acid

→Lactylic Esters of Fatty Acids

Stearyl Alcohol

Syn.: 1-Octadecanol; n-Octadecyl Alcohol
G.: Stearylalkohol; F.: 1-octadécanol

$CH_3–(CH_2)_{16}–CH_2OH$

m.w.: 270.5; m.p.: 58 °C b.p.; 214 °C (2.67 kPa/20 mm)

S. forms colorless crystals soluble in alcohol and ether. Pure C_{18}-alcohol (99.5%) is commercially available but is used only in specialties. Normally, derivatives are made from broader cuts (→detergent-range alcohols), also called s., which are used mainly in →surfactants.

Lit.: [112-92-5]

Stearylamine

Syn.: 1-Octadecylamine
G.: Stearylamin; F.: 1-octadécanamine

$CH_3–(CH_2)_{17}–NH_2$

m.w.: 269.52; m.p.: 53.1 °C; b.p.: 150–170 °C (0.5 kPa/3.75 mm)

S. is a skin-irritating, white, waxy material. For uses: →fatty amines.

Lit.: [124-30-1]

Sterculia urens

→Karaya Gum

Steroids

D.: Steroide; F.: stéroïdes

S. are compounds that contain a cyclopenta[a]phenanthrene ring, which, when fully hydrogenated, is called steran or gonan (5-α-steran skeleton):

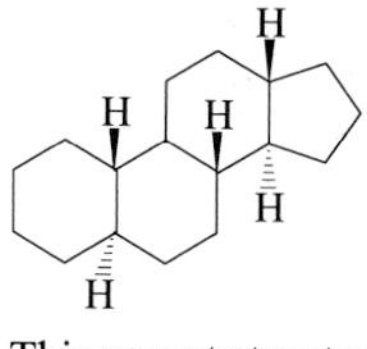

This parent structure is modified in nature in many ways, e.g., by ring contraction/enlargement or even ring opening.
The term s. comprises the following important classes:

- the sterols (e.g., →cholesterol, →stigmasterol and related compounds, vitamin D and methyl steroids);
- the bile acids (e.g., cholic acid and chenodeoxycholic acid);
- the steroid hormones: a) sex hormones, such as estrogens, gestagens and androgens, b) adrenal

Formulas:

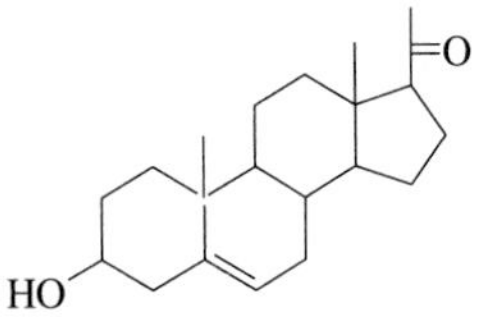

pregnenolone (A)

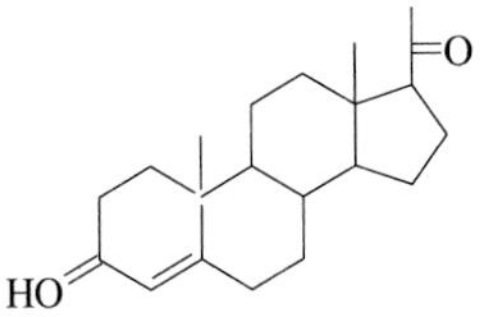

progesterone (B)

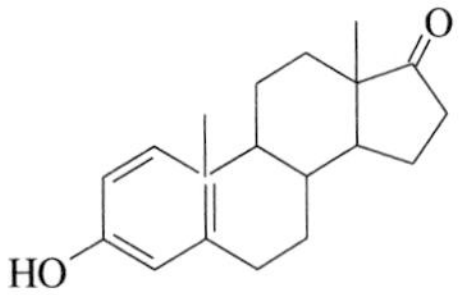

estrone (C)

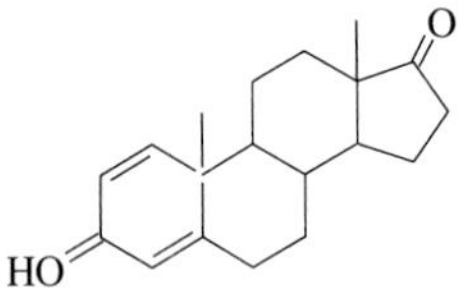

androstadienedione (D)

For other formulas:
→cholesterol, →diosgenin, →stigmasterol.

steroids (glucocorticoids and mineralcorticoids), c) calcium-regulatating sterols (→vitamin D) and d) insect skin-shedding hormones;
● the sapogenins (plant glycosides, called →saponins, that form a soapy foam in water; generally composed of an aglycon, sapogenin, and one or more sugars, →digitogenin);
● the steroid alkaloids and steroid lactones/cardenolides; and
● the bufadienolides, which are the aglycons of the cardiac glycosides).
Steroids (e.g., →diosgenin [512-004-9] from →Dioscoreaceae, →stigmasterol [83-48-7] from →soybean oil, β-sitosterol [83-46-5] from →wheat germ oil, →soy bean oil, other cereal germ oils and sugarcane wax, and →cholesterol [57-88-5] from wool grease, distillation residues of animal fatty acids, spinal cord and brain are valuable raw materials for the synthesis of steroid intermediates. Thus, stigmasterol, β-sitosterol and cholesterol are transformed (the key step is performed by suitable microorganisms) into androstadienedione, leading

diosgenin   stigmasterol

dehydropregnenolone ⟶ pregnenolone (A)

androstenolone

estrone (C)   progesterone (B)

androstadienedione (D)

cholesterol   β-sitosterol   stigmasterol

finally to estrone, which can also be prepared from diosgenin (via the dehydropregnenolon — pregnenolone — androstenolone route).

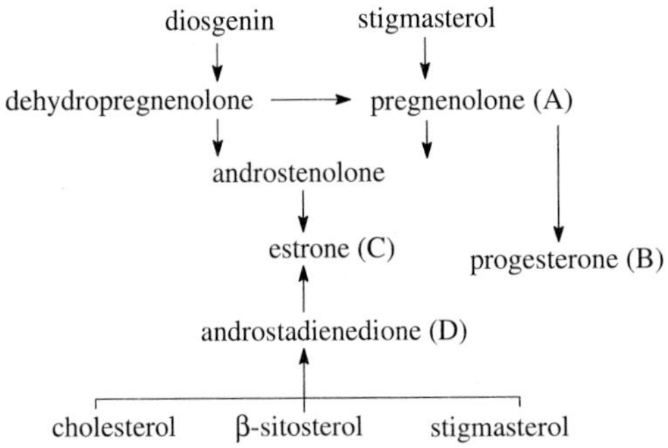

Lit.: Ullmann* (5.) **A13** 108 + **A25**, 309
Kirk Othmer* (3.) **21**, 645–729

Sterols
→Steroids

Stigmasterol
G.: Stigmasterin; F.: stigmastérol

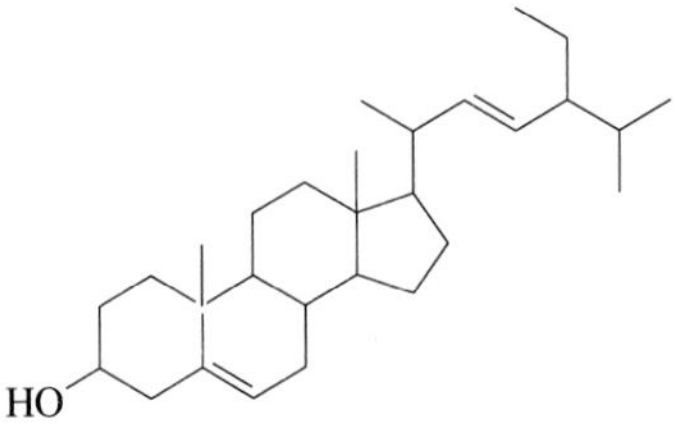

m.w.: 412.68; m.p.: 170 °C

S. is part (25%) of the nonsaponificable portion of soybean oil (→soybean) and is obtained from the deodorizer distillates, a by-product of soybean oil refining (→phytosterols), together with the →tocopherols. There are several processes to separate the tocopherols from the sterols by extraction, esterification, adsorption, molecular distillation and crystallization.
S. is an important starting material for steroid synthesis. The steroid is extracted together with β-sitosterol [83-46-5]), which differs from s. only by a double bond in the aliphatic side chain. This crude s. is subjected to microbiological degradation to yield 4-androsten-3,17-dione and 1,4-androstadien-3,17-dione (→steroids). The production of corticosteroids is an important outlet of s.

Lit.: Ullmann* (5.) **A13, 112**
Kirk-Othmer* **18**, 836
[83-43-7]

Stillingia Oil
→Chinese Tallow

Straw
G.: Stroh; F.: paille

S. is the dried stalks and leaves from cereals, oil crops and fiber plants; 30–45% is cellulose, the rest is lignin, pectin, pentoses and hexoses.

S. has hardly any commercial value, but its function to replenish organic matter of the soil is important. During recent years it has been considered as source of energy (1 kg of s. equals 0.5 kg coal or 0.4 kg oil) or as raw material for paper-making or as fiber material for other applications. Also, production of chemicals, such as glucose, xylose and single cell proteins, made from s., has been considered.

Due to logistic problems in collecting and handling of this low-weight, voluminous material (low density of energy) not many of such ideas have materialized yet.

Lit.: H.H.Leonhard, J.H.Martin "Cereal Crops" Macmillan NY (1968)

STS

→Sorbitan Esters of Fatty Acids

Styrax benzoin
Styrax tonkinensis

→Benzoin Resinoid

Styrax Resinoid/Oil

G.: Styrax Resinoid; F.: resinoïde de styrax

Crude styrax balsam is an exudate collected from incisions in the bark and sapwood of two wild-growing plants, *Liquidambar orientalis* and *L. styraciflua*, Hamamelidaceae. The first is a native of Asia Minor, the latter from Central America, where the balsam is collected in Honduras and Guatemala.

S. is produced by volatile solvent extraction of the dried and cleaned styrax balsam. The resulting product is a pale olive or greenish-brown viscous resinoid that possesses a sweet-balsamic, slightly spicy-cinnamic odor and good tenacity.

The typical constituents are cinnamic alcohol and esters of cinnamic acid, such as methyl and ethyl cinnamate. Styrax oil, steam-distilled from styrax balsam, is a pale yellow to almost water-white viscous liquid with a similar odor profile to s., except for greater diffusion and a more floral-spicy note.

Both products are used in floral (jasmine, hyacinth, lilac) notes and in floral-oriental (→odor description) fragrances as booster and as a pleasant addition to the →dry-down notes.

Lit.: Arctander*
       The H&R Book*

Suberic Acid

Syn.: n-Octanedioic Acid

G.: Korksäure, Suberinsäure; F.: acide subérique

$HOOC-(CH_2)_6-COOH$

m.w.: 174.19; m.p.: 144 °C; b.p.: 219.5 °C (1.33 kPa/10 mm)

S. is slightly soluble in water and soluble in alcohol.

It is produced by reaction of nitric acid with →castor oil or from cork. It can be used in the synthesis of drugs, dyes and polymers (→polyamide).

Lit.: Ullmann* (5.) **A8**, 531
       [505-48-6]

Succinic Acid

Syn.: Butanedioic Acid

G.: Bernsteinsäure; F.: acide succinique

$HOOC-CH_2-CH_2-COOH$

m.w.: 118.09; m.p.: 185–187 °C; b.p.: 235 °C

S. was observed by Agricolla in 1546 in the distillate of amber. S. occurs in fossils, fungi and lichens as a metabolite. It is obtained either by hydrogenation of maleic acid, oxidation of butane-1,4-diol, or from acetylene by oxo-synthesis. S. is used for the production of alkyd and polyester resins, surfactants(→sulfo-succinates), plasticizers, the manufacturing of lacquers and dyes and flavoring agents; its esters are used for perfumes. Its Fe, Mg, Ca or K salts are used as substitutes for sodium chloride in dietetic food.

Lit.: Ullmann* (5.) **A8**, 525
       [110-15-6]

Sucrose

Syn.: Saccharose

G.: Saccharose; F.: sucrose

S. is the most widespread disaccharide in the plant kingdom, occurring enriched in numerous plant organs, such as sweet fruits, stalks, roots and beets. It is industrially available as refined crystalline products or as aqueous solutions and syrups.

m.w. 342.303 ($C_{12}H_{22}O_{11}$); m.p. 185 °C;
$[\alpha]_D^{20}$: +65.5°

Colorless, monoclinic, water free crystals; solubility in 100 g water: 240 g at 20 °C; 487 g at 100 °C; solubility in Ethanol: 0.9 g ; insoluble in ether.

It is a nonreducing carbohydrate, consisting of one mole of →D-glucose and →D-fructose each, with the glucosidic C-atoms being glucosidically linked as an acetal compound. Total splitting of the glucosidic linkage by acid or →invertase leads to →invert sugars, an equimolecular mixture of both monomeric components (→inversion to $[\alpha]_D^{20}$: −22.2 °C).

S. is fully metabolized in the human and animal organisms. Digestion is started by the action of α-glucosidases and β-fructosidases on the brush borders of the mucosa, which split the disaccharide bond. Physiological energy content is 17.2 kJ/g.

The sweetness (r.s.: 100) is used as a standard value to compare with other caloric and noncaloric sweeteners.

Crystalline s. is nonhygroscopic. It is produced up to 99.7 % purity.

The →sucrose world market is supplied with 67 % →cane sugar and 33 % →beet sugar. Minor s. sources are (≈1 %): sweet sorghum (→sugary sorghum processing), maple sugar (→sugar maple processing), palm sugars (→sugar palm processing). They are limited to their growing territories.

The production of s. from cane, beet or sweet sorghum is characterized by aqueous extraction from the chopped or minced plant tissues, based upon the high solubility of s.

Precipitation of solubles, desludging and concentration are the following steps. The concentrated juice is further cooked in vacuum until crystallization starts. After crystallization, the massecuits are separated into crystalline crude sugar and mother liquor. Further washing, dissolution and recrystallization leads to various types of refined crystalline s.

Because of its nonreducing character, s. is highly stable on heating in neutral solutions as well as in the presence of amino acids or sulfur dioxide. The acetal linkage exhibits higher stability in alkaline than in acid solution. Acids cause formation of inversion products, which, on prolonged heating, undergo dehydration and formation of →5-hydroxymethyl furfural.

Based on 8 free hydroxyl groups, of which five exhibit similar chemical reactivity, typical organic reactions are possible, such as oxidation, dehydration, →esterification and →etherification.

→Hydrogenation after splitting the glucosidic linkages results in a mixture of →D-sorbitol and →D-mannitol. Hydrogenolytic splitting of both under forced conditions leads to mixtures of ethylene glycol, propandiol-1,2 and butandiol-1,2.

Heat treatment of alkaline solutions (pH >12) at >110 °C causes splitting of the glucosidic linkage, isomerization, enolization and destruction to C_3- and C_2-compounds: D,L→lactic acid, acetic acid, saccharinic acids or the corresponding aldehydes. Further aldolic condensation leads to →caramel color.

Biotransformations may be performed by microorganisms or by their isolated enzymes in different ways:

- →Sucrose, biotransformation to biopolymers (de-novo synthesis),
- →Sucrose, biotransformation to organic chemicals of low m.w.,
- enzymic transformation to →isomaltulose, →leucrose, →oligosylfructoses and →fructosylsucroses.

Despite numerous promising, well-known possibilities for reaction to chemicals and polymers, nonfood applications utilizes only about 2 % of the entire production.

S. food utilization constitutes the great preponderance of the current market, including some outlets in pharmaceutical and tobacco industry. The third outlet is s. →fermentation, but up to now the →sucrose molasses is preferentially used in the EC.

Annual world production in 1993/94: 110×10^6 mt.

Lit.: Rymon-Lipinski/Schiweck* 67–146 (1991)
Eggersdorfer* 211–224 (1992)
"Zucker in Zahlen", Wirtsch. Vereinig. Zucker, Bonn, (1994)

Sucrose, Biotransformation to Biopolymers

G.: Saccharose, Biotransformation zu Biopolymeren;

F.: sucrose, bioconversion en biopolymères

The production of biodegradable polymeric materials by biotechnological processing is a traditional nonfood use of sucrose, leading to some →homopolysaccharides: →Dextrans, formed by the action of cell-independent dextran sucrase from *Leuconostoc mesenteroides* on sucrose solutions or →molasses. The glucose moiety is used as the substrate donor, whereas the fructose remains unreacted; →Polyfructoses, formed by cell-immobilized enzymes from *Bacillus polymyxa* on sucrose solutions or molasses. The fructose moiety is used for building the polymeric chain, whereas the equivalent glucose and sucrose remain unreacted.

Sucrose, in these examples, is a specific substrate donor, which cannot be replaced by other sugars. Other biotransformations by living cells metabolize sugars as nonspecific C-sources.

Sucrose, glucose, maltose as well as lactose are equifunctional in this respect:

→Xanthan *Xanthomonas* sp. Heteropolysaccharide

→Pullulan *Pullularia pullulans* Homopolysaccharide

→sodium alginate *Acetobacter, Pseudomonas* Heteropolysaccharide

→Erwinia gum *Erwinia tahitica* Heteropolysaccharide.

For uses of these products: →specific keyword.

Most promising is the development of large-scale production of novel →biodegadable plastics: →polyhydroxy butyric acid and →polylactic acid. According to forecasts by the chemical industry, 200 000 mt sucrose or equifunctional sugars will be needed for EC production in the year 2000.

Lit.: Ruttloff* 485–513 (1991)
    Eggersdorfer* 287–290 (1992)

Sucrose, Biotransformation to Organic Chemicals

G.: Saccharose, Biotransformation zu organischen Chemikalien;

F.: sucrose, bioconversion en substances chimiques organiques

Production lines of numerous low-molecular organic chemicals from sucrose by cell-bound or cell-free fermentation processes are well known. Some of them have been running for many years. The small amount of pure sucrose used, compared to the entire production (EC, 1.2%), is indicative of the equifunctionality with other, lower-priced, C-sources, such as →sucrose molasses, →glucose, →glucose syrups and sugar-containing wastes. Rising consumption of sucrose-containing culture media will result from:

– the need to diminish use of imported molasses for environmental reasons (costy waste disposal);

– the change-over of limited traditional processing to large-scale production to exchange petrochemicals with RR;

– newly developed biotechnological routes for producing bulk chemicals.

Organic acids:

Present production: →Acetic acid, →citric acid, →gluconic acid, →lactic acid, →itaconic acid.

Possible production: →Aconitic acid, →butyric acid, →fumaric acid, →malic acid, →tartaric acid.

Solvents:

Present production: →Ethanol (for further chemical synthesis and energy).

Possible production: →Acetone, →butanol, →glycerol, →glycol.

Amino acids, feed proteins:

Present production: →Arginine, →glutamic acid, →lysine, →phenylalanine ; baker's yeast, single-cell proteins.

Possible production: →Asparagine, →tryptophane.

Other fermentation products:

Present production: →Ascorbic acid (vitamin C), →antibiotics, technical →enzymes, vitamin B12.

Possible production: →Biotransformation to biopolymers.

Prognostic numbers, issued by the chemical industry, amount to an additional 150 000 mt sucrose (or glucose) for citric acid, 100 000 mt for amino acids and for →lactic acid for →polylactic acid production.

Lit.: Nachwachsende Rohstoffe, Konzept zur Forschungsförderung 1990–1995, Bundesminister für Forschung und Technologie, Bonn (1990),

Sucrose, Chemical Modifications

G.: Saccharose, chemische Modifizierungen;

F.: sucrose, modification de méthodes chimiques

Sucrose in its crystalline form is a highly pure, chemically well-defined RR, a vital starting material for fine chemicals as well as a building block for syntheses. It is available in bulk quantities. The structure of this nonreducing disaccharide offers numerous possibilities for sucrose-specific chemical modifications:

– **Formation of sucrose esters and ethers** based on 8 hydroxyl groups, 5 of which are functionally nearly equivalent.

Examples include:

● →Sucrose fatty acid esters, s. phosphoric acid esters and ester chlorides with low DS for biodegradable surfactants and emulsifyers;

● Drying-oil esters with high DS for plasticizers and lubricants, surface coatings and additives for plastics[1];

● Functional disaccharides and reactive vinyl components for surfactants and polymerizable vinyl saccharides[2];

● Methacrylic esters for synthesis of sucrose methacrylate gels and chelating resins[3];

- Sucrose polyesters with natural fatty acids (DS 6–8) by transesterification for noncaloric fat substitutes with varying properties[4];
- Functional sucrose ethers, synthesized from acrylamide and sucrose in aqueous alkaline solution to cyanoethyl ethers.

Sucrose-β-amidoethyl ethers can be further reacted with formaldehyde as condensation component to yield formaldehyde resins. Higher sucrose ethers (by reaction with alkyl-isocyanates) can serve as starting materials for effective surfactants with good hydrolytical stability[5].

- **Hydrolytic splitting of the glucosidic linkage**.
  One example is the inversion reaction by acidic or →invertase treatment, which produces →invert sugar, a mixture of →glucose and →fructose. The components may be further reacted by hydrogenation to →sorbitol or →mannitol for direct use or as starting material.
- **Hydrogenolytic splitting of the sucrose molecule.**

Examples include:

- Conditions of hydrogenolysis can be directed to: formation of ethylene glycol, propanediol-1,2, butanediol-1,2 or other diols and tetrols, which, after transformation into epoxides, serve as building blocks for polymerization reactions[6];
- Polyol mixtures of varied composition are useful components in the production of →polyurethanes[7].
- **Transformation of the glucosidic bond by enzymic action.**
- The α-1,2-glucosidic bond is transformed into an α-1,6-glucosidic bond by immobilized enzymes from *Protaminobacter rubrum* to yield → D-isomaltulose or into α-1,5-glucosidic bond by action of dextransucrase from *Leuconostoc mesenteroides* to yield →leucrose[8].
- **→Dehydration of the molecule**.
- Acid catalyzed intramolecular elimination of three moles of water forms →5-hydroxymethyl furfural, a key substance for future polymer synthesis from RR to replace formaldehyde.
- **Thermolytic reactions, pyrolyses.**
- Formation of D-fructofuranosyl carboxonium (oxycarbenium) ions by reaction of free sucrose-OH-groups to a mixture of 41 % oligosaccharides and 46 % polysaccharides with hydrophilic, water-absorbing properties; with added alcohols, they produce glycosides, the properties of which are determined by the nature of the alcohols [9].

Possible drawbacks in these chemical modifications may arise from the tendency of inversion, which leads to reducing properties, acidic or alkaline degradation and color formation, depending on the environmental conditions of the reaction.

[1] Lichtenthaler* (1991), 37–42
[2] Lichtenthaler* (1991), 127–130
[3] Lichtenthaler* (1991), 93–111
[4] Lichtenthaler* (1991), 68–72
[5] Lichtenthaler* (1991),106–112
[6] Lichtenthaler* (1991), 57–64
[7] Eggersdorfer* (1992), 210–224
[8] Lichtenthaler* (1991), 183–195
[9] Lichtenthaler* (1991), 198–200

Sucrose, EC-Market

G.: Saccharose, EC-Markt;
F.: marché européen du sucrose

The sucrose market of the EC is regulated by the "Common Market Order" (1986). It was established to enable the EC sugar industry to provide the consumers with sucrose at appropriate, stabile prices and to ensure sufficient income for the producers.

Sucrose production of the member countries in 1993/94 is listed below (in 1000 mt white sugar equivalents):

Stock from past years	2.229
Production	16.235 (16.012 in 1992/93)
Imports from third countries	2.014 (1.977 „ „ „)
Exports into third countries	6.412 (5.810 „ „ „)
Actual consumption in EC	**11.875** (11.907 „ „ „)
Final stock	2.191

Sucrose production in EC is limited by the "system of quota", which regulates sale and payment of the produced sugars: A-quota (basis quota) has a limited price but unlimited sales are granted $(11.2 \times 10^6$ mt); B-quota (together with "A" forms the maximum quota), has a highly limited price, but an entire sales grant $(2.5 \times 10^6$ mt); C-quota, the "free sugar" without any security for price and sales, that can be offered to the →sucrose world market. The guaranteed prices are fixed every year by the " EC Council of Agriculture" to protect producers and consumers against extreme fluctuations. "Threshold prices" for imported sugars are intended to balance the difference between world market prices and the EC-market prices as well as to protect the sugar industry from low-priced imports. They represent the upper limit for the development of the EC-prices. The "intervention price" is to be payed for EC-sugar according to the "Quote Regulations" when governmental institutions buy from sugar factories or trade organizations. The "gross price" consists of the real costs of the raw material, transportation, processing plus storage cost and tax.

Main producer countries are: France, Germany, (which provide about one half of the entire production), Italy, Great Britain, Spain, The Netherlands, Belgium/Luxembourg (each $>1 \times 10^6$ mt). Utilization of sucrose is dominated by direct consumption as food (22%) as well as by processing in the food industry (78%). Nonfood applications in the EC (12 countries) in 1985/86 was 89000 mt.

The novel EC-market order enables the chemical industry and biotechnology to obtain sucrose at nearly world market prices. This is the starting condition for increased sucrose utilization. In 1988/89 it amounted to 200000 mt. The potential of further application is estimated at an additional 450000 mt for citric acid, amino acids and biopolymers.

Lit.: "Zucker in Zahlen","Zuckermarkt" 1992/93/94, Wirtschafts-Vereinigung Zucker, Bonn (1994)
Nachwachsende Rohstoffe, Konzept zur Forschungsförderung 1990–1995, Bundesminister für Forschung und Technologie, 27–32 Bonn (1990)

Sucrose, Energy Source

G.: Saccharose, Energieträger;
F.: sucrose, porteur d'énergie

Sucrose-bearing plants represent an appreciable potential for renewable energy. Utilization of such energy contributes less to global warming than does burning of the traditional fossil energy sources. See also: →Preface, →fuel alternatives →Biomass, such as leaves, →wood (waste) and →bagasse may be burned for direct energy gain; excess pulp and press mud from beet processing may be converted in reactors to biogas. Both are, however, of limited local significance. Large-scale energy gain is achieved by conversion into fuel alcohol by year-around continuous fermentation by immobilized yeasts or bacteria. See also: →fuel alternatives, →ethanol.

Comparison of energy input into beet production and conversion to energy output (leaves, fertilizers, pulp, press mud and ethanol) results in $\approx 63\%$ gain. The energetic value of the ethanol alone surpasses energy input.

According to EC directives an admixture of 5% ethanol to gasoline is allowed. This calculates to a need of 1.3×10^6 mt in Germany alone (equivalent to an area of 300000 ha sugar beet). The present

decisive drawback is the price of "bio-ethanol". It can presently only compete on the petrol world market if a permanent level of $\approx \$50$/barrel of crude oil is reached, or it must be subsidized.

The highest state of realization in the use of fuel ethanol from sucrose is evident in Brazil for the "ethanol motor" as well as by 10%–25% admixture to normal gasoline. Annual average used for ethanol production from cane exceeds 7×10^6 m^3. In addition 350 $\times 10^6$ liters were imported from USA. In the USA, fuel ethanol is produced from →corn starch-derived →D-glucose. The tendency to ethanol driven cars in Brazil has come to a standstill.

Lit.: "Nachwachsende Rohstoffe", Konzept zur Forschungsförderung 1990–1995, Bundesminister für Forschung und Technologie, Bonn (1990)
F.O.Lichts Internation. Melasse u. Alkohol.-Ber.31 (1994), 281,282.

Sucrose Fatty Acid Esters

Syn.: Fatty Acid Sucrose Esters
G.: Fettsäureester der Saccharose;
F.: esters sacchariques

R: H
or
$-CO-C_{5-19}$

(molecules contain 6, 7 or 8 fatty acid residues)

It seems to be rather reasonable to use cheap sugar (→sucrose) as a hydrophilic component in combination with →fatty acids to build nonionic →surfactants. The products are nontoxic, nonskin-irritating and completely biodegradable. Although known for many years, these surfactants are, however, not widely used due to difficulties in manufacturing. Sucrose is only soluble in rather toxic and expensive solvents, such as dimethylformamide and dimethylsulfoxide, which have to be used in reacting sucrose with →fatty acid methyl ester. The reaction is carried out below 100 °C, and methanol is stripped of in vacuum. The solvents have to be removed completely, especially for food applications. Depending on the ratio of methyl ester to sucrose, a mixture of mono-, di- and tri-esters results. The monoester is the most interesting product. There is also the possibilty to produce s. by transesterification of fats with sucrose without a solvent followed by separation of sucrose esters from glycerides. However, products are rather undefined.

Uses are mostly in the food industry as emulsifiers (baked goods and chocolate) and in cosmetic formulations because of their attractive physiological properties.

Lit.: Ullmann* (5.) **A 25,** 747

G. Schuster,"Emulgatoren in Lebensmittel" Springer Verlag Berlin/Heidelberg/NY/Tokyo (1985)

Sucrose, Food Utilization

G.: Saccharose, Lebensmitteleinsatz;
F.: sucrose, utilisation alimentaire

By far the largest share of the world sucrose production (99.5%) is consumed as food component, either directly (22%) or as an additive in the food industries (78%). For →nonfood utilization and as energy source, sucrose plays an inferior but increasing role. Average per-capita world consumption is about 15 kg/year, but the wealthiest countries are consuming >50 kg/capita, whereas the poorest countries consume 5 kg/a or less but consumption is increasing. In the high-consuming countries, the consumption is stagnating because of modern health considerations, the shift to other caloric bulk sweeteners, such as →isomerized sugars and the introduction of noncaloric sweeteners and "light products".

Numerous food applications are based on the following functional properties and overlapping characteristics:

It increases:
sweetness, viscosity, boiling point, moisture retention, osmotic pressure, freshness, prolonged shelf life;

enhances:
flavor, appearance (improved clarity), luster, gloss;

imparts:
plasticity, texture;

affects:
solubility of other accompanying ingredients;

provides:
energy, body, mouthfeel, bulk;

penetrates:
fruits and vegetables;

depresses:
freezing point (ice-creams, frozen foods);

ferments:
alcoholic beverages, substrate for microbiological processing;

melts:
glassy state, candy production;

microcrystallizes:
fondants, mashmallows, chewing gum, chocolates.

Sucrose producers provide numerous qualities of standard, adapted and tailored products. Frequently applied are blends of sucrose with →invert sugars or →glucose syrups. They can be shipped and stored in higher concentrations of d.m. than pure sucrose syrups.

Lit.: Salunkhe/Desai* 1–4 (1988)

Rymon-Lipinski/Schiweck* 67–146 (1992)

"Zucker in Zahlen", Zuckermarkt 1992/93/94, Wirtschafts-Vereinigung Zucker, Bonn (1994)

Sucrose Industry, By-products

G.: Saccharoseindustrie, Nebenprodukte;
F.: industrie sucrière, produits secondaires

Sugar crops of world-wide significance are →sugar cane, →sugar beet; future significance will include sweet sorghum (→sugary sorghum).
By-products of their processing are part of the RR complex.

- **Beet pulp** represents the residue left after sucrose extraction, including the beet tops; it is sold as wet pulp (85–90% water), as pressed pulp (80–85% water), dried pulp (10% moisture, 4.5% from 100 kg fresh beet). Dried pulp may be prepared as "molassed pulp" if disposal of surplus molasses is useful. Utilization is mainly for animal nutrition (ruminants), including silage products with 58–60% carbohydrate equivalent and 5–6% digestible protein.

- The sweetened off pressure-treated **filter cake** of the carbonization mud from beet as well as cane processing is to be used as agricultural soil fertilizer. Cane filter cake is also used for fuel, manure, cane wax, dye-stuffs, metal polish.

- **Molasses**: →sucrose molasses.

- →Bagasse, is the most promising by-product for industrial uses in the pulp- and →paper-industry. It will provide a significant alternative to reduce the destruction of woods.

Similar considerations are valid for the extracted stems of sugary sorghum (26% α-cellulose, 17% pentosans, 10% lignin).

Lit.: Salunkhe/Desai*, 27, 68–69, 88, 203–204 (1988)

Sucrose Molasses

Syn.: Molasses

G.: Melasse, Zuckerrübenmelasse; F.: mélasse

S. are the noncrystallizing liquors of the final low-grade massecuite crystallization of sucrose from →sugar beet processing or →sugar cane processing, as well as from sweet sorghum (→sugary sorghum processing). S. are dark-brown, heavy (d=1.6), viscous syrups; beet sugar molasses are not suited for human nutrition. The composition of beet molasses and cane molasses is somewhat different:

% dry basis	Beet molasses (3–4% of beet weight)	Cane molasses (2.5–4% of cane weight)
Sucrose	50	30–40
Inverted sugar	1.0	10–25
Other carbohydrates	0.5	2–5
Raffinose	0.5–2	traces
Sulfate ash	12	–
Carbonate ash	–	7–15
Nitrogen	1.8	0.4–0.7
Amino acids	2.2	0.5–1.5
Organic acids	4.3 (lactic acid)	1.5–6
Betain	5.5	–
Moisture	20	17–25

M. for long periods have been the least expensive carbohydrate sources for chemical, pharmaceutical and biotechnological uses in Europe and Japan (beet) as well as in the USA (beet and cane) and South America (cane). Due to more intensive utilization of the sugar content, the quality of s. has decreased during recent years (less sugar and more useless by-products).

Consumption: In Germany (1993/94), 921 320 mt of s. were produced, (1 034 404 mt in 1992/93). The largest amount of it, 918 661 mt, was consumed in animal nutrition for cattle feed or as a dried product with beet pulp (molassed pulp). For industrial consumption (fermentation products and yeast), 154 879 mt of s. were needed. Therefore, imports are necessary.

Nonfeed use in EC in 1991/92 was 687 000 mt (≈ 329 000 mt fermentable sugars). From this, 487 000 mt were used as feedstock in the chemical and pharmaceutical industries for fermented products, such as →citric acid and its esters, →glutamic acid and monosodium glutamate, lysine and antibiotics. Minor amounts are applied for alcohol fermentation.

Despite of their direct and easy fermentability, s. are of limited value for large-scale ethanol production. Prices fluctuate considerably and the supply of domestic s. is not sufficient.

Transportation costs of imported s. are confronted with the expense of moving 2–3 mt s. for 1 mt of yield. A novel important drawback is in the costly waste disposal because of environmental problems. Exchange of s. for sucrose as feedstoff for large-scale fermentations may become significance in the future. Animal nutrition will remain the main field of utilization.

Lit.: Salunkhe/Desai* 27, 29–30, 68–70, 200–201 (1988)
U.Sommer: Zuckerindustrie, 117 ,381, ,388, (1992)
"Zucker in Zahlen", Zuckermarkt 1992/93/94, Wirtschafts-Vereinigung Zucker, Bonn (1994)

Sucrose, Nonfood Utilization

G.: Saccharose, Industrierohstoff;

F.: sucrose, matière brute industrielle

Sucrose is one of the most prominent products of world trade (→sucrose, world market). It is produced mainly for utilization in food. In Germany (1992/93) only 1.3% of sucrose consumption, 35 100 mt, was utilized in the chemical industry; this is typical for the →EC-sucrose market.

A promising development is foreseen for the coming years as a consequence of the EC Market Order (1986), which enables the sugar industry to provide sucrose for chemical and biotechnological industries at prices similar to those of the world market. The EC market increased from 89 000 mt in 1985/86 to ≈ 200 000 mt in 1988/89 for the following applications:

organic-chemical products	16.6%,
pharmaceuticals	19.6%,
plastics, cellulose derivatives	34.0%,
various chemicals, production aids	28.4%,
albumins, enzymes, glues	1.5%.

Fermentation for →ethanol (→sucrose, energy source) is another form of sucrose utilization, which is difficult to evaluate because of the widespread use of →molasses as C-source. Ethanol is used in edible and pharmaceutical applications. Fuel alcohol is mainly produced from →corn starch in USA and from sugar cane in Middle and South America.

Sucrose nonfood utilization in traditional as well as enhanced materials until the year 2000 in the EC is listed below:

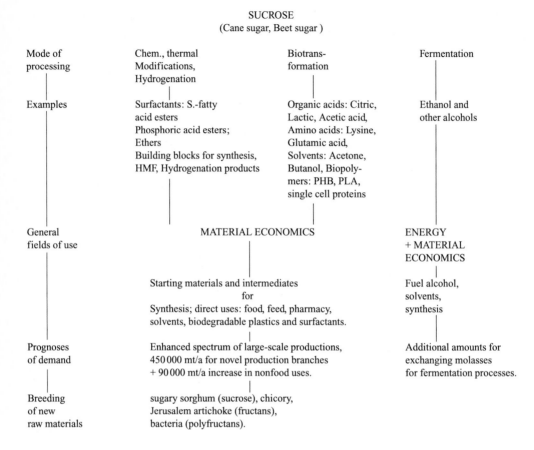

SUCROSE
(Cane sugar, Beet sugar)

Mode of processing	Chem., thermal Modifications, Hydrogenation	Biotrans-formation	Fermentation
Examples	Surfactants: S.-fatty acid esters Phosphoric acid esters; Ethers Building blocks for synthesis, HMF, Hydrogenation products	Organic acids: Citric, Lactic, Acetic acid, Amino acids: Lysine, Glutamic acid, Solvents: Acetone, Butanol, Biopoly-mers: PHB, PLA, single cell proteins	Ethanol and other alcohols
General fields of use	MATERIAL ECONOMICS		ENERGY + MATERIAL ECONOMICS
	Starting materials and intermediates for Synthesis; direct uses: food, feed, pharmacy, solvents, biodegradable plastics and surfactants.		Fuel alcohol, solvents, synthesis
Prognoses of demand	Enhanced spectrum of large-scale productions, 450 000 mt/a for novel production branches + 90 000 mt/a increase in nonfood uses.		Additional amounts for exchanging molasses for fermentation processes.
Breeding of new raw materials	sugary sorghum (sucrose), chicory, Jerusalem artichoke (fructans), bacteria (polyfructans).		

In numerous functions of nonfood utilization, starch-based saccharides are competitors on the market because of their equifunctionality of properties. Availability, world market conditions and possible yields of production may decide preferences. In organic reactions that are based on the specific peculiarities of its well-defined structure and conformation, sucrose is not exchangeable.

Lit.: "Nachwachsende Rohstoffe", Konzept zur Forschungsförderung 1990–1995, Bundesminister für Forschung und Technologie, Bonn (1990)

Sucrose World Market

G.: Zucker-Weltmarkt; F.: sucre, marché mondial

The annual world production of sucrose at present amounts to about 110×10^6 mt, of which 75% is utilized in the 111 producer countries and the rest is imported or exported.

One third of the production is regulated by quotas and guaranteed prices. The "free" world market includes about 19×10^6 mt, 17% of the world produc-

tion. Basis for all calculations is unrefined raw sugar; the transformation factor into white (refined) sugar is 0.92. Sugar production results for 67% from →sugar cane and for 33% from →sugar beet. Sweet sorghum, maple and sugar palms make up about 1% of the world production. Sweet sorghum is expected to be a more important source, if breeding for adaption to the European climate is successful. Sugar world production in 1991/92 was retrogressive for the first time since 1985/86, whereas consumption is increasing further. So, the stock of sucrose is slowly diminishing.

WORLD SUGAR BALANCE (in $\times 10^6$ mt)

	1991/92	1992/93	1993/94
Initial stock	34.981	39.549	39.098
Production	116.434	112.608	109.608
Imports	31.741	31.212	30.184
Exports	32.319	32.070	31.034
Consumption	111.228	112.201	112.955
Final stock	39.549	39.098	34.901
% of consumption	35.54	34.85	30.90

MAIN PRODUCERS OF SUGAR 1991/92 (in 10^6mt)

Beet sugar		Cane sugar	
FSU	6.700	India	13.500
France	4.423	Brazil	9.100
Germany	4.245	Cuba	6.500
USA	3.429	China, PR	6.145
Turkey	2.052	Thailand	5.100
China, PR	1.835	Mexico	3.500
Poland	1.641	USA	3.196
Italy	1.640	Australia	3.140
Great Britain	1.330	Pakistan	2.600
Netherlands	1.137	South Africa	2.418

The increasing consumption trend of annually 1.5 $\times 10^6$mt is caused by population growth and by increased consumption in South-East Asia and numerous developing countries.

Sucrose prices in the world market fluctuate between \$30–40/100 kg (1995).

SUCROSE CONSUMPTION, 1992,
in VARIOUS COUNTRIES
Entire consumption incl. nonfood and fermentation
(kg white sugar per capita)

Cuba	80.9	Hungary	54.4
Iceland	53.1	Netherlands	52.2
Denmark	51.6	Austria	49.8
Mexico	49.6	Ireland	47.7
Israel	47.0	Tchechia	46.5
Australia	41.0	Sweden	40.5
Great Britain	39.2	Norway	38.1
Canada	37.5	France	34.2
Germany	33.6	Greece	33.0
South Africa	31.1	USA*)	28.9
Middle America	45.7	South America	37.6
Europe	34.7	Africa	12.3
North America	30.0	Asia	11.7
World	19.0		

* not including starch sweeteners

Lit.: "Zucker in Zahlen","Zuckermarkt" 1992/93/94, Wirtschafts-Vereinigung Zucker, Bonn (1994)

Sugar Beet

Beta vulgaris L.ssp. *vulgaris* var. *altissima* Döll
Chenopodiaceae
G.: Zuckerrübe; F.: betterave sucrière

S. is a good example of drastic changes in crops as the result of →plant breeding methods. With selection methods the sugar content of the Mediterranean fodder beet was raised from 2% to 15–20%. Only this way commercially self-sufficient production of sugar was possible in Europe, which started with the first beet sugar factory in Prussia in 1801. Today, worldwide produced is 43×10^6mt of sugar from sugar beets (→sucrose world market).

Sugar beet; old drawing of the biennial plant with the beet and the inflorescence (Dalechamps, J., Historia generalis plantarum, Lyon, 1586).

Main production areas are in the moderate zones, where s. is grown as a spring annual, and in the subtropic zones, where it may be grown as a winter annual crop.

Botanically, s. is →biennial with seed production starting after a low-temperature shock. Seed production and sugar production therefore need to take place in different climates because frost resistance is poor.

As seeds are always formed in clusters of 2–4, which are not easy to separate, production of planting seed needs to take this into consideration, either by mechanically treating the seeds or by establishing genetically monogerm lines, which produce only one plantlet per seed cluster.

Also, planting the seeds needs to be done with single-seed planters to obtain the final spacing of the harvested beets of 30–50 cm. For germination, comparatively high temperatures of 20–25° C are needed, which means slow emergence and comparatively high →pesticide input in moderate climate, especially herbicides, to keep the soil free of weeds for about 1 month.

Nevertheless, at least one hand-hoeing cycle is required for good s. productivity; in earlier times, 3–4 hand-hoeings were necessary. The harvesting of s. is mostly done with fully mechanized s. harvesters, which first cut the leaves (topping), then pull the beets after slightly plaughing the soil and clean and load them into trucks.

The beets are transported during a harvest campaign of about 2 months to the sugar factory, slices of the beets are extracted and several cleaning and enrichment steps are carried out on the sirup. By-products of sugar manufacture are pulp and molasses (→sucrose, industry by-products) as well as waste lime, which may be used as → fertilizer. The topped leaves may be utilized as silo-feed. Depending on soil and climate about 50–100 mt of beets are produced per ha. Sugar yield is about 16% or 8–16 mt.

Lit.: Martin/Leonard* (1967)

Sugar Beet Processing

G.: Rübenverarbeitung;
F.: betteraves sucrières, traitement

The cells of →sugar beet (*Beta vulgaris* L.) contain 15%–19% →sucrose. The aim of processing is its entire extraction and isolation in crystalline form as well as in the recovery of the by-products. The beet sugar campaign in Middle Europe is concentrated in the period between October and January. Duration of the harvest is shorter than the production period because storage of the beets.

The technological steps after delivery from the field or from the storage place are:

Beet preparation: removal of stones, herbs, leaves, damaged

beets and roots, washing them free from soil.

Extraction: After cutting the beet to chips (cossettes) they are treated with water at 70–75 °C in countercurrent flow in diffusion batteries or continuously in extraction drums or towers. Extracted chips (pulp) are pressed for dewatering (17–22% residual water) and finally dried for feed. The press liquid is recycled to extraction. Raw juice contains 12–15% sucrose and numerous dissolved and insoluble impurities.

Juice filtration and purification: Pulp particles are filtered off, and the raw juice is treated in a two-step process with slaked lime and then carbonated by introduction of carbon dioxide. This procedure serves to neutralize and remove acids and colloids, to destroy nonsucrose components and to remove excess lime. Finally, the carbonation mud is filtered off. The thin juice contains 11–14% su-

crose. Further purification may be achieved by ion exchange treatment.

Evaporation and crystallization: The thin juice is concentrated to 65–70% in multistep evaporators. The cooking is continued under stirring in vacuum until formation of a crystal suspension (90% d.s.). The mixture is transferred into horizontal crystallizers with screw stirrers in which crystallization is finished. The massecuits are separated by centrifugation into brownish raw sugar and syrup, which is further purified, concentrated, crystallized, and after dissolution, recycled into the thick juice.

The mother liquor of the last repeated crystallization step, which cannot be further worked up for sucrose, is →sucrose molasses. For white sugar production, the raw sugar in the centrifuge is stripped with hot water to remove the brown syrups and thereafter transferred for drying into air-drying cooling devices.

Refining: Moist white sugar after stripping may be dissolved again, then purified with silica, →activated carbon or decolorizing resins, then concentrated, brought to crystallization and dried. The product, refined sucrose, has the same sucrose content as white sugar but less moisture humidity and minerals (sucrose of highest purity and whiteness).

Commercial sucrose: Refined sugar (Raffinade), = EC-Category 1; White sugar (Kristallzucker) = EC-Category 2. They are further subdived into 5 standard granulation types between 1–2 mm (coarse) and 5–110 μm (powder). Specialized types of white sugar are lump sugar, hail sugar, brown or white candy sugar, various instant or agglomerated types, and a large variety of liquid sugars (tailored for industrial processing).

Yield: 100 kg fresh beet can be worked up to 12–15 kg sucrose, 3.5 kg molasses, 4.5 kg dried pulp (cossettes), and varying amounts of carbonization mud (filter cake).

Lit.: Salunkhe/Desai* 188–191 (1988)

Sugar Cane

Saccharum spp.
Gramineae, Poaceae
G.: Zuckerrohr; F.: canne à sucre

S. is a large, →perennial tropical grass (→C4 plants), cultivated for its tall, thick stems from which cane sugar is obtained. In Europe, sugar was not available until the beginning of the 18th century. During the end of the 19th century, a competitor to s. was established by German plant breeders with the introduction of the →sugar beet.

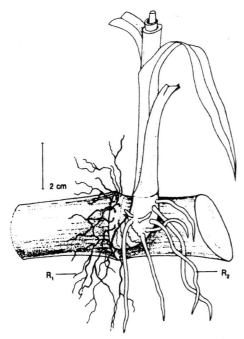

Sugar cane cutting after sprouting. R1: Short lived roots from the node, R2: Roots from the new shoot (from Rehm and Espig*, 1995, with permission).

Today, 60% of the world's annual sugar production comes from s. It may be grown between latitudes 35 degrees n. and s. of the equator. It requires temperatures warmer than 21 °C, with an optimum between 32° and 38° C. Annual rainfall should exceed 1500 mm.

S. produces remarkable quantities of d.s., more than 200 mt of cane per ha with about 12% →sucrose in it. The cane may be harvested one to two years after planting (density: about 12000–20000 stem cuttings per ha). The potential of regrowth would enable about 20 years of production per planting. It is, however, common to replant after two to three years due to →disease problems.

The most common species of s. is *S. officinarum* L., out of which interspecific hybrids have been developed in commercial sugar estates. *S. barberi* Jesw. and *S. sinense* Roxb. are still grown in restricted areas. *S. edule* Hassk. is only grown in New Guinea.

Sugar is generally removed from sugarcane in large factories by either the milling or the diffusion process (→Sugar cane processing).

Production of cane sugar in 1993 was 64.4×10^6 mt. (→sucrose world market).

Lit.: Cobley* (1976)

Sugar Cane Processing
G.: Zuckerrohr, Verarbeitung;
F.: canne à sucre, traitement

The stalks of native sugar cane (*Saccharum officinarum*, L.) contain about 15–20% sucrose. The aim of processing is its entire extraction and isolation in crystalline form as well as in the recovery of the by-products.

Production must be started immediately after harvesting because of the rapid deterioration and losses of sucrose during even a few storage days.

The technological steps after delivery from the field are:

Cane preparation: cleaning of the stalks by removing soil, leave residues and other plant tissues.

Extraction: after disintegration with rotating knives or shredders, the stalks are delivered to the mill battery or diffuser. Extraction is performed by countercurrent flow of hot water and diluted extraction juice and finally by squeezing out the mechanically bound liquid. The residue after extraction is termed →bagasse.

Juice clarification: the "mixed juice" contains the total sucrose and impurities, such as suspended insolubles, salts, nonsucrose saccharides and negatively charged colloids. It is treated with lime, followed by carbonization, in a two-step process.

Flocculation may be completed by doses of phosphate salts (superphoshate).

The muds are removed by pressure filtration or continuous precoat filtration, which leads to the by-product "press mud". Sucrose concentration of the clarified juice is 12–16%.

Juice concentration: is performed in multistep evaporators, which results in a thick juice of about 60% solids. Further concentration, until initiation of crystallization, is achieved in vacuum pans or stirred vacuum evaporators.

Crystallization and separation: the crystallizing syrup is transferred to coolers for either batch or continuous crystallization. After completion the massecuits (fillmass) are separated by centrifugation into A-sugar and A-molasses. The latter are separately purified, concentrated and crystallized. In this way B- and C-products, sugars and molasses, are produced. The C-sugar is returned into vacuum pans of the A-or B-step as seed grains; C-molasses is a by-product.

Refining: the first, crude sugar is shipped with 98–99% sucrose purity. For higher-grade products, it is dissolved in water, purified, concentrated and crystallized. The centrifuged and dried crystals are sold as refined cane sugar, nearly identical with refined beet sugar in purity and whiteness.

Yield: 100 kg sugar cane may be processed to 14–18 kg sucrose, 2–3 kg press mud (d.b.), 30–32 kg bagasse (d.b.) and 2.5–4 kg molasses.

Lit.: Salunkhe/Desai* 35–79 (1988)

Sugar Maple, Processing

G.: Zuckerahorn, Verarbeitung;

F.: érable à sucre, traitement

The sap from the xyleme tissue of the tree trunks of the genus *Acer* is rich in →sucrose and →invert sugar (*A. rubrum, A. saccharopherum, A. nigra, A. galchrum*). In eastern USA and adjacent Canada, it is exploited for the production of different types of maple syrup and sugar.

Processing: Mature stems are tapped for 2–5 weeks/a between February and April. Regularly drawing off sap does little harm to the trees or to the wood quality; the trees may reach an age of 70 years. The sap may contain 1–9% solids, 2–3% sugars on average, which is 98% pure sucrose.

Concentration: syrups require a 25–30-fold concentration under atmospheric pressure until a solids content of 65.5% is reached. Modern plants prefer to apply the principles of reverse osmosis, which reaches removal of 75% water. Syrups from the evaporation and finishing steps are passed through pre-filters, artificial fibers and flat-style filters.

Product formulations: standardized maple-syrups contain 65% solids; 92% sucrose, 5.5% reducing sugars, 1% minerals, with the remainder consisting of organic acids, phenolic compounds and proteinaceous material. Further concentration is performed to result in maple-cream and maple-butter, which exhibit creamy appearance and texture by formation of small sugar crystals. Maple candy is prepared by melting maple-cream. All these products possess high regional consumer acceptance as specially flavored sweeteners. They are not considered in →sucrose world market statistics.

Lit.: Salunkhe/Desai* 99–117 (1988)

Sugar Palms, Processing

G.: Zuckerpalmen, Zuckergewinnung;

F.: palmier à sucre, traitement

Different tropical and subtropical palms are exploited for the production of sucrose- and invert-sugar containing syrups by tapping the sap from inflorescences, trunks and stalks.

Palmyra palms (*Borassus flabellifer*) are found growing wild in West Africa, New Guinea and Asia, and are cultivated in India. Juice is collected after tapping the inflorescences. Large palms yield 11–20 liters/d. During its tapping life, a single palm may produce 120000 liters. The sap is fermented into intoxicants or vinegar, distilled into arrack or palm brandy, or evaporated to creamy palm sugar solids.

Caryota palms (*Caryota*) are from India to Queensland, (*Arenga*) in the rain forests, and (*Wallichia*) in eastern India and Indonesia. Juice is obtained by tapping the young inflorescences; a single inflorescence yields 7–14 liters, several on one trunk can produce 20–27 liters/d. The sap is utilized to produce sugar, palm wine and arrack.

Nipa palms (*Nipa fruticans* Wurmb.) are grown in Bangladesh, Andaman Islands and Saurashtra (Gujrath). Natural forms are found in the mangrove swamps and tidal forests. The sap is tapped from the stalks of the spadix, about 43 liters during a season of 2–3 months. One acre of nipa may yield 2800 kg sugar. Fresh nipa juice contains 17% sucrose and traces of reducing sugars. Natural fermentation prevents large-scale production of crystalline sucrose. Evaporation leads to brown syrups. Fermented sap is distilled for arrack or further reacted to vinegar.

Phoenix palms (*Phoenix sylvestris*) are commonly grown in India, Ceylon, West Africa, the Middle East, Canary Islands, China, Bangladesh and Pakistan, and they are the main sources of palm sugar. The sugar sap is obtained when it flows out of wounds that are inflicted on the active part of the trunk within the crown. A single tree may yield 4–19 liters/d. Averages are 250–1100 liters per season. Tapping can be performed for 40 years and more. The date palm (*P. dactylifera*) is tapped for the same purpose, but the whole central bud is cut out, and after 2 months the exhausted stem is dry and is used as firewood.

Amazonian fan palm (*Mauritia flexuosa*): the sap flows from the inflorescence stalks after scorching the palm with fire. Daily production is 4–20 liters from a single palm with the flow rate continued for 2–3 months. The sap may be evaporated into brown palm syrup (jaggery) or is fermented to palm wine, which may be further processed into arrack or vinegar.

Raphia palm (*Raphia vinifera*) originates with about 20 species in Africa. Extraction of juice is similar to those of *Phoenix* palms. Besides sugar sap production, the leaf stalks and sheaths are utilized -after retting in water- for fiber production (African bass).

The **sugar contents** of freshly prepared sap of all palms is between 10% (*Borassus*) and 17% (*Nipa*). Tender stem buds, peduncles and immature seeds, which are rich in starch, are eaten raw or after cooking.

Nonfood uses of most palms include thatching materials, coarse mats, baskets, bags and fibers. Leaflets contain up to 10% →tannins and 15% hardtans, which can be used for tanning of light leather. Their plant organs or their extracts are used for their traditional medicinal value.

Lit.: Salunkhe/Desai* 119–122 (1988)

Sugars, Inverted
→Invert Sugars

Sugars, Isomerized
→Isomerized Sugars

Sugary Sorghum Processing
Syn.: Sweet Sorghum
G.: Zuckerhirse, Verarbeitung;
F.: mil à sucre, traitement

Sweet sorghum, Sorgho (*Sorghum bicolor*, L., Gramineae) forms stalks, 1–5 cm in diameter, 40–600 cm in height, which contain 10–20% sucrose in the fully ripe state. It is grown in Middle and South America, Africa, Australia, India, China, Pakistan, South-East Asia and European Russia. S. is a typical short-day plant with an optimum photo period of 10–11 h. Under tropical and subtropical conditions, up to three crops per season are realistic. New cultivars in the USA yield about 50–70 mt/ha with two cuts. The juice contains 16–18% sugar. Breeders have developed an additional "syrup type", which contains sufficient →invert sugar as to prevent crystallization. The "sugar type" contains mostly sucrose.

Processing for syrup includes milling or shredding of the stalks, hot-water extraction (the remainder is →bagasse), clarification by storage in tanks for settling, assisted by heating and filter aid, concentration at 108–110 °C. A 100% purity syrup contains about 81% sucrose.

Crystallization is performed similarly to that of →sugar cane processing.

Both syrup and sugar serve to produce ethanol (1300- 1800 liters/ha).

S. is considered a highly promising future source for RR.

Breeding, to adapt it for Middle European growth conditions, has been started. The economy of this agro-industrial crop is highly positive: the whole plant may be utilized as an integrated food, feed, energy and biofertilizer system. Carbohydrate production is twice that of maize or beet sugar.

Lit.: Salunkhe/Desai* 81–98 (1988)

Sulfated Fats and Oils
G.: Sulfonierte Fette und Öle; F.: huiles sulfatés

Sulfated natural fats and oils are the first nonsoap →surfactants commercially produced in the middle of the 19th century.

Unsaturated triglycerides and those containing OH-groups react readily with sulfuric acid to form halfesters (sulfates) by addition of the sulfuric acid to the double bond and/or reaction with the OH-group. Neutralization is the final step in production.

Turkey red oil, derived from →castor oil, was one of the first products of this type. Other fats and oils of vegetable, animal or fish origin are used widely.

Carefully selected reaction conditions reduce →hydrolysis of the triglyceride to a minimum.

S. are still widely used as wetting and penetrating agents, emulsifiers and lubricants. The main outlet is as →textile and →leather auxiliary.

Lit.: Kirk Othmer* (3.) **22**, 359

Sulfation/Sulfonation
G.: Sulfatierung/Sulfonierung;
F.: sulfatation/sulfonation

Two important reactions introduce the SO_3-group into a fatty material:
– sulfation produces sulfates with a C-O-S bond and
– sulfonation yields C-S bonds.

The terms are often used interchangeably, which causes some confusion. Both are basic reactions to produce anionic →surfactants.

Sulfation:
Sulfates are half esters of sulfuric acid with OH-group- bearing oleochemicals such as →fatty alcohols, →fatty alcohol ethers, monoglycerides (→glycerides) and →ethanolamides as preferred starting materials. In contrast to sulfonates, sulfates are sensitive to →hydrolysis.

Sulfuric acid is introduced either as concentrated sulfuric acid/oleum, as chlorosulfonic acid, or in the more modern, large volume processes as SO_3.

$$R-OH + H_2SO_4/SO_3 \longrightarrow R-O-SO_3-H + H_2SO_4$$
$$R-OH + Cl-SO_3H \longrightarrow R-O-SO_3-H + HCl$$
$$R-OH + SO_3 \longrightarrow R-O-SO_3-H$$

Due to the importance of s. in modern surfactant manufacturing, there are batch and continuous processes that are carried out in many different types of sophisticated equipment. Because the reactions are highly exothermic, all process principles focus on temperature and color control. Subsequent neutralization also raises a problem of temperature control. Here again, modern continuous processes are offered by equipment manufacturers.

Oleum sulfation is applicable to all a.m. oleochemicals with the exception of fatty alcohol ethers. They are either sulfated by chlorosulfonic acid or SO_3, which is an agent useful for all these oleochemicals.

The sulfuric acid/oleum reaction can be carried out in simple equipment and is used mainly for the s. of tallow alcohol. There are continuous processes also, which focus on temperature control. The process is burdened by a large amount of waste sulfuric acid.

In SO_3-sulfation, the gas may be delivered by an outside manufacturer in tank cars, is produced in-house by stripping from oleum, or is generated close to the sulfation unit by burning sulfur and oxidizing of the SO_2 to SO_3 by air with vanadium oxide as catalyst. The latter is the preferred method in modern plants. The SO_3 is diluted with dry air to reduce reactivity. Falling film and cascade reactors are used to carry out s. Temperature control and adequate feed transport that respect the changes in viscosity during the process are the key criteria for these processes and for product quality (color).

Chlorosulfonic acid **sulfonation** is used for s. of short and medium chain fatty alcohols and their ethers ($\approx 2-3$ EO). Despite the enormous progress in SO_3-s., this process is still carried out for high quality sulfate manufacturing. Again, temperature control during reaction and neutralization is important. The by-product HCl is absorbed.

Sulfonation:

In contrast to sulfation, sulfonation is not of great importance to RR-based products. Sulfoacids are made mainly from linear alkylbenzenes, paraffins and olefins, which are all important feedstocks for bulk →surfactants.

With the exception of chlorosulfonic acid, the sulfonation agents are the same as in sulfation: sulfuric acid, oleum and SO_3. The same equipment for production and work-up procedures can be used. Sulfonation of unsaturated oleochemicals, e.g., oleic acid methyl ester with SO_3 and subsequent neutralization yield sulfonates with the following structural elements:

$$-CH=CH-CH_2-CH- \quad \text{and} \quad -CH-CH_2-CH_2-CH-$$
$$\qquad\quad |\qquad\qquad\qquad\qquad |\qquad\qquad\qquad |$$
$$\qquad\quad SO_3H \qquad\qquad\qquad OH \qquad\qquad SO_3H$$

Variations of sulfonation are sulfoxydation ($SO_2 + O_2$) and sulfochlorination ($SO_2 + Cl_2$), which are used in reactions with paraffinic feedstock. Neither reaction is applied in oleochemistry.

The sulfonation of fatty acid esters yields α-sulfo fatty acid esters, a modern sulfonate based on →oleochemistry.

Sulfitation is the addition of sodium sulfites and bisulfites to double bonds. →Sulfosuccinates are made in this way. Sulfitation is also applied to unsaturated fats and oils to introduce C-S bonds and yield fat liquors (→leather auxiliaries).

Another group of the few sulfonates based on RR are the →acyl amino alkanesulfonates (taurates) and the →acyloxy alkane sulfonates (isethionates). →Lignosulfonate is also a sulfonate based on natural feedstock.

The US production of sulfates and sulfonates was $1,4 \times 10^6$ mt in 1980. ($\approx 250\,000$ mt were sulfates, the remainder were sulfonates).

Lit.: Falbe* p. 54 (1987)
    A.Davidson and M.Milvidsky, "Synthetic Detergents", G. Godwin Ltd. London (1978)
    Kirk Othmer* (3.) **22,** 1

Sulfitation
→Sulfation/Sulfonation

Sulfobetains
→Amphoteric Surfactants

α-Sulfo Fatty Acid Esters
Syn.: Estersulfonates
G.: α-Sulfofettsäureester; Estersulfonate;
F.: sulfonates d'esters

$$R-CH-COO-CH_3 \qquad R: C_{14-16} \text{ alkyl}$$
$$\quad |$$
$$\quad SO_3^- Na^+$$

S. are anionic →surfactants that are stable to hydrolysis in a pH range of $4-9$ and insensitive to water hardness. They have good emulsifying and lime soap dispersing properties. Their aqueous solutions foam well and have a high textile detergency. The methyl esters are by far the most used version.

S. are made by SO_3 →sulfonation of the methyl esters of preferably C_{16-18} fatty acids. Reaction takes place in two steps, a rather rapid formation of a mixed anhydride, followed by a slower rearrangement. The sulfonation mixture is rather dark and must be bleached with H_2O_2 at $60-80\,°C$, and it is finally neutralized. By-product is the disodium salt of the α-sulfo fatty acid.

Uses are in →detergents and soap bars. S. is one of the RR-based alternatives to petrochemical surfactants in heavy-duty detergents.

Lit.: Ullmann* (5.) **A25**, 747
Falbe* 75–80 (1987)

Sulfonation

→Sulfation/Sulfonation

Sulfosuccinamates

→Sulfosuccinates

Sulfosuccinates/Sulfosuccinamates

G.: Sulfobernsteinsäureester; F.: sulfosuccinates

There are four categories of s.:

I: CH$_2$–COOR
        |
NaO$_3$S–CH–COOR

Dialkyl sulfosuccinates R: C$_{4-8}$ alkyl

II: CH$_2$–COO–CH$_2$–CH$_2$–NH–COR
        |
NaO$_3$S–CH–COO–CH$_2$–CH$_2$–NH–COR

Dialkyl sulfosuccinamates R: C$_{12-18}$ alkyl

III: CH$_2$–CO–(O–CH$_2$–CH$_2$)$_n$OR
        |
NaO$_3$S–CH–COONa

Monoalkyl sulfosuccinates
n: 3–12
R: C$_{12-18}$ alkyl, C$_{11-17}$ acyl

IV: CH$_2$–CO–(O–CH$_2$–CH$_2$)$_n$NHCOR
        |
NaO$_3$S–CH–COONa

Monoalkyl sulfosuccinamates
n: 3–12
R: C$_{11-17}$ alkyl

All these compounds are made by →esterification of the corresponding alcohols or ethoxylates with maleic acid anhydride (monoesters are obtained at $70-100\,°C$, while diester formation needs higher temperatures and an azeotropic elimination of water) and subsequent reaction with sodium hydrogen sulfite. All s. have one property in common: they are sensitive to →hydrolysis and can be used only in neutral media (pH 6–8).

Synthetic variability is high, and anionic →surfactants with a broad spectrum of properties can be made.

I: These are the largest-volume s. Because they are normally not RR-based, they are not covered further here.

II: They are used as collectors in ore flotation (→mining chemicals).

III: The monoalkyl s. are gaining importance as dermatologically suitable →surfactants and emulsifiers in cleaners, in →skin preparations, and in →hair preparations due to their detergency and foaming behavior. Because of their sensitivity to alkaline hydrolysis, they are not used in detergents. They are also used as polymerization emulsifiers (→polymerization additives).

IV: same as II.

Lit.: Ullmann* (5.) **A25**, 747
Falbe* 82–83 (1987)

Sunflower

Helianthus annuus L.
Compositae
G.: Sonnenblume; F.: tournesol; grand soleil

S. is one of the oldest endemic crop species in North America. It is known to have been used as a cultivated food crop by North American Indian tribes from 3000–2000 B.C. s. was introduced to Europe by Spanish explorers when returning from America in the early 16th century. During the next 300 years, it spread across Europe and into Russia. In Russia it was extensively grown as an oilseed crop during the middle of the 19th century. At the turn of the century in Russia, it became a major agricultural crop with breeding of local cultivars. With the detection of cytoplasmatic male sterility in sunflower by the end of the 1960's in France, and the identification of fertility-restoring genes in the beginning 1970's, efficient production of →hybrid sunflower seeds was enabled. As in →corn, this was also the beginning of the adaptation of sunflower production to moderate climates, such as in Central and North Europe. Due to its deep taproot with extensive side branching, s. is well adapted to dryland farming conditions, such as in Italy, Argentina or California.

Today's major production areas for s. seed and oil are in the FSU, Argentina and France.

Production in 1994 ($\times 10^6$ mt)	Sunflower seed	Sunflower oil
Argentina	4.74	1.65
FSU	4.49	1.12
USA	2.19	0.49
France	2.10	0.61
India	1.50	0.50
China	1.50	0.23
Spain	1.02	0.43
World	23.18	7.96

The fatty acid pattern of normal →sunflower oil is characterized by a high percentage of linoleic acid. In the last 15 years, new varieties of s. have been derived with the → hybrid breeding technique, resulting in high oleic acid contents, which start at 80% and today reach levels as high as 93% oleic acid (HOSO).

This development enables new applications, especially in →oleochemistry but also in food technology because of high stability and shelf life of derived products.

See also: →Taxonomy

The oil is normally derived by solvent extraction with by-products of 44% protein and 15–18% fiber. Seeds will need to be dehulled before extraction, otherwise fiber contents and protein contents would be 26% and 28%, respectively. The meal is well adapted for feeding animals.

Lit.: W.Dedio, E.D.Putt: "Sunflower" in Fehr/Hadley* p. 631 (1980)
G.N.Fick Genetics and Breeding of Sunflower J. Amer. Oil Chem. Soc. **60**, 1251–1253 (1983)

Sunflower Oil

G.: Sonnenblumenöl; F.: huile de tournesol

s.v.: 188–194; i.v.: 125–144

The pale yellow oil has a pleasant odor and the following fatty acid composition:

C16:0	C18:0	C18:1	**C18:2**	C20:0
3–9%	1–3%	14–43%	**44–70%**	1–4%

The oil is derived from the dried seeds of →sunflower by →expelling and subsequent →extraction. Filtration, deodorization, neutralization and bleaching may follow.

Main use for the various grades, including →hydrogenated and winterized types, is in margarine and salad oils.

The new high-oleic variety is used preferably for frying and salad oils.

S. is used technically in high gloss enamels (→coatings) and in →alkyd resins.

→Hydrolysis yields the fatty acids, which are used in →soap manufacturing and varnish production (→coatings).

Interesting applications are envisaged for the acids from the high-oleic variety, which gives almost pure →oleic acid with only one splitting operation, which is useful for old and new oleochemical processes. (→metathesis, →ozonolysis, →epoxidation, etc.)

World production (1993) was 7.32×10^6 mt with a strong increase during the last few years.

Lit.: Ullmann* (5.) **A10**, 226
[8001-21-6]

Sunflower Oil Fatty Acids

→Sunflower Oil

Super Absorbent

→Super Slurper

Superslurper

Syn.: Absorbents, Water-, Super Absorbent
G.: Superabsorber; F.: superabsorbants

Absorbents for water are polymeric substances, consisting of long-chain molecules, which are cross-linked at different sites to a three-dimensional network. They exhibit outstanding absorption power for aqueous liquids, 100 to 500 fold (distilled water) of its own volume.

Most applied products are synthetic polymers, such as crosslinked acrylate/acrylamide polymers. Possible problems of waste disposal, detoxification and demand for biodegradability may give preference to RR-based, biodegradable polymers. This has been successfully proven by crosslinked →carboxymethyl- and →hydroxyethyl-cellulose as well as gelatinized starches.

Starch-based superslupers are made by the following reaction:

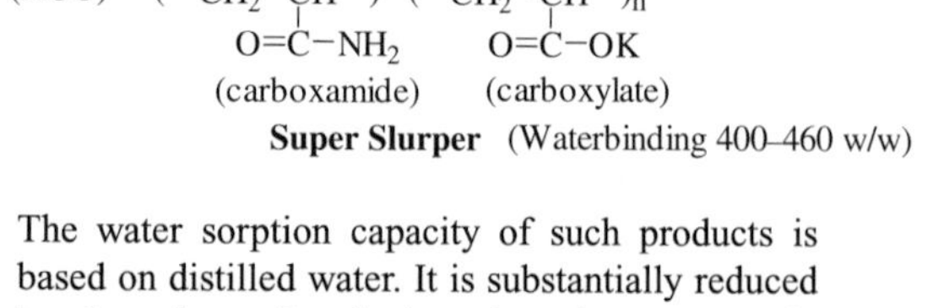

Gelatinized starch (AGU) + n $CH_2{=}CH{-}CN$

Grafting Reaction

(AGU) ····· $(-CH_2-CH-)_n$ Starch/Acrylonitril-copolymer
CN (Waterbinding 84 w/w)

Saponification + KOH + H_2O

(AGU) ··· $(-CH_2-CH-)-(-CH_2-CH-)_n$
O=C−NH$_2$ O=C−OK
(carboxamide)   (carboxylate)
Super Slurper (Waterbinding 400–460 w/w)

The water sorption capacity of such products is based on distilled water. It is substantially reduced by electrolytes. Standard products have to remain stable after absorption of 45 g urine and 84 g water. Preferred uses are hygienic products, such as napkins, diapers, baby clothes, sanitary towels and enhancers for water storage capacity of soil.

World production of all s. (including synthetic) was >200000 mt (1988) by 40 producers: EC, 70000 mt; Japan, 40000 mt and USA, 90000 mt.

Lit.: Tegge* 191–192 (1988)
   Eggersdorfer* 353–365 (1992)

Suppositories

G.: Suppositorien; F.: suppositoires

S. are semirigid, meltable or dissolvable masses that contain pharmaceuticals and are designed to be inserted into body cavities (rectum, vagina, urethra).

They are either based on →cocoa butter or on →glycerol/→gelatine blends, and sometimes contain sodium stearate (→soaps) as gelling agent.

Lit.: Hagers Handbuch der pharmazeutischen Praxis (4.)
   7a, 664

Surface-Active Agents

→Surfactants

Surfactants

Syn.: Surface-active agents
G.: Tenside; F.: agents tensioactifs, surfactants

S. are amphiphilic compounds, which means that they have a double character in one molecule: a hydrophobic (mostly long-chain) part and a hydrophilic counterpart. These molecules orient themselves in an aqueous solution with their hydrophobic and hydrophilic parts towards the corresponding surfaces or interfaces. Thus, hydrophobic liquids (oils) or solid particles (fatty dirt) are surrounded in aqueous systems by s. molecules whose hydrophobic molecule-segment is turned towards the suspended material, with the hydrophilic groups penetrating into the water phase. This principle is responsible for their detergency and their wetting and emulsifying (→emulsions) power. In depth information of the physical and chemical theoretical background of the rather complicated actions and interactions is given in the literature below.

S. are classified as follows:

Anionic s.: the hydrophobic residues are linked to an anionic group with small cations (Na,K,NH$_4$). The following RR-derived compounds belong in this group:

→soaps,→carboxymethylated fatty alcohol ethoxylates, →sarcosinates, →fatty acid/protein condensates, →α-sulfo fatty acid esters, →sulfosuccinates, →acyl aminoalkyl sulfates, →acyloxy alkane sulfates, →fatty alcohol sulfates, →fatty alcohol ether sulfates, →fatty alcohol phosphates, →sulfated fats and oils.

Nonionic s.: they are not able to dissociate into ions. The hydrophilic part is not an ion but a (poly)ethyleneglycol or glucose residue.

The following compounds are important:
→fatty alcohol ethoxylates, →fatty acid ethoxylates, fatty acid methyl ester ethoxylates (→fatty acid methyl ester) →fatty acid ethanolamides, →terminally blocked fatty alcohol ethoxylates, →mono/diglycerides and their many esters with acids, such as citric, lactic or tartaric acid, →sucrose fatty acid esters, →alkyl polyglucosides.

Cationic s.: they are cationic N-compounds with a hydrophobic residue and counterions such as chloride or sulfate. The following s. belong to this group:
→quaternary ammonium compounds, →imidazoline derivatives, →fatty alcohol amine oxides, →fatty amine ethoxylates

→**Amphoteric s.**: they have both cationic and anionic groups linked to one hydrophobic molecule (betaines, sulfobetains, imidazolinium betains).

→**Biosurfactants**

Lit.: Ullmann* (5.) **A25**, 747
   Kirk Othmer* (3.) **22**, 332
   M.J.Rosen "Surfactants and Interfacial Phenomena"
   J.Wiley & Sons, NY (1989)

Sweet Oil

→Olive Oil

Sweet Sorghum

→Sugary Sorghum Processing

Sweetzyme

→Isomerization

Systematics

→Taxonomy

T

Tall Oil

Syn.: Liquid Rosin
G.: Tallöl; F.: tallol

s.v.: 145–175; i.v.: 120–188

T. is an oily, resinous mixture of →rosin acid (33–47%), →fatty acids (42–55%), sterols, unsaponifiables, colored and odorous materials. The composition varies within broad limits, depending on the origin .

It is derived as a by-product from the Kraft (sulfate) process to make →paper pulp. Pinewood is the most common source. After separation of the pulp, the remaining "black liquor" is concentrated by evaporation, whereby the crude soap is salted out and collected. Acidulation with sulfuric acid and several refining steps gives crude tall oil. Production of one ton of paper pulp yields 30–40 kg of tall oil. Fractional steam-vacuum distillation (200–285 °C; 13.3 kPa/100mm; steam must be present to lower the b.p. and avoid decomposition) separates the oil from the pitch and low-boiling components. The oil is purified by additional fractionation procedures to separating the →rosin and to produce various grades of purities of tall oil fatty acids (TOFA).

Typical TOFA consists of 1–3% saturated acids; 45–65% →linoleic/→linolenic acid and 25–45% →oleic acid. Some TOFA grades still contain rosin up to 8%.

For further details about the rosin fraction: →rosin.

The largest use of TOFA used to be in coatings, primarily in →alkyd resins. Today, more TOFA is consumed for the production of →dimer acids and of →epoxides. Other areas of application are →mining chemicals and →oilfield chemicals, →soaps and →detergents.

In the early 1990's 1.5×10^6 mt crude t. were produced worldwide.

The major producer of t. and derivatives is the USA. In 1987, 900000 mt of tall oil and 220000 mt of TOFA were produced there.

Lit.: Kirk Othmer* (3.) **22**, 531 + (4.) **5**, 187
[8002–26-4]

Tall Oil Fatty Acids

→Tall Oil

Tallow (Beef)

G.: (Rinder)Talg; F.: suif (de bœuf)

s.v.: 190–200; i.v.: 32–47; m.p.: 40–50 °C

Beef tallow is the main t. used. There are other types, such as mutton t., which is used only industrially and has similar properties as beef tallow. Neatsfoot oil is a low-melting fat rendered from the feet of cattle; it is used as →lubricant and in fat liquors (→leather auxiliaries).

There are many grades and types of beef tallow, depending on the part of the animal from which it originates, the season of slaughtering, the ffa content, color, odor and many other factors.

Stability of t. is poor if no antioxidant is added. The fatty acid composition is:

C14:0	**C16:0**	C16:1	**C18:0**
2–4%	**23–29%**	2–4%	**20–35%**
C18:1	C18:2		
26–45%	2–6%		

High-purity grades (<1% free fatty acids), rendered from clean and fresh tissues at 50–60 °C, are used as edible t. and are added to margarines and shortenings. Only a small amount goes into these applications. Lower grades are used in animal feed. Technical grades are the most important raw materials for the manufacture of fatty acids by →hydrolysis. →Distillation, →fractional distillation, →hydrogenation and →crystallization yield different saturated and unsaturated fatty acids (mixtures, e.g., →stearin and →olein). They are used in many different applications: →soaps, →rubber chemicals, →candles, derivatives, such as →fatty alcohols and esters.

Production in 1994 ($\times 10^6$ mt)	Tallow/Grease
USA	3.27
Australia	0.47
FSU	0.30
Argentina	0.28
Brazil	0.28
UK	0.22
Canada	0.21
France	0.20
World	6.85

Lit.: Ullmann* (5.) **A10**, 234
[61789-97-7]

Tallow (Mutton)

G.: (Schaf)Talg; F.: suif (de mouton)

s.v.: 192–198; i.v.: 31–47; m.p.: 44–55 °C

Mutton tallow (→sheep) is used in the same areas as beef tallow but only for industrial applications. Superior-quality grades are similar to beef →tallow.

The fatty acid composition is also similar to beef →tallow. However, it contains a relatively high amount (4%) of odd-numbered, branched and *trans* fatty acids.

Lit.: Ullmann* (5.) **A10**, 235

Tallow Fatty Acids

→Tallow

Tannins

Syn.: Tanning Agents
G.: Tannine; F.: tannins

The term t. is used as a synonym for vegetable tanning (→leather auxiliaries) but classifies also a group of polyphenolic macromolecules.

Their chemical structure is rather complex. The main characteristic feature is that glucose (→dextrose) is partly or completely esterified with aromatic polyhydroxy carboxylic acids, e.g., gallic acid (3,4,5 trihydroxy benzoic acid) or its esters, e.g., di-galloyl-gallic acid and its condensation products.

There are two classes: hydrolyzable and condensed t.

T. are extracted from various parts (wood, bark, fruits, gall apples) of trees (oak, birch, mimosa, quebracho, chestnut) with water and spray-dried. Tannin is a white to yellowish powder.

T. are mostly used for tanning leather. Due to their astringent properties, t. are used as hemostatic and antiseptic agents.

They are also a source for gaining gallic acid and pyrogallol.

Lit.: Haslam "Vegetable Tannins Revisited" Cambridge University Press (1989)
E.Heidemann "Fundamentals of Leather Manufacture" p. 372 E.Roether KG Darmstadt (1993)

Tapioca

→Cassava

Tartaric Acid

Syn.: 2,3-Dihydroxy Butanedioic Acid
G.: Weinsäure; F.: acide tartrique

$$
\begin{array}{c}
COOH \\
| \\
H-C-OH \\
| \\
HO-C-H \\
| \\
COOH
\end{array}
$$

m.w.: 150.09
m.p.: L- and D-form 169–170 °C; meso 159–160 °C; rac. 205–206 °C.

T. forms three stereoisomers (→isomerism) and a racemate: L (+)-form [87-69-4], the D (−)-form [147-71-7], the meso-form [147-73-9], and the racemate DL(±) [133-37-9].

Only the L(+)-form occurs in nature as free acid or as salts in many plants and fruits. The salts are called tartrates.

T. is a strong acid, which is soluble in water and alcohol and has a refreshing taste. By careful heating, it forms an anhydride. T. is able to form complexes with heavy metal ions, e.g., Fe, Cu and Pb. Ammoniacal silver solution is reduced with formation of a silver mirror.

L(+)-t. is obtained by transferring by-products of wine making, such as tartar (potassium hydrogen tartrate), into the Ca salt and decomposing this with sulfuric acid into t. and $CaSO_4$.

The racemate is obtained from fumaric or maleic acid by treatment with H_2O_2.

T. is used as →textile auxiliary (dyeing aid, finishing, reduction agent) and as a universal acidulator in food processing (→food additives). Other applications are in electroplating as a metal-complexing agent, metal coloring, →ceramics and plasticizers. →Diacetyl tartaric acid esters of mono/diglycerides are common emulsifiers in food processing (→food additives).

World market is estimated to be in the range of 58 000 mt and potassium bi-tartrate at 20 000 mt (acid basis).

Lit.: Kirk Othmer* (4.) **13**, 1071–1078
[526-83-0]

Tartrates

→Tartaric Acid

Taurates

→ Acylamino Alkane Sulfonates

Taurides

→ Acylamino Alkane Sulfonates

Taxonomy

Syn.: Systematics

G.: Taxonomie, Pflanzensystematik; F.: taxonomie

T. is the division of botanical (and zoological) science that deals with the classification and segregation of different individual plants. It was first implemented by the famous Swedish botanist Karl von Linné in 1753. He described plants according to the name of their genus and species in a binomial nomenclature. Many plants and animals were first described by him, which is indicated by the abbreviation L. behind the botanical name ($\rightarrow$e. g., borage, castor, coconut, flax, hemp, oats, peanut, sugar beet). Other famous botanists and authors of first taxonomic descriptions are De Candolle (D.C.), Humboldt, Bonplant and Kuhnt (H.B.K.), Lamarck (Lam.), Linné filius (L.f.) and Sprengel (Spr.).

T. needs to consider several traits or "taxa" of a species, such as geographic distribution, similarity of appearance, flower, leaf, pollen morphology, chromosome and genome characteristics, as well as chemotaxonomical traits, such as the appearance of proteins, fatty acids, sugars, starch, essential oils, alkaloids, etc. The more similar such taxa are, the better is the chance that 2 individuals belong to the same species, genus or family of plants. There are several other segregating categories of taxonomy. The different hierarchic categories are listed here in a rough way for the $\rightarrow$sunflower (*Helianthus annuus* L.):

Category	Ending	Taxonomic Unit (example)
Regnum	– ota	Eukaryota
Phylum	– phyta	Spermatophyta
Subphylum	– phytina	Angiospermae (= Magnoliophytina)
Classis	– phyceae – mycetes – aae	Dicotyledoneae
Order	– ales	Asterales
Familia	– aceae	Compositae (= Asteraceae)
Genus		*Helianthus*
Species		*H. annuus* L.

Interspecific $\rightarrow$hybrids are named by an "x" between two species. If more than one author is known, the former "first describer" is mentioned in brackets.

In this dictionary, the name of a botanical source of RR within the head of a keyword is followed by the family name (e. g., Theaceae for $\rightarrow$teaseed).

Lit.: P.Raven et al. "Biology of Plants". (5.) Worth Publish. NY (1992)

Teal Oil

$\rightarrow$Sesame Oil

Tea Oil Plant

$\rightarrow$Teaseed

Teaseed

Syn.: Tea Oil Plant

Camellia oleifera, Abel; *Camellia sinensis* (L.) O. Kuntze

Theaceae

G.: Teesamen, Teestrauch; F.: thé

The genus *Camellia* is not only known for the production of tea leaves (*C. sinensis*, so-called Chinese tea) but also especially in mainland China for the production of t. oil from *C. oleifera*. In Japan, *C. japonica* is also used for oil production (also named tsubaki oil). Another well-known representative of *C. japonica* is the ornamental Camelia or China rose. The history of t. is closely connected with the history of Chinese tea. Tea is said to have been used in China already since 2700 B.C. Reports about teaseed oil date back to the year 300 B.C. Today, the main producing areas of t. are in South-China, especially in the provinces of Hunan, Jiangxi, Fujian, Guangxi and Guangdong. In China about 4×10^6 ha of t. trees are reported; however, the yield is only 150 000 mt oil per year.

The t. plant is a slow-growing, evergreen tree with a height of 1–4 m. T. trees are grown in large plantations. The tree may become several hundred years old. It will bear flowers from age 5–6, with typical fruits similar to those of walnuts or chestnuts with a fleshy fruit-wall, including 1–4 seeds that often have a cubic- or tetraeder-like shape. The seeds carry ca. 30% t. oil. The fatty acid pattern of t. oil is about:

C16:0	C18:0	**C18:1**	C18:2
9%	2%	**80%**	8%

and shows many similarities to $\rightarrow$olive oil and high-oleic $\rightarrow$sunflower oil (HOSO).

It is mostly applied as food oil. The meal of t. contains a comparatively high amount of $\rightarrow$saponins (13%), which makes the application for animal feed difficult. For a long time, the press cake was used preferably as soap especially for shampooing hair.

Lit.: L.Tang, E.Bayer, R.Chuang; Fat Sci. Techn. 95, 23–27. (1993)

Terminally Blocked Fatty Alcohol Ethoxylates

→Fatty Alcohol Ethoxylates

Terpenes

G.: Terpene; F.: terpènes

T. are an important class of numerous natural products that are →oligomers of isoprene. Thus, their number of C-atoms is always a multiple of 5. Monoterpenes are isoprene dimers (C_{10}), then there are sesquiterpenes (C_{15}) and diterpenes (C_{20}), and natural →rubber is a polyterpene. The term t. includes also the alcohols, ketones, aldehydes and esters derived from the respecting hydrocarbons.

T. are widely distributed in nature. The main use is as flavor and fragrance raw material (→essential oils). They are also starting materials for vitamins and pharmaceuticals. Some t. have pesticidal and allelophatic properties. T. resins are polymers of mainly α- or β-pinene, often with other comonomers (styrene, isoprene), and are used as tackifiers in →adhesives.

T. are obtained by isolation from natural sources but also by a sophisticated chemical synthesis. A cheap and frequently used starting material is →turpentine oil.

The following t. derivatives are of importance:
mono: →menthol, →camphor, →citronellal, →citral →eucalyptol
di: phytol (→tocopherol), vitamin A, →gibberelin, abietinic acid (→rosin);
tri: →steroids;
tetra: →carotene.

Lit.: Conolly and Hill (ed.) "Dictionary of Terpenoids"
Chapman & Hall, London (1992)
Kirk Othmer* (3.) **22**, 709

Tertiary Amines

→Fatty Amines

Tetraclinis articulata

→Resins, natural

Tetracosanoic Acid

→Lignoceric Acid

Tetracycline

G.: Tetracycline; F.: tétracyclines

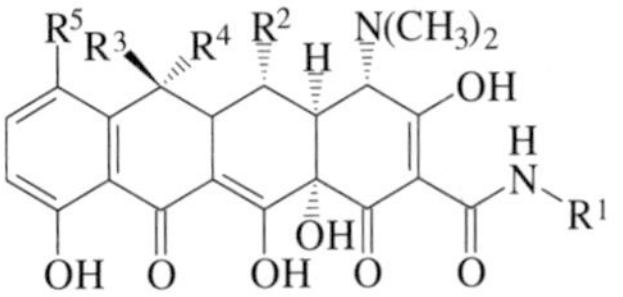

Name	R^1	R^2	R^3	R^4	R^5
tetracycline	H	H	OH	CH_3	H
oxytetracycline	H	OH	OH	CH_3	H
Auromycine	H	H	OH	CH_3	Cl

m.w.: 444.43; m.p.: 165 °C (swelling), 170–175 °C (d.)

T. is an antibiotic substance produced by *Streptomyces* spp., e.g., *Streptomyces viridifaciens*; fermentation with *Streptomyces rimoses* leads to 5-hydroxy-t. (= oxycycline), with *Streptomyces aureofaciens*, 7-chlorotetracycline (= aureomycine) is obtained. Its hydrochloride, the phosphate complex and its lauryl sulfate complex are used as antibacterials against gram-positive as well as gram-negative bacteria; t. acts as an anti-amebic and anti-rickettsial agent.

Lit.: The Merck Index* (11.) 9130
[64-75-5]

Tetradecanoic Acid

→Myristic Acid

1-Tetradecanol

→Myristyl Alcohol

Tetrahydrofuran

G.: Tetrahydrofuran; F.: tétrahydrofurane

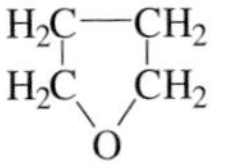

m.w.: 72.11; m.p.: –21.5 °C; b.p.: 66 °C

T. is a colorless liquid with an etherlike smell.
It is produced, aside from several other technical possibilities, by →hydrogenation of →furan, which is derived from RR.
It is used as solvent for PVC and other vinyl polymers and for →adhesives and →coatings. It functions also as an intermediate.

Lit.: Kirk Othmer* (3.) **18**, 645
[109-99-9]

Tetrahydrofurfuryl Alcohol

Syn.: 2-Hydroxymethyl-tetrahydrofuran
G.: Tetrahydrofurfurylalkohol;
F: alcool tétrahydrofurfurylique

$$H_2C\!-\!CH_2$$
$$H_2C \quad CH\!-\!CH_2OH$$
$$O$$

m.w.: 102.13; b.p.: 178 °C

T. is a colorless liquid. It is miscible with water and many organic solvents. T. is biodegradable and low in toxicity.

It undergoes all typical reactions of a primary alcohol, e.g., the formation of esters, such as tetrahydrofurfuryl acrylate, which is an interesting monomer.

Uses are mainly as solvent for resins, lacquers, insecticides and as industrial cleaner. It also serves as an organic intermediate.

Lit.: Ullmann* (5.) **A12,** 128
        [97-99-4]

Textile Auxiliaries

G.: Textilhilfsmittel; F.: produits auxiliaires textiles

Many (formulated) chemicals are necessary to produce fibers and to manufacture finished textiles from them. The base polymers of the →fibers may be natural, e.g., →cotton, →silk and →wool, or modified, e.g., →viscose. The raw materials of fully synthetic fibers are, with a few exceptions (→polyamide), of petrochemical origin. The chemicals and formulations necessary to make or process natural as well as synthetic fibers are called textile auxiliaries. Fiber manufacturers use spin finishes and coning oils. The fiber, whether natural or synthetic, is processed in a textile mill, which needs chemicals, such as sizes, pretreatment agents, auxiliaries for dyeing and printing, and finally, finishes. Some of them remain in or on the finished textile goods, others are only used during processing and are not contained in the finished products. Most of the products are surfactants or water-soluble polymers and are therefore frequently based on RR. Especially for →surfactants, it is not possible to mention all the various areas of application because they are broadly used in all kinds of textile processes as emulsifiers, detergents, wetting, levelling and dispersing, antifoaming and foaming agents.

Auxiliaries for Fiber Production:

Two groups of auxiliaries are used in fiber manufacturing:

- Additives for spinning solutions and baths:

In context with RR, only additives for the production of →viscose are of interest. →Sulfated oils, →fatty acid condensation products, laurylpyridinium salts and mainly ethoxylated products, especially fatty amines, are used as modifiers to produce high-tensile, high-modulus modal fibers (→viscose).

- Spin finishes:

Man-made, synthetic fibers need spin finishes, which enable production and processing of the fiber, such as drawing, warping and texturizing. They simultaneously act as lubricants and impart antistatic properties. They are applied onto the surface of the fiber and are formulations that contain lubricants (natural oils, e.g., →coconut or →peanut oil, esters and the ethoxylates of fatty alcohols, or →castor oil), emulsifiers (soaps, sulfonated oils, →fatty alcohol ether phosphates, →fatty amines either ethoxylated or quaternized, and →fatty alcohol ethoxylates). They contain also antistatic agents such as fatty alcohol ether phosphates, →quaternary ammonium compounds and betains.

Coning Oils and Spinning Lubricants:

These mostly paraffin-based formulations contain also →surfactants to assist washability after use. Sometimes, →fats and oils as well as ester oils (→lubricants) are used as such or in mixture with the paraffin.

Sizes:

Sizes make warp yarns smoother and more slippery to protect them during the weaving process. They increase the abrasion resistance by adhering the protruding fibers to the yarn body of a staple fiber yarn, respectively, by glueing the individual filaments of a filament yarn to each other. They are applied from aqueous solution and finally removed by desizing. A large volume of sizing agents is based on the following RR and are mainly used for sizing of staple fiber yarns:

- The main product group is →starch-based and is used for cotton and its blends with other fibers. Native, →pregelatinized starch, which has the disadvantage of →retrogradation, is increasingly substituted by starch derivatives, such as →starch phosphates or →starch acetates, as well as starch ethers, such as →hydroxyethyl starch or →hydroxypropyl starch and →carboxymethyl starch. The DS of these derivatives is ≈ 0.1.

- →Carboxymethyl cellulose is another important sizing agent, used for both natural and synthetic staple fiber yarns. The DS is ≈ 0.7, and purified types are required.

- Galaktomannans, derived from natural gums ($\rightarrow$guar, cassia and $\rightarrow$locust beans), are gums of minor but probably increasing importance due to their inherent biodegradability.

700 000 mt of sizes are used annually, 70% of which is starch-based.

There are also some additives used in formulating sizes Fatty substances, e. g., $\rightarrow$fatty acid esters, in combination with nonionic or anionic emulsifiers ($\rightarrow$surfactants), and $\rightarrow$sulfonated oils plasticize the size film, impart antistatic properties, help in foam control and improve smoothness. $\rightarrow$Fatty alcohol ethoxylates are sometimes added to increase wettability of the yarn.

It is important to remove sizes after weaving to guarantee smooth further processing (e. g., dyeing and printing). Water-soluble material is washed off, less-soluble sizes are removed by oxidation (H_2O_2 or persulfates) or by desizing $\rightarrow$enzymes, e. g., bacterial or pancreatic amylases or malt diastases. Nonionic and anionic wetting and dispersing agents enhance the desizing process.

Pretreatment Agents:

It is necessary to pretreat, especially natural fibers, prior to dyeing or printing to achieve even coloration.

- During the alkaline extraction (boiling, kiering and scouring) to remove noncellulosic impurities from cotton, alkaline- and electrolyte-resistant surfactants (e. g., $\rightarrow$fatty alcohol ethoxylates) are used as wetting and dispersing agents. $\rightarrow$Gluconic acid is used as complexing agent.
- Bleaching is mainly done with H_2O_2 today and needs agents for complexing heavy metals (among others, $\rightarrow$gluconic acid) and $\rightarrow$surfactants, which emulsify and remove impurities and reaction products. $\rightarrow$Fatty alcohol ethoxylates and sulfates are used in this application.
- Mercerizing is a method to improve dyeability, gloss (luster) and tensile strength of cotton by treatment with caustic. Surfactants, such as $\rightarrow$fatty alcohol (C_6–C_{12})sulfates and ether sulfates, as well as fatty alcohol ethoxylates (terminally blocked) and fatty amines improve wettability and foam control.
- $\rightarrow$Fatty alcohol ethoxylates are used in washing, carbonizing and bleaching of $\rightarrow$wool.

Auxiliaries for Dyeing and Printing:

Dyestuffs and pigment dispersions need dispersing agents (sulfonated oils and esters, fatty acid condensation products and ethoxylates). They are used

in combination with protective colloids [cellulose derivatives, alginates ($\rightarrow$sodium alginate), $\rightarrow$casein and $\rightarrow$lignosulfonate].

Wetting during these processes is improved by $\rightarrow$fatty alcohol sulfates and phosphates as well as $\rightarrow$sulfosuccinates.

Levelling agents that promote the distribution of dyestuff on the textile are sulfonated oils ($\rightarrow$sulfated fats and oils), $\rightarrow$fatty alcohol sulfates, $\rightarrow$fatty acid esters or amides and $\rightarrow$fatty acid/protein condensates. $\rightarrow$Fatty amine ethoxylates are used in wool dyeing. The formation of creases is prevented by fatty acid derivatives, fatty alcohols and their ethoxylates, and sulfonated oils. In padding, alginates ($\rightarrow$sodium alginate) and $\rightarrow$guar derivatives reduce the undesired migration of the dyestuff, while $\rightarrow$fatty alcohol ethoxylates function as antifrosting agents.

Printing thickeners are based on alginates ($\rightarrow$sodium alginate), tamarind, $\rightarrow$guar and $\rightarrow$locust bean gum products, $\rightarrow$starches and $\rightarrow$cellulose ethers. Some of these polymers are also used as printing and edge adhesives.

During dyeing, protein fibers are protected by $\rightarrow$protein hydrolyzates, $\rightarrow$fatty acid/protein condensates and $\rightarrow$lignosulfonates.

Finishing Agents:

RR-based products are not much used in finishing, an area dominated by formaldehyde and methylol compounds, which are used as easy care finishes, mainly for cotton.

- Within this process, hand builders (give a firmer, bulkier hand to fabrics), based on low-viscosity $\rightarrow$starch derivatives, are used.
- Oleochemically based softeners are mainly amides of fatty (stearic) acid with polyamines (e. g., diethylenetriamine), in which free amine functions are neutralized with mineral or organic acids. Also, cationic softeners ($\rightarrow$quaternary ammonium compounds), such as distearyl dimethylammonium or quaternized esters with tallow acids, are in use. Nonionic softeners are based on esters of stearic acid and ethoxylates of stearic acid or stearyl alcohol. As anionic softeners, sulfonated oils of fatty esters are known.
- Agents that make fabrics water repellent or water proof are based on fats, oils and waxes, and for higher durability, quaternary $\rightarrow$fatty acid amide derivatives are used. Others are fatty acid chromium complexes, resin-based reaction products of methylolated melamine with fatty acids, and the most modern fluorochemical repellents, which impart also oil repellency and

antisoiling properties. Copolymers of highly fluorinated alkyl acrylates (→fluoroalkyl compounds) are mainly used.

Finishes that reduce electric charges on fiber material are called →antistatic agents.

For wool finishing: →wool

There are several special finishes for technical textiles, which are used, in addition to the a.m., to increase fiber bonding. They are mainly polymeric formulations. From RR-based products, cellulose derivates, natural rubber latex and many oleochemically based emulsifiers and dispersants are used. This market is increasing.

A rough estimate is that 3×10^6 mt of t. are used worldwide annually.

Lit: Ullmann* (5.) **A26**, 227
Schlüter/Mattis*

Thebaine

Syn.: Paramorphine
G.: Thebain; F.: thébaïne

m.w.: 311.37 ($C_{19}H_{21}NO_3$); m.p.: 193 °C

T. is an alkaloid from *Papaver bracteatum*. It crystallizes into orthorhombic, rectangular plates, which upon consumption produce strychnine-like convulsions rather than narcosis. T. derives from →opium. It is used as a valuable starting material for the synthesis of the antitussive →codeine.

Lit.: Ullmann* (5.) **A1**, 903
[23979-17-1]

Theobroma cacao
→Cocoa Butter

Theobroma Oil
→Cocoa Butter

Thermoplastic Starches
→Starch Plastics

Thin-boiling Starches

Syn.: Acid-modified Starches
G.: Dünnkochende Stärken;
F.: amidon fluide, amidon soluble

T. are →modified starches in the granular state, insoluble in cold water, the paste of which, when warm, is more fluid than the paste of native starch in equal concentration.

Modification is achieved by treating of →starch granules with small amounts of sulfuric, hydrochloric or phosphoric acid (0.5–3%) at 48–55 °C in <40% aqueous slurries, filtering, washing and drying the starch granules.

The reaction is principally hydrolytic, but it is run to such a small extent that mainly the viscosity-causing →amylopectin is fragmented, and the gelling →amylose remains virtually unchanged. When the starch granules are gelatinized, they dissolve easily to give thin, clear solutions, which rapidly set back to white rigid gels on cooling.

Large quantities are used for textile sizing and finishing (→textile auxiliaries), for sizing of paper and board (→paper additives) to enhance printability and protection against abrasion. In the food industry, the rapid setting to gels in high concentrations is utilized for the manufacture of starch-based gum confectionary.

Products of other chemical treatment, such as →oxidized starches by treatment with sodium hypochlorite as well as white dextrins (→dextrins), are frequently also termed t. The term →soluble starch, frequently used, is incorrect with respect to the typical properties of this group of modified starches.

Usually, t. are not classified as →starch hydrolysis products because of their typical macromolecular starch properties.

A special type of t. by acid modification is **Lintner Starch**[1,2], which is used in the laboratory as a colorimetric indicator. It is obtained by moderate treatment of →potato starch with mineral acid. It retains the ability to give a blue color in the presence of iodine.

It is prepared by treating starch with 7.5% aqueous hydrochloric acid or 15% sulfuric acid over several days at r.t. The acid-free washed and dried product dissolves easily in boiling water to give a perfectly clear, stable fluid at a concentation of 2%. It is termed "soluble starch" or "Amylum solubile".

Beside its function as iodine indicator it is frequently used to determine amylase activity.

Another special type of t. by thermal modification is **Zulkowsy Starch**[2], which is used in the same applications as Lintner starch.

It is prepared by heating granular starch in pure anhydrous →glycerol for 30 min at 190 °C or until solubility in cold water is reached, precipitating by methanol, washing with methanol and vacuum-drying. It is an odorless, white, amorphous powder, soluble in cold water and gives clear, stable solutions. In this case, the frequently applied term "soluble starch" is correct.

Lit.: Tegge* 167–168 (1988)
      Blanchard* 306–307 (1992)
      Ullmann* (5.) **A25**, 13–14

[1] van Beynum/Roels* 89 (1985)
[2] M.Richter et al. "Ausgewählte Methoden der Stärkechemie" 49–50, VEB Fachbuchverlag, Leipzig und Wissenschaftliche Verlagsgesellschaft, Stuttgart (1969)

Thromboxanes

G.: Thromboxane; F.: thromboxanes

T. are compounds, derived from →prostaglandin endoperoxides, that cause platelet aggregation, contraction of arteries and other biological effects.

Lit.: B. Samuelsson, R. Paoletti (ed.) "Advances in Prostaglandin and Thromboxane Research", Raven Press, New York, Vol. 1–8, (1976–1980)

Tobacco

Nicotiana tabacum L., *N. rustica* L.
Solanaceae
G.: Tabak; F.: tabac

T. is a plant genus of American tropical origin of the nightshade family with hairy, sticky foliage and long-tubed, white, yellow, greenish, or purple flowers. T. is grown for its cured leaves. Its ability to produce large quantities of seed and ease of culture frequently make it the organism of choice for research on tissue culture (→genetic engineering). It is cultivated worldwide in areas with a frost-free period that is sufficiently long to permit crop maturity.

Due to the small seeds, t. needs to be grown in seedbeds, with transplantation of seedlings to the field (about 40 days later). For better leaf formation shoot inhibitors may be used, which are formulated from short-chain →fatty acid methyl esters and react in the t. plant as hormones. Leaves are hand-harvested when reaching the desired color (light greenish for cigars, yellow-green for cigarettes). Harvest may take place 4–5 times per week. Leaf yields reach 500 kg (oriental t.), 1000 kg (virginia t.) to 3000 kg (heavy cigar t.).

T. of commerce is classified according to the method of leaf curing. T. cured in heated barns is flue-cured (bright) or fire-cured. Air-cured t. includes light air-cured (burley and Maryland), dark air-cured, and cigar tobaccos, filler, binder, and wrapper. Flue-cured and burley are the principal t. used in cigarette manufacture and occupy the predominant hectarages.

The major effect of smoking t. is inhaling the alkaloid →nicotine. Its content varies around 3–5% (*N. rustica*), the latter reaching as much as 12% in special breeds for pharmaceutical applications and as →insecticide. Other sources of industrial nicotine are waste t. from the cigarette industry.

Main producing countries are China (803 000 mt), USA (793 000 mt), India (409 000 mt), Brazil (255 000 mt), and USSR (240 000 mt). Best t. qualities are produced in Cuba and Sumatra.

Lit.: Fehr/Hadley* (1980)

Tobacco Industry

G.: Tabakwarenindustrie; F.: industrie de tabacs

The leaves of the →tobacco plant (*Nicotiana tabacum* L.) are dried and fermented before cigarettes and cigars or other tobacco goods (chewing t. and snuff) are produced.

Several nature-based products are used to impart desired properties, and auxiliaries are necessary to manufacture the finished goods:

● →Glycerol and →sucrose find use as moisturizer;
● Blends of natural fragrances (→clove bud or →vanilla) are used in saucing tobacco;
● →Shellac, →gum arabic, →gelatine, →alginates, →cellulose ethers, →locust and →guar gum are used as thickeners and binders;
● Tobacco sheets, made from t. dust, and binders based on tobacco pectins or →methylcellulose, have to contain at least 75% tobacco and are used in cigar rolling;
● →Starch-based adhesives dominate as cigarette seam glues;
● Cigarette filter tows are made from →cellulose acetate (secondary), which is plasticized by →glyceryl triacetate

Lit.: Ullmann* (4.) **22**, 367

D-α-Tocopherol
Syn.: Vitamin E
G.: D-α-Tocopherol; F.: D-α-tocophérol

$R_1 = R_2 = R_3 =$ Methyl	d-α-Tocopherol
$R_1 = R_3 =$ Methyl, $R_2 =$ H	d-β-Tocopherol
$R_1 =$ H, $R_2 = R_3 =$ Methyl	d-γ-Tocopherol
$R_1 = R_2 =$ H, $R_3 =$ Methyl	d-δ-Tocopherol

m.w.: 430.7; m.p.: 2.5–3.5 °C

Naturally occurring t. is a mixture of the above-shown structures with various degrees of methylation. Only the trimethyl compound (α-) shows full vitamin activity. It is a yellowish oil that is insoluble in water but soluble in organic solvents. It is able to form esters such as acetates and succinates.

T. occurs in plants, especially in vegetable oils, from which it is isolated. Richest in t. is wheat germ oil (2500 ppm). Other sources are →soybean oil (1400 ppm), →cottonseed oil (1000 ppm) and → sunflower oil (600 ppm). Because wheat germ oil is only available in small quantities, soybean oil is normally used for commercial production. This, however, contains only 10% of α-t., which makes it necessary to methylate the other t. to the trimethyl compound. Various methods are available to accomplish this. The product obtained is completely identical with D-α-t. Mainly the deodorizer distillate of soybean oil is used as starting material, which contains 8% t. Sometimes, other plant oils (e.g., →palm oil) are used or considered as possible sources. There are several methods of concentration and isolation.

T. can be and is produced also synthetically. However, there are three assymmetric C-atoms, which results in 8 possible isomers (→isomerism), only one of which is identical with natural t. in structure and activity. Synthetic t. is used mainly for animal feed improvement.

T. is a powerful, lipid-soluble antioxidant and radical scavanger. Its natural function is to avoid oxidation (rancidity) of unsaturated oils in cell membranes in plants and in animals or human body.

Research studies have shown some activity against:

- a variety of cancers,
- artherosclerosis and other forms of heart disease,
- premature aging,
- cataract formation, and
- arthritis.

Due to antioxidant properties, it enhances the action of the immune system, and controls platelet aggregation of blood lipids.

The vitamin E market is estimated (early 1990's) at 6000–7000 mt/a, 2000 mt for human applications, shared equally by synthetic and natural t.

Lit.: Ullmann* (4.) **23**, 643
Kirk Othmer* (3.) **24**, 214–226
Haiyaishi, Mino "Clinical and Nutritional Aspects of Vitamin E", Elsevier, Amsterdam (1987)

TOFA
→Tall Oil

Topinambur
Syn.: Jerusalem Artichoke
Helianthus tuberosus L.
Compositae
G.: Topinambur, Erdbirne, Jerusalemartischocke; F.: topinambour

T. is a new world plant that was already under cultivation by North American Indians and had some importance in Europe, up to the middle of the 18th century, when it was superseded by the →potato. It has still some importance in South West Germany and limited importance in France, North America, USSR, Asia and Australia.

T. is a →perennial crop, which grows from a tuber into a ca. 2 m tall plant that is similar to the →sunflower, also in its flower morphology. However, the flower head is only 4–8 cm in diameter. For flower formation, it needs short-day conditions. At low parts of the sprout, shoots will grow from the leaf axils and form tubers at their end (under the soil). These tubers are about the size of →potato tubers, but they form roots and have a yellow, brown or red skin, depending on the variety. Their inner tissue is white. The crop may be harvested twice a year, if a first above-soil harvest for green fodder is desired, followed by the tuber harvest. Yields range from 100–250 dt/ha of tubers.

T. tubers contain about 78% water, 2.4% protein and 15.4% carbohydrates (7–8% of which is →inulin). They may be used similar to potatoes as a steamed, boiled or fried vegetable. A disadvantage of the tubers is their limited storage capacity, otherwise it would be worthwhile to isolate its inulin, e.g., for the manufacture of →fructose and similar derivatives. T. tubers are well adapted for the manufacture of →ethanol (yield 8–10 liters from 100 kg of tubers).

Lit.: Martin/Leonard* (1967)

Topnote
→Odor Description

Tragacanth
Syn.: Gum Tragacanth; Gum Dragon
G.: Tragant(h), Tragacanth;
F.: gomme tragacanthe, gomme adragante

T. is the dried, gummy exudation obtained by incision from *Astragalus gummifer* Labill. (Leguminosae, Papilionaceae) and other Asiatic *A.* species.
The shrub is native to Turkey/Greece and is not cultivated. Harvest of the air-dried gum particulates is carried out after cuts have been made in the bark of stem, branches and roots. T. consists of flattened, lamellated, frequently curved fragments or straight or spirally twisted linear pieces.
It is a white to weak yellow, translucent, odorless substance with an insipid mucilaginous taste. It is a mixture of two polysaccharides, bassorine (60–70%, consisting mainly of L-arabinose, D-galactose, L-rhamnose and methyl galacturonate) and tragacanthic acid (consisting of galacturonic acid, D-xylose, L-fucose, and D-galactose). In its powdered form, it appears white to yellowish. T. gum is about 6–10 times more expensive than →gum arabic.
Its application is thus limited today to pharmaceutical formulations (emulsifier and suspending agent).
Market trend in 1988–1995 has been -6%/a.

Lit.: Kirk-Othmer* (3.) **12**, 56
      [9000-65-1]

Train Oil
→Whale Oil

Transesterification
G.: Umesterung; F.: transestérification
T. is the reaction of an ester with an alcohol, whereby the alcohols are exchanged according to the formula:

$$R^1{-}COO{-}R^2 + R^3{-}OH \longleftrightarrow R^1{-}COO{-}R^3 + R^2{-}OH$$
R^1, R^2, R^3 = different alkyls.

The reaction is catalyzed by alkali, and the equilibrium is shifted to the right side when the alcohol on the left side is in excess and/or the alcohol on the right side is removed continuously.

T. is frequently used, especially in →oleochemistry. The reaction is used on a large scale to produce →fatty acid methyl esters from →fats and oils (methanolysis).
Other applications are the manufacture of →alkyd resins and of →glycerides, especially mono- and diglycerides.

Lit.: Ullmann* (5.) **A9**, 569 + **A10**, 188

Triacetin
→Glyceryl Triacetate

Triacontanoic Acid
→Melissic Acid

Triglycerides
→Fats and Oils
→Glycerides

1,7,7-Trimethylbicyclo (2.2.1)-2-Heptanone
→Camphor

Triolein
→Glycerides

Tripalmitin
→Glycerides

Tristearin
→Glycerides

Triticum spp.
→Wheat

Tsubaki Oil
→Teaseed

Tuberose
Polianthes tuberosa L.
Amaryllidaceae
G.: Tuberose; F.: tubéreuse

T. is a tall, slender, long-stemmed plant with several small, white flowers alongside the top part of the stem. The flowers are extremely fragrant, especially at night, and continue to be so for several days after being picked. This makes them an ideal material for the production of a floral →pomade, which then is further extracted to yield tuberose absolute.

The plant originates in Central America and grows wild in a multitude of tropical and subtropical countries. It is cultivated in France, India, Morocco, Egypt and the Comoro Islands for the production of →tuberose absolute.

Lit.: Gildemeister*
      The H&R Book*

Tuberose Absolute

G.: Tuberose Absolut; F.: absolue tubéreuse

T. is produced by alcohol extraction of either the →concrete or the →pommade, both extraction products of the →tuberose flower. The resulting product is dark orange to brown and ranges from a liquid (for tuberose from concrete) to a viscous or paste-like (for tuberose from pommade) consistency. The odor is extremely rich, sweet and heavy floral, with a slightly green, medicinical →topnote and a creamy, fruity, almost coconut-like →body note with a recognizable fatty undertone for t. from pommade. Due to the low yield, combined with a labor-intensive production process (1 mt of flowers yield about 300 g t.), the cost of t. is high, which restricts its uses to high-class fragrances for perfumes of the heavy-floral, floral bouquet or sweet-oriental (→odor description) types, where it adds tremendous richness and volume.

Lit.: Gildemeister*
      The H&R Book*

Tung

Syn.: Tung Oil Tree, Wood Oil Tree
Aleurites cordata (thunb.) R. Br. ex Steud.; *A. fordii* (Hemsl.); *A montana* (Lour.) Wils.;
Euphorbiaceae
G.: Tungölbaum, Holzölbaum;
F.: alévrite, bois de chine

The genus Aleurites consists of only 6 species, all of which are native to tropical and subtropical Southeast Asia. They are trees of varying size. However, only *A. cordata* and *A. montana* are important sources of → t. oil. Both species are native to China, where *A. montana* is found in the warmer, subtropical regions. It has been introduced to several tropical countries, with best productivity only at higher elevations.

The tree may be up to 20 m tall. It is adaptable to poor soils. The first crop may be produced 4 years after planting with a maximum life span of about 30 years. The →inflorescence is a raceme with varying numbers of flowers that are produced terminally during the growth of a new season.

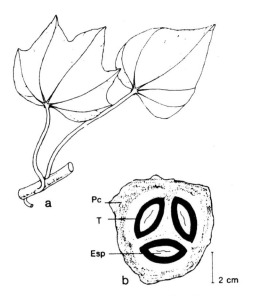

Tung; a: Leaves. b: Cross section of fruit (Pc: pericarp, T: testa, Esp: endosperm) (from Rehm and Espig*, 1995, with permission).

While some trees predominantly produce male flowers, others prefer to produce female flowers. The fruit is a drupe with a hard, woody pericarp; it is egg-shaped (up to 6 cm in diameter) and contains 3 broad, oval seeds of about 3 cm in length. For good productivity, at least 1000 mm of rainfall is needed annually as well as a cool, dry season as found in the tropics only at altitudes above 600 m. The fruits may be harvested from the trees or collected from the ground. The woody pericarp and the testas are removed mechanically or by hand, and the kernels are crushed and pressed to obtain the oil. The oil content is about 60% of the mature kernel weight.

China supplies more than half of the total world production of t. oil of about 120 000 mt per year.

Lit.: Purseglove* (1974)

Tung Oil

Syn.: Chinese Wood Oil; Abrasin Oil
G.: Tungöl; F.: huile de tung, huile d'abrasin

s.v.: 189–195; i.v.: 150–165

T. is a yellow, fast-drying oil because of its 80% content of →eleostearic acid. The residual fatty acids are small amounts (1–8%) each of C16:0; C18:0; C18:1 and C18:2.

It derives from the →tung tree.

Main uses of t. are in →coatings, especially in air-drying systems, in the →foundry industry and in the manufacture of →linoleum. The fatty acids resulting from →hydrolysis are used in →alkyd resins and →coatings.

World production in 1984 was 95 000 mt.

Lit.: Ullmann* (5.), **A10**, 232
       [8001-20-5]

Tung Oil Fatty Acids
→Tung Oil

Turkey Red Oil
→Sulfated Fats and Oils

Turnip Rape
Syn.: Colza; Field Mustard; Bird Rape
Brassica rapa L. *emend.* Metzg. ssp. *oleifera* DC.
Cruciferae
G.: Rübsen, Ölrübsen; F.: navette

Often also named as →rapeseed or canola, t. is actually a close relative of rapeseed being one of its two ancestors; the other is cabbage. T. often is grown because of its special winter hardiness in northern Europe or Canada or because of its short vegetation period as a spring crop in arid continental climates. The oil content of the seed and the fatty acid pattern of the oil are similar to that of →rapeseed. The oil may be used as cooking vegetable oil as well as for industrial and technical applications, especially the high →erucic acid varieties. The meal is used as animal feed. Young leaves are used as vegetable.

Lit.: G. Röbbelen et al. "Brassica species" 339–362, in:
       Röbbelen* (1989)

Turpentine Oil
Syn.: Spirit of Turpentine
G.: Terpentinöl; F.: essence de térébenthine

T. is an almost colorless liquid with a woody odor. It is gained from different pines (→rosin) by extraction and (steam-) distillation. It is the largest-volume essential oil. Many varieties are available in the market, depending on the type of tree and country of origin. These various grades are boiling between 160–170 °C.

Main constituents are α-pinene ($\approx 65\%$) and β-pinene ($\approx 33\%$).

T. is used as solvent for →coatings, paints, floor and shoe polishes. It is an important starting material for many perfumery raw materials that are synthesized from α- and β-pinene (e.g., terpineol, citronellol, isobornyl esters and →camphor). T. itself is used in perfumery for its refreshing, warm-balsamic odor, which contributes greatly to low-cost pine fragrances.

Distillation of the wood of various turpentine-producing pine trees yields →pine oil, which is often referred as "steam-distilled wood turpentine" and therefore confused with t., which is distilled from the resin.

Production, especially for industrial applications, is declining due to labor-intensity but is still practiced in the USA, Russia, Scandinavia and in other wood-processing countries.

Lit.: Kirk Othmer* (3.) **16**, 328
       [8006-64-2]

Tyrian Purple
→Dyes, natural

U

Undecenoic Acid
→Undecylenic Acid

Undecylenic Acid
Syn.: 10-Undecenoic Acid; C11:1 (Δ 10)
G.: Undecylensäure; F.: acide 10-undécénoïque

$CH_2=CH-(CH_2)_8-COOH$

m.w.: 184.27; m.p.: 24.5 °C; b.p.: 275 °C
U. derives from decomposition of →ricinoleic acid
esters at 500–600 °C:

$CH_3-(CH_2)_5-CHOH-CH_2-CH=CH-(CH_2)_7-COOCH_3$

$$\downarrow$$

$CH_3-(CH_2)_5-CHO + CH_2=CH-(CH_2)_8-COOCH_3$

Oenanthaldehyde (→oenanthic acid) is the by-product.
U. is used in organic synthesis of →fragrance raw materials, pharmaceuticals and →lubricants.
However, the main outlet of u. is the manufacturing of →polyamide 11.
U.-ethanolamide is a fungicide, antimycotic and bactericide. Because of its excellent skin behavior it is therefore used in cosmetic formulations.
Lit.: Ullmann* (5.) **A10**, 271
          [112-38-9]

Umbellatine
→Berberine

V

Vanilla Absolute/Resinoid
G.: Vanille Absolut; F.: absolue de vanille

V. is produced by hydrocarbon solvent extraction from the cured pods of *Vanilla planifolia,* Orchidaceae, native of Central America. In its natural habitat, the plant is pollinated by a local insect with a particularly long proboscis. After the orchid was brought to Europe by Fernando Cortez and subsequently to other tropical countries, it was discovered that the plant needed to be hand-pollinated to bear fruits (early 19th century). Vanilla is now cultivated in Indonesia, Madagascar, Reunion (i. e., Bourbon) and the Caribbean Islands.

The green, odorless fruits are subjected to a month-long curing (fermentation) process that consists of alternately moistening and sun-drying the pods, which then produce the chocolate-brown to black half-dried sticks called vanilla pods. Extraction with hydrocarbon solvents of these pods yields vanilla resinoid, which is either used as such or further purified to remove resinous matters and render the extract alcohol-soluble. The resulting product is vanilla absolute, a dark brown, clear and viscous liquid with a rich, warm note that consists of a sweet culinary aspect, surrounded by a deep, spicy, balsamic body and touches of woody and animalic notes.

Due to its high cost v. is not used in big quantities (it is often combined with synthetic vanilline), the note being extremely useful in all warm, sweet perfume types, such as oriental (→odor description) fragrances, and lending depth to floral and fruity notes. The flavor applications are equally numerous, ranging from baked goods to ice creams, beverages, liqueurs, and especially candies and chocolate-based products

Lit.: Arctander*
      Gildemeister*

Vanilla planifolia
→Vanilla Absolute/Resinoid

Varnishes
→Coatings

Vegetable Protein, Hydrolyzed
→Hydrolyzed Vegetable Protein

Veratrine
→Insecticides

Vernalization
→Biennial

Vernolic Acid
Syn.: 13-Epoxy-,*cis,cis*-9,12-Octadecenoic Acid
G.: Vernolsäure; F.: acide vernolique

$$CH_3-(CH_2)_4-\underset{\underset{O}{\diagdown\diagup}}{CH-CH}-CH_2-CH=CH-(CH_2)_7-COOH$$

m.w.: 296.44; m.p.: 30–31 °C

V. is an epoxy fatty acid, which is contained in the oil of →*Euphorbia lagascae* and *Vernonia* spp. The oil can be used in →epoxy plasticizers for PVC (→plastics additives) and in coatings.

Lit.: Kirk Othmer* (3.) **5**, 154
      [31263-20-4]
      [503-07-1]

Vernonia
Syn.: Indian Iron Weed, Purple Fleabane
Vernonia anthelmintica, V. galamensis
Compositae
G.: Vernonia; F.: vernonie

V. is produced on a limited scale commercially in Zimbabwe, Kenya and Pakistan. Old world species are distributed in India and Africa. It has been identified as potentially interesting →annual crop for the production of →vernolic acid, an epoxy fatty acid, present in v. at a level of about 80% of the oil contained in the seeds (24%). The residual fatty acids are C16:0, C18:1 and 18:2

Germplasm of v. have been selected that are neutral to day-length and thus suitable for the development of a temperate industrial crop.

Lit.: F.O.Ayorinde et al. "Vernonia galamensis: a rich Source of Epoxy Acids". J. Amer. Oil Chem. Soc. **67**, 844 (1990)

Vetiveria zizanoïdes
→Vetiver Oil

Vetiver Oil

G.: Vetiveröl; F.: essence de vétyver

V. is produced through steam-distillation of the roots of *Vetiveria zizanoïdes* Stapf (Graminae/Poaceae), a grass growing wild in many tropical countries and cultivated for essential oil production in Indonesia, Haiti, India, Reunion, Angola and Brazil. The oil is an amber-colored to olive or dark-brown viscous liquid, whose odor is sweet, mild and woody-earthy with leathery undertones.

Rarely used in flavors, v. and its derivatives find extensive use in perfumery, where its woody and earthy qualities and tenacity are appreciated in men's fragrances or as a complement to →oakmoss and other woody notes in women's fragrances.

Important isolates are vetiverol, which in turn is acetylated to produce vetiveryl acetate. Another quality of vetiveryl acetate is produced by acetylation of v. with acetic anhydride. Vetiverol displays more of the earthy-leathery aspects of the oil, whereas vertiveryl acetate is more clearly woody and therefore more versatile in use.

Lit.: Arctander*
The H&R Book*

Vicia faba

→Broadbean Starch

Viscose

Syn.: Rayon, Cellulose Regenerate
G.: Viskose; F.: viscose

The viscose process is used to produce fibers (viscose, rayon) and films (→cellophane).

The principle of the process is to convert cellulose to the soluble xanthate, which can be spun into fibers or transformed into a film, followed by a reconversion to cellulose by treatment with acid.

The following describes the different steps of fiber production (Specific process variations for making film: →cellophane):

Steeping:

Wood pulp (→paper) or, in rare cases, linters (→cellulose) is treated with 18–22% sodium hydroxide solution at 17–45 °C (steeping). Shredding is only necessary if one starts from pulp sheets. The resulting alkali cellulose is pressed to remove excess solution of caustic together with some soluble hemicelluloses.

Alkalization can be done also in two steps one for elimination of most hemicellulose, and the second for alkalization. Thus the hemicellulose, which would absorb a lot of CS_2, causing economical and ecological problems, is removed first before alkalization.

This tandem steeping reduces CS_2 consumption (step 3) into a range of 20–30%.

Aging:

Maturing and aging follows. During this period, atmospheric oxygen degrades the cellulose chain (at 20–45 °C) to a →DP of 300 for regular fiber and to 450 for high-tenacity yarn, resulting in decreased viscosity in the following steps of the process.

Xanthation:

When the material has aged sufficiently, it is treated with carbon disulfide (CS_2) and sodium hydroxide. Orange crumbs of cellulose xanthate are formed:

$$Cell-ONa + CS_2 \longrightarrow Cell-O-C{\overset{S^-}{\underset{S}{}}} \quad Na^+$$

Xanthation is rapid and exothermic (40–50 °C; 90 min). It is stopped at a DS of 0.5–0.6. Xanthate groups are not uniformly distributed along the chain. The batch process, used earlier has been abandoned. (Semi)continuous techniques are used today. Handling of CS_2 must be carried out carefully due to its explosivity.

Dissolving:

The orange crumbs are disolved in NaOH, and auxiliaries are added (→fiber finishes). Another maturing period follows, which makes the product more uniform.

Spinning:

After filtration and deaeration, the viscose solution is forced through spinneret openings into a bath that contains sulfuric acid (7–12%) and salts, e.g., sodium (12–24%) and zinc (0.5–3%) sulfate.

The body of the spinnerets consists of gold or platinum alloys. The orifices have a diameter between 50 and 250 μm, depending on the desired thickness of the fiber. In filament spinning, the number of holes is equal to the desired number of single filaments in the yarn (18–1000). In staple fiber production, the number of holes may surpass 90 000.

During dissolving and spinning, modifiers are added (→textile auxiliaries).

Regeneration:

The salts cause the xanthate to gel immediately to form a fiber of sufficient strength so that it can be drawn through the bath under tension. During this process, the salts coagulate the polymer while the sulfuric acid decomposes the xanthate to convert it back to cellulose, which is washed and dried. The CS_2 is recycled.

The overall regeneration process consists of three different mechanisms: the chemical regeneration (elimination of CS_2), the coagulation (salting-out effect), and the compound formation (by insoluble Zn-salts).

Types of fibers and uses:

The various steps and the many parameters of this process as well as the proper selection of the regeneration mechanism provide the possibility to tailor the final properties of the fiber to specific needs. Modifiers are normally added (→textile auxiliaries).

- Regular fibers, made after the process described above, are produced as staple and filament. They have a rather low wet modulus, low wet strength and high water absorption. They are used for outerware and nonwovens.

- Crimped fibers can be made from relatively "unripe" viscose solution under modified spinning conditions. They are used alone or in combination with polyester and acrylics for knitted and woven outerware.

- There are two varieties of modal fiber. Polynosic fibers require special spinning equipment. They are more cotton-like than regular viscose and show high tenacity, low elongation, dimensional stability, high alkali resistance and high wet modulus. They are used for knitted underwear, outerwear and decoration textiles. KHWM (high wet modulus) fibers can be spun in normal equipment by modified parameters. It has also high tenacity but higher elongation than polynosics. It is most sucessfully used in blends with cotton and synthetics for shirts, blouses, dresses and bed linen.

- High-tenacity viscose rayon is made by modifications that lead to more crystallinity of the fiber. This is accomplished by reduced aging (higher m.w.) of the alkali cellulose, additional CS_2, higher salt concentration and less acid in the spinning bath. Thus, greater tension (50–150% stretch) can be applied on the fiber during spinning, while slowed-down precipitation leads to higher crystallinity and tenacity. This fiber continues to play an important role in cord for tires and conveyor belts.

Fibers, grafted with different monomers (styrene, acrylonitrile, etc.), are in an early stage of commercial use.

For many years world production of regenerated cellulose fibers has been constant in volume at about 3×10^6 mt/a. However, the share of man-made cellulosic fibers has however decreased from 43% in 1970 to 20% in 1985 due to the strong growth of synthetic fibers.

The outlook is rather optimistic. Regenerated cellulose fibers are popular in many clothing, home textile and industrial applications for various reasons. They are based on RR, and the raw material is therefore available in the long term. It is however necessary to increase research efforts to improve the existing production processes in terms of their economical and ecological aspects and to find entirely new principles for regeneration and fiber formation.

There are already encouraging alternatives. Well advanced is the modification process with N-methyl morpholine oxide/water as a solvent for cellulose (→cellulose fibers). Under these circumstances, the importance of regenerated or modified cellulose fibers may increase once again.

Lit.: Ullmann* (5.) **A5**, 400
     Encycl.Polym.Sci.Engng.* (2.) **6**, 691

Vitamin B$_2$

Syn.: Riboflavin

G.: Vitamin B$_2$; F.: vitamine B$_2$

m.w.: 376.4

V. is a yellow powder and is contained in milk. It forms the prosthetic group in important enzymes such as oxidoreductases.

The main volume is produced by fermentation with *Ashbya gossypii* or modified *Bacillus subtilis* starting from →glucose. Improved yields can be obtained with lipids or mixtures of glucose and inosit or →chitin.

Lit.: Perlman, "Microbial Process for Riboflavin Production" in Peppler and Perlman (eds) "Microbial Technology", 2.edition vol 2, 263–302, Academic Press, London (1978)

Vitamin C

→L-Ascorbic Acid

Vitamin D

G.: Vitamin D; F.: vitamine D

V. is a →steroid and is contained in cod liver. Its precursors (provitamin D_{1-3}) are found in plant oils (phytosterols). Upon solar irridation these compounds are able to cure rickets. Today, it is possible to synthesize similar antirachitic compounds.

Lit.: Ullmann* **A25,** 311
      Kirk Othmer* **21,** 645–729

Vitamin E

→D-α-Tocopherol

Vitellaria paradoxa

→Shea Butter

Viverra civetta

→Civet Absolute

Vulcanized Fiber

G.: Vulkanfiber; F.: fibres vulcanisées

Known since 1855, this horn-like sheet material is produced by soaking unsized and unloaded paper sheets in a warm zinc chloride solution, and winding upon a roll under slight pressure until the desired thickness (0.5–50 mm) is obtained. It is cut, and the zinc salt is washed out, a process which takes a long time (2 weeks to several months) depending on the thickness of the sheet.

Chemically, v. is almost pure cellulose hydrate.

The grey material can be dyed and pressure-molded. It is oil-proof and absorbs 20–50% water on prolonged exposure, but dries (6–10% water) to its original size.

It is used for manufacture of trunks, gears, valves, seals, brake linings, and electrical insulation material.

Today, due to its the time-consuming manufacture, it is almost entirely substituted by synthetic plastics.

Because of its good properties and the advantages of being based on RR, a revitalization is not unlikely, especially if production technology can be improved.

In 1968, ≈20 000 mt were produced in USA by seven manufacturers.

Lit.: Encycl. Polym. Sci. Technology **14,** 757–767 (1971)

Vulcanized Oil

→Factice

W

Wallpaper Paste
→Adhesives

Washing Agents/Washing Powder
→Detergents

Water-Soluble Polymers
Syn.: WSP; Hydrocolloid
G.: Wasserlösliche Polymere;
F.: polymères hydrosolubles

W. are a group of water-soluble or swellable, some-times hydrogel-forming, RR-based (with a few exceptions) polymers, which act as →adhesives, binders, →film formers, thickeners, gelling agents, suspension aids, crystallization inhibitors, emulsion stabilizers, protective colloids, or water absorbents by giving certain rheological properties to aqueous systems.

These rheological properties depend on polymer composition, size (m.w.), shear rate, concentration, temperature, pH and the presence of organic solvents and salts.

The nature-based products are generally tasteless, odorless, colorless and nontoxic. With the exception of starch and starch derivatives they are noncaloric. W. are sensitive to microbiological degradation and to acid-catalyzed hydrolysis.

In context with RR, the following w. should be mentioned: alginates (→sodium alginate); →carboxymethyl cellulose; →carrageenan; →guar gum ; →locust gum; →gum arabic; →hydroxyethyl cellulose; →hydroxypropyl cellulose; →methylcellulose; →mixed ethers of cellulose; →pectin; →starch and derivatives, →dextran, →scleroglucan, →maltodextrins and →xanthan.

Lit: Encycl.Polym.Sci.Engng.* (2.) **17**, 730

Waxes
G.: Wachse; F.: cires

There are two definitions of w.: an old, chemical definition says that w. are →esters of →fatty

acids and →fatty alcohols, both with long chains. The other one defines w. as substances that are plastic solids at ambient temperature and, on being subjected to moderately elevated temperatures, they become low-viscosity melts. The chemical compositions range from pure hydrocarbons to vegetable- and animal-based products. The paraffin waxes represent a large-volume market today $(1.5 \times 10^6$ mt/a). W. gained from lignite and peat are on the borderline between fossile and RR products. Only the plant- and animal-derived w. are of interest in the context with RR (total market $\approx 20\,000$ mt/a).

The following are of major interest and therefore treated under separate keywords: →beeswax, →spermaceti, →lanolin, →carnauba, →candelilla and hydrogenated →jojoba oil.

There are some less important vegetable w., such as **Japan wax**, which is obtained from *Rhus succedanea* (Anacardiaceae), a small tree that is native in Japan and China. The wax melts at 53 °C, consists mainly of a triglyceride, and is used in special lubricants, food-related applications and as an additive to resins.

Ouricury wax (m.p. 82.5 °C) is obtained from a Brazilian palm and is a substitute for →carnauba wax when its darker color is tolerable.

Some commercial hope was directed towards **Douglas-fir wax**, which is extracted from the bark. It is mainly an ester wax (m.p. 59.2 °C). A small plant in the northwestern United States was shut down in ≈1970 because of poor economy.

Castor wax is hydrogenated →castor oil.

Rice-bran wax is extracted from →rice-bran oil, which is then purified and bleached (m.p. 79–83 °C). It consists of esters of →lignoceric and →behenic acid with $C_{22}-C_{36}$ alcohols. It is approved as a mold →release agent in plastics for food packaging and is used as a lipstick base. [8018-60-3]

In general, waxes are used in →polishes, in cosmetics, as coating for food (→food additives), and in →textile printing (batik), →pencil making and →candles.

Lit.: H.Bennett, Industrial Waxes, Chemical Publishing Co.,Inc., New York (1975)
Kirk Othmer* (3.) **24**, 466

Waxy Starches

Syn.: Glutinous Starches

G.: Wachsige Stärken; F.: amidons cireux

W., consisting of pure →amylopectin (99–100%), are obtained from certain cereal cultivars. Based on the plant source, the terms waxy corn (waxy maize), waxy rice, waxy sorghum starches are used.

Only waxy corn starch *(Zea mays,* var. *glutinosa)* has gained technical significance, especially in the USA, as the purest amylopectin, which needs no further isolation.

The corn kernels are subjected to →wet milling in the same way as regular →corn. →Starch granules are identical in size, shape and distribution.

M.w. ($\times 10^6$): amylopectin 76.9 (amylopectin of regular corn 112.2)[1], crude lipids 0.2%, crude proteins 0.25%, minerals 0.07%. The rheological properties during gelatinization, cooking and cooling deviate from regular corn starch. They resemble those of maniok starch : high consistency gain after gelatinization, low stability of the peak viscosity against cooking and/or shearing forces, formation of clear to slightly opaque pastes of long texture, very low susceptibility to →retrogradation.

For technical application as adhesives and for food thickening purposes, w. are commonly stabilized by chemical modification (→modified starch, →cross-linked starch).

Enzymic hydrolysis is performed by bacterial α-amylases to yield →maltodextrins of 5–20% DE with solutions of high clarity and stability.

Lit.: Van Beynum/Roels* 20–27,34,39 (1985)
Tegge* 157–158 (1988)

[1] Th. Aberle et al.: starch/stärke **44**, 331.(1994)

Wet Milling

G.: Naßvermahlung;

F.: extraction en milieu liquide

W. represents processes for the separation of starch-bearing materials into starches and their by-products. Modern wet milling is a continuous process with minimum modification or alteration of the final products used in production of →corn starch, →pea starch, →potato starch, →wheat starch, →starch by-products.

W. may be started from the whole grain, bulb or tuber, which are disintegrated or crushed after hydration and extraction of solubles (steeping), or from flour after initial →dry milling and starch (endosperm) enrichment in the case of cereal grain or seeds (legumes). All main processes are physical in nature.

Auxiliary chemical operations, such as the use of SO_2 or other compounds, are applied to stabilize pH level, suppress fouling and microbial growth (by *Lactobacillus* spp.), softening of corn kernels and endosperms, preventing oxidation and browning.

Lit.: Blanchard* 69–79 (1992)

Whale Oils

Syn.: Train Oil, Blubber Oil

G.: Walöl, Tran; F.: huile de baleine

s.v.: 185–205; i.v.: 110–135

W. are mainly triglycerides and are graded according to their ffa content. Good qualities are light-colored and have a fishy odor.

In contrast to →fish oils, w. do not contain major amounts of long-chain, highly unsaturated fatty acids:

C16:0	C16:1	C18:0	**C18:1**	C20:1
10–18%	13–18%	2–3%	**33–38%**	11–20%

W. is obtained from various whale species (28 mt of oil are gained from a whale weighing 120 mt), which are (were) processed in modern factory ships. Whale catching and processing is almost entirely abandoned due to international laws and agreements.

W. can be hydrogenated to different degrees of saturation. Soft types melt at 33–38°C and were used in shortenings, while the higher-melting (40–45°C) types were used in canned food or special margarines.

The fishy smell and taste disappear entirely after →hydrogenation and refining.

W. was formerly used also for technical applications, such as for fat liquors (→leather auxiliaries) and →lubricants and in metal processing. These areas are covered by substitutes today.

A special whale oil is →sperm oil.

A large volume of w. (140000 mt in 1974) was processed in former times. Whale catching today is limited and only done for the meat, e.g., in Japan.

Lit.: Ullmann* (5.) **A10,** 236

Wheat

Triticum aestivum L.

Gramineae, Poaceae

G.: Weizen; F.: froment, blé

Worldwide, w. is the most important →cereal crop, with 552 $\times 10^6$ mt of grain, which is 60% more than →rice. Wild forms of w. were developed into crops by simple selection methods (→plant breeding) already 8000 years ago in Eurasia.

During this process, several wild species of w. were combined, of which *T. durum* Desf. (hard w.), *T. dicoccoides* (wild emmer) and *T. spelta* L. (spelt w.) still have significance in world agriculture (10% of all the produced w.). W. is a good example of a rising degree of →polyploidy in a highly developed crop.

W. is a crop of the moderate climates with winter and spring varieties. It is well adapted to combine →harvesting and probably the highest yielding cereal, more than 100 dt/ha of seed, which is a consequence of its high demand for heavy nutritious soils.

The comparatively high protein content of w. grain makes it the most important source of human nutrition, and its several by-products from the food industry have opened several outlets for technical and industrial applications.

(→wheat gluten, →wheat germ oil, →wheat starch production).

Production 1995 ($\times 10^6$ mt)	Wheat
FSU	72.57
US	61.52
India	60.00
France	32.00
Canada	24.50
Turkey	16.00
Pakistan	16.70
Germany	17.70
UK	14.00
World	547.78

Lit.: Martin/Leonard* (1967)
       N.W.Simmonds "Evolution of Crop Plants" Longman London/NY (1976)

Wheat Germ Oil

G.: Weizenkeimöl; F.: huile de germes de blé

s.v.: 180–189; i.v.: 115–126

W. is a yellow oil that is contained at 8–12% in the wheat germ, which itself is 2% of the total grain weight. It is isolated by expelling and extracting.

The fatty acid composition of w. is:

C16:0	C18:0	C18:1	**C18:2**	C18:3
11–20%	1–6%	13–30%	**44–65%**	2–13%

The 3.5–6% of unsaponifiable material consists of →steroids (1700 mg/kg) and →tocopherols

(2500 mg/kg), 60–70% of which is the most active D,α-tocopherol.

Due to its valuable composition, w. is used as a dietetic oil and in cosmetic preparations.

Lit.: Ullmann* (5.) **A10,** 226

Wheat Gluten

G.: Weizengluten, Weizenkleber;
F.: gluten de froment

W. is the water-insoluble protein complex in the wheat endosperm. It is the commercial term for industrial products that are extracted as by-products of →wheat starch production by →wet-milling processes (cereal gluten).

Two main products are produced on an industrial scale:

Vital w., consisting of dry w. proteins in their native state, which exhibit -after rehydration- cohesive and elastic dough properties like freshly prepared wheat flour dough.

Devitalized w. (sweet gluten powder) consist of wheat gluten proteins that have irreversibly lost their native properties. During →wheat starch production, w. is obtained as a highly hydrated (70% water), cohesive, elastic dough after separation of the wheat starches by starch extractors and decanter centrifuges.

W. is further washed and dewatered in screen centrifuges and perforated conical screw presses.

For vital gluten production, drying must be done carefully to avoid devitalizing the proteins by overheating (<60 °C). Suitable drying systems are ring dryers, special types of flash dryers or ultrarotors, combinations of high-speed fine milling and drying in a hot-air stream. Vital gluten is separated by air cyclones as a white-greyish powder (6–8% water).

For preparation of sweet w., the dewatered dough (35% d.b.) is fed to roller-dryers that are steam-heated to 130–150 °C. The dry w. film is removed from the surface by scraper blades and then milled to fine granules or powder that contains 8–10% water.

W. is used for production of →hydrolized vegetable protein and →glutamic acid by acid hydrolysis. It may be prepared from freshly dewatered w. as well as from dried gluten preparations.

Commercial w. has the following composition (d.b.):

70–80%   crude protein,
 6–8%   crude lipids (neutral fat, phospholipids),
10–14%   carbohydrates (starch, fibers),
0.8–1.4% minerals .

The w. yield from 1 mt wheat flour (12% d.b.) may reach up to 143 kg (8% water, 85% crude protein) = 74.5% of the protein input (MARTIN process). Somewhat higher results can be achieved with whole wheat or whole wheat meal if the "homogenized slurry processing" with three-way centrifugal decanter separation is applied.

The current enhancement of wheat processing in the starch industry leads to a rising output of w., which exceeds traditional utilization of →gluten in food and feed.

Novel forms of nonfood applications, being developed at pilot plant and production scale, have led to promising results:

- Cobinders in paper-spreading colors after chemical modification to →gluten derivates;
- Rheology-labelling adjuvants in paper making;
- Nitrogen component in urea-formaldehyde or melamin-resin glues in wood processing (less formaldehyde emmissions);
- Setting-regulating additive in concrete and mortar mixtures;
- Additive bulking component for glues and spreading masses in pulp and paper processing;
- Component of masterbatches for extrusion or casting of edible or biodegradable foils and coatings.

Sometimes chemical modification of the cheaper devitalized sweet gluten is more convenient for the above-cited applications.

Lit.: J.M.Hesser Getreide, Mehl, Brot **49**, 21–25 (1994) "Stärke im Nichtnahrungsbereich" Reihe A/Heft 388, 189–196, 230–238, Schriftreihe des Bundesministers für Ernährung, Landwirtschft, Forsten, Landwirtschaftsverlag GmbH, Münster-Hiltrup (1990)

Wheat Starch Production

G.: Weizenstärkegewinnung;
F.: fabrication d'amidon de blé

W. leads to starch for food and industrial purposes as well to the by-products →wheat gluten and wheat feed in continuous →wet-milling processes with wheat flour as starting material.

There are two principal process routes:

- Formation of a dough by simply kneading wheat flour and water, which is exposed to rotating drum starch extractors, resulting in crude starch milk (and solubles) and wet gluten dough, (**Martin process**);
- Formation of an aqueous slurry from starch and small gluten particles by high energy input; the liquid system is distributed into product streams

of starch, protein concentrate and solubles by different techniques (**slurry process**).

In both technologies, fresh gluten remains a highly hydrated, coherent elastic dough of about 30% d.s., which is washed and dewatered by conical screw presses to 35% d.b. (→wheat gluten production).

The raw starch milk streams are freed from insolubles — small gluten particles, fibers and insoluble pentosans by rotating screens, DSM screens or jet refiners. These by-products are collected and shipped directly as slurry for feeding or may be concentrated and/or dried together with concentrated solubles and bran as wheat gluten feed.

Solubles may appear in the B-starch stream or in the light phase effluent of a three-phase decanter centrifuge. They contain the soluble pentosans, low-molecular carbohydrates and peptides, minerals and the d.s. of recycled production liquids.

A special problem in w. is caused by the bimodal distribution of granules with a small granule (B-starch) fraction, 2–15 μm in diameter (15–20%), and a large granule (A-starch) fraction, 20–35 μm (80–85%). The first one is highly contaminated with pentosans, fine fibers, lipids and protein. Therefore, it is sometimes advantageous to work up both fractions separately. This may lead to two final products: high-purity A-starch and low-purity B-starch. Otherwise, both starch streams are recombined after washing and concentration to one final product. This is usually done by nozzle separators and concentrators or by hydrocyclones. The resulting concentrated starch milk is ready for further processing.

If dry wheat starch is desired, the concentrated starch milk is dewatered by channel separators or rotary vacuum filters to a wet cake of about 42% humidity, which may be fed directly to the flash dryers or -in case of higher water content- after blending with dry returned wheat starch from the sifting station.

The concentrated B-starch slurry may be fed to drying cylinders for production of →pregelatinized starch, which is used in corrugated paper production, in the →foundry industry as binder, or in wheat gluten feed preparations.

Recent innovations focus on improved overall yield and reduced fresh water input. The high-pressure disintegration technique is applied to prepare a highly sheared slurry, which is fed to a three-phase decanter centrifuge to split the slurry into three fractions: A-starch plus fine fiber, B-starch plus gluten and solubles, plus finely suspended solids.

Other approaches start from whole grain or whole grain meal as raw materials from which the solubles are first removed by aqueous leaching.

Comparison of starch yield, biomass recovery and fresh water needs in different wheat processing techniques

Technique	Starch (A+B) yield, % d.b.	Biomass %	Water input /mt raw mat.
MARTIN	84.2 (74.2 +10)	87	8 – 10 mt
"Slurry"	90 (76 +14)	98	3 – 4 mt
Whole grain meal	90.3 (84.5 + 5.8)	99	3 – 4 mt
Whole grain	91.7 (82.1 + 9.6)	99	>3 mt

World wheat starch production has reached the second position after corn starch. In EC, w. exceeds now 1×10^6 mt/a.

Lit.: P.A.Seib J.applied Glycoscience (Jap.) **41**, 49–69 (1994)
"Stärke im Nichtnahrungsbereich" Reihe A/Heft 388, 41–55, 125–141, Schriftreihe des Bundesministers für Ernährung, Landwirtschaft, Forsten, Landwirtschaftsverlag GmbH, Münster-Hiltrup (1990)

Whey Acid
→Orotic Acid

White Dextrins
→Dextrins

Whole-crop Harvesting
G.: Ganzpflanzenernte;
F.: récolte de la plante complète

W. means that all parts of the plant are taken from the field for further processing, either for thermal utilization as fuel (→fuel alternatives) or for further separation into useful products.
For →cereal w., this means that straw and spikes are harvested in the field and transported to an on-farm "agro-refinery", where a separator segregates the harvested material into kernels, empty spikes, nodes and internodes (the segments of the straw). The grain can be used for food, spikes and nodes may be used in the production of cardboard, insulation material, etc., and the nodes could be used for the production of particle board.

Winterization
→Crystallization

Wood
G.: Holz; F.: bois
W. is one of our most important RR. It is available year round and can be stored for some time.

The original water content is 40–60%, which is reduced to 15–18% by air drying. It consists of 45–50% →cellulose, 25–30% →lignin and 15–20% pentoses and hexoses (→hemicelluloses). There are a few other minor ingredients, such as resins (→rosin), waxes, →terpenes and minerals.
All these figures fluctuate with the kind of tree, geographic origin, season of harvesting and processing technology.
W. is extremely versatile:

Functional uses:
W. is the oldest building and construction material used by human beings. It is or has been used for houses (partially or totally), bridges, railroad sleepers, construction in mines, telefone poles, ship building, furniture and works of art, to name only a few examples. There are also some composite materials to mention: ply w., w. flake- and particleboard, and fiber board. All require adhesives, some of which are →starch- or →protein-based.

Energy and fuel uses:
According to FAO, 2/3 of all human beings still use w. as their only source of household energy. It is estimated that 2×10^9 m^3 of w. is burned worldwide. The available energy is rather low (15 MJ/kg) due to the high oxygen content. W. residues are fired even today for generating process energy in saw mills and related industries.
In connection with the recent concerns about alternative energy, more use of waste w. and of fast growing trees, such as eucalyptus, has been considered for local power stations. Some rather large pilot projects are already on stream in Europe.
→Charcoal is used for barbecueing.

Use as a source for chemicals:
In former times, w. gasification and simultaneous →charcoal production was a source for →methanol and acetic acid.
Today, the gases resulting from charcoal manufacturing are burned for generation of process energy.
→Charcoal is used on a large scale in metallurgical processes, e. g., in Brazil for iron production. Fast-growing w.(eucalyptus introduced from Australia) is grown in well-organized plantations and is then carbonized. Some →activated carbon products are derived from w. via charcoal.
Many attemps were made in the past to use the large carbohydrate potential of w. by hydrolysis to obtain lower-m.w. sugars, which can be used as biotechnological feedstock, e. g., for ethanol production. Starch, used for this purpose today, is readily degradable. Hydrolysis of w. with either strong or week acids causes problems with the

large amount of acid consumed and costly work-up procedures. Enzymes are expensive and are not able to attack lignin-encrusted cellulose. Rather large plants have been on stream and have been shut down again for economical reason.

The most important chemical industry based on w. is →paper and pulp, which is used to produce fiber material ranging from newsprint to pure →cellulose, which is the raw material for many chemical products. The by-products of pulping are →lignosulfonate, →rosin acids and →tall oil fatty acids.

W. is also the source of naval stores: →resins and →turpentine.

An area of 4×10^9 ha (30% of world surface) is covered by w. The total volume of w. is estimated to be 330×10^9 m^3 worldwide. Round wood production in 1992 was 3.4×10^9 m^3 [1]. There are serious discussions on how to use our wood resources reasonably: 11.3×10^6 ha of tropical forest were destroyed in 1980 and 17×10^6 ha in 1990. In the long term, our w. resources can only be exploited to a certain extent, which is in balance with reforesting activities. Sustainable concepts need to be reduced to practice.

Lit.: Kirk Othmer* (3.) **24,** 579
      Encycl.Polym.Sci.Engng.* (2.) **17,** 843

[1] Vital Signs 1994, World Watch Inst. 80 W.W. Norton

Wood Polyoses
→Hemicelluloses

Wool
G.: Wolle; F.: laine

Wool is a protein fiber that is obtained from the fleece of →sheep. Aside from normal domestic sheep, there are other varieties, such as merino. The term w. covers also the hair of lamas (alpaca), camels, rabbits (angora) and goats (mohair, cashmere).

Chemically, wool is a →protein, based on keratin, and consists of 24 different →amino acids. Most important is the content of S-containing amino acids (cystine), which are responsible for the disulfide cross-linking.

W. is a hygroscopic fiber whose physical properties depend on the water content, which is about 15% at 65% R.H. and 21 °C. The fiber is highly extensible (elongation 25%) and has a relatively low breaking strength. It is degraded by acids and alkali rather rapidly.

Raw w. consists of $\approx 50\%$ fiber, $\approx 16\%$ grease/wax, 26% suint (i.e. perspiration salt) and dirt and $\approx 10\%$ moisture. The fiber is classified according to fineness (16–40 μm), length (5–35 mm) and crimp (≈ 6 waves/cm) of the fiber.

The processing of wool starts with scouring, which is the removal of impurities by washing with nonionic →surfactants and →soap or by treatment with strong acids (carbonizing). Wool grease or wax is recovered (→lanolin).

Some of the many mechanical steps of further processing are carding, combing, spinning, twisting, knitting, weaving. They are accompanied by bleaching, dyeing and finishing (setting, felting, treating for shrink, insect and flame resistance).

Aside from all kinds of textile applications, w. is used for making felt and wadding (milling with →soap, →amphoterics and →amine oxides).

World production in 1984 was about 2×10^6 mt, with Australia, New Zealand, Russia and China being the main producers. The volume is rather constant. The relative share of w. in the fiber market is decreasing due to a strong increase in other fibers. However, its excellent properties guarantee a stable future.

Lit.: Kirk Othmer* (3.) **24,** 612
      Encycl.Polym.Sci.Engng.* (2.) **6,** 689

Wool Wax
→Lanolin

Wrinkled-Pea Starch
→Pea Starch

WSP
→Water-Soluble Polymers

X

Xanthan Gum

Syn.: *Xanthomonas* Polysaccharide
G.: Xanthan; F.: xanthane

X. is the extracellular anionic heteropolysaccharide that is produced by the action of *Xanthomonas campestris* on culture media that contain →glucose as C-source (→microbial gums).

The main chain is structured of 1,4-β-linked AGU (principally, a →cellulose backbone). The side chains are trisaccharide units, bound in O-3 positions. They are constructed of the groups:

β-D-mannopyranosyl-, β-D-glucopyranosyl-, α-D-mannopyranosyl-, and pyruvic acid as an α-4,6-di-O-acetale, linked with half of the β-D-mannopyranosylic groups.

m.w. $\approx 2 \times 10^6$ (much higher figures have been reported).

Variations in molecular and functional properties occur with different bacterial strains, conditions of fermentation or composition of the culture medium with respect to nitrogen source and concentration of minerals.

Production is run in batches or continuously by aerobic growing of *X. campestris* in well-aerated submerged cultures. After separation of the culture medium from the cell biomass, sterilization and clarification by centrifugation or filtration, x. is isolated by solvent precipitation with isopropanol. Further processing is done by drying, milling, sifting, and eventually blending.

X. swells easily in water; after dissolution, it forms pseudoplastic solutions with unique viscosity stability against heating, changes of pH and electrolytes. Precipitation is achieved by polyvalent cations in slightly alkaline media. Noteworthy is the high synergistic viscosity enhancement by interaction with other hydrocolloids, such as galactomannans (→guar, →locust bean gum, tara gum) and →methylcellulose. In the pure state, non-gelling x. forms thermally reversible gels with these gums by physical interaction. Good compatibility between these polysaccharides favors gel forming with alginates and starch. Double-helix formation is considered to be the basis of structure formation. According to its structure, x. is a noncaloric and nontoxic food component, which is approved for humans.

Applications are based on the unique rheological properties; the market growth of 10.7% for food applications alone is the highest, compared with all other synthetic and natural hydrocolloids.

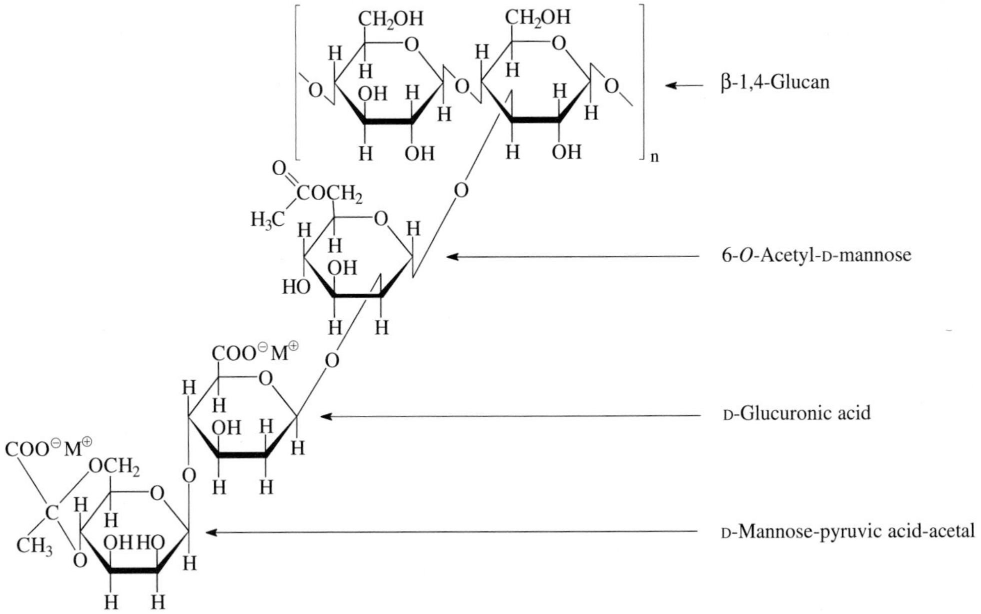

β-1,4-Glucan

6-*O*-Acetyl-D-mannose

D-Glucuronic acid

D-Mannose-pyruvic acid-acetal

Main food applications (→food additives) include salad dressings, low-calory sauces and dressings, chocolate syrup, dry-mix beverages, canned fruit juices, whipped and other toppings, creams, egg substitutes, snack foods, cakes and breakfast slices. The excellent freeze-thaw stability of pastes and gels make x. an effective stabilizer for different types of frozen food as well as ice cream thickener. The synergistic action of gum blends is utilized in canned frostings, cheese dressings, pizza creams, and starch-based gum candies.

The major nonfood application is in oil field (→oilfield chemicals). Other applications are in stabilizing suspensions, nondripping paints and →paint removers, textile print pastes, stabilizing agro-chemical suspensions, →metal-working fluids. In all these products, the a.m. excellent properties compensate for the high price and give a bright outlook for the future.

Annual production is estimated with 20 000 mt.

Lit.: Encycl.Polym.Sci.Engng.* (2.) **17,** 901–918
Ruttloff* 500–506 (1991)
Ullmann* (5.) **A25,** 51–54

Xylans
→Hemicelluloses

D-Xylose
→Hemicelluloses
→Pentoses

Y

Yam

→*Dioscorea*

Yellow Dextrins

→Dextrins

Ylang-Ylang Oil

G.: Ylang Ylang Öl; F.: essence d'ylang-ylang

Y. is steam-distilled from hand-picked flowers of the →ylang-ylang tree. This oil is commercially available in four grades: extra, I, II and III. Y. extra is the first distillate (30–45% of total), and the other grades are collected at increasing length of distillation time. Due to separation into low- and high-boiling constituents, the different grades have distinct odor characteristics. Y. extra is the most expensive and appreciated grade and has an extremely floral and diffusive, somewhat jasmine-like note with spicy and sweet undertones. Typical constituents are: methyl benzoate and methyl p-cresol.

At the other end of the scale, y. III is oilier in appearance, is more tenacious and has a mild floral character, combined with a balsamic woody note. Its uses are in lower-cost fragrances or wherever a mild, woody-floral background is required (→odor description).

Y. extra is used in all types of →fragrances, especially in florals, or rich orientals to boost the richness of the floral notes and add diffusion and liveliness.

Y. products are occasionally used in →flavors of the fruity or wintergreen type.

Lit.: Arctander*
      The H&R Book*

Ylang-Ylang Tree

Cananga odorata, Hook.f.&Thoms. *forma genuina*
Annonaceae

G.: Ylang-Ylangbaum; F.: arbre d'ylang-ylang

Originating from Indonesia and the Philippines, this tree is found in many tropical countries and is mainly cultivated in Madagascar and the Comoro Islands as well as in Reunion, Indonesia, Zansibar and some West Indian islands. Y. is a medium-sized tree and has flowers with white, long, almost feather-shaped petals, which are picked and further processed to yield →ylang-ylang oil.

Another form of y. is *Cananga odorata, forma macrophylla,* which grows in the same areas and whose flowers are extracted to produce cananga oil, which is similar to ylang-ylang oil, but less rich and as fine in odor.

Lit.: Arctander*
      The H&R Book*

Z

Zea mays
→Corn

Zein
G.: Zein; F.: zéine

m.w.: 10 000–35 000; m.p.: 180–200 °C

Z. (a mixture of prolamines) is the main part of the proteins of →corn gluten meal. It is extracted from high-grade meal with aqueous alcohol at 60–70 °C and precipitated by addition of water.

It is a granular, straw- to pale-yellow, amorphous powder or fine flakes with properties similar to the gliadins of →wheat gluten. For composition: →corn gluten meal.

Bland in taste and of a characteristic odor, it is insoluble in water and acetone but readily soluble in ethanol/water or acetone/water mixtures. It forms films, a property used in protective food coatings. It is used also as a wet granulation binder, as an remote release agent and for the manufacture of adhesives as well as a substitute for →shellac, for the production of laminated boards and in solid color printing.

Lit.: R. Mosse, Ann. Physiol. Vegetable **3,** 105 (1961).
Ullmann* (5.) **A22,** 293
J.B.S. Braverman "Introduction into the Biochemistry of Food", 126–129 Elsevier Publ. Co, Amsterdam/London/NY (1963)

Zinc Soap
→Metallic Soaps

Zulkowski Starch
→Thin-boiling Starch

Murphy, D.J. (ed.)

Designer Oil Crops

Breeding, Processing and Biotechnology

1993. XVI, 320 pages, 29 figures,
9 tables. Hardcover.
DM 195.-/öS 1424.-/sFr 189.-
ISBN 3-527-30040-6

This book is the first to address directly the exciting new vistas opened up for oil crops following the molecular biology 'revolution' of the 1980s and 1990s.

Its wide-ranging coverage includes:

- the major oil crops
- biochemistry of oil synthesis
- transformation of oil crops
- release of transgenic oil crops
- future prospects for oil crops

We are all facing the ominous problem of finding alternative renewable resources to the slow but sure depletion of non-renewable fossil oils. With its invaluable information on genetically engineered oil crops and their use in consumer, food and non-food products, 'Designer Oil Crops' is a vital contribution to solving this problem.

Denis Murphy is Head of the Brassica and Oilseeds Research Department at the John Innes Centre, Norwich (UK), one of the largest plant biotechnology institutes in Europe.

Date of information: December 1996

VCH, P.O. Box 10 11 61, D-69451 Weinheim, Fax 06201 - 60 61 84

VCH
A Wiley company